T0259012

Datacenter Connectivity Technologies:
Principles and Practice

RIVER PUBLISHERS SERIES IN OPTICS AND PHOTONICS

Indexing: All books published in this series are submitted to the Web of Science Book Citation Index (BkCI), to CrossRef and to Google Scholar.

The "River Publishers Series in Optics and Photonics" is a series of comprehensive academic and professional books which focus on the theory and applications of optics, photonics and laser technology.

Books published in the series include research monographs, edited volumes, handbooks and textbooks. The books provide professionals, researchers, educators, and advanced students in the field with an invaluable insight into the latest research and developments.

Topics covered in the series include, but are by no means restricted to the following:

- Integrated optics and optoelectronics
- Applied laser technology
- Lasers optics
- Optical Sensors
- Optical spectroscopy
- Optoelectronics
- Biophotonics photonics
- Nano-photonics
- Microwave photonics
- Photonics materials

For a list of other books in this series, visit www.riverpublishers.com

Datacenter Connectivity Technologies: Principles and Practice

RIVER PUBLISHERS SERIES IN OPTICS AND PHOTONICS

Series Editors:

MANIJEH RAZEGHI
Northwestern University
USA

KEVIN WILLIAMS
Eindhoven University of Technology
The Netherlands

Indexing: All books published in this series are submitted to the Web of Science Book Citation Index (BkCI), to CrossRef and to Google Scholar.

The "River Publishers Series in Optics and Photonics" is a series of comprehensive academic and professional books which focus on the theory and applications of optics, photonics and laser technology.

Books published in the series include research monographs, edited volumes, handbooks and textbooks. The books provide professionals, researchers, educators, and advanced students in the field with an invaluable insight into the latest research and developments.

Topics covered in the series include, but are by no means restricted to the following:

- Integrated optics and optoelectronics
- Applied laser technology
- Lasers optics
- Optical Sensors
- Optical spectroscopy
- Optoelectronics
- Biophotonics photonics
- Nano-photonics
- Microwave photonics
- Photonics materials

For a list of other books in this series, visit www.riverpublishers.com

Datacenter Connectivity Technologies: Principles and Practice

Editor

Frank Chang

Inphi Corporation
USA

Published, sold and distributed by:
River Publishers
Alsbjergvej 10
9260 Gistrup
Denmark

River Publishers
Lange Geer 44
2611 PW Delft
The Netherlands

Tel.: +45369953197
www.riverpublishers.com

ISBN: 978-87-93609-22-8 (Hardback)
   978-87-93609-21-1 (Ebook)

©2018 River Publishers

To my parents, for their unfailing love and encouragements,
who sacrifice so much for my happiness

To my wife, Jenny, and our children Pam and Olivia,
for all their unconditional love and inspiration

Frank Chang

To my parents, for their unfailing love and encouragements,
who sacrifice so much for my happiness

To my wife, Jenny, and our children Pam and Olivia,
for all their unconditional love and inspiration

Frank Chang

Contents

3   Direct Modulation Laser Technology: Past, Present,
and Future   **87**

Yasuhiro Matsui

Demis D. John, Grant Brodnik, Sarat Gundavarapu,
Renan L. Moreira, Michael Belt, Taran Huffman and
Daniel J. Blumenthal

Steve Yao, Wajih Daab, Gang He and Daniel Gariépy

Preface

Welcome to *"Datacenter Connectivity Technologies: Principles and Practice"*. This book discusses relevant concepts and inherent technologies that should be taken into account when designing and implementing data center connectivity solutions. It was motivated by the desire to explore the future perspectives of high-speed data center connectivity technologies by presenting the state-of-the-art results in both new optical devices, and circuit implementations. It's well known that, in recent years, investments by cloud companies in mega data centers and associated network infrastructure has created a very active and dynamic segment in the optical components and modules market. Optical interconnect technologies at high speed play a critical role for the growth of mega or hyperscale data centers, which flood the networks with unprecedented amount of data traffic.

And yet, there is only a limited number of books in this area that give a coherent and comprehensive review of data center technologies in general. Most of the data center related research materials are scattered around in journals, periodicals, conference proceedings, and a number of technical standards by industrial forums. Therefore we feel it is timely to publish a book that covers the different aspects of the data center connectivity technology. One may reliably count on this book as the first of its kind to address various advanced technologies connecting data centers.

This book provides a comprehensive and in-depth look at the development of various optical connectivity technologies which are making an impact on the building of data centers. The technologies span from short range connectivity, as low as 100 meters with multi-mode fiber (MMF) links inside data centers, to long distances of hundreds of kilometers with single-mode fiber (SMF) links between data centers. It is one of the purposes of this book to provide a balanced coverage of networking technologies, fiber optics transmission technologies, and electronics/components involved in developing data center connectivity solutions.

One point I'd like to mention is that electronic power is the lifeline of the data center. Since power represents up to 70% of the total operating costs, many enterprise users and colocation operators focus on their technology and site selection on low-cost-power options. Very often, researchers in new technology such as data centers will be driven by the desire to demonstrate technical smartness and overlook practicality. This tends to result in a lot of literatures which eventually becomes irrelevant due to economic reasons. Data center connectivity needs to be very cost-conscious with low power consumption. Therefore, when selecting the materials for this book, we tried to balance research interest with practical economic and engineering considerations.

This book represents a collection of recent achievements and the latest developments from well-known industry experts and academic researchers active in this particular field. The book is organized in 16 chapters, covering different aspects of data center connectivity technologies.

Chapter 1 begins with a contribution by Alibaba distinguished researchers and discusses from its networking perspective the requirements and technologies for the optical interconnect technologies in datacenters, including intra-datacenter and inter-datacenter interconnects. Alibaba is the world's largest retailer with online sales and profits surpassing Walmart, Amazon, and eBay combined. Alibaba is expanding its cloud data center footprint from 14 data centers in operation and growing rapidly to support its global expansion.

Chapter 2 is written by semiconductor laser veteran who pioneered the Vertical cavity surface emitting lasers (VCSEL) from its early days of fabrication. VCSELs have been the primary laser sources for short-distance datacom and data centers over multimode fibers from 1 to 25 Gbps per channel in the form of either serial or parallel optical data links.

Chapter 3 and 4 review two critical transmitter sources: directly modulated laser diodes (DML) and electro-absorption modulated lasers (EML) from two industry leading suppliers by Finisar and NeoPhotonics separately. DML and EML are widely used by default for data centers and client side transceivers.

Chapter 5 provides an overview of optical fibers and connectors used for data center connectivity from the perspective of fiber manufacturer. Multimode fibers still has many advantages in short-reach applications that have been enhanced by the introduction of WDM technology. On the other hand, single-mode fibers have become more prevalent in hyper-scale cloud based datacenters that have much longer reach requirements.

Chapter 6 to 8 discusses various optical modulation technologies for direct detection: PAM4, DMT and Duobinary. PAM4 has been adopted by

IEEE802.3 as signaling standard by default for Ethernet inside data centers. DMT was introduced to boost the line rate with the limited bandwidth of optoelectronic devices. It may play an important role in Metro DCI. Three-level Duobinary may also be useful for power-efficient ultra-high serial rate optical links.

Chapter 9 describes the basic principles and presents industrial features of LiNbO$_3$ (LN) Mach-Zehnder Modulator (MZM) which has been successfully operated in long distance and high capacity optical fiber transmission systems for over 30 years. Compact size with higher bandwidth LN-MZM is strongly demanded especially in the field of application related to the data center. New technologies are expected to be introduced in the production of next generation LN-MZM.

Chapter 10 provides an overview of implementing the PAM4 and Si Photonics for achieving the 100 Gbit/s, DWDM datacenter interconnections that span regional distances of up to 120 km. The combination of Si photonics for the highly integrated optical components, and high speed Si CMOS for signal processing is critical for the implementation of low-cost, low-power, switch pluggable optical modules.

Chapter 11 is devoted to the ultra-low-loss photonics integrated circuits (PIC) and packaging, contributed by researchers of UC Santa Barbara. PIC is a device that integrates multiple (at least two) photonic functions in optical domain and as such is similar to an electronics integrated circuit. Compact PICs are critical to address cost, size, weight, and form factor. One PIC example is integrated dispersion compensation fabricated in a low loss silicon nitride platform mitigating PAM4 dispersion for multiple WDM channels while at the same time satisfying the strict OSNR requirements.

Chapter 12 is instructive in terms of the common measurement techniques used for data centers, especially for high speed optical measurements at longer distance. The authors discuss in details the polarization effects, OSNR measurements and characterization of optical vector modulated signals such as what's used in coherent detection.

Chapter 13 provide a concise summary of various digital signal processing (DSP) technique, while Chapter 14 describes in details the multi-dimensional polarization modulation. The research community has put tremendous efforts to mirror the achievements in long-haul coherent transceivers for developing DSP strategies suitable for short-reach datacenter connectivity. If readers are interested in designing solutions beyond 400Gb/s, it is recommended to read through those two chapters written by DSP experts and testing professionals.

Chapter 15 is contributed from network operator perspective to look into the high speed flexible coherent optical transport networks. Flexible coherent transport supports multiple baud rates, modulation orders, and data rates to serve more capacity for inter-datacenter applications. Coherent tends to move to shorter reaches as baud rate increases, for example 400 Gbit/s switch pluggable 16QAM coherent modules are being standardized to address next-gen DCI modules.

Chapter 16 is devoted to present ultra-low-power SiGe Driver IC for high-speed EMLs. It's emphasized that co-design of the driver IC with the EML could enhance the TOSA power efficiency at high baud rates. If readers are interested in designing analog ICs working at beyond 53Gbaud rates, it is recommended to read through this chapter written by SiGe device expert.

It is our wish to present this book as a comprehensive reference of data center connectivity technologies for those interested in developing and understanding this fast-growing area. I sincerely believe that it will provide a major boost in your understanding of the various latest technologies after reading through the chapters by the international experts in industry and academia. The discussion points range from system requirements to component/IC design. They are useful to understand current state-of-the-art technologies in these development fields. Also relevant standard activities and technical details behind those are addressed. The readers can grasp the future potential of these interconnect technologies and research trends by reading this book.

This book is intended as a general reference for researchers, senior and graduate-level college students working in the field of data center networks. It can be also used by engineer and managers to obtain a working knowledge of data center connectivity technologies. The book provide the breadth for people who need to gain a generic understanding of this specific field, and the depth, for those who would like to dig deeper into data center connectivity technologies, and relevant areas.

Frank Chang, Ph.D
Silicon Valley, 2018
fymchang@gmail.com

Acknowledgements

First, I would really like to thank all the authors who have made tremendous efforts to contribute to this book out of their extreme busy schedule. I will always remember their contribution and active exchange of their thoughts and ideas.

I would like to thank River Publishers and Rajeev Prasad for giving me the opportunity to start this project and edit this comprehensive and valuable book. Special thanks to Junko Nakajima, our production manager, who has been extremely patient with my constant slipping of the schedule, and also to her for her excellent project management.

Last but not the least, I want to express my gratitude to my wife, Jenny. She was the first to provide invaluable insights to my work in progress during my completing this work. Since this book was completed mostly in my spare time at home with endless weekends, she gave up a lot of her free time to support me working on this project.

Frank Chang

Acknowledgements

First, I would really like to thank all the authors who have made tremendous efforts to contribute to this book out of their extreme busy schedule. I will always remember their contribution and active exchange of their thoughts and ideas.

I would like to thank River Publishers and Rajeev Prasad for giving me the opportunity to start this project and edit this comprehensive and valuable book. Special thanks to Junko Nakajima, our production manager, who has been extremely patient with my constant slipping of the schedule, and also to her for her excellent project management.

Last but not the least, I want to express my gratitude to my wife, Jenny. She was the first to provide invaluable insights to my work in progress during my completing this work. Since this book was completed mostly in my spare time at home with endless weekends, she gave up a lot of her free time to support me working on this project.

Frank Chang

List of Contributors

Alan Pak Tao Lau, *Department of Electrical Engineering, The Hong Kong Polytechnic University, Hung Hom, Kowloon, Hong Kong SAR, China*

An Li, *Futurewei Technologies, Santa Clara, CA, USA*

Chao Lu, *Department of Electronic and Information Engineering, The Hong Kong Polytechnic University, Hung Hom, Kowloon, Hong Kong SAR, China*

Chongjin Xie, *Alibaba Infrastructure Service, Alibaba Group, Sunnyvale, CA, USA*

Daniel J. Blumenthal, *University of California Santa Barbara, California, USA*

Daniel Gariépy, *EXFO Inc., Quebec, Canada*

Demis D. John, *University of California Santa Barbara, California, USA*

Di Che, *The University of Melbourne, Melbourne, Australia*

Frank Chang, *Inphi Corp, California, USA*

Gang He, *EXFO Inc., Quebec, Canada*

Grant Brodnik, *University of California Santa Barbara, California, USA*

Glenn A. Wellbrock, *Verizon, Richardson, TX, USA*

Gordon Ning Liu, *Huawei Technologies Co., Ltd., Shenzhen, China*

Guy Torfs, *imec - Ghent University, IDLab, Gent, Belgium*

Hirochika Nakajima, *Waseda University, Tokyo, Japan*

Jiahao Huo, *University of Science and Technology Beijing, Beijing, China*

Johan Bauwelinck, *imec - Ghent University, IDLab, Gent, Belgium*

John Kamino, *OFS Fitel LLC, Norcross, GA, USA*

Jung Han Choi, *Fraunhofer Heinrich-Hertz-Institute, Berlin, Germany*

Kangping Zhong, *MACOM Technology Solutions Inc, Shenzhen, China*

Li Zeng, *Huawei Technology Ltd, Shenzhen, China*

Liang Zhang, *Huawei Technologies Co., Ltd., Shenzhen, China*

Mark Filer, *Microsoft Corporation, Redmond, WA, USA*

Michael Belt, *Honeywell Inc., California, USA*

Qian Hu, *Nokia Bell Labs, Stuttgart, Germany*

Radhakrishnan Nagarajan, *Inphi Corporation, Santa Clara, CA, USA*

Renan L. Moreira, *University of California Santa Barbara, California, USA*

Sarat Gundavarapu, *University of California Santa Barbara, California, USA*

Steve Yao, *General Photonics Corporation, California, USA*

Taran Huffman, *GenXComm Inc., Austin, TX, USA*

Tianjian Zuo, *Huawei Technologies Co., Ltd., Shenzhen, China*

Trevor Chan, *Neophotonics, California, USA*

Tiejun J. Xia, *Verizon, Richardson, TX, USA*

Wajih Daab, *General Photonics Corporation, California, USA*

Wenbin Jiang, *WJ Technologies LLC, California, USA*

William Shieh, *The University of Melbourne, Melbourne, Australia*

Winston Way, *Neophotonics, California, USA*

Xi Chen, *Nokia Bell Labs, Holmdel, New Jersey, USA*

Xian Zhou, *University of Science and Technology Beijing, Beijing, China*

Xin (Scott) Yin, *imec - Ghent University, IDLab, Gent, Belgium*

Yasuhiro Matsui, *Finisar Corporation, Fremont, CA, USA*

Yi Sun, *OFS Fitel LLC, Norcross, GA, USA*

Yuya Yamaguchi, *National Institute of Information and Communications Technology (NICT), Tokyo, Japan*

List of Figures

List of Tables

List of Abbreviations

16QAM	16-ary quadrature amplitude modulation
2-D	2-dimensional
5G	5th generation wireless services
AAF	Anti-Aliasing Filter
ADC	Analog-to-Digital Converter
AGC	Automatic Gain Control
Al_2O_3	Aluminum Oxide
APC	Auto power control
APD	Avalanche Photodiode
AR	Anti-Reflection
ASE	Amplified spontaneous emission
ASIC	Application-specific Integrated Circuit
AWG	Arrayed Waveguide Grating
AWG	Arbitrary Waveform Generator
AWGR	Arrayed Waveguide Grating Router
BC	Baseband component
BCB	Benzocyclobutene
BER	Bit error rate
BI-MMF	Bending Insensitive MMF
BTB	Back to back
BTJ	Buried tunneling junction
BW	Bandwidth
CAGR	Compound Annual Growth Rate
CC	Cross-beating component
CD	Compact disk
CD	Chromatic Dispersion
CDC	Colorless, directionless, and contentionless
CDC-F	CDC and flexible
CDM	Complex direct modulation
CDR	Clock and data recovery
CFP	C form-factor pluggable

CHIL	Channel Insertion Loss
CIN	Cloud Integrated Network
CL	Conversion Loss
CL	Co-location
CM	Complex modulation
CMA	Constant modulus algorithm
CMOS	Complementary metal–oxide–semiconductor
CO	Central office
CoC	Chip-on-carrier
COHD	Coherent detection
CP	Cyclic prefix
CSPR	Carrier-signal power ratio
CTLE	Continuous Time Linear Equalization
CW	Continuous wave
CWDM	Coarse Wavelength Division Multiplexing
DAC	Digital-to-Analog Converter
DB	Duobinary
DBR	Diffractive Bragg reflector
DBR	Distributed Bragg reflector
DC	Datacenter
DC	Direct current
DCF	Dispersion Compensating Fiber
DCI	Datacenter interconnect
DCM	Dispersion-compensated module
DD	Direct detection
DD-MZM	Dual-driver Mach–Zender modulator
DEMUX	Demultiplexer
DFB	Distributed feedback laser
DFE	Decision Feedback Equalizer
DFT	Discrete Fourier transform
DGD	Differential Group Delay
DH	Double heterojunction
DM	Direct modulation
DMD	Differential Modal Delay
DML	Direct Modulated Laser
DMT	Discrete Multi Tone
DOP	Degree of Polarization
DPP	Dispersion power penalty
DPSK	Differential-phase-shift keying

DQPSK	Differential QPSK
DSB	Double sideband
DSL	Digital subscriber line
DSO	Digital Sampling Oscilloscope
DSP	Digital Signal Processing
DUT	Device under Test
DWDM	Dense Wavelength Division Multiplexing
DWDM	Dense WDM
EAM	Electro-absorption modulator
ECL	External cavity laser
ED	Error detector
EDB	Electrical duobinary
EDC	Electrical dispersion compensation
EDFA	Erbium-doped Fiber Amplifier
EF	Encircled Flux
EO	Electro optical
EM	Electromagnetics
EM	External modulation
EMB	Effective modal bandwidth
EML	Electro-absorption modulated DFB Laser
EO	Electronic-Optical
EON	Elastic optical network
ER	Extinction Ratio
ETDM	Electrical time-division multiplexing
EVM	Error Vector Magnitude
FEC	Forward error correction
FFE	Feed forward equalizer
FFT	Fast Fourier transform
FIR	Finite impulse Response
FMF	Few-mode Fiber
FSR	Free-Spectral Range
FWHM	Full-Width at Half-Maximum
FWM	Four-Wave Mixing
GbE	Gigabit Ethernet
Gbps	Gigabit per second
GC-SOA	Gain-Clamped Semiconductor Optical Amplifier
GI-MMF	Graded-index multimode fiber
GRINSCH	Graded index separate confinement heterostructure
HD-FEC	Hard decision forward error correction

HHI	Heinrich-Hertz Institute
High-Q	High Quality Factor
HPC	High performance computing
HR-DMD	High-resolution DMD
IFFT	Inverse fast Fourier transform
III-V	Alloys of elements from group Three and Five of the Periodic Table
IL	Insertion Loss
IM	Intensity modulation
IMDD	Intensity modulation and direct detection
InP	Indium Phosphide
IoT	Internet of Things
IP	Internet protocol
IQ-MZM	In-phase/quadrature Mach–Zehnder modulators
ISI	Inter-symbol interference
LAN	Local area network
LC	Liquid crystal
LCoS	LC on silicon
LED	Light-Emitting Diode
LER	Line-Edge Roughness
LIDAR	Light Detection and Ranging
$LiNbO_3$	Lithium niobate
LMS	Least-mean-squared algorithm
LO	Local oscillator
LP	Linearly polarized
LPCVD	Low-pressure chemical Vapor Deposition
LPF	Low-pass filter
LSB	Least Significant Bit
LW	Long wavelength
MBE	Molecular beam epitaxy
MCF	Multicore Fiber
MCS	Multi-cast switch
MDI	Medium dependent interface
MDIO	Management Data Input/Output
MFD	Mode field diameter
MIMO	Multi-input-multi-output
MLSE	Maximum likelihood sequence estimation
MMA	Multi-modulus algorithm
MMF	Multi-mode fiber

MN	Modal noise
MO	Magneto Optic
MOCVD	Metal organic chemical vapor deposition
MPD	Mode Power Distribution
MPI	Multi-Path Interference
MPLS	Multiprotocol label switching
MPN	Mode partition noise
MPW	Multi-Project Wafer
MQW	Multiple Quantum Well
MSA	Multi-source agreement
MSB	Most Significant Bit
MUX	Multiplexer
MZI	Mach–Zehnder Interferometer
MZM	Mach–Zehnder Modulators
NA	Numerical aperture
NC	Noncoherent
NRZ	Non-Return-to-Zero
OBO	On Board Optics
OBTB	Optical Back-to-back
ODB	Optical duobinary
OE	Optical-Electronic
OFDM	Orthogonal frequency-division multiplexing
OFL-BW	Overfilled Launch Bandwidth
OLSP	Optical-line-section protection
OM	Optical multimode
OMA	Optical Modulation Amplitude
OMB	Overfilled Modal Bandwidth
OMSP	Optical-multiplex-section protection
ONT	Optical network tester
OOK	On-Off Keying
OSA	Optical Spectrum Analyzer
OSNR	Optical-to-signal-noise ratio
OTN	Optical transport network
OUT	Optical transport unit
P/S	Parallel to serial
PAM	Pulse-amplitude modulation
PAM4	4-level pulse-amplitude modulation
PAPR	Peak-to-average power ratio
PBC/S	Polarization beam combiner/splitter

PBS	Polarization-beam splitter
PD	Photo-detector
PD	Photodiode
PDG	Polarization Dependent Gain
PDL	Polarization Dependent Loss
PDLC	Polarization Dependent Loss Compensation
PDM	Polarization Division Multiplexing
PDR	Polarization Dependent Response
PDS	Polarization Dependent Sensitivity
PDW	Polarization Dependent Ceneter Wavelength
PECVD	Plasma-Enhanced Chemical Vapor Deposition
PER	Polarization Extinction Ratio
PHY	Physical Layer
PIC	Photonic Integrated Circuit
PIN	Positive-intrinsic-negative
PIN	P-I-N Photodiode
PLC	Planar Lightwave Circuit – usually referring to a passive PIC of glass materials
PLL	Phase-locked loop
PM	Phase modulator
PM	Polarization-multiplexed
PMD	Polarization Mode Dispersion
PMDC	Polarization Mode Dispersion Compensation
PMG	Principal Mode Group
POL	Polarization
POL-(DE)MUX	Polarization (de)multiplexing
Pol-Muxed	Polarization Multiplexed
POP	point of presence
PPG	Pattern generator
PRBS	Pseudo-Random Binary Sequence
PSA	Polarization State Analyzer
PSG	Polarization State Generator
PSM	parallel single mode
PSM4	Parallel Single Mode 4 Channel
PSP	Principal State of Polarization
PZT	Lead Zirconium Titanate, a common piezo-electric material
QAM	Quadrature-amplitude modulation
QCSE	Quantum Confined Stark Effect

QPSK	Quadrature Phase Shift Keying
QW	Quantum well
RC	Raised-cosine
RF	Radio-Frequency
RHEED	Reflectoin high energy electon diffraction
RIE	Reactive ion etching
RIN	Relative intensity noise
RM	Rotation matrix
RMS	Root Mean Square
ROADM	Reconfigurable optical add/drop multiplexer
ROP	Received Optical Power
ROSA	Receiver optical sub-assembly
RTA	Rapid thermal annealing
RU	Rack unit
RX	Receiver
S/P	Serial to parallel
SDM	Space-division multiplexing
SDN	Software defined network
SE	Spectral efficiency
SECQ	Stress Eye Closure Quaternary (SECQ)
SerDes	Serializer/Deserializer
Si	Silicon
Si_3N_4	Silicon Nitride
SiP	Silicon Photonics
SLC	Single-layer capacitor
SLPM	Standard Litres per Minute
SM	Single-mode
SMF	Single-mode fiber
SMSR	Side-Mode Suppression Ratio
SNR	Signal-to-noise ratio
SOA	Semiconductor Optical Amplifier
SOC	System-On-Chip
SD-FEC	Soft decision forward error correction
SONET	Synchronous optical network
SOP	State of Polarization
SOPMD	Second Order Polarization Mode Dispersion
SRS	Stressed Receiver Sensitivity
SSB	Single sideband
SSBI	Signal to signal beating interference

SSBN	Signal-signal beat noise
SSM	Stokes-space modulation
SSMF	Standard single-mode fibers
SSPR	Short Stress Pattern Random
SSPRQ	Short Stress Pattern Random - Quaternary
SVR	Stokes vector receiver
SWDM	Shortwave wavelength-division multiplexing
TbE	Terabit Ethernet
TCO	Total cost of ownership
TDC	Tunable Dispersion Compensator
TDECQ	Transmitter and Dispersion Eye Closure Quaternary
TDM	Time-division multiplexing
TDP	Transmitter Dispersion Penalty
TE/TM	Transverse Electric/Magnetic Field
TEC	Temperature controller
TEC	Thermo-Electric Cooler
TEM	Transmission electron microscopy
TIA	Transimpedance Amplifiers
TLS	Tuanble Laser Source
TOSA	Transmitter optical sub-assembly
TW	Traveling-wave
TWA	Traveling-wave amplifier
TX	Transmitter
UCSB	University of California Santa Barbara
UHV	Ultrahigh vacuum
ULL	Ultra-Low Loss
ULLW	Ultra-Low Loss Waveguide
VCSEL	Vertical-cavity surface-emitting lasers
VIN	Visual Networking Index
VOA	Variable Optical Attenuator
WAN	Wide area networks
WBMMF	Wideband MMF
WDM	Wavelength Division Multiplexing

1

Optical Interconnect Technologies for Datacenter Networks

Chongjin Xie

Alibaba Infrastructure Service, Alibaba Group, Sunnyvale, CA, USA

1.1 Introduction

In 1965, Kao with Hockham found that by removing the impurity in glass material the fundamental limitation for light attenuation in glass is below 20 dB/km [1]. In 1970, Schultz, Keck, and Maurer in Corning made the first low-loss optical fibers, with a loss coefficient of 17 dB/km [2]. These started the era of fiber-optic communications. The first field experiment of fiber-optic communication systems was conducted in 1976, at a bit rate of 45 Mbps [3]. Since then, significant progress has been made in the field. With the advances in technologies such as time-division multiplexing (TDM), wavelength-division multiplexing (WDM), polarization-division multiplexing, digital coherent detection, etc., the capacity of fiber has increased by more than 10^6 times. Single-mode fiber (SMF) capacity has reached 100 Tbps [4], and by using multimode and multi-core technologies, over 2-Pbps capacity on a single fiber has been achieved [5, 6].

In the past, optical communications had mostly been used by telecom operators in their long-haul, metro, and access networks, to connect their central offices with long-haul and metro networks and bring their customers to central offices with access networks. In 2008, the demand for optical communications from hyper-scale datacenters exceeded that of telecom operators in the United States [7]. Today optical interconnect demand from datacenters has become a major driving force for optical communications.

Internet services have become part of our daily lives. Almost all the internet services such as web-browsing, e-mail, e-commerce, video streaming,

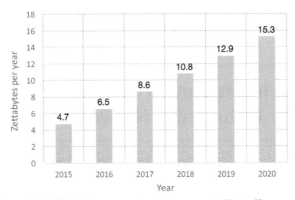

Figure 1.1   Global datacenter internet protocol (IP) traffic growth.

social networking, and cloud computing run in datacenters, where a massive number of servers are connected and work together like a supercomputer [8], which generates a huge amount of internet traffic both inside and outside of datacenters. Most internet traffic has originated or terminated in a datacenter since 2008, and datacenter traffic will continue to dominate internet traffic in the foreseeable future. As shown in Figure 1.1, the annual global datacenter internet protocol (IP) traffic will reach 15.3 ZB (Zettabytes, Zetta = 10^{21}) in 2020, up from 4.7 ZB in 2015, a three-fold increase in 5 years with a compound annual growth rate (CAGR) of 27% in the 5-year time frame, driven mainly by video streaming and cloud computing [9]. The distribution of global internet traffic by destinations forecasted by Cisco is given in Figure 1.2 [9], which shows that by 2020 non-datacenter-related internet traffic only accounts for less than 1% of the total IP traffic, and more than 99% of internet traffic will be originated and/or terminated in a datacenter. Datacenters have replaced telecom center offices to become the center of internet traffic.

A typical architecture of datacenter networks is shown in Figure 1.3. In general, a worldwide hyper-scale datacenter operator has datacenters distributed around the world. They divide the world into many regions, and these regions are connected with optical mesh networks. Each region has one or a few gateways connecting to public internet, through which users can access their datacenters. In each region, there are one or a few metro areas, depending on customer distributions. Typically, a few datacenters are set up in a metro area with distances among them limited mostly to 80 km, which is the consideration of both disaster prevention and the latency requirement set by synchronous replication among different datacenters. To improve user

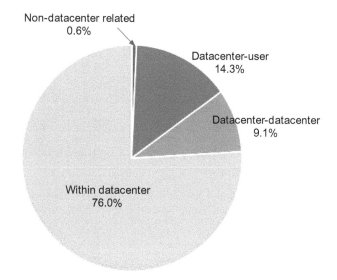

Figure 1.2   Global internet protocol (IP) traffic by destination in 2020.

experience and reduce the latency for customers connecting to their networks, many points of presence (POPs) may be built close to customers. The POPs have direct connections to datacenters so that they can quickly bring customer traffic to datacenters. These POPs can also be used as edging computing sites where data can be computed and processed locally. This brings two benefits: (1) the computation is close to data sources so that intelligence can be provided locally with reduced latency; (2) most data are processed locally so that the bandwidth requirement on the connections from POPs to datacenters can be significantly reduced. Inside datacenters, servers are connected through intra-datacenter networks.

From Figure 1.3 and the above descriptions, one can see that a datacenter network infrastructure includes access networks (connecting POPs to datacenters), backbone networks (connecting datacenters in different metro areas), metro networks (connecting datacenters within a metro area), and intra-datacenter networks (connecting servers inside a datacenter). Each of these networks has its own characteristics and requirements. The distance ranges for these networks are quite different. Intra-datacenter distances are less than 2 km and metro-network links are less than 80 km, but for backbone and access networks, hundreds to thousands of kilometers are needed. The different length requirements of different networks result in different

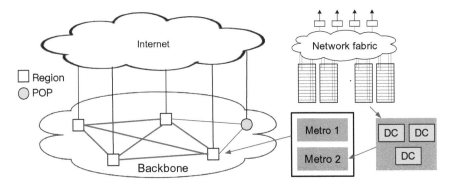

Figure 1.3 Architecture of a datacenter network, including intra-datacenter networks, metro-networks, backbone networks, and access networks. POP: point of presence; DC: datacenter.

optical interconnect technologies. In the following sections, we will discuss the details on optical technologies for intra-datacenter interconnects and inter-datacenter interconnects in metro networks and wide-area-networks (WANs).

1.2 Intra-datacenter Interconnects

The computing and storage powers needed by internet services and cloud computing are provided by a massive number of servers located in each datacenter. Servers in each datacenter are connected with intra-datacenter networks so that they can communicate and exchange information with each other. The ideal solution is to use one big switch to connect all these servers, but this solution is technically difficult and cost prohibitive. In practice, an intra-datacenter network is built with many small switches and they are architected in multiple layers. Lots of network architectures have been proposed and implemented for datacenters, with the aim to reduce cost and increase scalability and efficiency for a given bisection bandwidth and number of connected servers [10, 11]. Figure 1.4 shows a typical architecture of an intra-datacenter network. It has three layers of switches. The edge switches are directly connected to servers, and the aggregate switches and core switches are used to increase the size and the scalability of the network. The incoming and outgoing traffic is handled by the routers and transmitted to other datacenters in metro networks or wide-area networks by optical transport networks. Due to the massive number of servers in a datacenter, there are lots of links

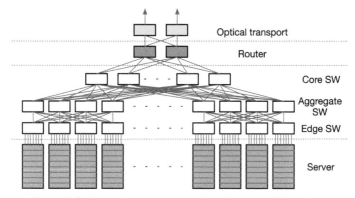

Figure 1.4 Intra-datacenter network architecture. SW: switch.

in a datacenter network, as well as switches. For a datacenter with 10,000 servers, the numbers of links and switches can be from 15,000 to 30,000 and 1,000 to 3,000, respectively, depending on the network architecture [10].

At low speeds, these switches and servers are connected with copper cables. With the increase of network speeds and sizes, copper cables may not be able to meet most of the interconnect requirements. From 1 Gb-Ethernet, optical technologies start to be widely used in datacenters. In today's datacenters, optical interconnects are used in almost every connection outside of servers, providing high-bandwidth channels between connected objects. At the interface of an electrical switch, optical transceivers are used to convert the outgoing signals from the electrical domain to the optical domain and the incoming signals from the optical domain to the electrical domain, and optical fibers are used as media to transmit optical signals from one location to another, as shown in Figure 1.5 [12]. Optical interconnects are used as big "information pipes" in the networks.

There are three different speeds in datacenter networks, server speed (network-interface-card speed of a server), switch speed, and router speed, and datacenter network speed usually refers to the switch speed. Datacenter

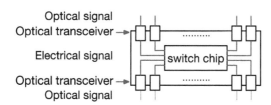

Figure 1.5 Diagram of optical transceivers in an electrical switch.

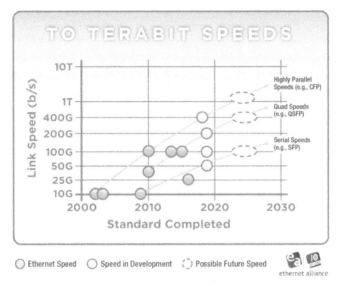

Figure 1.6   Ethernet speed roadmap issued by Ethernet Alliance.

traffic is mostly Ethernet traffic. Figure 1.6 gives the Ethernet speed evolution published by Ethernet Alliance, from 10G to Tb [13]. There are three speed evolution curves. One curve is for "Serial Speeds", which are determined by the speeds of SERDES and are usually the same as the speeds of servers. One way to get to higher speeds is to use multiple parallel lanes, and four lanes are widely used for switches, as the "Quad Speeds" curve shows. For example, in 40G networks, server speed is 10G, and the switches use four 10G lanes to get to 40G. The next server speed after 10G is 25G and the corresponding switch speed is 100G, which can be supported by 25G serial speed and 100G quad speed that has already been standardized and deployed, as shown in Figure 1.6. The datacenter speeds after 100G is either 200G or 400G. 200G networks can be supported by 50G serial speed for servers and 200G quad speed for switches as illustrated in Figure 1.6. The best way to realize a 400G datacenter network is to use a 100G serial speed for servers and a 400G quad speed for switches, but due to the limitations of SERDES, 100G serial speed may not be realized until 2025 according to Ethernet Alliance Roadmap. Before that, 400G networks can be achieved by using higher parallelism, for example, $2 \times 50G$ for 100G servers and $8 \times 50G$ for 400G switches. The "Highly Parallel Speeds" curve in Figure 1.6 is in general for routers and optical transport equipment, which typically demand higher speeds than switches and can be realized with more than four parallel lanes. For example, early-day 100G and 400G are realized with ten 10G lanes and sixteen 25G lanes, i.e., $10 \times 10G$ and $16 \times 25G$, respectively.

Due to short reaches and large quantities, there are different requirements on connections inside datacenters from those used in long-haul, metro, and access networks, so different technologies have been developed. Most links inside datacenters are in the range of a few meters to a few hundred meters. There are some long links between different buildings in a campus, with the longest link limited to 2 km. Bandwidth cost is the primary requirement for intra-datacenter connections. When calculating cost, the cost of both optical transceivers and fibers need to be included. The second requirement is bandwidth density or faceplate density, which can be quantified by the capacity that the front panel of a 19-in 1-RU (482.60mm $\times$ 44.45mm) switch can accommodate. The bandwidth of switch I/O still evolves in Moore's law, as shown in Figure 1.7 [14]. To fully utilize the switch bandwidth, the bandwidth density of optical interconnects must be able to scale in the same speed. Figure 1.7 shows that before 2010, the bandwidth density of pluggable optical transceivers is higher than that of switch I/O, but it is not anymore after 2010. The bandwidth density is determined by not only the speeds, but the form factors of transceivers as well. Figure 1.8 compares the port density of different form factors, i.e., how many transceivers of different form factors can fit in a 1-RU face plate. There are more than a dozen optical transceiver form

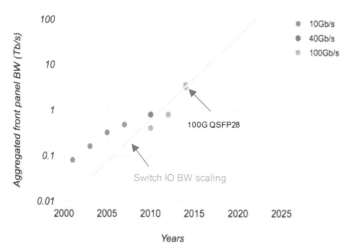

Figure 1.7 Historical view of front panel bandwidth of switch I/O and pluggable optics per RU. RU: rack unit. Courtesy of X. Zhou.

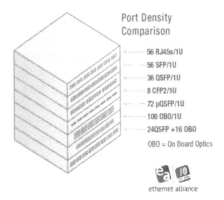

Figure 1.8   Port density comparison for different form factors.

factors, and SFP and QSFP have become the main transceiver form factors in datacenters. The third requirement is the power consumption of optical transceivers. Although the power consumption of optical transceivers or even networks is a small portion of the total power consumed by a datacenter, power-efficient optical transceivers are critical for high bandwidth-density switches as low power consumption optical transceivers can help power dissipation and increase bandwidth density.

1.2.1 40G Optical Interconnect Technologies

Up to 40G, most intra-datacenter optical interconnects use multimode technology, i.e., vertical-cavity surface-emitting lasers (VCSELs) combined with multimode fiber (MMF). VCSELs not only have a vertical structure so that thousands of VCSELs can be processed simultaneously on a single wafer, but are of low power consumption as well. Compared with SMF, MMF has a larger core size and numerical aperture so that it is much easier to couple light in and out MMF than SMF. Due to these two reasons, multimode technology is widely used in short-reach applications. However, light launched into MMF excites different modes, which travel at different speeds and limit the bandwidth of MMF, as shown in Figure 1.9. To increase the bandwidth of MMF, four generations of MMF have been developed, from OM1 to OM4. Table 1.1 gives the characteristics of different MMF. The bandwidths and link distances have been significantly improved from OM1 to OM4. For example, the link distance of a 10G system has been increased from 33 m for OM1 to 550 m for OM4. Note that the link distances listed in Table 1.1 are defined by IEEE standard [15, 16], and the link distances can be further increased using some distance-enhancement technologies.

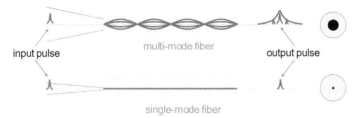

Figure 1.9 Schematic of pulse propagation in multimode and single-mode fibers.

Table 1.1 Characteristics of different multimode fibers

Fiber	Core Diameter (μm)	EMB (MHz.km)	Link Distance (m)			
		850 nm	1G	10G	40G	100G
OM1	62.5	N/A160–200	275	33	N/A	N/A
OM2	50	N/A400–500	550	82	N/A	N/A
OM3	50	2000	N/A	300	100	70
OM4/OM5	50	4700	N/A	550	150	100

EMB: effective mode bandwidth

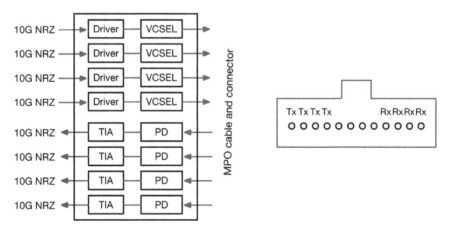

Figure 1.10 Schematic of 40G SR4 and MPO connector layout.

There are mainly two types of 40G optical transceivers – one uses multi-mode technology, called SR4, and the other single-mode technology, called LR4. Both are standardized by IEEE [15]. The schematic of a 40G SR4 optical transceiver is depicted in Figure 1.10. It consists of four electrical lanes and optical lanes in each direction, with each lane working at 10G. The lasers are VCSELs with wavelength at 840 nm ~ 860 nm, and the four optical lanes are carried by four MMFs. The optical transceiver uses 12 fiber multi-parallel optics (MPO) cables and connectors, with four fibers not connected

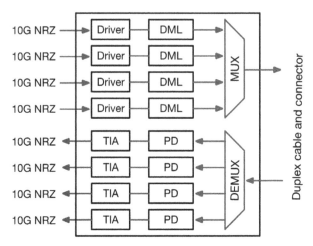

Figure 1.11   Schematic of 40G LR4.

or empty, as shown in Figure 1.10. Although according to IEEE standard, 40G SR4 can only reach 100 m and 150 m over OM3 and OM4 fibers, respectively, the distance of 40G can be significantly improved using some distance-enhancement technologies such as equalization and low-linewidth VCSELs. Up to 300 m and 550 m over OM3 and OM4 fibers can be achieved, respectively, using enhanced 40G SR4 technologies, which cover most of the links in datacenters.

For links of distances exceeding the range of multi-mode technology, one can use 40G LR4 transceivers and SMF. The schematic of a 40G LR4 transceiver is illustrated in Figure 1.11. It also has four electrical and four optical lanes with each lane working at 10G. Unlike SR4, it uses either directly-modulated lasers (DMLs) or electro-absorption-modulated lasers (EMLs) and the four optical lanes have different wavelengths according to coarse wavelength division multiplexing (CWDM) specifications, which are multiplexed into a single fiber with a wavelength multiplexer. In the receiver side, the four optical lanes are separated with a wavelength demultiplexer. Instead of an MPO cable, a duplex fiber cable is used for an LR4 optical transceiver, and the link distance can reach 10 km. The wavelengths of the four lanes are given in Table 1.2. The wavelength spacing of the four lanes is 20 nm and wavelength range is $\pm 6.5\ nm$, which allows the use of uncooled lasers.

As no optical amplifier and dispersion compensation are used, one of the most important parameters for short-reach optical interconnect technologies

Table 1.2 Wavelength assignment of 40G LR4 (CWDM)

Lane	Center Wavelength	Wavelength Range
0	1271 nm	1264.5 to 1277.5 nm
1	1291 nm	1284.5 to 1297.5 nm
2	1311 nm	1304.5 to 1317.5 nm
3	1331 nm	1324.5 to 1327.5 nm

Table 1.3 Link power budget of 40G SR4 and LR4 (in dB)

	SR4 (for OM3)	LR4
Power budget	8.3	9.3
Channel insertion loss	1.9	6.7
Allocation for penalties	6.4	2.6

is the link power budget. Table 1.3 gives the link power budget for 40G SR4 and LR4 [15]. The total power budget for SR4 and LR4 are 8.3 and 9.3 dB respectively. Most of the power budget is allocated to the transmitter and dispersion penalties for SR4 due to large-mode dispersion of MMF. For LR4, most of the power budget is allocated to insertion loss because of the long transmission length. Note that the channel insertion loss is calculated using the maximum distances specified in IEEE standard and cabled optical fiber attenuation, which is 3.5 dB/km for MMF at 850 nm and 0.47 dB/km for SMF at 1264.5 nm, plus an allocation for connection and splice loss, which is 1.5 dB and 2 dB for MMF and SMF, respectively.

1.2.2 100G Optical Interconnect Technologies

While multi-mode technologies can cover most of the links in datacenters, they are not any more at 100G due to larger mode dispersion penalties at higher bit rates. A large number of links have to use single-mode technologies in 100G as IEEE only standardized one single-mode 100G technology, i.e., 100G LR, which covers up to 10 km over SMF. To reduce the chromatic dispersion penalties, 100G LR4 uses local area network (LAN) WDM, whose wavelength assignment is given in Table 1.4 [15]. The channel spacing of LAN WDM is about 4.5 nm (800 GHz) and the wavelength range about $\pm 1.0\ nm$, so uncooled lasers cannot be used and temperature controllers (TECs) are required. Considering these facts, 100G LR4 does not meet the bandwidth cost and power consumption requirements for datacenter applications. Therefore many different 100G single-mode technologies for datacenter optical interconnects are developed by multi-source agreements (MSAs).

Table 1.4 Wavelength assignment of 100G LR4 (LAN WDM)

Lane	Center Frequency	Center Wavelength	Wavelength Range
0	231.4 THz	1295.56 nm	1294.53 to 1296.59 nm
1	230.6 THz	1300.05 nm	1299.02 to 1301.09 nm
2	229.8 THz	1304.58 nm	1303.54 to 1305.63 nm
3	229.0 THz	1309.14 nm	1308.09 to 1310.19 nm

One single-mode 100G transceiver MSA is CWDM4 [17]. The required reach of CWDM4 is up to 2 km on standard SMF (SSMF). Compared with LR4, two changes are made in 100G CWDM4 to reduce cost. First, it uses course WDM wavelengths instead of LAN WDM. The wavelengths are the same as those of 40G LR4, i.e., from 1291 nm to 1331 nm with a 20-nm channel spacing. Therefore, uncooled lasers can be used. Second, the optical link is specified to operate at a bit error ratio (BER) of 5×10^{-5}, not 1×10^{-12}, which relaxes the requirements on transmitter power and receiver sensitivity, but Reed–Solomon forward error correction (FEC) RS(528,514) is required to be implemented by the host in order to ensure reliable system operation of the host system. Another similar 100G transceiver technique is CLR4 [18]. The only difference between CLR4 and CWDM4 is that CLR4 supports two operation modes, an FEC-enabled mode for an increased link margin and an FEC-disabled mode for lower power consumption and lower latency.

Another single-mode 100G transceiver MSA is PSM4 (PSM stands for parallel single mode), targeting a low-cost solution capable of reaching up to 500 m over a parallel SMF infrastructure [19]. For PSM4, each optical lane is carried not by an optical wavelength but an optical fiber and the laser wavelength range is from 1295 nm to 1325 nm, which further reduces the requirements on lasers compared with CWDM4. In addition, when external modulators are used, which is the case for silicon photonics technologies, lasers can be shared by two or four optical lanes. Figure 1.12 shows the diagrams of two different PSM4 transceivers, where one uses four DML and the other one has one laser shared by four external modulators. Note that in the case of four lasers, external modulators can also be used. Like CWDM4, PSM4 specifies the optical link to operate at a BER of 5×10^{-5} too.

Table 1.5 lists the main characteristics of four 100G technologies, including the reaches, single-mode or multimode techniques, fiber cables types, BER requirements and power budget. Note that the reaches listed here are defined by IEEE and MSAs, and real products may have longer reaches.

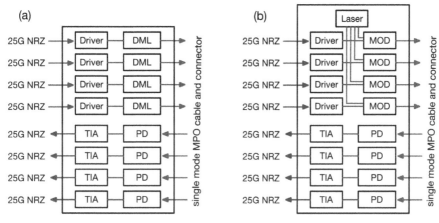

Figure 1.12 Diagrams of two 100G PSM4 transceiver architectures. (a) Using DML, (b) using external modulators and one laser.

Table 1.5 Main characteristics of different 100G technologies

	SR4	CWDM4	PSM4	LR4
Reach (m)	2–100	2–2,000	2–500	2–10,000
Mode	multimode	single-mode	single-mode	single-mode
Fiber	MMF MPO	SMF, Duplex	SMF, MPO	SMF, Duplex
BER	5×10^{-5}	5×10^{-5}	5×10^{-5}	1×10^{-12}
Power budget (dB)	8.2	8.0	6.2	8.5

When choosing technologies, one of the main factors to consider is the Total Cost of Ownership (TCO). Optical interconnect cost includes the cost of optical transceivers and link fibers. SR4 has the least expensive transceivers but most expensive fibers, whereas for CWDM4, the cost of fibers is the lowest, but the cost of transceivers is the highest. The TCO of optical interconnects depend on many factors, including datacenter sizes, the lifetime of datacenters, and the cost of transceivers, and has to be calculated based on the parameters of each datacenter to get the most cost-effective solutions. Figure 1.13 depicts the link cost versus distance for the three different 100G technologies, SR4 (assuming it can reach 300 m with distance-enhancement techniques), PSM4 and CWDM4 [20]. In the figure, we assume that the PSM4 transceiver price is 1.5 times SR4, and CWDM4 two times PSM4. It shows that for small datacenters with most link distances less than 100 m, multimode technology is the most cost-effective, but for mega-datacenter with link distances large than 100 m, PSM4 technology has the lowest cost.

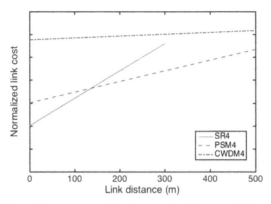

Figure 1.13   Link cost versus link distance for different 100G technologies.

The perspective of the technologies can be changed by some disruptive technologies in devices and components. For example, if the transceiver cost of CWDM4 is the same as that of PSM4, CWDM4 can become the dominant technology for most datacenters.

1.2.3 400G and Beyond Optical Interconnect Technologies

400G transceivers can be realized by $4 \times 100G$, i.e., to scale up the speeds of both electrical and optical lanes from 25G, currently used by 100G transceivers, to 100G. 100G per optical lane has been experimentally demonstrated in real time using external modulators including Mach-Zehnder modulators and electro-absorption modulators with various modulation formats such as 4-level pulse-amplitude modulation (PAM4), duobinary and NRZ [21–23]. For DML and VCSEL, 100G per channel was only demonstrated with offline processing using massive DSP due to their limited bandwidths [24–26]. Alternatively, one can also use the scale-out method to achieve 400G by increasing the number of lanes with lower lane speeds, such as $16 \times 25G$ or $8 \times 50G$. Optical lanes can be either fiber lanes or wavelength lanes, as shown in Figure 1.14. According to the speeds of electrical lanes (SERDES speeds) and optical lanes, 400G optical transceiver technologies can be divided into four generations, as illustrated in Table 1.6, with lane speeds evolving from 25G for the first generation to 100G for the fourth generation.

In March 2013, the IEEE 802.3 400 Gbps Ethernet Study Group was formed to explore development of a 400 Gbps Ethernet standard [27]. IEEE has specified four 400G transceiver technologies so far, which are

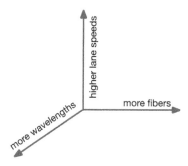

Figure 1.14 Methods to increase transceiver speeds.

Table 1.6 Generations of 400G transceiver technologies

			Multimode Technology		Single-mode Technology	
Generation	SERDES	Elec. I/O	Opt. I/O	Name	Opt. I/O	Name
1st	25G	16×25G	16×25G	SR16	8×50G	FR8, LR8
2nd	50G	8×50G	8×50G	SR8	8×50G	FR8, LR8
3rd	50G	8×50G	4×100G	SR4	4×100G	DR4, FR4, LR4
4th	100G	4×100G	4×100G	SR4	4×100G	DR4, FR4, LR4

SR16, DR4, FR8, and LR8. SR16 is a multimode technology, which uses 16 MMF lanes with a lane speed of 25G and can reach up to 100 m over OM4 fiber. It requires 32-fiber MPO cables and connectors. DR4, FR8, and LR8 are single-mode technologies, with reach up to 500 m, 2 km, and 10 km over SSMF, respectively. DR4 is equivalent to PSM4, which uses four parallel SMF lanes with the lane speed of 100G. FR8 and LR8 use eight wavelength lanes with the lane speed of 50G. An additional four wavelengths shorter than LAN WDM are assigned to FR8 and LR8, and the wavelength assignments are the same for FR8 and LR8 as shown in Table 1.7. The wavelength spacing in the upper and lower four lanes is 800 GHz, same as LAN WDM, but the spacing between the upper four wavelengths and lower four wavelengths is 1.6 THz. Temperature control may have to be used due to the narrow wavelength range requirement.

To reduce the requirements on the bandwidths of electronic and optical components, IEEE suggested using PAM4 for both electrical and optical signals when lane speeds are larger than 25G. PAM4 has a higher requirement on the power and relative intensity noise (RIN) of transmitters than non-return-to-zero (NRZ) modulation format [28]. Higher-gain FEC are suggested by IEEE for 400G, which has an FEC threshold at BER of 2.4×10^{-4}.

Table 1.7   Wavelength assignment of 400G FR8 and LR8

Lane	Center Frequency	Center Wavelength	Wavelength Range
0	235.4 THz	1,273.54 nm	1,272.55–1,274.54 nm
1	234.6 THz	1,277.89 nm	1,276.89–1,278.89 nm
2	233.8 THz	1,282.26 nm	1,281.25–1,283.27 nm
3	233.0 THz	1,286.66 nm	1,285.65–1,287.68 nm
4	231.4 THz	1,295.56 nm	1,294.53–1,296.59 nm
5	230.6 THz	1,300.05 nm	1,299.02–1,301.09 nm
6	229.8 THz	1,304.58 nm	1,303.54–1,305.63 nm
7	229.0 THz	1,309.14 nm	1,308.09–1,310.19 nm

Table 1.8   Characteristics different 400G transceiver form factors

	CFP8	OSFP	QSFP-DD	QSFP
Elec. I/O	$16 \times 25G/$ $8 \times 50G$	$8 \times 50G$	$8 \times 50G$	$4 \times 100G$
Size (mm)	$102 \times 40 \times 9.5$	$107.7 \times 22.6 \times$ 9	$78.3 \times 18.4 \times$ 8.5	$72.4 \times 18.4 \times$ 8.5
Capacity/RU	6.4T	12.8T	14.4T	14.4T

For datacenter operators, it is desirable to have four optical lanes for 400G single-mode solutions, either four fiber lanes (DR4) or wavelength lanes (FR4), same as 100G technologies. To meet the power and cost requirements, a larger channel spacing than that of FR8 may be needed so that TECs are not required for lasers. But with the lane speed at 100G, chromatic dispersion-induced penalties may become an issue at larger channel spacing, and some equalization techniques may need to be used.

As fiber cost is a significant part for the cost of multimode solutions, it is preferred not to use more than four fiber lanes in real implementation. Before the maturity of 100G VCSELs, one can reduce the fiber lanes of multimode solutions with either shortwave wavelength division multiplexing (SWDM) technique [29], which may transmit two VCSEL wavelengths in one fiber, or bidirectional technique, where two optical lanes counter-propagate in one fiber.

Before 400G QSFP becomes available, other form factors may be used, including CFP8 [30], OSFP [31], and QSFP-DD [32]. The main difference between these form factors are the size, and thus power dissipation capabilities and bandwidth density. The characteristics of these form factors are listed in Table 1.8. A form factor of a bigger size has a larger power dissipation capability and will be available earlier, and it may also be able to accommodate more complex modules such as coherent modules.

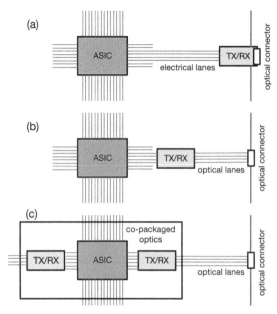

Figure 1.15 Diagrams of pluggable optics (a), mid-board optics (b), and co-packaged optics (c).

Going beyond 400G, mid-board optics [33] and even electro-optical co-packaging [34] may be needed. The benefits of mid-board optics and co-packaged optics can be seen in Figure 1.15. For pluggable optics, the long electrical lanes between a switch application-specific integrated circuit (ASIC) and optical transceivers can cause bigger signal distortions and penalties for higher speed signals. By moving optical transceivers close to switch ASIC, not only can the signal integrity be significantly improved due to the reduced length of electrical lanes, but the front panel port density can be increased as well since only optical connectors are in the front panel. Electro-optical co-packaging requires more optical integration and can achieve a higher bandwidth density. As optics are close to the hot switch ASIC in mid-board co-packaged optics, thermal control of optical chip could be a big challenge.

1.3 Inter-datacenter Interconnects

Dense-wavelength-division-multiplexing (DWDM) optical transport technologies are predominantly used for inter-datacenter interconnects in both metro networks and WANs. Figure 1.16 depicts the evolution of capacity carried by a single fiber. Due to advances in technologies such as TDM, WDM, digital coherent detection, digital signal processing (DSP), optical fibers, and optical amplifiers, optical transport systems can today achieve a capacity of 100 Tbps in a single SMF, over trans-oceanic distances with a capacity over 70 Tbps, and single-channel bit rate has reached 1 Tbps [4, 35, 36]. We are close to the fundamental limit of fiber-optic communication systems that is determined by non-linear Shannon limit [37, 38], and further significant increase of the system capacity might be only achievable with space-division-multiplexing (SDM) technologies [39].

Due to its ability to achieve high receiver sensitivity and spectral efficiency, and compensate most of the optical impairments in the electrical domain with DSP, digital coherent detection technology, which emerged in late 2000, has become the main technology for high-speed and high-capacity optical transport networks and is predominantly used in today's inter-datacenter interconnects [40, 41]. The block diagram of a digital coherent optical communication system is illustrated in Figure 1.17. For simplicity, only one channel of a WDM is shown in the figure. In a coherent optical communication system, both polarization and phase of light wave are used to carry information to increase spectral efficiency and system capacity. In the transmitter, a continuous wave (CW) from a low-linewidth laser such as an external cavity laser (ECL) is split into two parts, one for each polarization, and each part is modulated with an inphase/quadrature (I/Q) modulator by electrical signals. The electrical driving signals to the modulators are

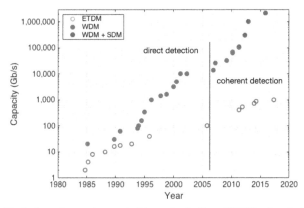

Figure 1.16 Evolution of capacity carried by a single fiber. ETDM: electrical time-division multiplexing; WDM: wavelength-division multiplexing; SDM: space-division multiplexing.

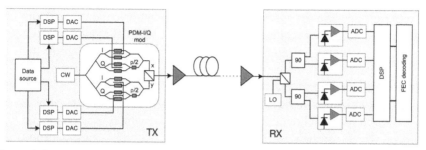

Figure 1.17 Digital coherent optical transmission system. TX: transmitter; RX: receiver; DSP: digital signal processing; DAC: digital-to-analog convertor; CW: continuous wave; LO: local oscillator; ADC: analog-to-digital convertor.

preferably generated using digital-to-analog converters (DACs) fed by an ASIC, which can generate signals with specific waveforms and spectral shapes for purposes such as to improve non-linear tolerance and/or spectral efficiency.

During the propagation in fibers, signal polarization is not maintained but randomly rotated. At the polarization and phase diversity receiver, the signal is split by a polarization-beam splitter (PBS). Each polarization of the signal after the PBS is combined with a local oscillator (LO) in a 90° hybrid. The four tributaries (x and y polarizations, I/Q branches) of the combined signal and LO after the hybrids are detected by four detectors (or four pairs of balanced detectors) and then are sampled and converted to digital form by analog-to-digital converters (ADCs). The signals are then processed by DSP, including retiming and re-sampling, CD compensation (non-linearity compensation if allowed by the complexity of the DSP), polarization demultiplexing and PMD compensation, carrier frequency and phase estimation, and symbol detection. FEC can be part of the DSP, for example in the case of coded modulation. In a real system, the DSP is typically included in an ASIC.

Using coherent detection, both inphase and quadrature can be used to encode information and information can be carried in both polarizations, which effectively increases the spectral efficiency by four times compared with intensity modulation and direct-detection (IM-DD) technology. In addition, spectral efficiency can be further increased by using high-order quadrature-amplitude modulation (QAM). Note that higher spectral efficiency modulation formats require a higher signal-to-noise ratio (SNR). For example, compared with quadrature-phase-shift keying (QPSK), 16-ary

QAM (16QAM) and 256-ary QAM (256QAM) can double and quadruple the spectral efficiency, but they require about 3.7-dB and 12.5-dB higher SNR per bit than QPSK, as shown in Figure 1.18 [42]. This means that to upgrade the capacity of a system with more spectral efficient modulation formats, some techniques have to be used to accommodate higher required SNRs.

Coherent polarization-division-multiplexed (PDM) QPSK has become a de facto standard for 100G optical transport networks, which uses 28-Gbaud or 32-Gbaud QPSK signals on both polarizations with 7% hard-decision FEC or 20% soft-decision FEC and can achieve more than 3,000 km in terrestrial networks. Unlike 100G, there is no single solution for 400G transport networks, and many solutions have been proposed. One can use one-carrier higher-order modulation formats such as 16QAM or 64QAM, or two-carrier lower-order modulation formats such as QPSK or 8QAM, as shown in Table 1.9. Different solutions have different requirements on occupied bandwidths and different target reaches as well. For example, 400G using QPSK modulation occupies a 150-GHz bandwidth and is applicable to long-haul networks, whereas 400G using 64QAM only occupies a 50-GHz bandwidth but its reach may be limited to a couple of hundreds kilometers. Super-channels may be used for 1-Tb and beyond transport networks. A super-channel can be comprised of a few wavelength channels which are treated as a single channel in propagation and routing [43]. It can also be based on space channels, where sub-channels are in different space modes but have the same wavelength. These space channels are transmitted, routed, and received together [44].

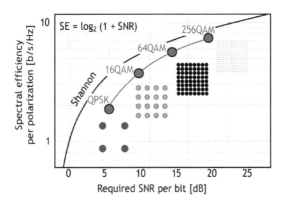

Figure 1.18   Spectral efficiency and required SNR for different modulation formats.

Table 1.9 400G solutions for optical transport networks (two polarizations for each solution)

Modulation	QPSK	8QAM	16QAM	16QAM	32QAM	64QAM
Carriers	2	2	2	1	1	1
Bandwidth	150 GHz	100 GHz	75 GHz	75	62.5	50

1.3.1 Inter-datacenter Interconnects in Metro Networks

Compared with a carrier metro network, a datacenter metro network is much simpler. It has fewer access points and the services are virtually all Ethernet. Due to the latency requirement, link distances in a datacenter metro network are limited to 80 km, much shorter than those in a carrier's metro network. It is not necessary to use reconfigurable optical add-drop multiplexers (ROADMs) and point-to-point optical links are sufficient for most datacenter metro networks. As synchronous replications require large data transfer among datacenters, high-capacity optical links are needed in a datacenter metro network. Table 1.10 lists the differences between a carrier metro network and a datacenter metro network.

Figure 1.19 depicts a typical architecture of an inter-datacenter optical interconnect in a metro network. It is a point-to-point optical link with $1 + 1$ protection. Optical signals of different wavelengths from transponders

Table 1.10 Differences between a carrier metro network and a datacenter's metro network

	Carrier Metro	Datacenter Metro
Distance	<600 km	<80 km
Access point	Many	A few
Network topology	Ring and mesh	Point-to-point
ROADM	Many	Few
Services	Voice, mobile, data, enterprise	Ethernet
Rate granularity	Many	Few
Capacity	Small and medium	Large

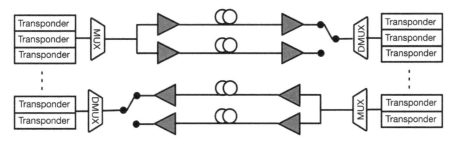

Figure 1.19 Architecture of an inter-datacenter optical interconnect in a metro network.

are multiplexed together with a WDM multiplexer, split into two copies, and sent to two booster amplifiers. At the receiver side, an optical switch is used to select signals after the pre-amplifiers from two different optical paths. The signals are demultiplexed into different wavelengths and sent to the corresponding transponders. The protection shown in Figure 1.19 is called optical-multiplex-section protection (OMSP), which protects both optical amplifiers and transmission fiber. Optical-line-section protection (OLSP) may also be used, which only protects transmission fiber.

Due to its ability to achieve high spectral efficiency and its low requirements on line systems, such as having no need for optical dispersion compensation and low requirement for optical-to-signal-noise ratios (OSNRs), coherent-detection technology is predominantly used in datacenter metro networks. Note that in optical communication systems, instead of SNR, OSNR is usually used to quantify system performance, which is defined as the ratio of optical signal power to optical noise power in 0.1-nm bandwidth. Since link distances in metro networks are short and there is only one span, delivered OSNR is typically high and high-level modulation formats such as 64QAM can be used to increase fiber link capacity.

To reduce the cost and size of metro datacenter interconnects, many other solutions have also been investigated, including simplified coherent solutions and direct-detection solutions [46–51]. Direct-detection solutions are particularly attractive for datacenter operators who have abundant fibers. The simplest direct-detection solution is to use NRZ intensity modulation, but it requires expensive high-bandwidth optics and electronics. Hence, many other direct-detection solutions using advanced modulation formats combined with DSP have been proposed. Out of these, PAM4 [48, 49] and discrete multi-tone (DMT) [50, 51] are heavily discussed, and the 100G direct-detection solution based on PAM4 has already been commercialized.

The diagram of an IM-DD PAM4 system is given in Figure 1.20. Compared with a coherent detection system, the transmitter and receiver of an IM-DD system is much simpler, which potentially can significantly reduce the cost and size of optical transponders. The drawback is that optical dispersion compensation is required, as shown in the line system in Figure 1.20, and the required OSNR is much higher, which puts a high requirement on fibers and amplifiers. Currently, the spectral efficiency of a commercial PAM4 system is limited to 1 b/s/Hz, which gives about 4 Tbps link capacity per fiber in C-band.

DMT modulation is a special implementation of orthogonal frequency-division multiplexing (OFDM) and widely employed in digital subscriber

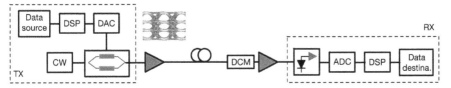

Figure 1.20 Diagram of an intensity-modulation direct-detection PAM4 system. DCM: dispersion compensation module.

lines (DSLs). Unlike OFDM, in DMT the modulation format and power of each subcarrier is in general different and is adapted to the frequency-dependent transmission characteristics of the channel, thus being able to make the maximum use of the bandwidth of a transmission link, as shown in Figure 1.21. The diagram of an IM-DD DMT system is similar to Figure 1.20, except that a dispersion compensation module (DCM) may not be needed. For a given bandwidth, DMT in general can achieve a higher bit rate than PAM4, but the DSP of DMT is much more complex than PAM4. Although chromatic dispersion of fiber induces power fading in an IM-DD DMT system, DMT can in principle adapt to the power fading by adjusting the modulation on each subcarrier and thus does not need optical DCM. However, the power fading can cause performance degradation and bit rate reduction, which limits the transmission distance of an IM-DD DMT system. To solve this problem, one can either put optical DCM in the line system to compensate for chromatic dispersion or use single-sideband (SSB) DMT. It has been shown that 100G SSB-DMT can transmit over 80-km SSMF without any optical DCM [51].

1.3.2 Inter-datacenter Interconnects in WANs

Inter-datacenter interconnects in WANs uses long-haul optical transport technology, covering distances from a couple of hundred kilometers between datacenters in adjacent cities to more than 10,000 km between datacenters

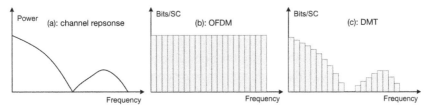

Figure 1.21 Diagram of a channel response, OFDM modulation and DMT modulation.

in different continents. The advancement on long-haul transport technology is the progress on battles against attenuation, dispersion, and non-linearities in optical fibers, with the goals to achieve the longest distances without electrical regeneration and the largest capacity in a single fiber. Figure 1.22 shows the evolution of single-channel bit rates for commercial systems. Today inter-datacenter interconnects in WANs predominantly use coherent 100G per channel technology, even in submarine systems. With Flex-grid technology, the system can achieve 12 Tbps in C-band. When per-channel speeds of optical transport systems were increased from 2.5G to 10G and to 100G, the spectral efficiency was increased by four and 40 times, respectively, but there was no sacrifice in transmission distances, thanks to dispersion management technology for 10G and digital coherent detection technology for 100G. Note that there was distance sacrifice for 40G systems using direct-detection differential-phase-shift-keying (DPSK) and differential-quadrature-phase-shift-keying (DQPSK) modulation formats and they were a transition technology and not widely deployed.

Today 100G coherent transport systems use the PDM-QPSK modulation format. To further increase the spectral efficiency and fiber capacity may need higher level modulation formats such as 16QAM and 64QAM, which require much higher OSNR, as Figure 1.18 indicates. Theoretically, doubling the capacity of the current PDM-QPSK 100G system with PDM-16QAM modulation format requires an $\sim$7-dB higher OSNR, which means that the un-regenerated distances will be approximately five times shorter than current 100G systems to achieve the same BER on the same line systems, which will be difficult to implement in real systems. The challenge for datacenter interconnects in WANs is how to increase the capacity and spectral efficiency without sacrificing un-regenerated transmission distances.

One way to increase the transmission distances is to increase the delivered OSNR. For a transmission link with equal span losses, the delivered OSNR (in dB) after N number of spans can be expressed as

$$OSNR = 58 + P_s - NF - 10log_{10}N - L \qquad (1.1)$$

where P_s is the launch power in dBm, NF is the noise figure of optical amplifiers, L is the loss of each span. It shows that OSNR can be increased by increasing the launch power, decreasing the NF of amplifiers and span losses. The launch power, which is limited by fiber non-linearities, can be increased by using large-core optical fibers and non-linearity compensation techniques. The effective core area of new large-core fibers has reached 150 μm^2, with fiber non-linearity $\sim$50% of SSMF, which can effectively increase the

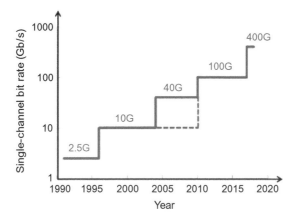

Figure 1.22 Evolution of single-channel bit rates for commercial optical transport systems.

launch power by ~3 dB. Span loss can be decreased by reducing fiber loss. Since Kao's prediction of low-loss silica glass fibers in 1966, continuous efforts have been made to lower fiber losses. The lowest loss fiber with the loss coefficient of 0.142 dB/km has been demonstrated recently [52], which can increase the delivered OSNR by more than 5 dB at an 80-km span length compared with SSMF. Due to their intrinsic lower noise figure and distributed gain profile along transmission fibers, effectively Raman amplifiers can significantly reduce the noise figure. With new fibers and Raman amplifiers, the 100G PDM-QPSK transport networks can be upgraded to 400G PDM-16QAM, with the same un-regenerated distances and a two-time increase in the spectral efficiency, and the spectral efficiency can be further increased by using Flex-grid technology [53, 54]. Another way to increase the transmission distances is to lower the required OSNR. Using advanced FEC and coded modulation techniques, close to Shannon limit performance can be achieved [55].

In addition to the capacity and distances, flexibility, agility and automation are necessary features for optical transport networks. With the fiber capacity approaching non-linear Shannon limit [37, 38], increasing network efficiency through re-configurability and programmability of optical transport networks becomes more and more important. Open optical transport networks, which consist of open line systems and open control and management platforms, are highly desirable [56]. As shown in Figure 1.23, open line systems decouple optical line systems from optical terminal equipment, which allows datacenter operators to choose the best technologies for

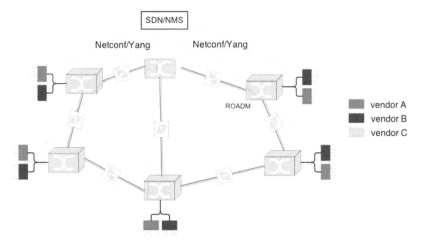

Figure 1.23 Open optical transport networks.

terminal optics and line systems in terms of performance, features and cost to meet their network requirements. A unified control and management platform that allows management and control of optical elements from multiple providers requires standard interfaces and vendor-neutral data models [57].

1.4 Summary

In this chapter, we discussed the optical interconnect technologies for datacenters, including intra-datacenter and inter-datacenter interconnects. For intra-datacenter interconnects, we started with 40G and showed how optical interconnect technologies evolve when the network speeds move to 100G, then discussed technologies for 400G and beyond networks. For inter-datacenter interconnects, we first described digital coherent optical transmission technology, which has become the main technology for optical transport networks. Then we discussed the requirements and technologies for inter-datacenter optical interconnects in metro networks. Finally, inter-datacenter interconnect technologies in WANs are described.

References

[1] Kao, K. C., and Hockham, G. A. (1966). "Dielectric-fibre surface waveguides for optical frequencies," *in Proceedings of the IEE,* 113, 1151–1158.

[2] Schultz, P. C. (2010). Making the first low-loss optical fibers. *Opt. Photon. News* 21, 30–35.

[3] Shumate, P. W., Chen, F. S., and Dorman, P. W. (1978). Special issue on Atlanta fiber system experiment. *Bell Syst. Tech. J.* Vol. 57:6.

[4] Qian, D., Huang, M.-F., Ip, E., Huang, Y.-K., Shao, Y., Hu, J., et al. (2011). "101.7-Tb/s (370×294-Gb/s) PDM-128QAM-OFDM transmission over 3×55-km SSMF using Pilot-based Phase Noise Mitigation," *in Proceedings of the OFC'2011*, Los Angeles, CA.

[5] Puttnam, B. J., Luís, R. S., Klaus, W., Sakaguchi, J., Delgado Mendinueta, J. -M., Awaji, Y., et al. (2015). "2.15 Pb/s transmission using a 22 core homogeneous single-mode multi-core fiber and wideband optical comb," *in Proceedings of the ECOC'2015*, Valencia, Spain.

[6] Soma, D., Igarashi, K., Wakayama, Y., Takeshima, K., Kawaguchi, Y., Yoshikane, N., et al. (2015). "2.05 Peta-bit/s Super-Nyquist-WDM SDM transmission using 9.8-km 6-mode 19-core fiber in full C band," *in Proceedings of the ECOC'2015*, Valencia, Spain.

[7] Lam, C. F., Liu, H., Koley, B., Zhao, X., Kamalov, V., and Gill, V. (2010). "Fiber optic communication technologies: what's needed for datacenter network operations," *in Proceedings of the IEEE Communications Magazine*, 48, 32–39.

[8] Barroso, L. A., and Hölzle, U. (2009). *The Datacenter as a Computer – an Introduction to the Design of Warehouse-Scale Machines.* Williston, VT: Morgan & Claypool Publishers.

[9] Cisco (2016). "*Cisco Global Cloud Index: Forecast and Methodology, 2015–2020*". Cisco White Paper.

[10] Al-Fares, M., Loukissas, A., and Vahdat, A. (2008). "A Scalable, Commodity Data Center Network Architecture," *in Proceedings of the SigComm'2008*, Seattle, WA, 63–74.

[11] Vahdat, A., Al-Fares, M., Farrington, N., Mysore, R.-N., Porter, G., and Radhakrishnan, S. (2010). "Scale-out Networking in the Data Center," *in Proceedings of the IEEE Micro*, 30, 29–41.

[12] Papen, G. (2017). "Optical Components for Datacenters," *in Proceedings of the Optical Fiber Communications Conference and Exhibition*, Los Angeles, CA.

[13] The Ethernet Alliance (n.d.). Available at: http://www.ethernetalliance. org/

[14] Zhou, X., Liu, H., and Urata, R. (2017). Datacenter optics: requirements, technologies, and trends. *Chin. Opt. Lett.* 15:120008.

[15] IEEE 802.3 Ethernet Working Group (n.d.). Available at: http://www.iee
     e802.org/3/ba/index.html
[16] IEEE 802.3 Ethernet Working Group (n.d.). Available at: http://www.iee
     e802.org/3/bm/index.html
[17] The 100G CWDM Multi-Source Agreement (n.d.). Available at:
     www.cwdm4-msa.org
[18] The 100G CRL4 Alliance (n.d.). Available at: www.clr4-alliance.org
[19] The 100G PSM4 Multi-Source Agreement (n.d.). Available at:
     www.PSM4.org
[20] Xie, C. (2015). "Optical interconnects in data centers," *in Proceedings
     of the ACP'2015*, Hong Kong.
[21] Lee, J., Shahramian, S., Kaneda, N., Baeyens, Y., Sinsky, J., Buhl,
     L., et al. (2015). "Demonstration of 112-Gbit/s optical transmission
     using 56GBaud PAM-4 driver and clock-and-data recovery ICs," *in
     Proceedings of the ECOC'2015*, Valencia, Spain.
[22] Verplaetse, M., Lin, R., Van Kerrebrouck, J., Ozolins, O., De Keu-
     lenaer, T., Pang, X., et al. (2017). Real-time 100 Gb/s transmission
     using three-level electrical duobinary modulation for short-reach optical
     interconnects. *J. Lightwave Technol.* 35, 1313–1319.
[23] Verbist, J., Verplaetse, M., Srivinasan, S. A., De Heyn, P., De Keulenaer,
     T., Pierco, R., et al. (2017). "First Real-Time 100-Gb/s NRZ-OOK
     transmission over 2 km with a silicon photonic electro-absorption
     modulator," *in Proceedings of the OFC'2017*, Los Angeles, CA.
[24] Xie, C., Dong, P., Randel, S., Pilori, D., Winzer, P., Spiga, S., et al.
     (2015). "Single-VCSEL 100-Gb/s short-reach system using discrete
     multi-tone modulation and direct detection," *in Proceedings of the
     OFC'2015*, Los Angeles, CA.
[25] Yang, C., Hu, R., Luo, M., Yang, Q., Li, C., Li, H., et al. (2016).
     "IM/DD-Based 112-Gb/s/lambda PAM-4 transmission using 18-Gbps
     DML," *in Proceedings of the IEEE Photonics Journal*, 8.
[26] Lavrencik, J., Varughese, S., Thomas, V. A., Landry, G., Sun, Y.,
     Shubochkin, R., et al. (2017). "4λ × 100Gbps VCSEL PAM-4
     Transmission over 105m of Wide Band Multimode Fiber," *in Proceed-
     ings of the OFC'2017*, Los Angeles, CA.
[27] IEEE 802.3 Ethernet Working Group (n.d.). Available at: http://www.iee
     e802.org/3/bs/index.html
[28] Sadot, D., Dorman, G., Gorshtein, A., Sonkin, E., and Vidal, O. (2015).
     Single channel 112 Gbit/sec PAM4 at 56 Gbaud with digital signal
     processing for data centers applications. *Opt. Exp.* 23, 991–997.

[29] SWDM Alliance (n.d.). Available at: www.swdm.org

[30] The CFP Multi-Source Agreement (n.d.) Available at: www.cfp-msa.org

[31] The OSFP Multi-Source Agreement (n.d.). Available at: www.osfpmsa. org

[32] The QSFP-DD Multi-Source Agreement (n.d.). Available at: www.qsfp-dd.com

[33] The Consortium for On-Board Optics (n.d.). Available at: http://cobo.az urewebsites.net

[34] Sun, C., Wades, M. T., Lee, Y., Orcutt, J. S., Alloatti, L., Georgas, M. S., et al. (2015). Single-chip microprocessor that communicates directly using light. *Nature* 528, 534–538.

[35] Cai, J.-X., Batshon, H. G., Mazurczyk, M. V., Sinkin, O. V., Wang, D., Paskov, M., et al. (2017). "70.4 Tb/s capacity over 7,600 km in C+L band using coded modulation with hybrid constellation shaping and nonlinearity compensation" *in Proceedings of the OFC'2017*, Los Angeles, CA.

[36] Schuh, K., Buchali, F., Idler, W., Eriksson, T. A., Schmalen, L., and Templ, W., et al. (2017). "Single carrier 1.2 Tbit/s transmission over 300 km with PM-64 QAM at 100 GBaud," *in Proceedings of the OFC'2017*, Los Angeles, CA.

[37] Essiambre, R.-J., Kramer, G., Winzer, P. J., Foschini, G. J., and Goebe, B. (2010). Capacity limits of optical fiber networks. *J. Lightwave Technol.* 28, 662–701.

[38] Richardson, D. J. (2010). Filling the light pipe. *Science* 330, 327–328.

[39] Winzer, P. J., Ryf, R., and Randel, S. (2013). "Spatial multiplexing using multiple-input multiple-output signal processing," in *Optical Fiber Telecommunications VIB*, eds I. P. Kaminow, T. Li, and A. E. Willner (San Diego, CA: Academic Press).

[40] Kikuchi, K. (2016). Fundamentals of Coherent Optical Fiber Communications. *J. Lightwave Technol.* 34, 157–179.

[41] Xie, C. (2011). Impact of nonlinear and polarization effects in coherent systems. *Opt. Exp.* 19, B915–B930.

[42] Winzer, P. J. (2012). "Optical network beyond WDM," *in Proceedings of the ACP'2012*, Guangzhou, China.

[43] Renaudier, J., Muller, R., Schmalen, L., Tran, P., Brindel, P., and Charlet, G. (2014). "1-Tb/s PDM-32QAM superchannel transmission at 6.7-b/s/Hz over SSMF and 150-GHz-grid ROADMs," *in Proceedings of the ECOC'2014*, Cannes, France.

[44] Nelson, L. E., Feuer, M. D., Abedin, K., Zhou, X., Taunay, T. F., Fini, J. M., et al. (2014). Spatial superchannel routing in a two-span ROADM system for space division multiplexing. *J. Lightwave Technol.* 32, 783–789.

[45] Winzer, P. J. (2014). Making spatial multiplexing a reality. *Nature Photon.* 8, 345–348.

[46] Xie, C., Dong, P., Winzer, P., Grus, C., Ortsiefer, M., Neumeyr, C., et al. (2013). 960-km SSMF transmission of 105.7-Gb/s PDM 3-PAM using directly modulated VCSELs and coherent detection. *Opt. Exp.* 21, 11585–11589.

[47] Xie, C., Spiga, S., Dong, P., Winzer, P., Bergmann, M., Kgel, B., et al. (2015). 400-Gb/s PDM-4PAM WDM system using a monolithic 2×4 VCSEL array and coherent detection. *J. Lightwave Technol.* 33, 670–677.

[48] Yin, S., Chan, T., and Way, W. I. (2015). "100-km DWDM transmission of 56-Gb/s PAM4 per λ via tunable laser and 10-Gb/s InP MZM," *in Proceedings of the IEEE Photonics Technology Letters*, 27, 2531–2534.

[49] Eiselt, N., Wei, J., Griesser, H., Dochhan, A., Eiselt, M., Elbers, J.-P., et al. (2016). "First real-time 400G PAM-4 demonstration for inter-data center transmission over 100 km of SSMF at 1550 nm," *in Proceedings of the OFC'16,* Anaheim, CA.

[50] Yan, W., Li, L., Liu, B., Chen, H., Tao, Z., Tanaka, T. et al. (2014). "80 km IM-DD transmission for 100 Gb/s per lane enabled by DMT and nonlinearity management," *in Proceedings of the OFC'2014*, San Francisco, CA.

[51] Randel, S., Pilori, D., Chandrasekhar, S., Raybon, G., and Winzer, P. (2015). "100-Gb/s discrete multitone transmission over 80-km SSMF using single-sideband modulation with novel interference-cancellation scheme," *in Proceedings of the ECOC'2015*, Valencia, Spain.

[52] Tamura, Y., Sakuma, H., Morita, K., Suzuki, M., Yamamoto, Y., Shimada, K. et al. (2017). "Lowest-Ever 0.1419-dB/km loss optical fiber," *in Proceedings of the OFC'2017*, Los Angeles, CA.

[53] Xie, C., Zhu, B., and Burrows, E. (2014). "Transmission performance of 256-Gb/s PDM-16QAM with different amplification schemes and channel spacings," *in Proceedings of the IEEE/OSA Journal of Lightwave Technology*, 32, 2324–2331.

[54] ITU-T Standard (2012). *ITU-T G.694.2.* Available at: https://www.itu.int/rec/T-REC-G.694.1-201202-I

[55] Cai, J.-X., Zhang, H., Batshon, H. G., Mazurczyk, M., Sinkin, O. V., Sun, Y., et al. (2014). "Transmission over 9,100 km with a capacity of 49.3 Tb/s using variable spectral efficiency 16 QAM based coded modulation," *in Proceedings of the OFC'14*, San Francisco, CA.

[56] Vusirikala, V., Zhao, X., Hofmeister, T., Kamalov, V., Dangui, V., and Koley, B. (2015). "Scalable and flexible transport networks for inter-datacenter connectivity," *in Proceedings of the OFC'2015*Los Angeles, CA.

[57] Shaikh, A., Hofmeister, T., Dangui, V., and Vusirikala, V. (2016). "Vendor-neutral network representations for transport SDN," *in Proceedings of the OFC'2016*, Anaheim, CA.

2

Vertical Cavity Surface Emitting Lasers

Wenbin Jiang

WJ Technologies LLC, California, USA

2.1 Introduction

Optical sources for fiber optic communications commonly operate at wavelengths near 850 nm, 1,300 nm, or 1,550 nm, which coincide with the transmission window of optical fibers. Vertical cavity surface emitting lasers (VCSELs) operating at wavelengths between 780 and 980 nm have been well developed and in large deployment in data centers and enterprise networks. Substantial efforts have been devoted to the research and development of 1,310 nm and 1,550 nm long-wavelength VCSELs. Red visible VCSELs have also been developed, which may be applied for transmission over plastic optical fibers. In this chapter, we will introduce some fundamentals of VCSEL technology, primarily for fiber optic data communications. We will start with the operating principles behind the VCSEL technology, followed by the fabrication and manufacturing of the VCSEL devices.

2.2 Technology Fundamentals

We begin this section with a brief review of semiconductor properties, and then describe the fundamentals of laser diodes that include the VCSELs. For the range of wavelengths used in data communications, the efficiency of the laser sources is determined by the bandgap energy. Intrinsic semiconductor materials are characterized by an electrical conductivity much lower than pure metals, which increases rapidly with temperature. The Fermi level lies between the conduction and valence bands of the material. These materials may be doped with various impurities as either n-type or p-type, so that the Fermi level moves either closer to the conduction or valence bands. Common room-temperature semiconductors include germanium (Ge) and silicon (Si) from group IV of the periodic table and binary compounds such as gallium arsenide (GaAs) and indium phosphide (InP) from groups III and V of the periodic table.

Silicon and germanium can be made n-type by doping with donor impurities such as P, As, or other group-V elements, and p-type by doping with acceptor impurities such as B, Ga, or other group-III elements. These dopant concentrations are small, from 10^{14} to 10^{21} cm^{-3} in Si. Typically, the Fermi level is uniform throughout the doped material. When an abrupt junction between p-type and n-type materials is formed, the conduction band, valence band, and Fermi band bend to accommodate the difference. Electrons tend to accumulate on the n-side of the junction, and holes on the p-side. A thin depletion region is formed at the junction itself, which is effectively depleted of both hole and electron carriers.

An external voltage applied across the junction will have an effect of raising or lowering the potential barrier between the two materials, depending on its polarity. When the p-side is connected to a positive voltage, the junction is forward biased, while the opposite condition is known as reverse bias. In a forward-biased junction, excess carriers are injected into the depletion region. The thickness of the depletion layer decreases under forward bias conditions. There is thus a balance of carrier flow – injection of carriers over the depletion region, which gives rise to an excess carrier concentration, is balanced by both diffusion away from the junction and carrier recombination.

Recombination is a process for the electrons and holes to recombine. Several types of recombination processes exist, but we are only interested in radiative recombination process that releases photons, which is a direct band-to-band transition across the bandgap. Direct bandgap semiconductor materials (such as GaAs and other III–V compounds) have an energy (E) – momentum (k) relationship such that the conduction band minimum and the valence band maximum occur at the same value of k on the E–k diagram, as shown in Figure 2.1. With direct bandgap materials, the holes and electrons recombine directly. In an indirect bandgap material such as silicon, by contrast, the conduction band minimum does not line up with the valence band maximum in the k-space, and it is more complicated for the electrons and holes to recombine while satisfying the requirements of both conservation of energy and conservation of momentum. This leads to favoring non-radiative transition, which involves intermediate energy levels that trap carriers deep within the crystal lattice. During the non-radiative recombination process, energy is lost to thermal energy within the crystal.

In direct bandgap materials, the radiative recombination process is proportional to the excess minority carrier concentration, and gives rise to the

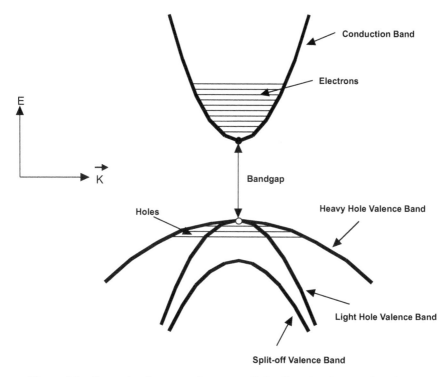

Figure 2.1 Energy band structure (energy vs. k) for direct bandgap semiconductor.

creation of radiation from the junction. The process is known as injection luminescence, and is the fundamental mechanism for the creation of optical radiation, or photon, from semiconductor lasers. The operation of such sources is also influenced by the minority carrier lifetime, diffusion length near the junction, injection efficiency, and width of the depletion layer. Equivalent circuits for the *p–n* junction allow the modeling of junction capacitance and series resistance, which affect how rapidly the optical source can be modulated to carry information.

The recombination of electrons and holes in a forward-biased *p–n* junction produces photons at wavelengths of 800–1,700 nm that are suitable for fiber optical communications. When only a single photon of energy E is produced, the wavelength may be determined from Planck's Law:

$$\lambda = hc/E = 1.24/E \text{ (microns/electron volt)} \tag{2.1}$$

where h is Planck's constant and c is the speed of light. The probability that an electron of energy E_2 will recombine with a hole of energy E_1 is proportional to the concentration of electrons at E_2 and the concentration of holes at E_1. The spectrum of the radiated energy may be determined by integrating the product of these carrier concentrations over all values of E_1 or E_2, subject to the constraint that the difference between E_1 and E_2 is the desired photon energy E.

It is desirable to design devices with high internal quantum efficiency, defined as the ratio of the rate of photon generation to the rate at which carriers are injected across the junction. A related parameter is the external quantum efficiency, defined as the ratio of the number of emitted photons to the number of carriers crossing the junction. External quantum efficiency is smaller than internal quantum efficiency for several reasons. For instance, some light will be reabsorbed before it can reach the emitting surface. Only light emitted towards the semiconductor-air interface is useful for coupling into the fiber, and some of this light is reflected back from the surface. Further, only light reaching the surface at less than the critical angle will be coupled into an adjacent optical fiber.

There are three essential elements for any semiconductor lasers including VCSELs to operate:

1. means for optical feedback, provided by multilayer diffractive Bragg reflectors (DBRs);
2. gain medium, consisting of a properly doped or intrinsic semiconductor material;
3. pumping source, provided by injection current.

For a semiconductor material with gain g and loss a, placed in a cavity of length L between two reflective mirrors of reflectivity R_1 and R_2, the minimum requirement for lasing action to occur is that the light intensity after one complete trip through the cavity must at least be equal to its starting intensity. This is called the threshold condition, given by the relation:

$$R_1 R_2 \exp(2(g - \alpha)L) = 1 \qquad (2.2)$$

Thus, the laser diode begins to operate when the internal gain exceeds a threshold value. Two key parameters associated with the laser are its efficiency of converting electrical current into laser light, known as the slope efficiency dP/dI, and the amount of current required before the laser begins stimulated emission, known as the threshold current I_{th}. This information is summarized in the power vs. current (*P* vs. *I*) characteristic curve of a laser diode, as shown in Figure 2.2. Applying current, *I*, to a laser device initially gives rise to spontaneous emission of light. As the current is increased, the

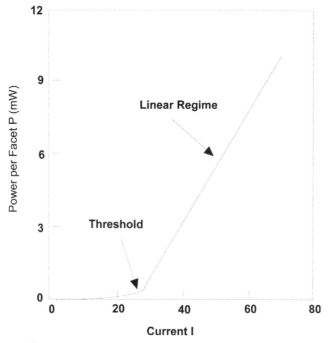

Figure 2.2 Power vs. current (*P* vs. *I*) characteristic curve of a laser diode.

cavity and mirror losses are overcome, and the laser passes the threshold and begins to emit coherent light. The optical power emitted above threshold, *P*, is given by

$$P = (I - I_{th})dP/dI \qquad (2.3)$$

Threshold current is found by extrapolating the linear region of the *P/I* curve. Low-threshold currents and high efficiencies are desirable in laser diodes. The slope of the linear region is called the slope efficiency, dP/dI; and it is related to the external differential quantum efficiency, η, by

$$\eta = (\lambda q)/(hc)(dP/dI) \qquad (2.4)$$

where λ is the wavelength, q is the charge of an electron, h is Plank constant, and c is the speed of light. The threshold current in a conventional edge-emitting semiconductor laser diode is a function of temperature, *T*, according to the relation

$$I_{th}(T) = I_{th}(T_1)\exp[(T - T_1)/T_0] \qquad (2.5)$$

where T_0 is the laser's characteristic temperature, and T_1 is room temperature (300 K). A large T_0 is desirable since it translates into a reduced sensitivity of threshold current to operating temperature. VCSEL, however, does not follow the same trend, which we will discuss in the later section. Usually, it is only possible to measure the case temperature of a packaged laser diode. The junction temperature can be estimated by taking into account the device thermal resistance, the input electrical power, and the output optical power.

Increasing the laser diode current above threshold allows the gain coefficient to stabilize at a value near the threshold point, just large enough to overcome losses in the media:

$$G = \alpha + ln(R_1 R_2)/2L \tag{2.6}$$

For lasers biased above threshold, the slope of the power vs. current characteristic curve in the spontaneous emission region corresponds roughly to the external quantum efficiency. The slope in the lasing region is related to the differential quantum efficiency, η_d, by:

$$\eta_d = q\Delta P/E\Delta I \tag{2.7}$$

In practice, this ideal characteristic curve is less well behaved, especially with VCSELs due to thermal effect and transverse mode competition.

2.3 VCSEL Device Structure

Semiconductor lasers were first reported in 1962 [1–4]. Those lasers were called edge-emitting lasers due to laser beam emitted from cleaved facets in parallel to the chip surface. In the late 1970s, Iga et al. [5] proposed to have a semiconductor laser oscillating perpendicular to the device surface plane, termed VCSEL. VCSELs have demonstrated many advantages over edge-emitting semiconductor lasers. First, the monolithic fabrication process and wafer scale probe testing as per the silicon semiconductor industry substantially reduces the manufacturing cost because only known good devices are kept for further packaging [6, 7]. Second, a densely packed two-dimensional (2-D) laser array can be fabricated because the device occupies no larger area than a commonly used electronic device [8]. This is important for applications in optoelectronic-integrated circuits. Third, the microcavity length allows inherently single longitudinal cavity mode operation due to its large mode spacing. Temperature-insensitive devices can therefore be fabricated with an offset between the wavelength of the cavity mode and the active gain peak [9, 10]. Finally, the device can be designed with a low numerical aperture (NA) and a circular output beam to match the optical mode of an optical fiber, thereby permitting efficient coupling without additional optics [7, 11].

A VCSEL needs both of its DBRs to be highly reflective to reduce the cavity mirror loss because its active gain layer is less than 1 μm thick. The first VCSEL was demonstrated with GaInAsP/InP in 1979, which operated pulsed at 77 K with annealed *Au* at both sides as reflectors [5]. A room-temperature pulsed-operating VCSEL was demonstrated with GaAs active region in 1984 [12]. Room-temperature CW-operating GaAs VCSELs succeeded by improving both the mirror reflectivity and current confinement [13].

Since then, output power of over 100 mW has been obtained from an InGaAs/GaAs VCSEL with GaAs/AlAs monolithic DBRs [14, 15]. VCSELs with lasing threshold of sub-100 μA [16, 17] or wall-plug efficiency of over 50% have been reported with lateral oxidized-Al confinement blocks [18, 19]. Room-temperature CW InGaAsP/InP VCSELs have been challenging to fabricate primarily due to low index difference between GaInAsP and InP, which causes difficulty in preparing highly reflective monolithic DBR [20]. Nevertheless, CW InGaAsP VCSELs at 1.5 μm have been reported using GaAs/AlAs DBR mirrors by wafer fusing [21, 22]. Nitride active materials-based 1.3 μm VCSELs with GaInNAs buried tunnel junction and AlGaAs/GaAs DBR lattice matched to GaAs substrate have been developed and may operate above 85°C [23]. Tremendous progress has been made on LW VCSELs with active tunnel junction structure and lattice-matched AlGaInAs/InP DBRs at both 1.3 μm and 1.55 μm, which operate up to 100°C [24]. High-speed tunnel junction AlGaInAs short cavity VCSEL has been demonstrated to operate at 35 Gbps over 10 km single-mode fiber [25].

Two-dimensional arrayed VCSELs [8, 26, 27] can find important applications in stacked planar optics, such as simultaneous alignment of large numbers of optical components used in parallel multiplexing lightwave systems and parallel optical logic systems, free space optical interconnects, etc. High-power lasers can be made with phase-locked 2-D arrayed VCSELs [28, 29]. Commercial applications of array VCSELs, however, have been on one-dimensional configurations such as 1×4 and 1×12.

The majority of the VCSELs that have been developed today are in the near-infrared wavelength range based either on GaAs/AlGaAs or strained InGaAs active materials. GaAs VCSELs at 850 nm are preferred as the light sources for short-distance optical communications because either Si or GaAs PIN detectors can be used as the receiver to reduce the total system cost. A typical GaAs VCSEL epitaxial layer structure is shown in Figure 2.3, and

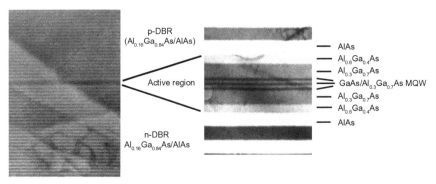

Figure 2.3　TEM photo of a GaAs vertical cavity surface emitting laser (VCSEL) epitaxial layer structure.

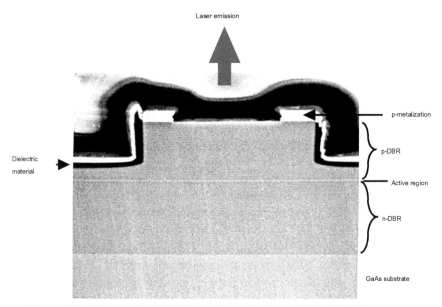

Figure 2.4　Cross section SEM photo of an etched mesa GaAs VCSEL structure.

the correspondent etched mesa-type device structure is shown in Figure 2.4. It includes three major portions: bottom DBR, active region, and top DBR.

Generally, the epitaxial material is grown by either MOCVD or MBE techniques. The bottom DBR is *n*-type doped, grown on an *n*-type doped *GaAs* substrate. Typically there is a *GaAs* buffer layer grown between the *n*-DBR and the substrate. Silicon (Si) and selenium (Se) are two commonly used n-type dopants. The n-DBR is comprised of 30.5 pairs [30, 31] of $Al_{0.16}Ga_{0.84}As/AlAs$, which starts and stops with the AlAs layer alternated with the $Al_{0.16}Ga_{0.84}As$ layer. Each DBR layer has an optical thickness equivalent to a quarter of the designed lasing wavelength. The intrinsic cavity region is comprised of two $Al_{0.6}Ga_{0.4}As$ spacer layers and three to four GaAs quantum-wells, with each quantum-well sandwiched between two $Al_{0.3}Ga_{0.7}As$ barriers. With the quantum-well width of 100 Å and the quantum barrier width of 70 Å, the lasing wavelength is around 850 nm. The $Al_{0.6}Ga_{0.4}As$ spacer can be replaced by $Al_xGa_{1-x}As$ spacer with x graded from 0.6 to 0.3 to form a graded index separate confinement heterostructure (GRINSCH). The total thickness of the spacers is such that the laser cavity length between the bottom and the top DBRs is exactly one wavelength or its multiple integer. The p-type doped DBR is grown on top of the active region. It consists of 22 pairs of $Al_{0.16}Ga_{0.84}As/AlAs$ alternating layers which starts with AlAs layer and stops with $Al_{0.16}Ga_{0.84}As$. Similar to the n-DBR, each layer has an optical thickness of a quarter wavelength. The most common p-type dopants are carbon (C), zinc (Zn) and Beryllium (Be). Typically, C is used for the p-doping in the top DBR when using MOCVD growth techniques with the top several layers doped by Zn for better metalization contact [32]. Finally, a GaAs cap is used to prevent the top AlGaAs layer of the p-DBR from oxidization. The cap is highly p-doped with Zn, and is kept to 100 Å thick as the material will be highly absorptive to the optical mode if it is too thick.

The DBR in a VCSEL is to reflect part of the laser emission back into the laser cavity and to transmit part of the laser emission as the output. A pair of low refractive index and high refractive index layers with optical thickness of each layer at one quarter of a specific wavelength will enhance the reflectivity to this wavelength. Many pairs of such alternating layers stacked together will form a mirror with its reflectivity reaching above 99% centered at the designed wavelength with a certain bandwidth. The mirror can be designed to achieve higher reflectivity and wider reflection bandwidth with a larger refractive index difference between the two alternating layers. $Al_xGa_{1-x}As$ is an ideal semiconductor material that can be monolithically grown on a GaAs substrate as alternating layers because the refractive index of this material can vary continuously from 3.6 to 2.9 with x varying from 0 to 1. For example, if one quarter-wavelength thick layer is made of GaAs (x = 0) and the other quarter-wavelength thick layer is made of AlAs (x = 1), 20 pairs of such alternating layers will provide a reflectivity of 99% at the desired wavelength with a bandwidth of larger than 70 nm, as long as the desired wavelength is

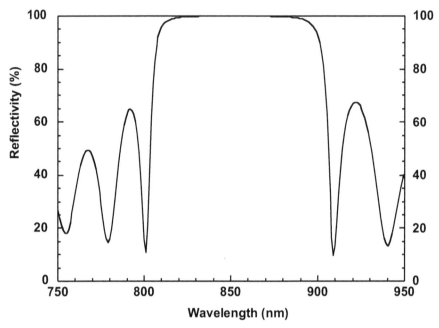

Figure 2.5   Reflectivity spectrum of 22 periods of $Al_{0.16}Ga_{0.84}As/AlAs$ quarter wavelength DBR mirror stacks centered at a wavelength of 850 nm.

longer than the GaAs absorption band-edge, which is around 875 nm at room temperature.

The reflectivity of a VCSEL DBR has to be more than 99% to reduce the cavity loss due to the extremely short gain length in the active region. For the example shown in Figure 2.3, the optical gain is provided by the three GaAs quantum-wells with each quantum-well thickness of only 100 Å. Due to the large refractive index difference between $Al_{0.16}Ga_{0.84}As$ and AlAs, the 22-pair DBR in this VCSEL has a reflectivity of 99.9% at 850 nm with a bandwidth of about 70 nm, as shown in Figure 2.5. The large bandwidth provides some tolerance to any variation in the lasing wavelength due to growth variation in quantum-well thickness and laser cavity length, the center wavelength mismatching between the top and the bottom DBRs, and the growth reactor/tooling variance. A typical GaAs VCSEL output power vs. input current is shown in Figure 2.6 for a mesa diameter of 10 μm and a laser emission aperture of 7 μm.

Strained InGaAs VCSELs operating at around 980 nm have also been extensively studied. The strain in the active region provides higher gain

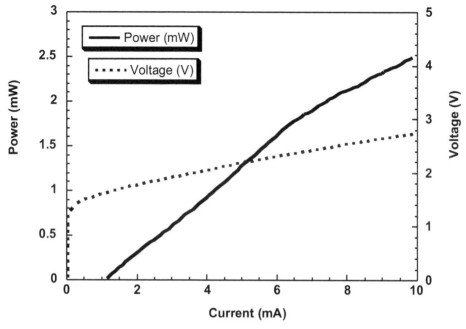

Figure 2.6 A GaAs VCSEL output power vs. input current for an etched mesa structure with a mesa diameter of 10 μm and an emission aperture of 7 μm. The laser wavelength is around 850 nm.

that allows lower lasing threshold and higher differential gain that allows larger intrinsic modulation bandwidth. In addition, the InGaAs VCSEL has a wavelength transparent to the GaAs substrate, allowing the light emission towards the substrate side and thus the epitaxial side down packaging scheme. In this way, heat generated in the *p*-DBR mirror and the active junction region can be dissipated more efficiently, resulting in lower junction temperature and higher output power. One disadvantage is that the low-cost Si or GaAs PIN detectors do not respond to the 980 nm light sources. In addition, OM3 an OM4 multimode fibers have been optimized in bandwidth for operation at a wavelength of 850 nm instead of 980 nm, making the 850 nm VCSEL more appropriate for short-distance high-speed data communications. A typical InGaAs VCSEL consists of two DBR mirrors with reflection bands centered approximately at 970 nm [10, 33]. Each mirror is composed of one-quarter-wavelength-thick layers alternating between AlAs and GaAs. The top *p*-doped DBR mirror contains 15 periods while the bottom *n*-doped DBR mirror contains 18.5 periods. The cavity between the mirrors is filled by spacer layers of $Al_{0.3}Ga_{0.7}As$ which are used to center three 8 nm $In_{0.2}Ga_{0.8}As$ quantum-wells (almost the thickness limit for coherently strained materials) separated by 10 nm GaAs barriers to form a one wavelength-long cavity. The $Al_{0.3}Ga_{0.7}As$ spacer on the *p*-DBR mirror side is *p*-type doped, and the $Al_{0.3}Ga_{0.7}As$ spacer on the *n*-DBR mirror side is *n*-type doped. Above the *p*-DBR mirror stack, a heavily *p*-doped (3×1019 cm^{-3}) GaAs phase-matching layer is deposited to provide non-alloyed Ohmic contact to the hybrid Au mirror, which also acts as a *p*-contact. The DBR mirrors are uniformly doped to 1×10^{18} cm^{-3} except for the digital grading region that is uniformly doped to 5×10^{18} cm^{-3}. The *n*-dopant is Si and the *p*-dopant is carbon. The epitaxial structure is grown by either MBE or MOCVD technique on an *n*-doped GaAs substrate. The device is designed to emit laser toward the substrate side.

Clearly, highly reflective semiconductor DBR mirrors are necessary to ensure low cavity loss, thus allowing the VCSEL to reach lasing threshold at a reasonable threshold carrier density level in the gain medium. While the refractive index difference between the two constituents of the DBR structures is responsible for high optical reflectivity, the accompanied energy bandgap difference which scales roughly linearly with the index difference results in electrical potential barriers in the heterointerfaces. These potential barriers impede the carrier flowing in the DBR structures and result in a large-series resistance, especially in the *p*-type doping case. The large-series resistance gives rise to thermal heating and thus deteriorates the laser performance.

The series resistance due to the heterojunctions in the DBR mirror can be minimized by grading and selectively doping the interfaces. In practice, the simplest approach to grade the interface is to introduce an extra intermediate-composition layer between the two alternating DBR constituents [34]. Instead of transiting directly from GaAs to AlAs, for example, a thin $Al_{0.5}Ga_{0.5}As$ layer can be grown in between to help smooth the interfaces. With the introduction of more $Al_xGa_{1-x}As$ layers of intermediate Al composition at each interface, greater performance over a single intermediate transition layer can be achieved. The advancement of MOCVD technology has allowed the continuous grading of an arbitrary composition profile. Very low resistance DBRs have been achieved using this technique [35]. In the structure shown in Figure 2.3, the hetero-interfaces are linearly graded from 15% Al composition to 100% over a distance of 12 nm. The total optical thickness of a pair of alternating layers is maintained to be one half wavelength. The effect of this interface linear grading on the DBR mirror reflectivity is minimal. Likewise, hetero-interface parabolic grading [36–38] and sinusoidal grading [39] have also been used in some cases to flatten the valence band and therefore reduce the *p*-DBR series resistance.

The graded interfaces may be heavily doped (5×10^{18} cm^{-3}), while the remainder of the mirror lightly doped (1×10^{18} cm^{-3}) to reduce scattering and free-carrier absorption loss in the mirror, in addition to a reduction in series resistance [40]. A simplified delta doping approach [41, 42] has worked successfully to reduce the series resistance in the *p*-DBR mirror. In this technique, *p*-doping is carried out at levels as high as the crystal can incorporate at interfaces where the nodes of the optical intensity are located. This heavy doping at the hetero-interfaces causes the valence band-edge to shift upward, and the thermionic emission current to increase. The excess resistance at the higher bandgap side (AlAs) of the hetero-interfaces is also reduced together with the relaxation of the carrier depletion in this region. Furthermore, the delta doping introduces a thinner potential barrier that allows for an increased tunneling current. Because the carrier density is increased only locally, the excess free-carrier absorption in the DBR mirror is minimized.

The intracavity metal contact technique [43–45] is an alternative approach to achieve low series resistance. This technique allows the electrical contact to bypass the resistive *p*-DBR mirror stack layers. Furthermore, the mirror stack above the metal contact does not require any doping, thereby reducing the intracavity free carrier absorption loss and the optical scattering loss.

The intracavity metal contact structure starts with a conventional VCSEL design similar to that shown in Figure 2.3 [43]. *P*-type and *n*-type bulk layers that are multiples of half wavelengths are inserted on either side of the active region to provide electrical paths for the current to reach the active region from the ring contacts that are deposited on top of the inserted layers. A current blocking region must be formed to force the current into the optical mode. This current blocking region can be formed by ion implantation, or undercutting using wet etching between the *p*-type insertion layer and the active region. A resistive layer is further introduced between the conductive current distribution layer and the active region to overcome any residual current crowding effects near the contact periphery so that the injected current can be more uniformly distributed in the optical mode area. The current blocking layer in the *p*-type region can also be a thin *n*-type GaAs or AlGaAs layer that is inserted into the top *p*-doped cladding layer [44]. A second growth is needed to complete the epitaxial structure after a current flow path is opened by either wet or drying etching. The final VCSEL device will have a

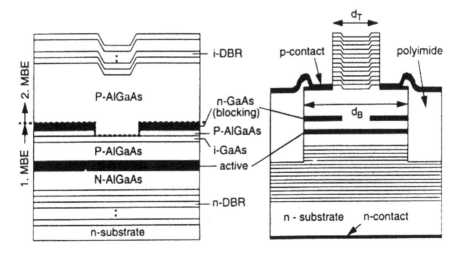

Figure 2.7   Schematic diagram of an intra-cavity contact VCSEL structure [44].

reverse-biased p–n junction inside the cavity in the p-doped region for current blocking, as shown in Figure 2.7.

One of the simple approaches to make intracavity metal contact is to start with a VCSEL with an only partially grown p-DBR mirror stack, as shown in Figure 2.8. After the metal contact has been deposited onto the p-DBR, a dielectric mirror stack consisting of quarter-wavelength thick alternating layers TiO_2/SiO_2 is used to complete the device [45]. In this instance, resistance of 50 Ω has been achieved with a laser emission aperture of 5 μm and a proton implantation aperture of 20 μm.

Within the microcavity structure of a VCSEL, only one Fabry Perot cavity mode exists in the designed DBR reflective bandwidth. Lasing can only be sustained at the wavelength of the cavity mode. Temperature-insensitive VCSEL can therefore be demonstrated by taking advantage of this micro-cavity mode characteristics [9]. Typically, the peak of the active gain profile shifts with the temperature at a rate of 3–5 Å/°C, and the cavity resonant mode shifts at a rate of 0.5–1 Å/°C [46]. If the resonant cavity mode is designed to initially sit at the longer wavelength side of the gain profile, the gain peak will gradually walk into the cavity mode with the rise in temperature (Figure 2.9). Conversely, the gain peak usually decreases in wavelength with the temperature. Together, the actual gain for the VCSEL cavity mode will vary little with temperature, and the VCSEL threshold current will stay almost constant within a certain temperature range. This temperature insensitivity

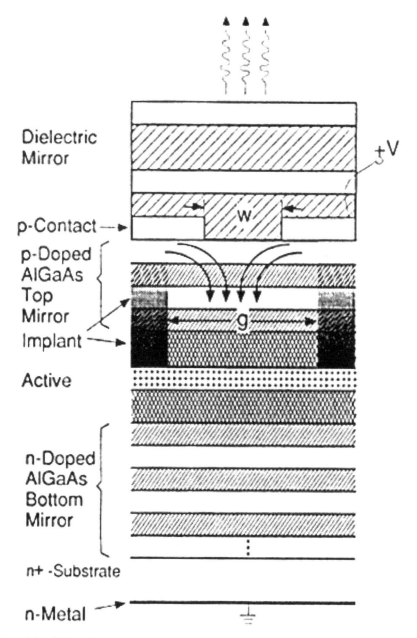

Figure 2.8   Schematic diagram of a VCSEL structure with partial monolithic semiconductor DBR and partial dielectric DBR on the *p*-doped contact side [45].

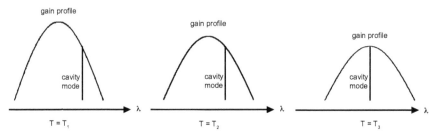

Figure 2.9   VCSEL cavity mode vs. gain profile for temperatures T_1, T_2, and T_3 with $T_3 > T_2 > T_1$.

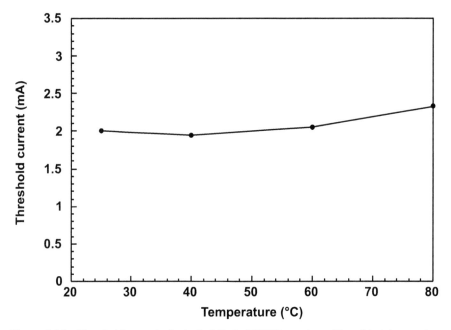

Figure 2.10   Threshold current of a typical GaAs VCSEL varying with ambient temperature with minimum threshold current at $40°$C.

allows the VCSEL to be designed to operate optimally at the system temperature mid-point region. For example, in an optical data link, the VCSELs can be designed to operate with minimum threshold current at around $40°$C in a required working range of $0–70°$C, as shown in Figure 2.10 [10, 47]. The system can be implemented without any auto power control (APC) circuitry, thereby simplifying the packaging and reducing the system cost [7]. This characteristic has worked well for VCSEL arrays in parallel optical data links, in which active power monitoring and power control becomes a challenge. The application of this method has also allowed the demonstration of VCSELs operating at a record high temperature of 200°C [31].

Apart from VCSELs at 830–870 nm based on GaAs multi-quantum-wells (MQWs) and VCSELs at 940–980 nm based on strained InGaAs MQWs, VCSELs operating at other wavelengths, such as 780 nm based on AlGaAs MQWs, 650–690 nm based on InAlGaP MQWs, and 1.3–1.5 μm VCSELs based InGaAsP MQWs, have all been extensively studied or commercialized.

Lasers at 780 nm are predominantly used for compact disk (CD) data storage and laser printing. A typical VCSEL at 780 nm has an epitaxial layer structure similar to that of a VCSEL at 850 nm [48–50]. The larger bandgap requirement for 780 nm drives the MQW active region to AlGaAs ternary system. The active region usually consists of three or four periods of $Al_{0.12}Ga_{0.88}As$ quantum-wells sandwiched between the $Al_{0.3}Ga_{0.7}As$ barriers. The DBR mirror stack consists of 27 pairs of *p*-type doped $Al_{0.25}Ga_{0.75}As/AlAs$ and 40 pairs of *n*-type doped $Al_{0.25}Ga_{0.75}As/AlAs$, with the bandwidth centered at 780 nm. The laser performance of a 780 nm VCSEL is similar to that of an 850 nm GaAs VCSEL, as shown in Figure 2.11. The increased aluminum concentration in both the active region and the DBR mirror stack over that used in the 850 nm VCSEL raises a concern with the 780 nm VCSEL device reliability, which could be inferior to that of the 850 nm VCSEL.

Red visible VCSELs are of interest for their potential applications in plastic fiber optical communications, bar-code scanners, pointers, etc. The epitaxial structure of a red visible VCSEL is grown on a GaAs substrate mis-oriented 6° off (100) plane toward the nearest |111> direction or on a (311) GaAs substrate [51–54]. It consists of three to four periods of In0.56Ga0.44P QWs with InAlGaP or InAlP as barriers, InAlP as both *p*-type and *n*-type cladding layers, and two DBR mirrors (Figure 2.12). The active QW layer is either tensile or compressive strained to enhance the optical gain. Typically, the QW thickness is 60–80 Å and the barrier thickness is 60–100 Å. The total optical cavity length including the active region and the cladding layers ranges from one wavelength or its multiple integer up to eight wavelengths. The DBR mirrors are composed of either InAlGaP/InAlP or $Al_{0.5}Ga_{0.5}As/AlAs$. The $Al_{0.5}Ga_{0.5}As/AlAs$ DBR mirror performs better because of a relatively larger index difference between the two DBR constituents, thus a higher reflectivity and a wider bandwidth. In general, because the index difference between $Al_{0.5}Ga_{0.5}As$ and AlAs is smaller than that used for the 850 nm VCSELs, more mirror pairs are needed to achieve the required

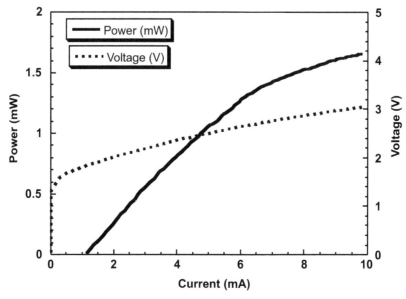

Figure 2.11   Etched mesa structure VCSEL output power vs. input current at a wavelength of 780 nm.

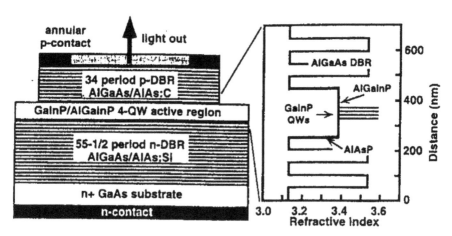

Figure 2.12   A visible VCSEL structure [53].

DBR reflectivity. Typically, 55 pairs are needed for the *n*-DBR and 40 pairs are needed for the *p*-DBR to ensure reasonable VCSEL performances. As a rule of thumb, the more pairs in the DBR mirror, the higher series resistance, thus more heat generated in the active region. This implies that the active junction temperature will be higher. So far, sub-mA threshold red VCSELs have been demonstrated. More than 5 mW output power from a red VCSEL has also been reported. The carrier confinement of the red visible VCSELs is poor because of the smaller bandgap offset between the quantum-well and the barrier, and between the active and the cladding. Therefore, the red visible VCSELs are more temperature-sensitive and more challenged for high-temperature operations.

Long-wavelength VCSELs at 1.3 μm and 1.55 μm have drawn attention because of their applications in data links over single-mode fibers. Traditionally, long-wavelength VCSELs are based on InP substrate with InGaAsP MQWs used as the active region. However, the lattice-matched monolithic InGaAsP/InP DBR mirrors do not have sufficient reflectivity for the long-wavelength VCSELs because of the small index difference between the two DBR mirror pair constituents, InGaAsP and InP. In addition, the Auger recombination-induced loss becomes evident due to smaller energy bandgap of the long-wavelength VCSEL active medium. To overcome the difficulty, dielectric mirrors with 8.5 pairs of MgO/Si multilayers and Au/Ni/Au on the *p*-side and six pairs of SiO$_2$/Si on the *n*-side have been used instead of the semiconductor DBR. A continuous wave 1.3 μm VCSEL has therefore been demonstrated at 14°C [55].

As the growth technology matures for AlGaInAs/InP DBRs on InP substrate, both 1.3-μm and 1.55-μm long wavelength VCSELs have achieved operation beyond 100°C using buried tunneling junction (BTJ) for improved carrier confinement in AlGaInAs-active QWs [24]. The long-wavelength VCSEL wafers are grown on InP (100) substrates in a multiple-wafer MOCVD reactor. The bottom DBR consists of over 40 pairs of InP and AlGaInAs lattice-matched to InP, which has a reflectivity of 99.9% at either 1.3 μm or 1.55 μm, depending on which wavelength is targeted for optimization. Six strain-compensated AlGaInAs QWs constitute the active region. Above the active layers, a p++-AlGaInAs/n++AlGaInAs tunnel junction is grown. Carbon is used as *p*-type dopant to achieve hole concentration of up to 7×10^{19} cm^{-3}. The bandgap of AlGaInAs in the tunnel junction is higher than the photon energy at the lasing wavelength to avoid optical absorption. The tunnel junction is thin and positioned at a node of the standing wave of the VCSELs for the same purpose. Because of this, the total optical absorption by the tunnel junction is negligible even if its layers have very high carrier concentrations. After the growth of the tunnel junction, the wafers were patterned with photolithography and circular mesas were formed by wet-etching down to an etch-stop layer. Finally, an *n*-type InP layer and n++-GaInAs contact layer were regrown on these mesas. This "buried tunnel junction" structure can block the current outside of the tunnel junction mesa. In addition, the difference in refractive index between AlGaInAs in the tunnel junction and InP outside creates optical confinement. The advantage of this structure is that the size and shape of the mesas is defined by photolithography. Therefore, it is more uniform and repeatable than the AlAs-oxidation structure for GaAs-based VCSELs. The top DBR consists of amorphous-Si (a-Si) and Al_2O_3 evaporated by e-beam evaporation. The reflectivity of 99.4% is intentionally lower than that of the bottom AlGaInAs–InP DBR to increase output power from the top side. This technology has allowed long-wavelength VCSEL to achieve over 10 Gbps operation at room temperature [56]. Relying on the same design architecture but shorter cavity mode through the use of dielectric DBR mirror with a larger index difference between the two layer constituents, larger bandwidth 1.55-μm VCSEL has been demonstrated to operate at 25 Gbps [25].

Dielectric DBR mirrors and monolithically grown quaternary semiconductor DBR mirrors do not have as good a thermal conductivity as binary semiconductor DBR mirrors. This limits the long-wavelength VCSEL operation temperature range and output power at higher operation temperature. Wafer-fusing techniques have been adopted to bond GaAs/AlAs binary DBR mirrors onto a structure with an InGaAsP MQW active layer sandwiched between the InP cladding layers that are epitaxially grown on the InP substrate [57, 58]. The InP substrate is removed to allow the GaAs/AlAs DBRs to be bonded onto one or both sides of the InGaAsP active region (Figure 2.13). As the DBR mirrors are either *n*-type or *p*-type doped, the completed fused wafer can be processed like a regular GaAs VCSEL wafer. In this way, a 1.5 μm VCSEL has been successfully fabricated that operates CW up to 64°C [21, 22].

Manufacturing yield and reliability has always been a concern with the VCSEL wafer fusion technique. To address this issue, monolithically grown long-wavelength VCSELs on GaAs substrate has been researched with GaInNAs BTJ active materials and AlGaAs/GaAs DBR [23, 59]. An *n*-type bottom mirror consists of 35 pairs of AlGaAs/GaAs DBR. Three GaInNAs QWs are sandwiched by the bottom DBR and a *p*-type spacer. A heavily doped tunnel junction is grown on the *p*-type spacer. To make the optical absorption loss

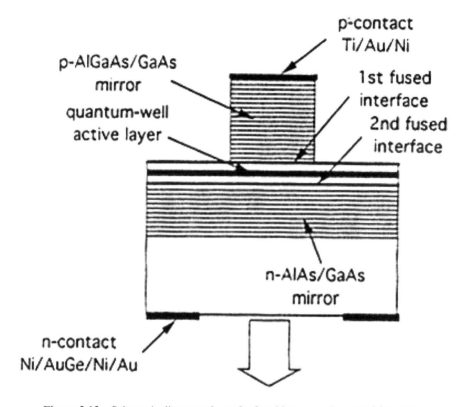

Figure 2.13 Schematic diagram of a wafer fused long wavelength VCSEL [21].

minimum, the tunnel junction is placed at the node of the optical field. A tunnel junction mesa is formed by conventional photolithography and etching techniques. An *n*-type current-spreading layer and a contact layer are grown to cover the tunnel junction mesa. The contact layer at the optical output aperture is removed to reduce the optical absorption loss. A circular mesa is formed to reduce the device capacitance. Finally, nine pairs of ZrO_2/SiO_2 dielectric DBR are deposited on the top of the device after forming a top electrode. The VCSEL operates at a threshold of 2 mA with wavelength at around 1.26 μm. Its bandwidth reaches 10 Gbps. The drawbacks of this approach are the limit of wavelength to 1.3 μm or shorter and the available output power at higher operation temperatures.

In applications, monitoring the output power of VCSEL is frequently required and has typically been achieved by placing a beam splitter on the path of the VCSEL output beam in the packaging. This complicates the

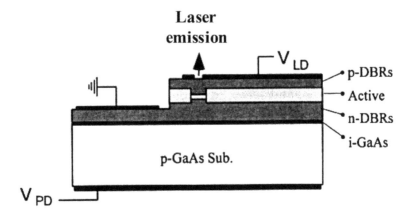

Figure 2.14   Schematic diagram of a VCSEL with integrated detector [50].

packaging architecture, in particular, for parallel optical data links when 850 nm VCSEL arrays are used. Due to the unique vertical stacking feature of VCSELs, a detector can be integrated underneath or above the VCSEL structure during the epitaxial growth [50, 60–63], as shown in Figure 2.14. For example, a VCSEL can start with a *p*-type GaAs substrate, with a PIN detector structure grown first on top of the substrate. The PIN detector has a GaAs intrinsic layer of around 1 μm and *p*-doped AlGaAs cladding of around 2,000 Å between the substrate and the intrinsic absorption layer. The detector structure stops at an *n*-type doped cladding layer of around 2,000 Å. A regular GaAs VCSEL epitaxial structure follows the PIN detector, with layers of *n*-DBR, *n*-cladding, active, *p*-cladding, and *p*-DBR grown in order. The detector cathode in this structure shares a common contact with the VCSEL cathode, with two independent anodes for both the PIN detector and the VCSEL. In practical applications, the anode of the detector can be either reverse biased or without any bias if detector speed is not a major concern. The VCSEL backward emission transmitted through the *n*-DBR is normally in proportion to the VCSEL forward emission. It will be received by the integrated PIN detector and generate a current. The VCSEL output power and the integrated PIN detector response is shown in Figure 2.15. There is a one-to-one relationship between the PIN detector current and the VCSEL output power up to a certain point when the VCSEL output power saturates, but the detector current keeps rising due to the effect of spontaneous emission. This device allows for VCSEL output power to be monitored and controlled

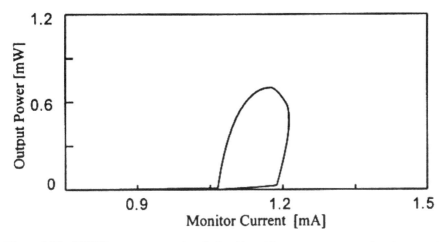

Figure 2.15   VCSEL output power in relationship with current response of an integrated detector [50].

by monitoring the current variation generated in the integrated detector when the VCSEL operates below the saturation [50, 60].

Super-low threshold micro-cavity-type VCSELs have been proposed that utilize the spontaneous emission enhancement due to more spontaneous emission being coupled into the lasing mode [64, 65]. Although a threshold-less laser is theoretically possible when the spontaneous emission coupling efficiency β is made approaching unity, the proposed structures are difficult to make in practice. One of the successful examples in research is to use oxidized lateral carrier confinement blocks by oxidizing an AlAs layer in the DBR or the cladding regions [16, 17, 66], as shown in Figure 2.16. Typically, sub-100 µA threshold can be achieved with this technique. A VCSEL with an extremely low threshold of 8.7 µA has been reported with an active area of 3 µm² [17]. VCSELs with oxidized mirrors have been demonstrated with extremely simple epitaxy layers [67, 68]. In this structure, only four to six pairs of GaAs/AlAs DBR stacks are grown on one or both sides of an active region that is made of strained InGaAs MQWs at 970 nm, as shown in Figure 2.17. The AlAs layers in the DBR mirrors are oxidized during the fabrication procedure. The extremely large index difference between GaAs and the oxidized AlAs layer makes it possible that only four pairs of GaAs/AlAs stacks will provide sufficiently high reflectivity with very large bandwidth for proper device operation. The VCSEL electrical contacts in this case will have to be made laterally inside the cavity as opposed to the top of the DBR

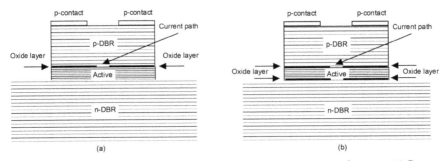

Figure 2.16   VCSEL with native aluminum oxide for lateral current confinement. (a) Current confinement on *p*-side, and (b) current confinement on both *p*-side and *n*-side.

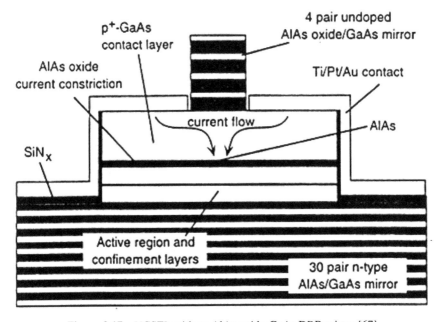

Figure 2.17   VCSEL with an AlAs oxide-GaAs DBR mirror [67].

mirror stacks because the electrical conduction through the DBR mirror is prohibited once the AlAs constituent of the mirror is oxidized.

High-speed data transmission requires that a VCSEL be modulated at tens of Giga-Hertz. The cavity volume of a VCSEL is very small, resulting in high photon density in the VCSEL cavity. The resonance frequency of a semiconductor laser typically scales as the square root of the photon density,

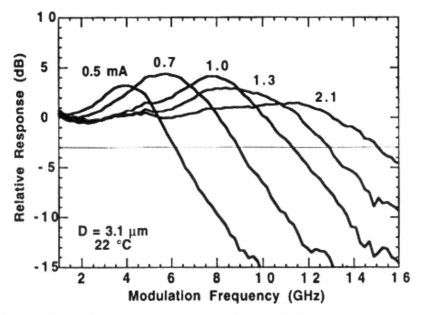

Figure 2.18 Small signal modulation response of a 3 µm VCSEL at various bias current. The maximum 3-dB bandwidth is about 15 GHz [70].

thus indicating that a VCSEL has a potential advantage in high-speed operation. However, the parasitic series resistance caused by the semiconductor DBR and the device heating limit the maximum achievable VCSEL modulation bandwidth. Nevertheless, modulation bandwidth of larger than 16 GHz VCSEL has been demonstrated with an oxide-confined VCSEL at a current of 4.5 mA [69]. Modeling results indicate that a gain compression limited oxide VCSEL with a diameter of 3 µm has an intrinsic 3-dB bandwidth of 45 GHz [70] and a measured 3-dB bandwidth of 15 GHz at 2.1 mA due to the parasitic resistance and device heating, as shown in Figure 2.18. So far, oxide VCSEL technology has been commercialized for supporting from 1 to 25 Gbps Ethernet over multimode fibers. Error-free transmission at above 40 Gbps over 50 m OM4 multimode fiber has also been demonstrated at room temperature [71].

2.4 VCSEL Material Growth

VCSEL technology is in essence an epitaxy growth technology because VCSEL structure may consist of more than 100 epitaxy layers. The thickness and the doping level of each layer have to be precisely controlled. Any slight deviation to the designed parameters may affect the VCSEL performance, which will result in manufacturing yield loss. In practice, the total variation of each layer thickness across the wafer cannot exceed 1% during the growth. Such precise thickness control is needed because of two reasons. First, the net optical gain in the VCSEL cavity is very small because of the thin active layer. Highly reflective DBR mirrors, typically in excess of 99.5%, are needed to ensure that the laser operates effectively. In order to achieve such a high reflectivity, each repetitive layer in the DBR mirror should be exactly the same, i.e., one quarter wavelength for each layer, to retain appropriate constructive interferences among those layers. Second, the cavity resonance peaks of a VCSEL are usually spaced widely apart due to the typical one-λ cavity. The requirement that the spectral position of this resonance fall within the optical gain region of the active MQWs places a tight constraint on the optical thicknesses of the layers making up the cavity.

Figure 2.19 shows the simulation results of net reflectivity of an 850 nm GaAs VCSEL structure grown on GaAs substrate, which consists of a top DBR, a bottom DBR, and an active region sandwiched in-between, as shown in Figure 2.3. Figure 2.19(a) is the reflectivity of an ideally grown structure. A deep Fabry–Perot resonate peak sits in the center of the reflective band, 850 nm. Figure 2.19(b) illustrates the reflectivity of a structure with the top DBR wavelength being longer than the bottom DBR wavelength by 1%. The resonate peak shifts toward the longer wavelength and becomes less pronounced, and a small glitch appears on the long wavelength end of the high reflectivity band. Figure 2.19(c) illustrates the reflectivity of a structure with the top DBR wavelength being shorter than the bottom DBR wavelength by 1%. The resonate peak shifts toward the shorter wavelength and becomes less pronounced, and a small glitch appears on the short wavelength-end of the high reflectivity band. Figure 2.19(d) illustrates the laser cavity being 1% longer than the designed wavelength of 850 nm. The center resonate peak moves toward the longer wavelength, and the depth of the resonate peak is reduced. Figure 2.19(e) illustrates the laser cavity being 1% shorter than the designed wavelength of 850 nm. Accordingly, the center resonate peak moves toward the shorter wavelength, and the depth of the resonate peak is also reduced. In order to achieve better than 1% growth control for the VCSEL structure, not only is the exact knowledge of the refractive index of the required alloy concentration needed, but also the precise growth rate has to be determined and maintained throughout the growth. Typically, the simulation results as shown in Figure 2.19 are used as references to judge

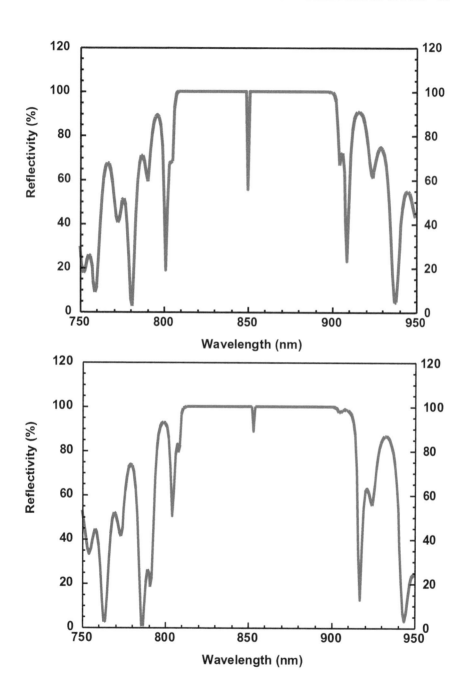

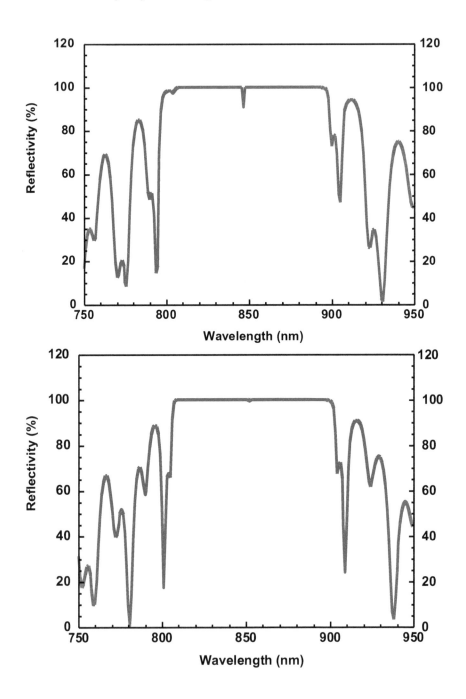

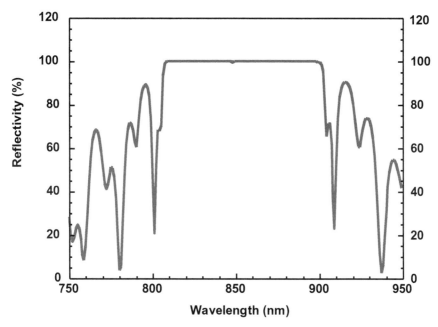

Figure 2.19 Net reflectivity simulation results of a complete GaAs VCSEL structure grown on a GaAs substrate for (a) a perfectly grown structure with identical top and bottom DBR mirrors, (b) the top DBR center wavelength being 1% longer than the bottom DBR center wavelength, (c) the top DBR center wavelength being 1% shorter than the bottom DBR center wavelength, (d) the VCSEL cavity being 1% longer than the designed wavelength, and (e) the VCSEL cavity being 1% shorter than the designed wavelength.

if the growth is successful, or how much deviation the growth is from a designed structure. The information drawn from the comparison will be used to make corrections for the subsequent growth runs. Successful growth requires an extremely stable epitaxial growth reactor that has the necessary long-term reproducibility and controllability. Two types of epitaxial techniques are used to grow VCSEL structures, molecular beam epitaxy (MBE) and metal-organic chemical vapor deposition (MOCVD).

An MBE growth system usually takes a 2-in or a 3-in GaAs wafer. The growth process takes 8–12 h to complete the entire VCSEL structure. Standard high temperature effusion cells provide group III sources such as Al, Ga, and In, and group V source such as As. The MBE system is also equipped with cells for *n*-type and *p*-type dopants, which are usually Si and Be, respectively. Modifications are usually made to the MBE system to reduce the growth rate transients to less than 1% during the repetitive switching of constituents for the VCSEL multiple layers. When substrate rotation is employed during the growth, 1% lateral thickness uniformity across entire 2 inch wafers can be achieved. This will lead to high yield and high uniformity across the wafer.

Reflection high energy electron diffraction (RHEED) oscillation measurements can be used to count the atomic layers being deposited, thus precisely monitoring the growth thickness [72, 73]. However, RHEED can only be used to monitor one particular spot on a wafer. When RHEED is used for *in-situ* growth monitoring, the wafer is not allowed to rotate, leading to roughly 5%/cm linear variation in layer thickness across the wafer. While such a variation may be advantageous in the research environment, it is not acceptable to the commercial application due to wafer growth non-uniformity affecting production yield. Therefore, RHEED oscillation calibrations are usually performed prior to the actual growth on a small test sample centrally mounted on a different sample block. Re-evaporation effect of group III species may affect the crystal quality grown by MBE. This effect is strongly substrate-temperature-dependent. To achieve good crystal quality, the GaAs–AlGaAs–AlAs layers are usually grown at a substrate temperature of 580–630°C, and InGaAs at 520–530°C.

The growth-rate calibration before the actual growth is only accurate to about 1%. It's not only time consuming, but also subject to drift. A modified RHEED *in situ* monitoring scheme is based on discrete substrate rotations [74]. Following the growth of each pair of quarter-wavelength Bragg layers, the growth is paused for a few seconds and the substrate is rotated by 180°. Any growth non-uniformity across the wafer will be compensated during the subsequent growth after the substrate rotation. The net effect of the discrete rotation will be equivalent to that of the continuous substrate rotation.

Optical reflectometry measurement is another effective *in situ* monitoring technique, where the bottom DBR mirror and most of the VCSEL active cavity, ranging between 94 and 100% of the cavity length, is grown first. The half-finished wafer is then taken out to evaluate the reflectivity spectrum. A comparison with the simulation will reveal information on how to complete the rest of the cavity and the top DBR [75].

A more rigorous *in situ* reflectivity measurement apparatus is shown in Figure 2.20 [76], where the measurement is conducted at room temperature through the viewpoint of the MBE transfer tube. The wafer remains in the ultrahigh vacuum (UHV) environment for the measurement so that any potential oxidation-induced defects can be avoided at the interface between the portions grown and portions of subsequent growth. In order to minimize the

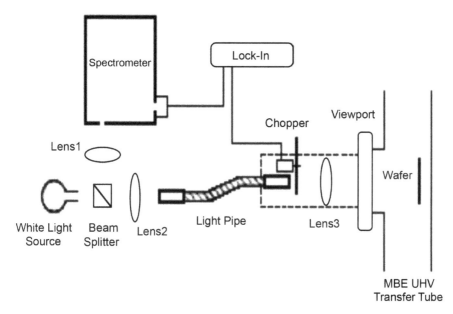

Figure 2.20 An *in situ* reflectivity measurement system for growing VCSELs by a MBE system [76].

weight and bulk of the equipment attached to the viewport, a light pipe is used to direct the white light toward and receive the reflected light from the wafer. A chart recorder synchronized with the scan rate of the spectrometer records the unnormalized reflectivity spectra. During the growth, the reflectivity spectrum is first measured after only a few periods of the bottom DBR mirror having been grown. Any relative error in the center wavelength determined from the measurement will be used to correct the subsequent growth. The second interruption and reflectivity measurement can be made just after the growth of the cavity. However, as the bottom DBR mirror reflectivity is very high and the cavity absorption is low, the Fabry–Perot resonant dip is very shallow and sometimes difficult to identify. To enhance the resonant dip visibility, two to three periods of the top DBR mirror can be grown before the second reflectivity spectrum measurement. Any cavity thickness correction can then be made by adjusting the thickness of the immediate one layer in the top DBR mirror when the growth resumes.

Other methods have also been demonstrated to be capable of *in situ* monitoring the VCSEL growth precisely to within 1%, such as multi-wavelength pyrometric interferometry measurements [77] and single-wavelength pyrometric interferometric measurements [78, 79].

While many research works on near-infrared VCSELs (780–980 nm) are based on MBE, MOCVD has its advantages in VCSEL applications due to its high growth rate, wide range of host and dopant species, and capability of continuously varying material composition.

In a high-volume manufacturing environment, VCSEL growth times of 8–12 h by an MBE reactor suffers from high cost of ownership and low throughput. On the other hand, a MOCVD reactor may achieve growth rates three times as high as MBE [80] while maintaining $\pm 1\%$ thickness uniformity across the entire wafer. Good yields have been achieved with multi-wafer MOCVD reactors which hold 15×2-inch wafer loads [81] or up to 7×6" wafers [82].

In order to maintain sharp doping profiles and prevent the dopant out-diffusion into the active region of the device, dopants with low diffusion coefficients are used. Zinc (Zn) is the most commonly used *p*-dopant in MOCVD growth [83], and is the dominant dopant in GaAs-based devices. It possesses, however, a high diffusion coefficient and is therefore not recommended for the VCSEL *p*-dopant from device performance consideration [84]. The other commonly used *p*-doping alternative is carbon (C) which has proven to be a reliable *p*-dopant with very low diffusivity and high solubility [85–88]. A commonly used carbon precursor for *p*-doping is carbon tetrachloride (CCl_4). The growth rate of C-doped AlGaAs is usually lower than that of undoped or *n*-doped AlGaAs with the effect of slow etching of AlGaAs growth surface while doping with CCl_4. It should be noted that CCl_4 is an ozone depleting material and its use has been discouraged for environmental considerations. Therefore, alternative C-doping method has been attempted. One candidate is the use of organometallic As sources such as tertiarybutylarsine (tBAs). Tertiarybutylarsine allows the incorporation of intrinsic carbon into the grown material due to enhancement of background carbon levels from organometallic decomposition and has the advantages of being roughly on an order of magnitude less toxic than arsine [89–91].

Low series resistance in the *p*-doped DBR mirror ensures high VCSEL performances, which is usually achieved by grading the hetero-interfaces between the alternative DBR mirror pairs. In MBE, the continuous interface grading can be accomplished by varying source cell temperature [39, 92], or by repeatedly switching the cell shutters to create a super-lattice grading [38]. Even though the techniques have been demonstrated with high reproducibility, the processes are not easy to implement from a manufacturing standpoint.

In comparison, the continuous grading of an arbitrary composition profile can be readily achieved with MOCVD, which has therefore become the primary tool for VCSEL manufacturing.

The MOCVD growth is typically carried out in a commercially available horizontal quartz tube reactor with graphite susceptor and fast-switching run-vent manifold. Sources used are trimethylindium (TMIn), triethylgallium (TEGa) or trimethylgallium (TMGa), trimethylaluminum (TMAl), arsine (AsH_3), and phosphine (PH_3). Dopants are Si from Si_2H_6 or Te from diethyltellerium (DETe) for *n*-type materials. Carbon (C) from CCl_4 for p-type AlGaAs alloys, Zn from diethylzinc (DEZn) for *p*-type GaAs cap, and Zn or Mg, from bis-cyclopentadienyl magnesium (Cp_2Mg) for *p*-type AlGaInP. Growth pressure inside the reactor is in the range of 80–200 mbar, and growth temperatures are typically in the range of 725–775°C. With AlGaAs DBR mirrors, the growth rate is around 3 μm/h, V to III ratio is 50:1, and the interface grading, whether linear or parabolic, can be accomplished by simultaneous ramping of TMGa and TMAl flows in very short increments, such as 0.1 s, controlled by the computer. Great yield and uniformity can be achieved with both 3-in wafers [7, 93], and 4-in wafers [94] by MOCVD.

The growth reproducibility of VCSEL structures is of a major concern as the DBR structures are sensitive to growth fluctuations. The statistics obtained for 150 VCSEL wafers grown by MOCVD and operating at 850 nm is shown in Figure 2.21, where the variation of the reflectivity center wavelength as well as the bandwidth of the reflectivity spectrum is shown in Figures 2.21(a) and (b), respectively. The standard deviation is 11 nm and 6 nm for the center wavelength and the bandwidth, respectively [95]. The application of *in situ* reflectometry growth monitoring [96–99] has allowed a better run-to-run thickness control across a 3-in wafer using MOCVD [100].

2.5 VCSEL Fabrication Process

The post-epitaxial VCSEL process involves photolithography, ion implantation, metal deposition, wet and dry chemical etching, oxidation, oxide and nitride deposition, etc. Each fabrication step is customized for compound semiconductor material systems.

From the emission direction standpoint, there are top emitting VCSELs and bottom emitting VCSELs. GaAs based VCSELs will not emit toward the bottom substrate side if the wavelength is shorter than 870 nm because of the substrate absorption, unless the substrate is removed. In this section, we will concentrate on the discussion of top emitting GaAs VCSEL processing.

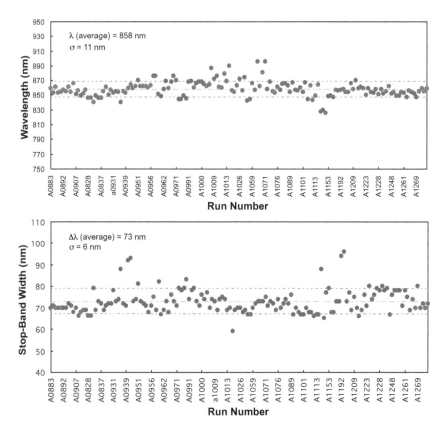

Figure 2.21   The statistics for 150 VCSEL wafers grown by MOCVD and operating at 850 nm. (a) The variation of the reflectivity center wavelength, and (b) the bandwidth of the reflectivity spectrum [95].

The same fabrication technology can easily be extended to bottom emitting VCSELs, such as strained InGaAs VCSELs at around 980 nm.

Vertical cavity surface emitting lasers are either gain guided or index guided. The gain-guided VCSEL has a planar top surface as shown in Figure 2.22. The current path of the gain-guided VCSEL is surrounded by the proton (H^+) implant buried in the top p-doped DBR mirror, which forms a high-resistance enclosure right above the active region. The optical mode is determined by the active region where the current flows through.

The fabrication of the gain-guided VCSEL is straightforward. Immediately after the photoresist patterning on top of the wafer that defines the

Laser emission

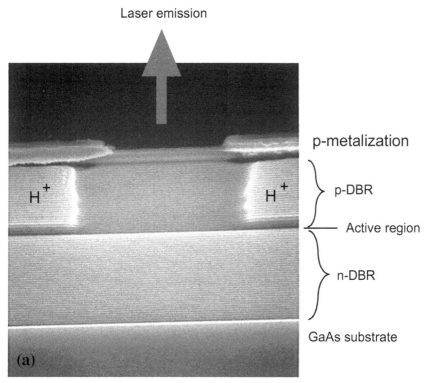

p-metalization

p-DBR

Active region

n-DBR

GaAs substrate

(a)

Figure 2.22   Cross section SEM photo of a planar GaAs VCSEL structure, where proton (H$^+$) implantation region is delineated by stain etching.

VCSEL emission aperture window, BeAu/Ti/Au three layer *p*-contact metal is deposited by thermal evaporation [101] for lift-off. The total contact metal thickness is around 1,500–3,000 Å. The emission window diameter typically varies from 5 to 50 μm, depending on the requirements for optical output power and transverse mode design. A thick layer of photoresist is patterned to cover the laser emission window with the diameter slightly larger than that of the emission window as a proton (H$^+$) implant mask. The photoresist implant mask thickness varies from 6 to 12 μm depending on the proton implant energy. The implant energy is such that the majority of the protons rest at 0.5 μm above the active region to create a highly resistive layer, leaving the topmost portion of the DBR mirror relatively conductive. For the top DBR mirrors of 18 to 30 quarter-wavelength stack pairs, the required implant energy varies from 300 to 380 kV. The proton implant dose is around

10^{14} to 10^{15}/cm^2. The implant straggle width is around 0.5 μm. The proton implant defines an annular high-resistance region underneath the p-metal contact and right above the active region in the p-DBR mirror, so that the injected current can funnel into the central active region underneath the laser emission aperture window. The diameter of the implant mask is chosen to be larger than the laser emission window defined by the p-metal contact. This helps lower the p-metal contact resistance because of a ring contact with an area unaffected by the implant. The smaller emission window also behaves like a filter that enhances the lowest order transverse mode [101, 102].

After the photoresist mask is stripped off, a new photoresist mask pattern is deposited for a shallow proton implantation to electrically isolate individual VCSELs between each other. The second implantation can be replaced by etched trenches if the requirement for planar surface is not essential [103]. Subsequently, an n-metal contact layer consisting of GeAu/Ni/Au is deposited by either thermal evaporation or e-beam evaporation on the bottom of the GaAs substrate. The wafer is then annealed at 450°C for 30 s. The anneal serves to alloy both contacts and to further shape the resistance profile due to proton implantation by further reducing the surface resistance while retaining a high-resistance layer under the metallized region [104]. A metal interconnect deposition on the top surface completes the device fabrication process. As an option, a layer of dielectric material, such as SiN or SiO$_2$, can be finally deposited by plasma-enhanced chemical vapor deposition (PECVD) for the purposes of laser device surface passivation.

The gain-guided VCSEL by proton implantation has been proven manufacturable with exceptionally high yield [7, 93]. There are, however, concerns on the devices' turn-on delay due to thermal lensing effect [105] and turn-on delay-induced timing jitter [106, 107]. These effects give rise to problems in high-speed fiber optical applications. The adverse effects of the turn-on delay have been alleviated via careful biasing of the laser at or above threshold [108, 109]. To further address this issue, index-guided VCSELs or ridge-waveguide VCSELs have been developed and achieved better performances than the gain-guided VCSELs [7].

A typical index-guided VCSEL structure is shown in Figure 2.4 [110]. The fabrication process starts with the deposition of 4,000 Å TiW metal alloy by sputtering and 4,000 Å SiN by PECVD. A photoresist mask is used to define the index-guided VCSEL mesa diameter, which usually ranges from 8 to 50 μm. The SiN and TiW outside of the photoresist mask is dry etched by fluorinated chemistry in a reactive ion etching (RIE) reactor to expose the AlGaAs semiconductor epitaxial layer. After the removal of the photoresist, the remaining SiN serves as a hard mask for the following mesa etching by chlorine chemistry in a RIE reactor. The mesa etching depth is *in situ* monitored using laser interferometry to a level above the active layer. Protons are then implanted into the remaining *p*-DBR mirror to further electrically isolate the mesa region. The implantation energy is 50 keV and the implantation dose is in the range of 10^{14} to 10^{15}/cm^2. After the implantation, a layer of 4,000 Å SiN is deposited by PECVD onto the entire wafer. A photoresist mask is then spun onto the SiN with the mesa top exposed for dry etching using fluorine chemistry to remove the SiN on the mesa. This etching process is not selective, so some TiW will be etched as well. After the photoresist is removed, 4,000 Å TiW will be sputtered onto the entire wafer. A new photoresist mask is then used to define a VCSEL emission aperture on the mesa. The diameter of the emission aperture is smaller than the mesa diameter and fluorine chemistry is used to selectively remove the TiW in the emission aperture. Gold interconnect pads can then be deposited on top of the TiW by a photoresist liftoff process. This completes the device fabrication on the *p*-side. A broad area *n*-metal contact formed by Ni/AuGe/Au is deposited on the bottom surface of the GaAs substrate. The wafer is then annealed in a rapid thermal annealing (RTA) chamber for 30 s at 420°C. Finally, a SiN passivation layer can be deposited on the VCSEL surface for protection. The wafer is then sawn into individual chips.

When the VCSEL is driven at a current well above threshold, multi-transverse modes develop partially due to spatial hole burning in the active region. The number of allowable transverse modes of a regular index guided VCSEL is largely determined by the waveguide composed of etched mesa and the materials surrounding the mesa, such as SiN, metal, or air. As a consequence of the large index difference between the semiconductor material and the surrounding materials, the diameter of the mesa has to be very small for the waveguide to support only the fundamental/single transverse mode. Single-mode VCSELs are preferred for an optical interconnect based on single-mode fibers, laser printing, compact disk data storage, bar code scanning, etc. Reducing the mesa diameter also helps reduce the VCSEL threshold. Super-low threshold VCSEL designs are appealing for some high-speed applications, as well as for some basic physics studies. However, the mesa-type index-guided VCSEL cannot go too small in diameter because of the optical scattering loss from the etched sidewall. The mesa sidewall damage due to exposing to the ion beam [111] may also become non-radiative recombination centers that limit the minimum achievable threshold and cause potential reliability problem.

Selectively lateral oxidation-defined index-guided VCSEL, as shown in Figure 2.16, is an alternative to the mesa-defined index-guided VCSEL [112–115] and has become the primary devices for today's datacenter applications due to its higher operation bandwidth to support beyond 40 Gbps per channel data transmissions [71]. Native oxide of Al_xO_y is stable enough to be utilized in the device fabrication for current confinement [116–119]. The effective index difference between the semiconductor and its native oxide also provides for an excellent waveguide for optical mode confinement. The diameter of the lateral oxide VCSEL waveguide can be made relatively larger for supporting only the single transverse mode, owing to the relatively moderate effective index difference. In addition, the *p*-metal has a larger area to make contact with for the oxide VCSEL, reducing the series resistance significantly even for a very small active area. Lower series resistance is effective in reducing the active junction temperature while operating, therefore more desirable from the reliability standpoint.

As an example, to fabricate the device structure shown in Figure 2.16(a), a VCSEL epitaxial structure is grown that includes $Al_{0.85}Ga_{0.15}As/Al_{0.16}Ga_{0.84}$ As DBR mirrors and a triple GaAs quantum-well active region in a graded AlGaAs cavity having the aluminum mole fraction of the AlGaAs mirror layer closest to the cavity increased from $x = 0.85$ to $x = 0.98$ on the *p*-side. Devices are formed in mesas created by etching away all the surrounding *p*-side mirror stacks and part of the cavity materials above the active layer to expose the edge of the cavity layers. A portion of the AlGaAs layers starting from the exposed edge is then converted to an electrically insulating aluminum oxide (Al_xO_y) by wet thermal oxidation at 425°C for a few hours. The aluminum oxide has a refractive index of ~ 1.55. The lateral oxidation is conducted in a furnace supplied with a flow of N_2 bubbled through deionized water heated to a temperature of 95°C. The $x = 0.98$ layer oxidizes more rapidly [120] and the resultant oxide protrudes further into the mesa to provide electrical and optical confinement about the active region. A backside substrate contact and an annular contact on the mesa complete the VCSEL [115]. This selective lateral oxidation is the most effective method so far in achieving record low-threshold VCSELs [16, 17]. Single transverse mode operation of up to 2.7 mW has also been demonstrated [121].

The same oxidation technology described above can be used to oxidize through the high aluminum mole fraction layer to create DBR mirrors with one layer of semiconductor material and the other layer of a native oxide Al_xO_y. The large index difference between the two alternative quarter-wavelength thick layers enables the demonstration of high reflective and wide bandwidth DBR mirror with only 4 or 5 mirror pairs [67, 68].

Oxide VCSEL has superb reported performances in single-mode power operation [121] and low threshold [16, 17]. Its long-term reliability, manufacturing uniformity across the wafer and yield are also proven through more than 10 years of field deployments. Other techniques for high-power single-mode emission include the leaky mode solution [122], which uses the regrowth technique to deposit higher index materials to surround the etched mesa pillar [123, 124]. A simpler leaky mode technique is to use a planar proton-implanted VCSEL with smaller current isolation diameter and emission aperture. This technique has allowed the generation of a single-mode output power up to 4.4 mW at 980 nm [125].

In systems that are polarization sensitive, such as some optical data storage systems, the precise knowledge of the VCSEL polarization is required. VCSELs generally polarize along $|110>$ and $|1–10>$ crystalline directions. In order to stabilize the VCSEL polarization along a predetermined direction, certain asymmetries have to be introduced into the laser structure. The most straightforward technique is to utilize the anisotropic transverse cavity geometry to select and maintain a dominant polarization state [126, 127]. Typically, the airpost or ridge waveguide mesa structure is elongated into a shape of rectangular, rhombus, or dumbbell. This technique works successfully in stabilizing the VCSEL polarization along the elongated mesa direction. Alternatively, the polarization can be stabilized by using a differential loss method obtained from coatings on the sides of the upper DBR [128], or by applying biaxial stress to the active layer [129]. A grating formed on the very top layer of the p-DBR, with metalization on the side walls of the grating structure, has also been demonstrated with good polarization extinguishing ratio [130].

The GaAs VCSEL manufacturing processes discussed above apply to InGaAs/GaAs VCSELs at 980 nm and InAlGaP/GaAs VCSELs at red visible wavelengths except that the 980 nm VCSELs can be made with laser emission toward the GaAs substrate side. Long-wavelength VCSELs at 1.3 μm and 1.55 μm are of interest to the telecommunication industries, but except for the 1.3-μm long-wavelength VCSEL with GaInNAs as the active materials, which follow the GaAs VCSEL manufacturing processes, the fabrication for long-wavelength VCSEL is typically more challenging due to the use of dielectric DBR mirrors, wafer fusing, and BTJ for active carrier confinement.

In one example of a monolithically grown long-wavelength VCSEL on InP substrate, dielectric mirrors with 8.5 pairs of MgO/Si multilayers and

Au/Ni/Au on the p-side and six pairs of SiO_2/Si on the n-side have been used instead of the semiconductor DBR mirrors [55]. Continuous wave VCSEL operation at 14°C has therefore been demonstrated at a wavelength of 1.3 μm. A monolithically grown long-wavelength VCSEL with BTJ active region normally has one of the DBRs as dielectric DBR [24, 56].

A more rigorous technique in the long-wavelength VCSEL fabrication is to use a wafer-fusion or bonding process to unite separately made GaAs/AlAs mirrors formed on a GaAs substrate to the InGaAsP double-heterostructure (DH) laser diode [57]. The higher reflectivity of the GaAs-system mirrors at either 1.3 or 1.55 μm make them more advantageous than InP-system mirrors. The wafer-fused VCSEL fabrication begins by fusing the GaAs/AlAs mirror wafer with the top of the InP DH wafer at 650°C for 30 min. in an H_2-ambient environment under a graphite load [131]. After the wafers have been fused together, the InP substrate is removed by a selective wet chemical etchant, such as $HCL:H_2O$ (3:1). The resulting fused structure, an InP DH active layer with a GaAs/AlAs DBR mirror on a GaAs substrate, is mechanically robust, and undergoes typical fabrication procedures without any special handling. The completed mesa structure VCSEL has a Pd/Zn/Pb/Au intracavity ring contact and a Si/SiO_2 dielectric mirror on the p-side, and an AuGe/Ni/Au contact on the n-type GaAs substrate [58]. This device operates pulsed at room temperature and CW at −45°C at a wavelength of 1.3 μm. When both sides of the InP DH wafer are fused with the GaAs/AlAs DBR mirrors, the VCSEL performances are further improved [21]. A double-fused VCSEL operating CW up to 64°C has been demonstrated at 1.55 μm in combining with the lateral DBR mirror oxidation technique [22].

2.6 Conclusion

Vertical cavity surface emitting lasers have been the primary laser sources for short-distance Datacom and data centers over multimode fibers from 1 to 25 Gbps per channel in the form of either serial or parallel optical data links. VCSELs are grown by MBE or MOCVD techniques. From a manufacturing standpoint, MOCVD offers the best growth throughput and excellent material quality, thus is predominantly used by all of the commercial VCSEL manufacturers.

Vertical cavity surface emitting lasers have only one longitudinal mode because of their extremely short laser cavity. Multi-transverse mode operation, however, is common due to the large active area relative to the laser cavity length. There are three basic types of VCSEL structures: proton implanted VCSELs, etched mesa VCSELs, and lateral oxidation confined VCSELs. Some applications require that the laser beam be diffraction limited. Single transverse mode VCSELs can be achieved by reducing the laser emission aperture, but the single-mode output power becomes limited as well [121, 123–125]. VCSEL beams have the advantage of being circular, and therefore offer a high coupling efficiency into an optical fiber.

Vertical cavity surface emitting lasers can be tested on wafer, and only known good dies will be packaged after being sawn into chips. From a manufacturing standpoint, VCSELs are more suitable for low-cost mass production.

Great progress has been made in VCSEL at 1.3 and 1.55 µm using wafer bonding technique, GaInNAs technology, and BTJ for carrier confinement. From commercial application perspective, their technology maturity lags far behind that of 850 nm GaAs VCSELs.

Vertical cavity surface emitting lasers have been proven reliable with extrapolated mean-time-to-failure (MTTF) of up to millions of hours at normal operation conditions without any burn-in screening [7, 132]. Hermeticity may not be necessary for VCSEL packaging. Such an advantage will allow VCSELs to be used in non-hermetic environment, thus substantially reducing the packaging cost.

References

[1] Hall, R. N., Fenner, G. E., Kingsley, J. D., Soltys, T. J., and Carlson, R. O. (1962). Coherent light emission from GaAs junctions. *Phys. Rev. Lett.* 9, 366.

[2] Nathan, M. I., Dumke, W. P., Burns, G., Dill, Jr., F. H., and Lasher, G. (1962). Stimulated emission of radiation from GaAs p-n junctions. *Appl. Phys. Lett.* 1, 62.

[3] Holonyak, Jr., N., and Bevacqua, S. F. (1962). Coherent (visible) light emission from Ga($Al_{1-x}P_x$)As junctions. *Appl. Phys. Lett.* 1, 82.

[4] Quist, T. M., Rediker, R. H., Keyes, R. J., Krag, W. E., Lax, B., McWhorter, A. L., et al. (1962). Semiconductor maser of GaAs. *Appl. Phys. Lett.* 1, 91.

[5] Soda, H., Iga, K. I., Kitahara, C., and Suematsu, Y. (1979). GaInAsP/InP surface emitting injection lasers. *Jpn. J. Appl. Phys.* 18, 2329.

[6] Iga, K., Koyama, F., and Kinoshita, S. (1988). "Surface emitting semi-conductor lasers," *in Proceedings of the IEEE Journal of Quantum Electronics*, 1845.

[7] Lebby, M. S., Gaw, C. A., Jiang, W., Kiely, P. A., Shieh, C. L., Claisse, P. R., et al. (1996). "Use of VCSEL arrays for parallel optical interconnects," *in Proceedings of SPIE – The International Society for Optical Engineering*, 2683, 81–92.

[8] Orenstei, M., Von Lehmen, A. C., Chang-Hasnain, C., Stoffel, N. G., Harbison, J. P., and Florez, L. T. (1991). Matrix addressable vertical cavity surface emitting laser array. *Electron. Lett.*, 27, 437–438.

[9] Shieh, C. L., Ackley, D. E., and Lee, H. C. (1993). "*Temperature insensitive vertical cavity surface emitting laser*," US Patent# 5,274,655.

[10] Young, D. B., Scott, J. W., Peters, F. H., Thibeault, B. J., Corzine, S. W., Peters, M. G., et al. (1993). "High-power temperature-insensitive gain-offset InGaAs/GaAs vertical-cavity surface-emitting lasers," *in Proceedings of the IEEE photonics technology letters*, **5**, 129–132.

[11] Tai, K., Hasnain, G., Wynn, J. D., Fischer, R. J., Wang, Y. H., Weir, B., et al. (1990). 90% coupling of top surface emitting GaAs/AlGaAs quantum well laser output into 8 μm diameter core silica fibre. *Electron. Lett.* **26**, 1628–1629.

[12] Iga, K., Ishikawa, S., Ohkouchi, S., and Nishimura, T. (1984). Room temperature pulsed oscillation of GaAlAs/GaAs surface emitting laser. *Appl. Phys. Lett.*, **45**, 348–350.

[13] Koyama, F., Kinoshita, S., and Iga, K. (1988). "Room-temperature CW operation of GaAs vertical cavity surface emitting laser," *in Proceedings of the IEICE Transactions (1976–1990)*, 1089–1090.

[14] Peters, F. H., Peters, M. G., Young, D. B., Scott, J. W., Thibeault, B. J., Corzine, S. W., et al. (1993). High power vertical cavity surface emitting lasers. *Electron. Lett.* **29**, 200–201.

[15] Grabherr, M., Weigl, B., Reiner, G., Michalzik, R., Miller, M., and Ebeling, K. J. (1996). High power top-surface emitting oxide confined vertical-cavity laser diodes. *Electron. Lett.* 32:1723.

[16] Huffaker, D. L., Shin, J., and Deppe, D. G. (1994). Low threshold halfwave vertical-cavity lasers. *Electron. Lett.* **30**, 1946–1947.

[17] Yang, G. M., MacDougal, M. H., and Dapkus, P. D. (1995). Ultralow threshold current vertical-cavity surface-emitting lasers obtained with selective oxidation. *Electron. Lett.* **31**, 886–888.

[18] Lear, K. L., Choquette, K. D., Schneider, R. P., Kilcoyne, S. P., and Geib, K. M. (1995). Selectively oxidised vertical cavity surface emitting lasers with 50% power conversion efficiency. *Electron. Lett.* **31**, 208–209.

[19] Jäger, R., Grabherr, M., Jung, C., Michalzik, R., Reiner, R., Weigl, B., et al. (1992). 57% wallplug efficiency oxide-confined 850 nm wavelength GaAs VCSELs. *Electron. Lett.* 33, 330–331.

[20] Iga, K. (1992). Surface emitting lasers. *Opt. Quantum Electron.* **24**, S97–S104.

[21] Babic, D. I., Streubel, K., Mirin, R. P., Margalit, N. M., Bowers, J. E., Hu, E. L., et al. (1995). Room-temperature continuous-wave operation of 1.54-μm vertical-cavity lasers. *IEEE Photon. Technol. Lett.* **7**, 1225–1227.

[22] Margalit, N. M., Babic, D. I., Streubel, K., Mirin, R. P., Naone, R. L., Bowers, J. E., et al. (1996). Submilliamp long wavelength vertical cavity lasers. *Electron. Lett.* **32**:1675.

[23] Onishi, Y., Saga, N., Koyama, K., Doi, H., Ishizuka, T., Yamada, T., et al. (2009). Long-wavelength GaInNAs VCSEL with buried tunnel junction current confinement structure. *SEI Tech. Rev.* **68**, 40–43.

[24] Nishiyama, N., Caneau, C., Hall, B., Guryanov, G., Hu, M. H., Liu, X. S., et al. (2005). "Long-wavelength vertical-cavity surface-emitting lasers on InP with lattice matched AlGaInAs-InP DBR grown by MOCVD", *in Proceedings of the IEEE Journal of Selected Topics in Quantum Electronics*, **11**, 990–998.

[25] Muller, M., Hofmann, W., Grundl, T., Horn, M., Wolf, P., Nagel, R. D., et al. (2011). "1550-nm high-speed short-cavity VCSELs", *in Proceedings of the IEEE Journal of Selected Topics in Quantum Electronics*, **17**, 1158–1166.

[26] Vakhshoori, D., Wynn, J. D., Zydik, G. J., and Leibenguth, R. E. (1993). 8 x 18 top emitting independently addressable surface emitting laser arrays with uniform threshold current and low threshold voltage. *Appl. Phys. Lett.*, **62**, 1718–1720.

[27] Uchiyama, S., and Iga, K. (1985). Two-dimensional array of GaInAsP/InP surface-emitting lasers. *Electron. Lett.* **21**, 162–164.

[28] Deppe, D. G., van der Ziel, J. P., Chand, N., Zydzik, G. J., and Chu, S. N. G. (1990). Phase-coupled two-dimensional $Al_xGa_{1-x}As$-GaAs vertical-cavity surface-emitting laser array. *Appl. Phys. Lett.* **56**, 2089–2091.

[29] Orenstein, M., Kapon, E., Stoffel, N. G., Harbison, J. P., Florez, L. T., and Wullert, J. (1991). Two-dimensional phase-locked arrays of vertical-cavity semiconductor lasers by mirror reflectivity modulation. *Appl. Phys. Lett.* **58**, 804–806.

[30] Hasnain, G., Tai, K., Wynn, J. D., Wang, Y. H., Fischer, R. J., Hong, M., et al. (1990). Continuous wave top surface emitting quantum well lasers using hybrid metal/semiconductor reflectors. *Electron. Lett.* **26**, 1590–1592.

[31] Morgan, R. A., Hibbs-Brenner, M. K., Marta, T. M., Walterson, R. A., Bounnak, S., Kalweit, E. L., et al. (1995). "200 degrees-C, 96-nm wavelength range, continuous-wave lasing from unbonded GaAs MOVPE-grown vertical cavity surface-emitting lasers," *in Proceedings of the IEEE Photonics Technology Letters*, **7**, 441.

[32] Zhou, P., Cheng, J. L., Schaus, C. F., Sun, S. Z., Zheng, K., Armour, E., et al. (1991). "Low series resistance high-efficiency GaAs/AlGaAs vertical-cavity surface-emitting lasers with continuously graded mirrors grown by MOCVD," *in Proceedings of the IEEE Photonics Technology Letters*, **3**, 591–593.

[33] Tan, M. R. T., Hahn, K. H., Houng, Y. M. D., and Wang, S. Y. (1995). Surface emitting laser for multimode data link applications. *Hewlett-Packard J.* 67–71.

[34] Tai, K., Yang, L., Wang, Y. H., Wynn, J. D., and Cho, A. Y. (1990). Drastic reduction of series resistance in doped semiconductor distributed Bragg reflectors for surface-emitting lasers. *Appl. Phys. Lett.* **56**, 2496–2498.

[35] Zhou, P., Cheng, J., Schaus, C. F., Sun, S. Z., Zheng, K., Armour, E., et al. (1991). "Low series resistance high-efficiency GaAs AlGaAs vertical-cavity surface-emitting lasers with continuously graded mirrors grown by MOCVD," *in Proceedings of the IEEE Photonics Technology Letters*, **3**, 591–593.

[36] Schubert, E. F., Tu, L. W., Zydzik, G. J., Kopf, R. F., Benvenuti, A., and Pinto, M. R. (1992). Elimination of heterojunction band discontinuities by modulation doping. *Appl. Phys. Lett.* **60**, 466–468.

[37] Peters, M. G., Young, D. B., Peters, F. H., Scott, J. W., Thibeault, B. J., and Coldren, L. A. (1994). "17.3-percent peak wall plug efficiency vertical-cavity surface-emitting lasers using lower barrier mirrors," *in Proceedings of the IEEE Photonics Technology Letters*, **6**, 31–33.

[38] Peters, M. G., Thibeault, B. J., Young, D. B., Scott, J. W., Peters, F. H., Gossard, A. C., et al. (1993). Band-gap engineered digital alloy

interfaces for lower resistance vertical-cavity surface-emitting lasers. *Appl. Phys. Lett.* **63**, 3411–3413.

[39] Lear, K. L., Chalmers, S. A., and Killeen, K. P. (1993). Low threshold voltage vertical cavity surface-emitting laser. *Electron. lett.* **29**, 584–586.

[40] Young, D. B., Scott, J. W., Peters, F. H., Peters, M. G., Majewski, M. L., Thibeault, B. J., et al. (1993). "Enhanced performance of offset-gain high-barrier vertical-cavity surface-emitting lasers," *IEEE Journal of Quantum Electronics*, 29, 2013–2022.

[41] Schubert, E. F., Fischer, A., Horikoshi, Y., and Ploog, K. (1985). GaAs sawtooth superlattice laser emitting at wavelength l > 0.9 μm. *Appl. Phys. Lett.* **47**, 219–221.

[42] Kojima, K., Morgan, R. A., Mullaly, T., Guth, G. D., Focht, M. W., Leibenguth, R. E., et al. (1993). Reduction of p-doped mirror electrical resistance of GaAs/AlGaAs vertical-cavity surface-emitting lasers by delta doping. *Electron. Lett.* **29**, 1771–1772.

[43] Scott, J. W., Thibeault, B. J., Young, D. B., Coldren, L. A., and Peters, F. H. (1994). "High efficiency submilliamp vertical cavity lasers with intracavity contacts," *in Proceedings of the IEEE Photonics Technology Letters*, **6**, 678–680.

[44] Rochus, S., Hauser, M., Rhr, T., Kratzer, H., Bhm, G., Klein, W., G. et al. (1995). "Submilliamp vertical-cavity surface-emitting lasers with buried lateral-current confinement," *in Proceedings of the IEEE Photonics Technology Letters*, **7**, 968–970.

[45] Morgan, R. A., Hibbs-Brenner, M. K., Lehman, J. A., Kaiweit, E. L., Walterson, R. A., Marta, T. M., et al. (1995). Hybrid dielectric/AlGaAs mirror spatially filtered vertical cavity top-surface emitting laser. *Appl. Phys. Lett.* **66**, 1157–1159.

[46] Dudley, J. J., Crawford, D. L., and Bowers, J. E. (1992). "Temperature dependence of the properties of DBR mirrors used in surface normal optoelectronic devices," *in Proceedings of the IEEE Photonics Technology Letters*, **4**, 311–314.

[47] Lebby, M., Gaw, C. A., Jiang, W. B., Kiely, P. A., Claisse, P. R., and Ramdani, J. (1996). "Vertical-cavity surface-emitting lasers for communication applications," *in Proceedings of the OSA annual '96, WR1*, Rochester, NY.

[48] Lee, Y. H., Tell, B., Brown-Goebeler, K. F., Leibenguth, R. E., and Mattera, V. D. (1991). "Deep-red CW top surface-emitting vertical-cavity AlGaAs superlattice lasers," *in Proceedings of the IEEE Photonics Technology Letters*, **3**, 108–109.

[49] Shin, H. E., Ju, Y. G., Shin, J. H., Ser, J. H., Kim, T., Lee, E. K., et al. (1996). 780 nm oxidised vertical-cavity surface-emittng lasers with $Al_{0.1}Ga_{0.89}$ As quantum wells. *Electron. Lett.* **32**, 1287–1288.

[50] Kim, T., Kim, T. K., Lee, E. K., Kim, J. Y., and Kim, T. I. (1995). "A single transverse mode operation of top surface emitting laser diode with a integrated photo-diode," *in Proceedings of the Lasers and Electro-Optics Society Annual Meeting (LEOS'95)* (San Francisco, CA: IEEE), **2**, 416–417.

[51] Schneider, Jr., R. P., Choquette, K. D., Lott, J. A., Lear, K. L., Figiel, J. J., and Malloy, K. J. (1994). "Efficient room-temperature continuous-wave AlGaInP/AlGaAs visible (670nm) vertical-cavity surface-emitting laser diodes," *in Proceedings of the IEEE Photonics Technology Letters*, **6**, 313–316.

[52] Choquette, K. D., Schneider, R. P., Crawford, M. H., Geib, K. M., and Figiel, J. J. (1995). Continuous wave operation of 640–660 nm selectively oxidised AlGaInP vertical-cavity lasers. *Electron. Lett.* **31**, 1145–1146.

[53] Schneider, Jr., R. P., Crawford, M. H., Choquette, K. D., Lear, K. L., Kilcoyne, S. P., and Figiel, J. J. (1995). Improved AlGaInP-based red (670–690 nm) surface-emitting lasers with novel C-doped short-cavity epitaxial design. *Appl. Phys. Lett.* **67**, 329–331.

[54] Crawford, M. H., Schneider, Jr., R. P., Choquette, K. D., and Lear, K. L. (1995). "Temperature-dependent characteristics and single-mode performance of AlGaInP-based 670–690-nm vertical-cavity surface-emitting lasers," *in Proceedings of the IEEE Photonics Technology Letters*, **7**, 724–726.

[55] Baba, T., Yogo, Y., Suzuki, K., Koyama, F., and Iga, K. (1993). Near room temperature continuous wave lasing characteristics of GaInAsP/InP surface emitting laser. *Electron. Lett.* **29**, 913–914.

[56] Hofmann, W., Muller, M., Bohm, G., Ortsiefer, M., and Amann, M. C. (2009). "1.55-μm VCSEL with enhanced modulation bandwidth and temperature range", *in Proceedings of the IEEE Photonics Technology Letters*, **21**, 923–925.

[57] Dudley, J. J., Ishikawa, M., Miller, B. I., Babic, D. I., Mirin, R., Jiang, W. B., et al. (1992). 144°C operation of 1.3 µm InGaAsP vertical cavity lasers on GaAs substrates. *Appl. Phys. Lett.* 61, 3095–3097.

[58] Dudley, J. J., Babic, D. I., Mirin, R., Yang, L., Miller, B. I., Ram, R. J., et al. (1994). Low threshold, wafer fused long wavelength vertical cavity lasers. *Appl. Phys. Lett.* **64**, 1463–1465.

[59] Onishi, Y., Saga, N., Koyama, K., Doi, H., Ishizuka, T., Yamada, T., et al. (2009). "Long-wavelength GaInNAs vertical-cavity surface-emitting laser with buried tunnel junction", *in Proceedings of the IEEE Journal of Selected Topics in Quantum Electronics*, **15**, 838–843.

[60] Shin, H. K., Kim, I., Kim, E. J., Kim, J. H., Lee, E. K., Lee, M. K., et al. (1996). Vertical-cavity surface-emitting lasers for optical data storage. *Jpn. J. Appl. Phys.* **35**, 506.

[61] Hasnain, G., and Tai, K. (1992). "Self-monitoring semiconductor laser device," *US Patent# 5,136,603.*

[62] Hasnain, G., Tai, K., Wang, Y. H., Wynn, J. D., Choquette, K. D., Weir, B. E., et al. (1991). Monolithic integration of photodetector with vertical cavity surface emitting laser. *Electron. Lett.* **27**, 1630–1632.

[63] Hibbs-Brenner, M. K. (1995). "Integrated laser power monitor," *US Patent# 5,475,701.*

[64] Bjork, G., and Yamamoto, Y. (1991). "Analysis of semiconductor microcavity lasers using rate equations," *in Proceedings of the IEEE Journal of Quantum Electronics*, 2386–2396.

[65] Ram, R. J., Goobar, E., Peters, M. G., Coldren, L. A., and Bowers, J. E. (1996). "Spontaneous emission factor in post microcavity lasers," *in Proceedings of the IEEE Photonics Technology Letters*, **8**, 599–601.

[66] Huffaker, D. L., Deppe, D. G., and Kumar, K. (1994). Native-oxide ring contact for low threshold vertical-cavity lasers. *Appl. Phys. Lett.* **65**, 97–99.

[67] MacDougal, M. H., Dapkus, P. D., Pudikov, V., Zhao, H. M., and Yang, G. M. (1995). "Ultralow threshold current vertical-cavity surface-emitting lasers with AlAs oxide-GaAs distributed Bragg reflectors," *in Proceedings of the IEEE Photonics Technology Letters*, **7**, 229–231.

[68] MacDougal, M. H., Yang, G. M., Bond, A. E., Lin, C. K., Tishinin, D., and Dapkus, P. D. (1996). "Electrically-pumped vertical-cavity lasers with Al_xO_y-GaAs reflectors," *in Proceedings of the IEEE Photonics Technology Letters*, **8**, 310–312.

[69] Lear, K. L., Mar, A., Choquette, K. D., Kilcoyne, S. P., Schneider, Jr., R. P., and Geib, K. M. (1996). High frequency modulation of oxide-confined vertical cavity surface emitting lasers. *Electron. Lett.* **32**, 457.

[70] Thibeault, B. J., Bertilsson, K., Hegblom, E. R., Strzelecka, E., Floyd, P. D., Naone, R., et al. (1997). "High-speed characteristics of low-optical loss oxide-apertured vertical-cavity lasers," *in Proceedings of the IEEE Photonics Technology Letters,* **9**, 11–13.

[71] Westbergh, P., Safaisini, R., Haglund, E., Gustavsson, J., Larsson, A., Geen, M., et al. (2013). "High-speed oxide confined 850-nm VCSELs operating error-free at 40 Gb/s up to 85°C", *in Proceedings of the IEEE Photonics Technology Letters*, **25**, 768–771.

[72] Neave, J. H., Joyce, B. A., Dobson, P. J., and Norton, N. (1983). Dynamics of film growth of GaAs by MBE from Rheed observations. *Appl. Phys. A.* **31**, 1–8.

[73] Walker, J. D., Kuchta, D. M., and Smith, J. S. (1991). Vertical-cavity surface-emitting laser diodes fabricated by phase-locked epitaxy. *Appl. Phys. Lett.* **59**, 2079–2081.

[74] Walker, J. D., Kuchta, D. M., and Smith, J. S. (1993). "Wafer-scale uniformity of vertical-cavity lasers grown by modified phase-locked epitaxy technique. *Electron. Lett.* **29**, 239–240.

[75] Chalmers, S. A., and Killeen, K. P. (1993). Method for accurate growth of vertical-cavity surface-emitting lasers. *Appl. Phys. Lett.* **62**, 1182–1184.

[76] Bacher, K., Pezeshki, B., Lord, S. M., and Harris, J. S. (1992). Molecular beam epitaxy growth of vertical cavity optical devices with *insitu* corrections. *Appl. Phys. Lett.* **61**, 1387–1389.

[77] Grothe, H., and Boebel, F. G. (1993). *Insitu* control of Ga(Al)As MBE layers by pyrometric interferometry. *J. Cryst. Growth* **127**, 1010–1013.

[78] Houng, Y. M., Tan, M. R. T., Liang, B. W., Wang, S. Y., Yang, L., and Mars, D. E. (1994). InGaAs(0.98 μm)/GaAs vertical cavity surface emitting laser grown by gas-source molecular beam epitaxy. *J. Cryst. Growth* **136**, 216–220.

[79] Houng, Y. M., Tan, M. R. T., Liang, B. W., Wang, S. Y., and Mars, D. E. (1994). *In situ* thickness monitoring and control for highly reproducible growth of distributed Bragg reflectors. *J. Vac. Sci. Technol.*, **B12**, 1221–1224.

[80] Lear, K. L., Schneider, R. P., Choquette, K. D., Kilcoyne, S. P., Figiel, J. J., and Zolper, J. C. (1994). "Vertical cavity surface emitting lasers

with 21-percent efficiency by metalorganic vapor phase epitaxy," *in Proceedings of the IEEE Photonics Technology Letters*, **6**, 1053–1055.

[81] Hibbs-Brenner, M. K., Schneider, R. P., Morgan, R. A., Walterson, R. A., Lehman, J. A., Kalweit, E. L., et al. (1994). Metalorganic vapour-phase epitaxial growth of red and infrared vertical-cavity surface-emitting laser diodes. *Microelectron. J.* **25**, 747–755.

[82] Christiansen, K., Luenenbuerger, M., Schineller, B., Heuken, M., and Juergensen, H. (2002). Advances in MOCVD technology for research, development and mass production of compound semiconductor devices. *Opto-Electronics Rev.* **10**, 237–242.

[83] Kawakami, T., Kadota, Y., Kohama, Y., and Tadokoro, T. (1992). "Low-threshold current low-voltage vertical-cavity surface-emitting lasers with low-Al-content p-type mirrors grown by MOCVD," *in Proceedings of the IEEE Photonics Technology Letters*, **4**, 1325–1327.

[84] Yoon, S. F. (1992). Observation of nonbiased degradation recovery in GaInAsP/InP laser diodes. *J. Lightwave Technol.* **10**, 194–198.

[85] Cunningham, B. T., Haase, M. A., McCollum, M. J., Baker, J. E., and Stillman, G. E. (1989). Heavy carbon doping of metalorganic chemical vapor deposition grown GaAs using carbon tetrachloride. *Appl. Phys. Lett.* **54**, 1905–1907.

[86] Cunningham, B. T., Baker, J. E., and Stillman, G. E. (1990). Carbon tetrachloride doped Al_xGa_{1-x} As grown by metalorganic chemical vapor deposition. *Appl. Phys. Lett.* **56**, 836–838.

[87] de Lyon, T. J., Buchan, N. I., Kirchner, P. D., Woodall, J. M., Scilla, G. J., and Cardone, F. (1991). High carbon doping efficiency of bromomethanes in gas source molecular beam epitaxial growth of GaAs. *Appl. Phys. Lett.* **58**, 517–519.

[88] Buchan, N. I., Kuech, T. F., Scilla, G., Cardone, F., and Potemski, R. (1989). Carbon incorporation in metal-organic vapor phase epitaxy grown GaAs from CH_xI_{4-x}, HI, and I_2. *J. Electron. Mat.* **19**, 277–281.

[89] Stringfellow. (1989). "Non-hydride group V sources for OMVPM," *Mat. Res. Soc. Symp. Proc.*, **145**: III–V Heterostructures for Electronic/Photonic Devices, 171.

[90] Lum, R. M., Klingert, J. K., and Stevie, F. A. (1990). Controlled doping of GaAs films grown with tertiary-butylarsine. *J. Appl. Phys.* **67**, 6507–6512.

[91] Kuech, T. F., Wolford, D. J., Veuhoff, E., Deline, V., Mooney, P. M., Potemski, R., et al. (1987). Properties of high-purity $Al_xGa_{1-x}As$

grown by the metal-organic vapor-phase-epitaxy technique using methyl precursors. *J. Appl. Phys.* **62**, 632–643.

[92] Chalmers, S. A., Lear, K. L., and Killeen, K. P. (1993). Low resistance wavelength-reproducible p-type (Al, Ga)As distributed Bragg reflectors grown by molecular beam epitaxy. *Appl. Phys. Lett.* **62**, 1585–1587.

[93] Hibbs-Brenner, M. K., Morgan, R. A., Walterson, R. A., Lehman, J. A., Kalweit, E. L., Bounnak, S., et al. (1996). "Performance, uniformity, and yield of 850-nm VCSEL's deposited by MOVPE," *in Proceedings of the IEEE Photonics Technology Letters*, **8**, 7–9.

[94] Lebby, M. L., Gaw, C. A., Jiang, W. B., Kiely, P. A., Claisse, P. R., and Grula, J. (1996). "Key challenges and results of VCSELs in data links," *in Proceedings of the Lasers and Electro-Optics Society Annual Meeting, 1996 (LEOS 96)* (Boston, MA: IEEE), WV2.

[95] Grodzinski, P., Denbaars, S. P., and Lee, H. C. (1995). From research to manufacture – the evolution of MOCVD. *J. Miner. Metal. Mater. Soc.* **47**, 25–32.

[96] Kawai, H., Imanaga, S., Kaneko, K., and Watanabe, N. (1987). Complex refractive indices of AlGaAs at high temperature measured by *in situ* reflectometry during growth by metalorganic chemical vapor deposition. *J. Appl. Phys.* **61**, 328–332.

[97] Breiland, W. G., and Killeen, K. P. (1995). A virtual interface method for extracting growth rates and high temperature optical constants from thin semiconductor films using *in situ* normal incidence reflectance. *J. Appl. Phys.* **78**, 6726–6736.

[98] Frateschi, N. C., Hummel, S. G., and Dapkus, P. D. (1991). *In situ* laser reflectometry applied to the growth of $Al_xGa_{1-x}As$ Bragg reflectors by metalorganic chemical vapour deposition. *Electron. Lett.* **27**, 155–157.

[99] Azoulay, R., Raffle, Y., Kuszelewicz, R., Leroux, G., Dugrand, L., and Michel, J. C. (1994). *In situ* control of the growth of GaAs GaAlAs structures in a metalorganic vapour phase epitaxy reactor by laser reflectometry. *J. Cryst. Growth* **145**, 61–67.

[100] Hou, H. Q., Chui, H. C., Choquette, K. D., Hammons, B. E., Breiland, W. G., and Geib, K. M. (1996). "Highly uniform and reproducible vertical-cavity surface-emitting lasers grown by metalorganic vapor phase epitaxy with *in situ* reflectometry," *in Proceedings of the IEEE Photonics Technology Letters*, **8**, 1285–1287.

[101] Hasnain, G., Tai, K., Yang, L., Wang, Y. H., Fischer, R. J., Wynn, J. D., et al. (1991). "Performance of gain-guided surface emitting lasers

with semiconductor distributed Bragg reflectors," *in Proceedings of the IEEE Journal of Quantum Electronics*, 1377–1385.

[102] Morgan, R. A., Guth, G. D., Focht, M. W., Asom, M. T., Kojima, K., Rogers, L. E., et al. (1993). "Transverse mode control of vertical cavity top-surface emitting lasers," *in Proceedings of the IEEE Photonics Technology Letters*, **5**, 374–377.

[103] Lee, Y. H., Tell, B., Brown-Goebeler, K., and Jewell, J. L. (1990). Top-surface-emitting GaAs four-quantum-well lasers emitting at 0.85 μm. *Electron. Lett.* **26**, 710–711.

[104] Tell, B., Lee, Y. H., Browngoebeler, K. F., Jewell, J. L., Leibenguth, R. E., Asom, M. T., et al. (1990). High-power cw vertical-cavity top surface-emitting gaas quantum well lasers. *Appl. Phys. Lett.* **57**, 1855–1857.

[105] Hasnain, G., Tai, K. C., Yang, L., Wang, Y. H., Fischer, R. J., Wynn, J. D., et al. (1991). "Performance of gain-guided surface emitting lasers with semiconductor distributed Bragg reflectors," *in Proceedings of the IEEE Journal of Quantum Electronics*, 1377–1385.

[106] Ding, G., Corzine, S. W., Tan, M. R. T., Wang, S. Y., Hahn, K., Lear, K. L., et al. (1996). "Dynamic behavior of VCSELs under high speed modulation," *in Proceedings of the OSA Annual* (Rochester, NY: IEEE), WR6.

[107] Law, J., and Agrawal, G. P. (1996). "Effect of carrier diffusion on modulation and noise characteristics of vertical-cavity surface-emitting lasers," *in Proceedings of the OSA Annual* (Rochester, NY: IEEE), WII8.

[108] Schwartz, D. B., Chun, C. K. Y., Foley, B. M., Hartman, D. H., Lebby, M., Lee, H. C., et al. (1995). "A low cost, high performance optical interconnect," *in Proceedings of the 45th Electronic Components and Technology Conference*, 376–379.

[109] Nordin, R. A., Buchholz, D. B., Huisman, R. F., Basavanhally, N. R., and Levi, A. F. J. (1993). "High performance optical data link array technology," *in Proceedings of the 43rd Electronic Components and Technology Conference*, 795–797.

[110] Shieh, C. L., Lebby, M. S., and Lungo, J. (1995). "Mothod of making a VCSEL," *US Patent# 5,468,656.*

[111] Scherer, A., Craighead, H. G., Roukes, M. L., and Harbison, J. P. (1988). Electrical damage induced by ion beam etching of GaAs. *J. Vac. Sci. Technol.*, 227–279.

[112] Lebby, M. S., Shieh, C. L., and Lee, H. C. (1994). "High efficiency VCSEL and method of fabrication," *US Patent# 5,359,618.*

[113] Huffaker, D. L., Deppe, D. G., and Kumar, K. (1994). Native-oxide ring contact for low threshold vertical-cavity lasers. *Appl. Phys. Lett.* **65**, 97–99

[114] Huffaker, D. L., Deppe, D. G., and Rogers, T. J. (1994). Transverse mode behavior in native-oxide-defined low threshold vertical-cavity lasers. *Appl. Phys. Lett.* **65**, 1611–1613.

[115] Lear, K. L., Choquette, K. D., Schneider, R. P., and Kilcoyne, S. P. (1995). Modal analysis of a small surface emitting laser with a selectively oxidized waveguide. *Appl. Phys. Lett.* **66**, 2616–2618.

[116] Dallesasse, J. M., Holonyak, N., Sugg, A. R., Richard, T. A., and Elzein, N. (1990). Hydrolyzation oxidation of $Al_xGa_{1-x}As$-AlAs-GaAs quantum well heterostructures and superlattices. *Appl. Phys. Lett.* **57**, 2844–2846.

[117] Sugg, A. R., Chen, E. I., Richard, T. A., Holonyak, N., and Hsieh, K. C. (1993). Native oxide-embedded $Al_yGa_{1-y}As$-GaAs-$In_xGa_{1-x}As$ quantum well heterostructure lasers. *Appl. Phys. Lett.* **62**, 1259–1261.

[118] Sugg, A. R., Chen, E. I., Richard, T. A., Holonyak, N., and Hsieh, K. C. (1993). Photopumped room-temperature continuous operation of native-oxide-$Al_yGa_{1-y}As$-GaAs-$In_xGa_{1-x}As$ quantum-well-heterostructure lasers. *J. Appl. Phys.* **74**, 797–801.

[119] Maranowski, S. A., Sugg, A. R., Chen, E. I., and Holonyak, N. (1993). Native oxide top-confined and bottom-confined narrow stripe p-n $Al_yGa_{1-y}As$-GaAs-$In_xGa_{1-x}As$ quantum well heterostructure laser. *Appl. Phys. Lett.* **63**, 1660.

[120] Choquette, K. D., Schneider, R. P., Lear, K. L., and Geib, K. M. (1994). Low threshold voltage vertical-cavity lasers fabricated by selective oxidation. *Electron. Lett.* **30**, 2043–2044.

[121] Weigl, B., Grabherr, M., Michalzik, R., Reiner, G., and Ebeling, K. J. (1996). "High-power single-mode selectively oxidized vertical-cavity surface-emitting lasers," *in Proceedings of the IEEE Photonics Technology Letters*, **8**, 971–973.

[122] Hadley, G. R., Choquette, K. D., and Lear, K. L. (1996). Understanding waveguiding in vertical-cavity surface-emitting lasers. *CLEO Technic. Digest Ser.* **9**:425.

[123] Wu, Y. A., Chang-Hasnain, C. J., and Nabiev, R. (1861). Single mode emission from a passive-antiguide-region vertical-cavity surface-emitting laser. *Electron. Lett.* **29**, 1861–1863.

[124] Wu, Y. A., Li, G. S., Nabiev, R. F., Choquette, K. D., Caneau, C., and Chang-Hasnain, C. J. (1995). "Single-mode, passive antiguide vertical cavity surface emitting laser," *in Proceedings of the IEEE Journal of Selected Topics in Quantum Electronics*, **1**, 629–637.

[125] Lear, K. L., Schneider, R. P., Choquette, K. D., Kilcoyne, S. P., Figiel, J. J., and Zolper, J. C. (1994). "Vertical cavity surface emitting lasers with 21% efficiency by metalorganic vapor phase epitaxy," *in Proceedings of the IEEE Photonics Technology Letters*, **6**, 1053–1055.

[126] Yoshikawa, T., Kosaka, H., Kurihara, K., Kajita, M., Sugimoto, Y., and Kasahara, K. (1995). Complete polarization control of 8x8 vertical-cavity surface-emitting laser matrix arrays. *Appl. Phys. Lett.* **66**, 908–910.

[127] Choquette, K. D., and Leibenguth, R. E. (1994). "Control of vertical-cavity laser polarization with anisotropic transverse cavity geometries," *in Proceedings of the IEEE Photonics Technology Letters*, **6**, 40–42.

[128] Shimizu, M., Mukaihara, T., Baba, T., Koyama, F., and Iga, K. (1991). A method of polarization stabilization in surface emitting lasers. *Jpn. J. Appl. Phys.* **30**:L1015.

[129] Mukaihara, T., Koyama, F., and Iga, K. (1992). Polarization control of surface emitting lasers by anisotropic biaxial strain. *Jpn. J. Appl. Phys.* **31**:1389.

[130] Ser, J. H., Ju, Y. G., Shin, J. H., and Lee, Y. H. (1995). Polarization stabilization of vertical-cavity top-surface-emitting lasers by inscription of fine metal-interlaced gratings. *Appl. Phys. Lett.* **66**, 2769–2771.

[131] Ram, R. J., Yang, L., Nauka, K., Houng, Y. M., Ludowise, M., Mars, D. E., et al. (1993). Analysis of water fusing for 1.3 μm vertical cavity surface emitting lasers. *Appl. Phys. Lett.* **62**, 2474–2746.

[132] Guenter, J. K., Hawthorne, R. A., III, Granville, D. N., Hibbs-Brenner, M. K., and Morgan, R. A. (1996). "Reliability of proton-implanted VCSELs for data communications," *in Proceedings of the SPIE*, **2683**, 102–113.

3

Directly Modulated Laser Technology: Past, Present, and Future

Yasuhiro Matsui

Finisar Corporation, Fremont, CA, USA

3.1 Introduction

Direct modulation of laser diodes is probably the oldest form of a modulation scheme for light from semiconductor lasers. It often refers to the modulation of the current supplied to the laser. Then, the carriers in the cavity of directly modulated lasers (DMLs) follow the current modulation, which generates the gain for the photons through the stimulated emission process. The photons then follow the carriers after some finite time delay. The energy conversion process through stimulated emission is efficient, and it is the reason for low power consumption, compact, and the low-cost nature of the DML transmitters.

Optical communication systems have been evolving from 10 Gb/s, 25 Gb/s, 4 × 25 Gb/s, 4 × 25 Gbaud PAM4 [1], to 100 Gb/s single channel, and even further to 400 Gigabit Ethernet [2]. As of today, DMLs seem to live up to the expectations for 4 × 25 Gbaud PAM4 in an un-cooled operation, which requires a modulation bandwidth (BW) of about 20 GHz for the transmitters. Under cooled condition, the BW of the fastest DML is presently 55 GHz and is capable of 56-Gbaud PAM4 (112-Gb/s) single channel operation [3]. In coarse wavelength division multiplexing (CWDM) systems in the O-band (1260 nm–1360 nm), four lasers with a 20-nm wavelength spacing are coupled into a single fiber. In this case, the dispersion tolerance is also an important consideration for high bit-rate transmission systems especially for longer wavelength channels. In an LAN-WDM system with 800 GHz (~4.5 nm) channel spacing, or future multi-core fiber (MCF) systems, the chirp for DMLs will be less of an issue because choices for a smaller fiber dispersion are available. High-speed DML technology is expected to play a crucial role in such future high-capacity transmission systems.

Figure 3.1 shows an overview of the progress on the BW of DMLs. In the late 1980's, the gain properties of the multiple quantum wells (MQWs) were extensively studied both in theory and experiments [4, 5]. The MQW structures confine the injected electrons in very thin layers with a thickness on the order of a few nanometers, which is close to the de Broglie wavelength of the electrons. When the electrons are confined in such a thin space, the quantum confinement effect discretizes the allowed energy of the electrons in the perpendicular direction to the thin MQWs, making it easier to fill up the energy states of electrons that contribute to the gain. This was supposed to increase the differential gain and therefore the intrinsic speed of DMLs [6].

In spite of the predicted superior gain property of MQWs compared to the bulk lasers which had already reached to ∼24 GHz BW [7–9], lattice-matched InGaAs/InP MQW structures at 1.55 μm showed somewhat compromised performances due to the difficulty in smooth carrier injection into MQWs with large numbers of wells [10–12]. At longer wavelengths, the dipole moment of the transition is smaller, and the rates of inter-valence band absorption (IVBA) [13] and Auger recombination [14] are higher.

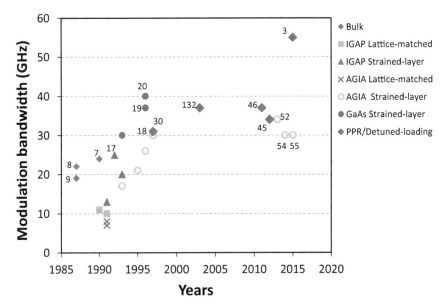

Figure 3.1   Evolution of DML modulation bandwidth.

The increase in the threshold carrier density as a result of the above adverse effects degraded the differential gain and therefore the speed of the DMLs. In order to overcome such difficulties, the introduction of strain in the MQW was proposed [5]. Strain effect in the MQW can separate the energy states between the heavy and light holes in the valence band, thus suppressing the band-mixing effect. This reduces the density of states (DOS) for the carriers to fill up, leading to the improvement in the gain properties. However, the well thickness was further reduced to $\sim$3 nm for compressively-strained InGaAs ternary wells to achieve 1.5 μm operation [15, 16]. The problem of the thin wells was solved by using InGaAsP quaternary wells [17], instead of InGaAs ternary wells, which provided a degree of freedom in choosing the bandgap and strain independently. Further improvement was achieved by using InGaAlAs barriers, which simultaneously suppressed the electron overflow in the conduction band (CB) and the hole carrier transport effects in the valence band (VB) [18]. This allowed the use of a large number of wells to increase the optical confinement factor and reduced the threshold carrier density per well without causing a hole transport issue. A BW of 30 GHz was reported based on this design with 20 pairs of InGaAlAs/InGaAsP MQWs at 1.55-μm wavelength [18].

At shorter wavelengths around 980 nm using AlGaAs/GaAs MQWs material systems, much fewer numbers of wells around four were often used. Thanks to superior gain properties, a BW exceeding 30 GHz was reported [19, 20]. The design approaches for high-speed DMLs in each operation band at 1.55 μm, 1.3 μm, and 850 nm are reviewed in Section 3.3.

The DMLs in a 1.55-μm regime suffer not only from the compromised gain property at longer wavelength, but also from the large dispersion of fibers over 20 km for Passive Optical Network (PON) applications, or 80 km for metro applications. The large linewidth enhancement factor of DMLs, typically reported to be around 2–4, is often held responsible for the limited transmission reach. However, to be more precise, this chapter points out that it is the transient chirp that should be suppressed, but that the adiabatic chirp can even be an advantage in extending the reach. According to this concept, even a proper amount of the carrier transport effect can be exploited since it can effectively suppress the transient chirp and increase the adiabatic chirp while maintaining a high differential gain [21]. Using this design scheme, a 12-nm tunable distributed Bragg reflector (DBR) laser with a proper gain compression property was successfully used for 20-km Time- and Wavelength-Division Multiplexed (TWDM) PON applications, which was also capable of burst-mode operation through fast-speed thermal chirp compensation by current tuning [21]. The gain compression phenomenon and the chirp control scheme are discussed in Section 3.4.

The reach of DML was further extended to 300 km at 10 Gb/s by combining optical filtering for vestigial sideband (VSB) generation, the extinction ratio (ER) enhancement by FM-AM conversion, and a minimum shift keying (MSK) chirp generation, which is the special case of continuous-phase frequency-shift keying (FSK) [22], to generate a format equivalent to an optical duobinary (ODB) [23]. This type of transmitter is called a chirp managed laser (CML) [24, 25] and has a common feature as the dispersion supported transmission (DST) [26]. CML applications from 10 Gb/s to 56 Gb/s NRZ formats are summarized in Section 3.5.

On the frontier of the high-speed performance, the concept of a high FM laser, as described above, was applied for the short-cavity distributed reflector (DR) lasers in order to enhance the speed by the detuned-loading effect [27], in-cavity FM-AM conversion to counter the RC limitation [28], and the photon-photon resonance (PPR) effect [29]. The detuned-loading effect can effectively enhance the differential gain and the modulation BW of DMLs when the lasing mode is located on the long wavelength side of the transmission profile of the DBR mirror. In this case, the frequency up-chirp under modulation of DMLs can reduce the mirror loss as the mode shifts toward the reflection peak of the DBR mirror. This increases the effective differential gain, and therefore the speed of DMLs improves. Photon-photon resonance (PPR) can improve the modulation response coinciding with the frequency difference between the main and side modes when the modulation sidebands of the main mode can couple into the side mode and are resonantly amplified [30]. Together with a record-wide 55 GHz BW, achieved by combining the above three effects, a single channel transmission of 112 Gb/s PAM4 was successfully demonstrated without an equalizer [3]. The detuned-loading and PPR effects are discussed in Section 3.6.

Throughout this chapter, the emphasis will be on the importance of the chirp in DMLs, which not only affects the transmission performance in the battle against the fiber dispersion, but also pushes the envelope of the high-speed performances of DMLs beyond 50 GHz BW. In the following section, we first build an intuitive picture of the dynamics of DMLs and then explain the important device parameters for high-speed operation of DMLs. Then, we review the design approaches for each operation wavelength at 1.55 μm, 1.3 μm, and 850 nm.

3.2 Intuitive Picture of the Dynamics of Directly Modulated Lasers

It might be beneficial first to look at the dynamics of DMLs and important parameters for high-speed modulation of DMLs before going further. Mathematically, the dynamics of DMLs can be described by the rate equations for the carriers and the photons. The commonly-used form of the rate equations can be stated as follows;

$$\frac{dN}{dt} = \frac{I}{e.vol} - g(N) \cdot S - \frac{N}{\tau_N}, \qquad (3.1(a))$$

$$\frac{dS}{dt} = \Gamma_g(N)S - \frac{S}{\tau_p} \cdot \beta - \frac{N}{\tau_N}. \qquad (3.1(b))$$

where N is the carrier density, I is the current, e is the electron charge, *vol* is the volume of active region, g is the gain, S is the photon density, τ_n is the carrier lifetime, Γ is the optical confinement factor, τ_p is the photon lifetime, and β is the spontaneous coupling coefficient. Numerically, such a set of equations can be solved by tracing the small changes in each step of time on the right-hand side of the equations, and through plotting the time evolution by integrating the equations. What makes the laser oscillator a non-linear system comes down to the fact that the stimulated emission depends on the product of N and S. This is a *nonlinear* first-order differential equations, also known as Lotka–Volterra equations, which was developed to model the dynamics in the natural population system consisting of the predator and prey. The most popular example is the population of the snowshoe hare and the lynx. However, in the interpretation in this chapter, the carriers are the prey and the photons are the predator.

Figure 3.2 shows the case when the bias is modulated from 0 bit to 1 bit, starting from the initial condition for the steady state of N and S under the bias for 0 bit. The dynamics of the carriers and photons can be likened to the predator-prey population cycle which shows the boom-and-bust pattern. As the current is suddenly increased (food supply to the prey), the prey (or N) starts to increase after some finite time delay. As the prey increases, the predators (or S) starts to increase after some finite time delay. The increase in the predators (S), in turn, depletes the prey (N) and, therefore, N starts to decrease after showing the overshooting. This interplay results in the oscillations of N and S (population cycles); however, such oscillation in laser systems often damps and asymptotically approaches to the steady condition.

The frequency of such oscillation is called a relaxation oscillation frequency (F_r) and determines the intrinsic speed of DMLs.

The damping of the relaxation oscillation is a result mainly due to two reasons. One is the non-linear gain compression phenomenon, which stems from the fact that the gain diminishes as the photon density increases when the supply of the carriers falls behind the rate of the stimulated emission under higher photon density. An empirical expression for the gain compression can be written as:

$$g\left(N\right) = \frac{g_0\left(N\right)}{1 + \varepsilon S},$$ (3.2)

where ε is the gain compression factor. The detailed physical origin of the gain compression is discussed in Section 3.4.3.

Another source of the damping can be a linear phenomenon, similar to a friction loss for a harmonic oscillator. One way to illustrate the hypothetical "linear laser" equation in simple terms would be to apply it to the dynamics of the relationship between Romeo and Juliet, which could be written in a form of the first-order ordinary differential equation [31];

$$\frac{dJ}{dt} = a \times J + b \times R,$$ (3.3(a))

$$\frac{dR}{dt} = c \times J + d \times R.$$ (3.3(b))

In this version of the "linear laser rate equation", we used R for the love of Romeo and J for the love of Juliet. First, we set $a = d = 0$, meaning there is no "friction of love". If $b < 0$, Juliet's character is that of a fickle lover; the more Romeo loves her, the more she feels like wanting to walk away from him. As for Romeo, if we set $c > 0$, Romeo is discouraged when the love of Juliet withers, but he lightens up when Juliet loves him. In this scenario, the story never ends, but the lovers' emotions are in perpetual oscillation, called a "limit cycle", *i.e.*, the love of Juliet oscillates as the sinusoidal function, and that of Romeo as the delayed sine function (Figure 3.3(a)). Damping of oscillation can still be introduced in this linear system by setting $a < 0$, and/or $d < 0$ (Figure 3.3(b)). These terms are, again, similar to the "friction loss" in the equation of the harmonic oscillator. In the laser rate equations, this corresponds to the photon and carrier lifetimes, which makes the laser a dissipative oscillator.

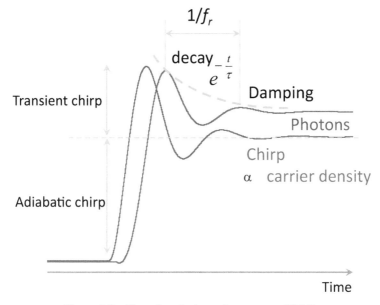

Figure 3.2 Time-domain dynamic response of DML.

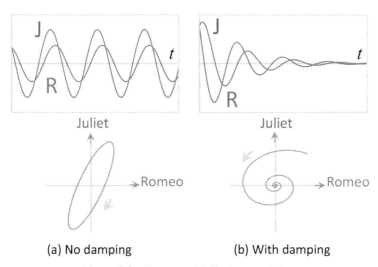

(a) No damping   (b) With damping

Figure 3.3 Romeo and Juliet love evolution.

In the modulation response of lasers (or S21 response), the damping factor determines the sharpness of the resonance and, therefore, the flatness of the S21 response. The weaker damping may create peaking in the S21 response and increase the 3-dB BW; however, the non-flatness in the S21 response and the group delay ripples associated with F_r can distort the eye diagram in the large signal modulation. Therefore, proper damping should be realized.

If the frequency shift of the light, or the chirp, is important, then one more equation for the phase of the light can be added to Equation (3.1). The chirp is the time derivative of phase ϕ in rate equations and is directly associated with the change in the carrier density in and around the MQW region, which changes the refractive index inside the laser cavity due to the plasma effect of the free carriers. Namely;

$$\frac{d}{dt}\phi = \Delta v = \frac{v_g}{\lambda} \sum_i \Gamma_i \frac{dn}{dN_i} \Delta_{N_i}, \tag{3.4}$$

where the suffix denotes the layers consisting of the MQW region, including the wells, barriers, and separate confinement hetero (SCH) layers [32]. It is important to consider the carriers not only in the wells, but also in the barriers and SCH. Using a rate equation for photons,

$$\Gamma v_g g - \frac{1}{\tau_p} = \frac{1}{S}\left(\frac{d}{dt}S - \Gamma R_{sp}\right), \tag{3.5}$$

and expanding the gain around g_{th} ($=1/\tau_p$),

$$g = \frac{1}{\tau_p} + dg(N - N_{th}) - a_p S, \tag{3.6}$$

the formula for chirp can be derived by substituting Equation (3.5) and Equation (3.6) into Equation (3.4).

$$\Delta v(t) = \frac{\alpha}{4\pi}\left(\frac{1}{P}\frac{d}{dt}P + \frac{2\Gamma\varepsilon P}{\eta h v \cdot vol}\right). \tag{3.7}$$

where R_{sp} is the spontaneous emission rate, α_p is defined as $-\partial g/\partial PS$, α is the linewidth enhancement factor (or called α-parameter), P the output power, η_i the internal quantum efficiency, and hv the photon energy. The first term is the transient chirp and the second term is the adiabatic chirp. It shows that a large gain compression factor leads to suppression of the transient chirp by damping the relaxation oscillation and, *at the same time,* increases the second term for an adiabatic chirp. The larger gain compression denotes the imperfect clamping of the carrier density above the threshold, and that more carriers have to be stored for 1 bits relative to 0 bits. This reduces the refractive index in the cavity of DMLs for 1 bits due to the plasma effect and, therefore, the frequency for 1 bits increases. This formula is useful for extracting the α-parameter and the gain compression factor under large signal modulation [33]. For the traveling wave type of devices, for example, an EAM or a semiconductor optical amplifier (SOA), only transient chirp can be generated.

Figure 3.4 shows the chirp waveform including the contributions from the carriers in the wells and the SCH. The carriers in the wells can show an overshooting or undershooting, called "transient chirp". On the other hand, the SCH behaves like a capacitor for the carriers, charged or discharged as the DML is modulated. This results in the net change in the stored carriers in the cavity of the laser, and the frequency of 1 bits shifts to higher frequency relative to that for the 0 bits due to the refractive index reduction caused by the plasma effect of the free carriers. This type of the shift in the frequency is called "adiabatic chirp".

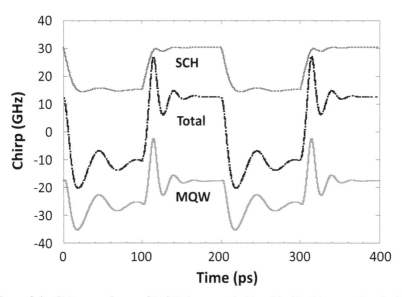

Figure 3.4 Chirp waveforms of MQW laser: total chirp (black), chirp associated with the wells (green), and SCH (red).

To understand the important parameters for the high-speed laser design, it would be good to derive the expressions for F_r from the rate equations by applying a small signal analysis. It is easy to find many variations for the expressions of F_r, and it might give different impressions; however, of course, those are essentially all the same.

$$(2\pi \cdot F_r)^2 = v_g \cdot dg \frac{S}{\tau_p}, \tag{3.8(a)}$$

$$= \frac{\lambda}{hn_{eq}} dg \frac{\Gamma}{vol} \frac{\alpha_{int} + \alpha_m}{\alpha_m} P_{out}, \tag{3.8(b)}$$

$$= v_g \cdot dg \frac{\Gamma}{vol} \frac{\eta_i}{\eta_d} \frac{P_{out}}{hv}, \tag{3.8(c)}$$

$$= v_g \cdot dg \cdot \eta_i \frac{\Gamma(I_b - I_{th})}{e \cdot vol}, \tag{3.8(d)}$$

where v_g is the group velocity of light, dg the differential gain (dg/dN), η_i the internal quantum efficiency, Γ the optical confinement factor, I_b the bias of current, I_{th} the threshold current, e the electron charge, vol the volume of active region, P_{out} the output power, v the frequency of light, η_d the slope efficiency of the output power, τ_s the carrier lifetime, J_{th} the threshold current density, n_{eq} the equivalent refractive index of material, λ the wavelength of light, α_{int} the internal loss of the laser cavity, α_m the mirror loss, τ_p and τ_s are the photon and carrier lifetimes, respectively, and S the photon density. Some parameters in the expressions are inter-dependent; therefore these expressions should be interpreted with some caution. For example, in the expression (3.8(a)), the reduction of the photon lifetime seems to increase the speed. However, the photon density, S, changes with the photon lifetime, and therefore, those two parameters are interdependent. The following expressions are useful for converting the expressions for F_r from one to another in Equation (3.8).

$$P_{out} = hv \frac{S}{\tau_m} vol = hv \frac{S}{\tau_p} \frac{\alpha_m}{\alpha_{int} + \alpha_m} vol = hv \frac{S}{\tau_p} \frac{\eta_d}{\eta_i} vol = \eta_d \frac{hv}{q} (I_b - I_{th}). \tag{3.9}$$

In fact, the variations in the expression for F_r comes from the various ways to express the P_{out}. Among those expressions, Equation (3.8(d)) is easier to interpret because it contains lesser degree of interdependency among the parameters used. In this expression, it can be seen that the gain length or the mirror reflectivity does not affect F_r in a direct manner if a linear dependence of the gain on the carriers is assumed. It can *indirectly* affect F_r if the logarithmic functional form of the gain is assumed. In this case, the smaller mirror reflectivity results in higher α_m and I_{th}, which degrades the differential gain, and therefore the speed *implicitly*. It also shows that short-cavity laser does not necessarily imply a faster laser since the maximum bias is limited in reality by the thermal effect that causes the rollover in the output power.

Equation (3.8(d)) also shows that the higher ratio between the optical confinement factor and the volume can improve the speed. Therefore, a strong optical confinement is an advantage. This will be discussed later in this section in Figure 3.6.

The unit for the gain may sometimes be confusing at first glance, but it is good to recall the expression below for the linear gain as a function of the carrier density:

$$g(N) = dg \times (N - N_{tr}) \times v_g, \tag{3.10}$$

where the corresponding units are:

$$s^{-1} = cm^2 \times cm^{-3} \times cm/s. \tag{3.11}$$

The gain in the unit of s^{-1} on the left side describes the gain for the photons in unit time, while $dg \times (N–N_{tr})$ is for the gain in the unit of cm^{-1} as the light travels over the distance L, and grows according to the exponential form e^{gL}. The differential gain, $dg \times v_g$, is expressed in a unit of cm^3/s.

The designs of high-speed lasers and approaches are somewhat different depending on the wavelength regions due to the difference in the available material system and the gain property. In the next section, we review separately the progress of FP and DFB lasers and the design approaches for wavelengths at 1.55 μm, 1.3 μm, and short-wavelength around 850 nm. The performance of high-speed DMLs exploiting the cavity effects including the PPR, detuned-loading, and FM-AM conversion in DBR mirror will be discussed in Section 3.6.

3.3 Progress of High-Speed FP and DFB Lasers

3.3.1 1.55 μm DML

As mentioned earlier, the gain properties at 1.55 μm suffer from the smaller dipole moment, larger IVBA, and Auger rates, and this affects the design approaches of high-speed DMLs at 1.55 μm. Some of the shortcomings in gain properties can be overcome by lowering the threshold carrier density in each well. This can reduce not only IVBA and Auger recombination, but also can increase the differential gain because of the logarithmic functional form of the gain vs. the carrier density. Therefore, increasing the optical confinement factor by using a large number of wells is effective. However, non-uniform hole injection can increase the hole density and therefore the IVBA. In early times, when the MQW structures were proposed and replaced the bulk gain medium, the speed of MQW lasers was not any faster than that of bulk lasers which had a modulation bandwidth of 24 GHz [7, 8, 9]. At 1.55 μm, InGaAs wells were often used, and the well width became very thin (~3 nm) in order to achieve lasing at 1.55 μm, especially for compressively strained InGaAs wells having a smaller band gap [15, 16]. As a result of thin wells, it was difficult to achieve a high optical confinement factor, which increased the threshold carrier density per wells; therefore the IVBA and Auger rates increased. To avoid the adverse effects of thin wells, a large number of wells were often grown [10–12, 34, 35]. However, due to a large valence band offset for the strained InGaAs/InP material system, it was difficult to inject holes uniformly over large numbers of wells. Uomi reported that the uniform injection was difficult to achieve for more than 10 wells, and no improvement in the speed was obtained [12]. The use of thicker wells using InGaAsP quaternary material was then proposed. Morton used 7nm-thick 8 InGaAsP wells to achieve a higher optical confinement factor and demonstrated 25 GHz BW [17]. The MQW structure was doped with Zn to reduce the carrier transport effect. However, the small conduction band offset for InGaAsP MQW caused an electron overflow issue, and the dipole moment was compromised due to a poor wave-function overlap between the electrons and the heavy holes.

Further improvement in speed was achieved at 1.55 μm by using InGaAlAs for the barriers and InGaAsP for the wells. The band diagram of this material system is shown in Figure 3.5. This material system provides much higher conduction band discontinuity of ~250 meV compared ~120 meV for InGaAsP MQWs, which improves the dipole moment and suppresses the electron overflow [18]. In the valence band, a small valence band offset for the material system can be further reduced by the tensile strain in the barriers, which lowers the light hole energy in the barriers. As a result, a uniform hole injection is expected due to the faster thermionic emission process of holes in the wells, while the strain compensation enables a stable growth of large numbers of MQW pairs. However, the smaller valence band discontinuity results in a smaller energy separation between the heavy

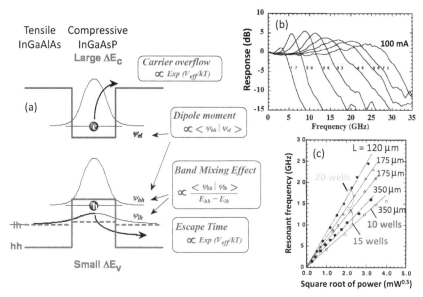

Figure 3.5 (a) Band diagram of InGaAlAs/InGaAsP strain compensated MQW, showing a flat light hole band, (b) S21 response, and (c) F_r vs. square root of output power. (after Reference 18).

and light holes. This increases the off-diagonal element in Luttinger-Kohn Hamiltonian [6]:

$$H_{nm} \propto \frac{\langle \psi_{hh} | \psi_{lh} \rangle}{E_{hh} - E_{lh}}. \tag{3.12}$$

which is responsible for the band mixing effect. This can increase the density of states in the valence band and degrades the differential gain by making the Fermi level less sensitive to the injection of the carriers [36]. In strain-compensated InGaAlAs/InGaAsP MQW, the confinement energy for light holes can be reduced to zero or even negative (i.e., type-II), and the band mixing effect can be suppressed by reducing the wave-function overlap between the heavy and light holes [37, 38]; namely, the numerator in Equation (3.12). Based on this design concept, an F_r of 25 GHz and 3-dB BW of 30 GHz was achieved for 20-pair strain-compensated InGaAlAs/InGaAsP MQW at 1.55 μm. A remarkably low K factor of 0.135 ns was obtained. The speed was limited by the Joule heating associated with the mushroom structure, which suffered from the thermal bottleneck around the active stripe [38].

Such high-speed DML was used to demonstrate very short pulse generation based on gain-switching technique. A 4-ps gain switched pulse was generated after linear compression of the gain-switched chirped pulse by injection seeding of FP laser and achieved 40 nm tuning [39]. Injection seeding also suppressed the occurrence of chaos for the DML laser with a reduced gain compression [40]. The measured chirp parameter was 4. The pulse was further compressed to 1 ps by high-order soliton compression, and then further down to 260 fs by high-quality adiabatic soliton compression in a 6m-long anomalous dispersion Er-doped fiber with a gain of 7 dB. Then the gain-switched pulse was compressed to 20 fs (4 optical cycles) by cascading a 4m-long dispersion-decreasing fiber for adiabatic soliton compression, followed by a 60cm-long dispersion-flattened fiber for an effective dispersion-decreasing profile as a result of Raman frequency shift [41]. The pulse was tunable over 160 nm by simply adjusting the EDFA output power to control the Raman frequency shift.

Akram also reported a uniform current injection over 20 pairs of InGaAlAs/InGaAsP MQW as well as superior gain property [42, 43]. Möhrle demonstrated a 3-dB BW of 12.5 GHz at $90°C$ using fewer number of InGaAlAs/InGaAsP MQW pairs [44]. The growth of InGaAlAs/InGaAsP MQW is easier at 1.55 μm wavelength range compared to 1.3 μm due to lower Al content. As Al content increases in the InGaAlAs growth, the suppression of As vacancy and oxygen incorporation by a use of high AsH_3 pressure and higher growth temperature become more critical, which makes it difficult to switch V-group gases between the InGaAlAs barriers and InGaAsP wells in MOVPE growth for 1.3 μm InGaAlAs/InGaAsP MQWs.

The above approaches for high-speed DML designs relied on the improvement on the material gain properties. In another approach, a dynamic mirror loss modulation in the DBR mirror was utilized in the detuned-loading effect to achieve a 30 GHz modulation BW [30]. The PPR effects improved the speed of passive-feedback lasers (PFLs) to 37 GHz at 1.55 μm [45] and 32 GHz [46] at 1.3 μm. This will be discussed in more detail in Section 3.6.

A unique approach to enhance the modulation BW and reduce the power consumption was proposed based on the realization of an extremely high optical confinement factor in a small volume of a membrane DFB laser as shown in Figure 3.6. A wafer bonding process was used to attach an MQW onto the SiO_2/Si substrate. After selective regrowth of InP on the side of the mesa stripe, Zn diffusion and Si ion implantation were then used to form a PN junction for a lateral current injection. In the vertical direction, the

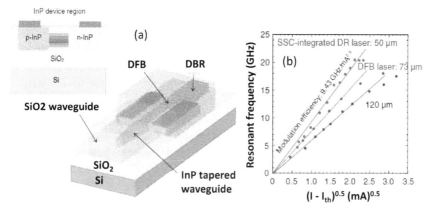

Figure 3.6 Structure of SSC-integrated membrane laser and (b) modulation efficiency. (after Reference 48).

light was strongly confined by the air due to thin InP clad layers (0.1 μm) at the top and bottom of the MQW region. As the expression for F_r shows in Equation (3.8), the higher ratio of Γ over the volume of an active region can increase the modulation speed. A DFB laser with a 73-μm length and 1 μm width showed an F_r of 18 GHz obtained at only 7 mA bias at 25°C [47]. Also, a DR laser with a DFB length of 50 μm was used, and achieved an F_r of 20 GHz at a bias of 4 mA [48]. This corresponds to a modulation efficiency of 9.43 GHz/mA$^{0.5}$. The importance of the strong optical confinement was fully displayed in a photonic crystal (PhC) structure used to strongly confine the light both in the lateral and vertical directions in an ultra-compact active region of 2.5 $\times$ 0.3 $\times$ 0.15 μm^3, and achieved a 10 Gb/s operation at an extremely low bias of 25 μA [49, 50].

3.3.2 1.3 μm DML

High-speed DFB lasers at 1.3 μm have recently been studied more intensively for the data center applications. Transmitters for 100G Ethernet needs to support 4 wavelengths $\times$ 25 Gbit/s with an 800 GHz spacing, called LAN-WDM, which is cooled to around 50°C, or 20-nm spaced CWDM, which is supposed to be uncooled up to commercial temperature (70°C) or industrial temperature (85°C). Also, non-hermetic DMLs are preferred to reduce the cost of packaging. For the next generation 400 Gigabit Ethernet (400GbE) [2], it is critical to develop transmitters that can be used for PAM4 formats at 8 $\times$ 56 Gbit/s or 4 $\times$ 112 Gbit/s.

InGaAlAs/InGaAlAs MQW is a popular choice at 1.3 μm because it provides higher electron confinement. Even higher electron confinement can be achieved for GaInNAs/GaAs MQW for which $\Delta E_c > 400$ meV is expected. Kitatani demonstrated a high T_0 of 100 K and 10 Gbit/s operation at 100°C [51]. However, it seemed difficult to achieve an operation close to 1.3 μm wavelength or longer due to the required higher nitrogen content for GaInNAs wells.

While the BH structure has gained widespread adoption at 1.55 μm wavelength, both BH structure and ridge structure are used at 1.3 μm. Kobayashi obtained a 3-dB BW of 34 GHz and an F_r of 26 GHz at 25°C for a 150-μm DFB ridge structure using eight pairs of InGaAlAs MQWs [52]. The advantages of ridge structures were low leakage current, a small capacitance of 0.12 pF achieved by using benzocyclobutene (BCB), and a corresponding large RC cutoff of 30 GHz. The challenge for ridge structures is a high electrical and thermal resistance; however, Nakahara demonstrated 10 Gbit/s operation at 120°C [53]. On the other hand, the advantages of BH structure are low thermal and electrical resistances, a lateral confinement of injected carriers, and a strong optical confinement using a narrower stripe width that also increases the injection current density. However, processing InGaAlAs MQWs into BH structures is more challenging because the sidewalls of an Al-MQW stripe are exposed to the air after forming a mesa stripe before the regrowth of blocking structures. The challenge seemed to be overcome by several groups where high-performance Al-MQW BH lasers have been reported in the literatures by careful treatment on the Al-MQW surface prior to the regrowth step. Using a ridge-shaped BH structure with InGaAlAs MQW buried with Fe-doped blocking layer, Nakahara achieved a 3-dB BW of 29.5 GHz at a bias of 85 mA at 25°C, and 22 GHz at a bias of 74 mA at 85°C for a 120-μm long DFB laser [54]. Nakahara also demonstrated 50 Gb/s transmission at 80°C [54]. Matsui reported a 29 GHz 3-dB BW for a 150-μm DFB laser using InGaAlAs MQW in a BH structure at a bias of 60 mA at 50°C. The parasitic capacitance of PN blocking was reduced to 0.8 pF (RC cutoff of 22 GHz) by forming a double-channel stripe filled with BCB. Wide open eyes at 56 Gb/s NRZ and 28 Gbaud PAM4 were obtained using a 1-tap Feed Forward Equalization (FFE) circuit on the Tx side. A successful transmission over 12 km of a single mode fiber was demonstrated at 1308 nm [55, 56]. Also, using the same DFB laser, 112-Gb/s 16-QAM Nyquist Subcarrier Modulation [57] and 112 Gb/s PAM-4 [58] transmission were demonstrated successfully. The DR laser reported in Reference [59] showed a 3-dB BW of 30 GHz at 25°C. The involvement of the PPR and detuned-loading effects was not mentioned. PFL laser was implemented at 1.3 μm and achieved 34 GHz BW [46] by taking advantage of the PPR effect. Matsui reported a 55 GHz BW for the DR laser by using the PPR, detuned-loading, and FM-AM conversion effects [3, 28]. The details of PPR, detuned-loading, and in-cavity FM-AM conversion effects will be discussed in Section 3.6.

3.3.3 Short-Wavelength DML

For short-wavelength edge-emitting high-speed lasers, very different approaches were used from those for long-wavelength DMLs. Sharfin first observed a very high gain compression factor for a single quantum well laser [60], which could not be explained by a framework of the spectral hole burning theory of gain compression [61]. Rideout proposed a "well-barrier hole burning" effect where the concept of "hole burning" was extended to the carrier transfer between the wells and barriers through the capture and escape, which have finite relaxation times [62]. Tessler developed a model that included the transport effects in the SCH with drift and diffusion, and well-barrier capture and emission [63]. Non-uniform distribution of holes creates an internal electric field which promotes the electron overflow into p-side SCH. This effect was evidenced by an observed large FM modulation associated with p-side SCH [21]. By Be doping to the barrier layers in four MQW pairs grown by molecular beam epitaxy (MBE), the K factor was significantly reduced to 0.14 ns and a 3-dB BW was extended to 30 GHz [19]. The group later reported 40 GHz BW even for un-doped InGaAs/GaAs MQW, although the K factor was increased to 0.24 ns [20]. As is often the case, a much lower number of wells is used for short-wavelength DMLs.

For short wavelength applications today, high-speed VCSELs at around 850 nm are far more important over the edge emitting laser counterpart for data-center applications over multi-mode fibers due to the low cost packaging capability. It is possible to close the link of a 300-m OM4 fiber with a 25.8 Gbit/s NRZ format under uncooled condition [64]. Due to the very small volume of the active region, VCSELs require very low drive current for modulation, which makes it possible to use FFE driver ICs based on a 130-nm BiCMOS directly wire-bonded to the VCSEL. A 56 Gb/s NRZ transmission was demonstrated using a VCSEL with a limited modulation bandwidth of only 18 GHz [65]. Also, 71 Gb/s NRZ transmission was demonstrated using a VCSEL with a modulation BW of 26 GHz [66].

In the other approach to extend the intrinsic modulation BW of VCSELs, the PPR effect was exploited in a passive transverse-coupled cavity [67]. A modulation BW of 26 GHz was reported by including the PPR effect created by an optical feedback effect from the transverse-coupled cavity while the BW was only 9 GHz without the PPR effect.

3.4 Reach Extension of DML for PON and Metro Applications

3.4.1 Principle of Reach Extension of DML by Tailoring Chirp

Applications of DMLs at 1.55 μm are limited partly due to a large dispersion of a fiber at this wavelength. At 10 Gbit/s, the possible transmission distance of DFB lasers is typically ∼5 km while 40 km – 80 km is required for Metro and 20 km for Ethernet Passive Optical Network (EPON) or Next-Generation Passive Optical Network 2 (NGPON2) applications. It is widely believed that the reason for poor dispersion tolerance is due to the large alpha parameter for DMLs; in fact, this statement is not very accurate. To be more precise, it is the transient chirp that creates an issue in anomalous dispersion regime of fibers, however, the adiabatic chirp can even improve the dispersion tolerance.

Figure 3.7(a) shows a 010 bit pattern with transient chirp corresponding to the positive alpha parameter of DMLs. In the leading edge, the positive spiky (blue) chirp travels faster in an anomalous dispersion regime of a fiber while a negative spiky (red) chirp in the trailing edge travels slower. Therefore, 1 bit broadens and degrades the bit error rate (BER) after transmission. On the other hand, in a normal dispersion of a fiber, the 1 bit initially compresses after transmission, and the BER can improve for DMLs before the 1 bit re-broadens for even larger dispersions. In the case where there is only adiabatic chirp, as shown in Figure 3.7(b), the blue-chirped 1 bit as a whole travels faster in the anomalous dispersion or slower in the normal dispersion regime. If the adiabatic chirp between 1 bit and 0 bit is $\Delta\lambda$, then the difference in the arrival times between the 1 bit and the 0 bit, Δt, is given by [26];

$$\Delta t = \Delta\lambda DL, \tag{3.13}$$

where D is the fiber dispersion. If Δt is smaller than the bit period, the distortion due to dispersion effect is moderate, and the results are the smooth trailing edge as well as some interference effects between the energies for the 1 bit and the 0 bit in the leading edge in the anomalous dispersion regime.

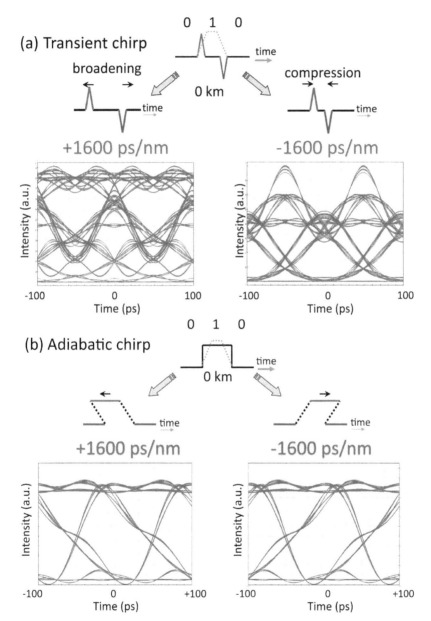

Figure 3.7   Transmitted eyes for 10-Gb/s NRZ signal with (a) transient and (b) adiabatic chirp over positive (anomalous) and negative (normal) dispersions.

In the normal dispersion regime, the eye is simply the time-inversion of the eye in an anomalous dispersion regime. Since the transmitted eyes in the positive and negative dispersions are symmetric inversion in time, there is no difference in the BER. According to the concept of the DST, the BER improves due to FM-AM conversion by a fiber dispersion when Δt is close to a bit period [68]. For example, 180-km transmission is possible at 10 Gb/s if the FM is 4 GHz for NRZ format according the Equation (3.13). However, low ER (< 2 dB) before the fiber transmission can degrade the receiver sensitivity in the back-to-back (BtB). Also, the DST format cannot be received with a standard detection scheme for NRZ; rather it needs the encoder on the transmitter and the decoder on the receiver.

The issues with the lower ER for the DST and the need for non-standard detection are less of an issue when proper AM and FM modulations are combined. By simulation, Binder showed that the 80 km transmission with 10-Gb/s DMLs is possible if the modulation spectrum is narrowed by canceling the AM and FM sidebands on the higher frequency side [69], as shown in Figure 3.8(a)–(c). The condition for the sideband cancellation for the ASK-FSK format is given below:

$$ER = \frac{1 + 2\Delta f/B}{1 - 2\Delta f/B}, \tag{3.14}$$

where B is the bit rate, Δf is the amount of adiabatic chirp, and the ER is the extinction ratio in linear scale. Figure 3.8(e) shows the transmitted eyes after 80 km and 180 km for a combination of an FM of 3 GHz and an AM of 6 dB (modulation index $m_{IM} = 0.6$). Due to the narrowed spectrum by AM-FM sideband cancellation, the eyes are open after transmissions. Is it possible to combine this effect with the DST effect? For 4 GHz of FM, which is the optimum chirp for 180 km transmission by the DST condition, the required ER for sideband cancellation needs to be 10 dB. Such high ER is difficult to achieve by DMLs in reality.

In the following sections, we first discuss the DML designs for extending the reach. The reduction of transient chirp is achieved through the increase of the gain compression factor by material and cavity designs.

In Section 3.5, we, then, discuss the technique called chirp managed laser (CML) where ER is enhanced by FM-AM conversion on the edge of an optical band-pass filter [24, 25], which achieved > 250 km transmissions. In the other approach, Erasme proposed the idea of co-modulation of EML where both DFB and EA sections were modulated to generate a single-sideband (SSB) ASK/FSK format with a high ER [70].

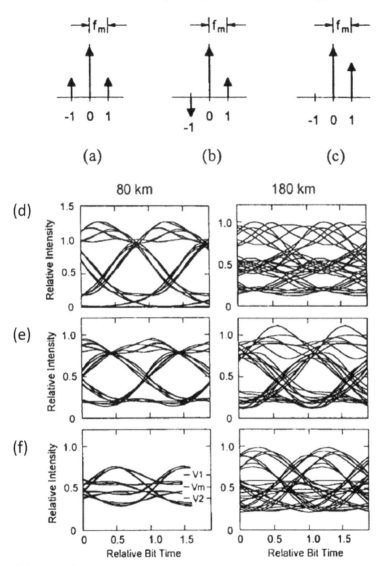

Figure 3.8 (top) Schematic frequency-domain representation of different modulation schemes: (a) AM, (b) narrowband FM, and (c) single-sideband large carrier modulation. (bottom) Calculated 10-Gbit/s eye diagrams for 80-km and 180-km length of standard single-mode fiber at 1.55 μm wavelength. Modulation formats: (d) Pure IM with $m_{IM} = 1$ (80 km) and $m_{IM} = 0.6$ (180 km); IM and simultaneous FM with (e) $m_{IM} = 0.6$ and $\Delta f_{pp} = 3$ GHz, and (f) $m_{IM} = 0.1$ and $\Delta f_{pp} = 3$ GHz. (after Reference 69).

3.4.2 10 Git/s Transmission Performance of Adiabatic- and Transient-Chirp Dominant DMLs

Figure 3.9 describes by simulations the transmission performance of an adiabatic-chirp dominant DML and a transient-chirp dominant DML for the 10-Gbit/s NRZ transmission over 20 km at 1.55 μm. Both cases used a linewidth enhancement factor of 4. The difference in the rate equation parameter is *not* the α-parameter, but rather the gain compression factor, which controls the ratio between the adiabatic and transient chirps according to Equation (3.7). The physics of the gain compression factor will be discussed in Section 3.4.3. In the chirp waveform in Figure 3.9(a.1) for the adiabatic-chirp dominant case, a larger gain compression factor is used, which results in a large adiabatic chirp (FM) of 25 GHz, as defined by the frequency difference between 1 bits and 0 bits. The corresponding optical spectrum for the adiabatic-chirp dominant case shows the two clear peaks separated by 25 GHz (Figure 3.9(a.2)). The transient chirp component, defined for the chirp associated with the overshooting (undershooting) in the leading (trailing) edge, is 8 GHz. The eye diagram after 20 km transmission is shown in Figure 3.9(a.3). A clear eye opening is confirmed for the adiabatic-chirp dominant case, even though we assumed a large linewidth enhancement factor

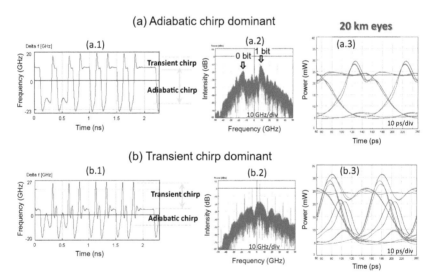

Figure 3.9 Chirp waveforms (a.1, b.1), optical spectrums (a.2, b.2), and eye diagrams (a.3, b.3) after 20-km transmission of 10-Gb/s NRZ signals for (a) adiabatic-chirp dominant DML, and (b) transient-chirp dominant DML.

of 4. Contrary, the chirp waveform for the transient-chirp dominant case is shown in Figure 3.9(b.1), created by using a smaller gain compression factor. The FM modulation amplitude is only 8 GHz while the transient chirp is as large as 22 GHz. The optical spectrum for the transient-chirp dominant case shows very strong carrier spikes with 10-GHz spacing (Figure 3.9(b.2)). The eye diagram after 20 km transmission is severely distorted by a fiber chromatic dispersion due to the strong transient chirp (Figure 3.9(b.3)).

Figure 3.10 shows the AM waveforms and the eye diagrams with and without Bessel-Thomson (BT) filter for an adiabatic-chirp dominated DBR laser after 25 km at 10 Gb/s [21]. Since the adiabatic chirp is ~25 GHz (0.14 nm), after a propagation over 25 km of a fiber, the 1 bit travels ~80 ps faster than the 0 bit, given by Equation (3.13). As a result, the energy in the leading edge of the 1 bit interferes with the energy of the preceding 0 bit, creating a high beat-frequency oscillation at 25 GHz in the leading edge. This is shown in Figure 3.10(b) for the waveform and the eye diagram after 25 km of a fiber without a BT electrical filter. The ER was 6.5 dB. High frequency oscillation can be seen, however, with the BT filter on, the eye becomes open due to the bandwidth limitation effect of the BT filter (Figure 3.10(a)). In CML, as discussed in Section 3.5, such high-frequency oscillation is not observed because the FM is reduced to half the bit rate (MSK condition), and higher ER (> 10 dB) is used; therefore, the energy for 0 bits is less.

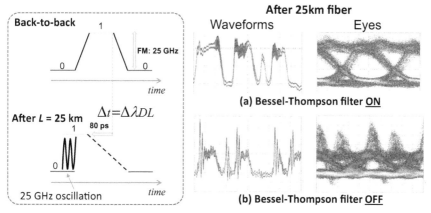

Figure 3.10 Schematic of chirp waveforms for 10-Gb/s NRZ 010 bit pattern before and after 25 km fiber transmission (left), measured waveforms (middle), and eye diagrams (right) after 25 km fiber (a) with Bessel-Thompson (BT) filter on, and (b) BT-filter off.

3.4.3 Gain Compression Phenomena

The impact of the gain compression factor is two-fold. It can damp the relaxation oscillation (therefore, suppress the transient chirp) and enhance the adiabatic chirp. The damping phenomenon of DMLs consists of two parts. One is the "dissipation terms" in the rate equations, which acts like "friction" for a harmonic oscillator: namely, the photon and the carrier lifetimes. The other phenomenon relates to the non-linear material properties stemming from a finite time delay between the injection of the carriers (input) and stimulated emission (outcomes). In principle, any time delays of the carriers during the transport process can lead to gain compression.

i) Cavity effect on damping

The damping of the relaxation oscillation is proportional to f_r squared;

$$\gamma_d = \frac{1}{\tau_{dN}} + K f_r^2, \tag{3.15}$$

where γ_d is the damping factor, τ_{dN} is the differential carrier lifetime, and the proportional factor, K, is called a K factor, which can be written as:

$$K = \frac{4\pi^2}{v_g} \left(\frac{1}{\alpha_m + \alpha_{\text{int}}} + \frac{\varepsilon}{\Gamma_l(dg/dN)} \right), \tag{3.16}$$

where v_g is the group velocity, α_m and α_{int} are the mirror and internal loss of the cavity, ε is the gain compression factor, Γ_l is the longitudinal optical confinement factor, and dg/dN is the differential gain. The first term is the dissipation term for photons, similar to a friction for a harmonic oscillator, which is a linear effect. The gain compression in the second term is a non-linear phenomenon (the ratio between the 1st and 3rd order susceptibilities); therefore, the effect happens only where the gain interacts with the photons. On the other hand, the differential gain is a linear effect (first-order susceptibility), and can be diluted when the passive section is integrated in the cavity. Therefore, DBR lasers shows higher damping compared to DFB lasers due to the factor, Γ_l, which can be less than the unity for DBR lasers. In Equation (3.16), it is shown that the cavity loss also affects the K factor.

As discussed in Section 3.6.1, the detuned-loading effect can also affect the effective gain compression factor [71] and chirp parameter [72]. On the long-wavelength side of the DBR mirror, the gain compression factor is reduced. The other sources of gain compression include the cavity standing wave effect [73] and the spatial hole burning effect [74], in which the carrier depletion, carrier heating, and spectral hole burning effects reduce the gain locally at the antinodes of the cavity standing wave [73] or the grating phase shift location for $\lambda/4$-shifted DFB lasers [74].

ii) Material effect on damping

When the gain compresses as the number of the photons increases, the relaxation oscillation is damped. This is opposite to the self-pulsation or mode-locking where the absorption saturates as the photon number increases. In this case, the effective gain compression factor becomes negative, and the initial small ringing or noise develops into pulsation due to un-damping. The gain compression phenomenon for MQW gain material stems from many physical origins. From the material point of view, well-known mechanisms include carrier density pulsation (CDP) [75], carrier heating (CH) [76], and spectral hole burning (SHB) [61], as shown in Figure 3.11. The carrier density pulsation compresses the gain by creating an index grating optically generated by the beating between two lights (or two FP modes), which scatters the light. The carrier heating is a process where the stimulated emission steals the carriers with lower energy than the statistical average, leaving heated carriers that reduce the gain. Spectral hole burning is the case where, again, the stimulated emission depletes the carriers of the corresponding transition energy, leaving a hole in the energy distribution due to the finite scattering time of the carriers to fill the hole. The gain compression factor is related to the third-order susceptibility (χ_3) and is also responsible for the four-wave mixing (FWM) process where two incident lights with different frequencies, ω_1 and ω_2, generate new frequencies, $\omega_1 + (\omega_1 - \omega_2)$ for up-conversion and

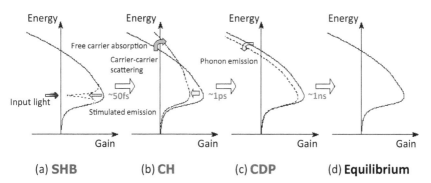

Figure 3.11　Carrier distribution depicting the non-linear gain phenomena for (a) spectral hole burning, (b) carrier heating, (c) carrier density pulsation, and (d) for equilibrium state.

$\omega_2 + (\omega_2 - \omega_1)$ for down-conversion [77]. For MQW structures, the down-conversion efficiency in FWM is more efficient because all three sources of the gain compression, CDP, CH, and SHB effects, are in phase and these add up constructively. The gain compression phenomenon also plays a role in the mode hop behavior and the hysteresis of DBR lasers [78].

In the framework of the SHB theory of the gain compression based on a density matrix formula, ε is written as [61]:

$$\varepsilon \propto \frac{\Gamma_3 g^3}{\Gamma_1 g^1} \propto \langle R_{cv} \rangle^2 (\tau_c + \tau_v) \tau_{in}, \qquad (3.17)$$

where g^1 (g^3) is the imaginary part of the first-order (third-order) susceptibility, respectively, Γ_1 (Γ_3) is the corresponding optical confinement factor, $<R_{cv}>^2$ is the dipole moment, $\tau_c(\tau_v)$ is the carrier energy relaxation time in the conduction (valence) band, and τ_{in} is the intraband carrier relaxation time. The gain compression factor is the ratio between the 1st and 3rd order gains in the SHB theory; therefore, it is mainly determined by the dipole moment and the intraband relaxation times. It does not show a strong dependence on the strain effect or the temperature effect [37].

iii) Carrier transport effect

Figure 3.12 illustrates the carrier transport effects considered for MQW structures. A common picture of the above gain compression phenomena is that any time delay of the carriers in filling the non-equilibrium energy distribution can lead to the gain compression. This concept was applied to the quasi-equilibrium populations of the carriers in the barrier and well states, which occur due to the finite capture time by phonon emissions and the escape time by thermionic emissions, leading to an additional gain compression of the MQWs. This phenomenon is called "well-barrier hole burning" [62]. The effect of damping due to the carrier transport effect in SCH was extensively studied [62, 79–82]. SCH is especially important not only for damping phenomenon, but also for chirp dynamics. Nagarajan studied a model in which the carrier diffusion and capture times across the SCH (τ_{sch}) and the thermionic emission time from the wells (τ_{ther}) were taken into account. The model predicted a degradation in the effective differential gain by a factor of $1+(\tau_{sch}/\tau_{ther})$ [82]. The gain compression factor is unaffected by this transport effect, although SCH creates a roll-off in S21 response similar to an RC roll-off (first order Bessel function). Also, an effective linewidth enhancement factor is increased due to the transport effect [80]. The conclusion on the gain compression phenomenon appears to differ from

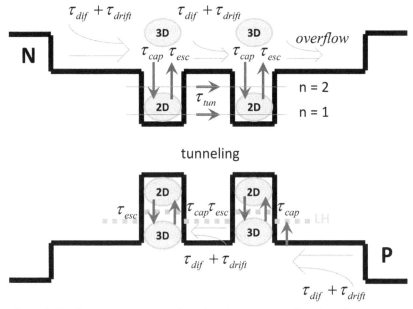

Figure 3.12   Schematic representation of carrier transport effects in MQW structure.

the "well-barrier hole burning" model proposed by Rideout [62] where the gain compression increases due to "well-barrier hole burning".

The effect of the carrier transport on the chip was shown in Figure 3.4, in which a set of rate equations based on the well-barrier hole burning and the SCH transport effect was numerically solved to depict the importance of the transport effects [37, 63]. The green trace is the chirp associated with the carriers in the wells and the barriers, and the red trace is the chirp associated with the carriers in the SCH. The black trace is the sum of those two. A slow hole transport time in the p-side SCH creates an internal electric field that attracts the electrons by drift and promotes the electron overflow into the p-side SCH. Therefore, the p-side SCH is more important for the chirp behavior than the n-side SCH, due to the electron overflow into the p-side SCH [21]. As can be seen in Figure 3.4, the SCH creates only the adiabatic chirp because the SCH acts like a capacitor that is charged or discharged as the laser is modulated.

iv) Lateral carrier diffusion effect

Besides the carrier transport effect in the vertical direction, the transport effect in the lateral direction can affect the chirp property of DMLs. For BH lasers, the carriers are laterally confined, and therefore can generate an adiabatic chirp efficiently. On the other hand, the carriers for ridge lasers can laterally diffuse, and, therefore, suppress the adiabatic chirp. Such lateral diffusion poses a practical upper limit on the numbers of wells and thickness of the SCH for ridge lasers.

v) Interface doping on damping

The carrier transport effect on the p-side SCH can be strengthened by n-type interface doping, for example Si, at the interface of p-side SCH and p-InP clad. This is confirmed in Figure 3.13(a) by the stronger roll-off in the simulated S21 response with Si interface doping using the CrossLight simulator. The simulation was in agreement with the measured S21 response, as shown in Figure 3.14(a), obtained for a DBR laser with a Si doping at the p-side SCH/InP clad interface. In comparison, Figure 3.14(b) shows the S21 response for the MQW *without* interface Si doping. Contrary to what is believed for the carrier transport effect, here F_r was almost unaffected (22 GHz for both cases), while the K factor could be increased from 0.23 ns to 0.34 ns by the interface Si doping. Therefore, it was possible to create a strong damping effect without significantly affecting F_r.

3.4.4 Experimental S21 Response and Transmission Performance of a Highly-Damped DBR Laser

In order to extend the reach of DMLs, it is effective to suppress the transient chirp by the design of MQWs and the laser cavity, then equalize the S21 response afterward by using a high-pass filter (HPF) for the drive waveform. As shown in Figure 3.14(a), the 3-dB BW for the DBR laser was limited to 13 GHz, due to the strong roll-off effect caused by a combination of the cavity damping effect of the DBR laser, the carrier transport effect in the SCH layer, and the Si doping at the interface between the p-SCH/InP cladding layers. Figure 3.14(c) shows the S21 response with an additional simple HPF circuit to achieve a flat S21 response for the highly-damped DBR laser. The transient chirp is well suppressed by equalizing a slightly over-damped S21 response, making this suitable for extending the reach of DMLs.

In Figure 3.15(a), the avalanche photodiode (APD) receiver sensitivity at a BER of 1E-3 was plotted as a function of the ER by changing the amplitude of the current modulation for the DBR laser described above. The DC bias was set at 90 mA, and the average fiber-coupled power was 10 mW, which is favorable for high-power PON applications. The gain and DBR lengths for the DBR laser were 300 µm and 80 µm, respectively.

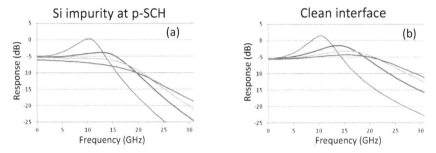

Figure 3.13 (a) Crosslight simulations for S21 responses of MQW with Si impurity at the p-SCH/clad interface, (b) with clean interface.

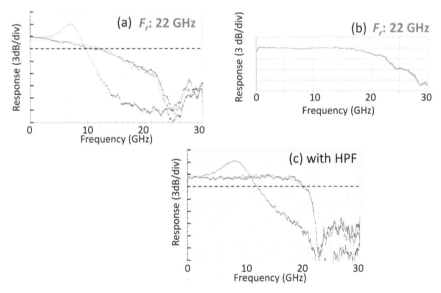

Figure 3.14 Experimental S21 responses for MQW (a) with interface Si impurity, (b) clean interface, and (c) RC equalization applied to Figure 3.14(a) with Si impurity.

The DBR laser was tunable over 12 nm, and therefore suitable for TWDM PON applications. Figure 3.15(b) shows the eye diagrams for the BtB, 10.5 km, and 21 km. The sensitivity in the BtB increases as the ER increases. A good eye opening after 21 km was obtained for an ER lower than 7.5 dB. Under higher ER conditions, the transient chirp becomes stronger, and eventually the transmitted eye fell apart above 7.5 dB of ER. According to the DST condition, the maximum reach of the FSK format can be determined by the distance when the 1 bit travels faster than 0 bit by one bit period, as

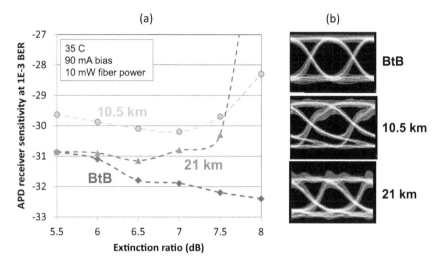

Figure 3.15   (a) APD receiver sensitivity at a BER of 1E-3 for back-to-back, 10.5 km, and 21 km transmissions plotted vs. ER changed by adjusting the modulation current, and (b) the corresponding eye diagrams.

shown in Equation (3.13). Indeed, by lowering the modulation current and reducing the adiabatic chirp (ER: 5 dB), it was possible to open the eye after 40 km. The eye after 10 km is more closed compared to the eye after 20 km. The reason can be understood by observing the 101 bit sequence where the energy for the 1 bits before and after the 0 bit spreads into the 0 bit and interferes destructively at 20 km, resulting in the opening of the eye. This is similar to the principle of the ODB.

3.4.5 Thermal Wavelength Drift Stabilization for the Burst-Mode NGPON2 Application

i) Sources for low-frequency FM response non-flatness

In TWDN-PON systems, multiple channels with a spacing of 100 GHz are multiplexed by an arrayed waveguide (AWG) filter; therefore, the lasing frequency stabilization under burst-mode operation is a critical requirement. After turning on the DMLs, the Joule heating causes the lasing wavelength to shift to a longer wavelength. This can cause a crosstalk between the channels as the signal passes through the AWG filter. This requires a stabilization of the wavelength within ~250 ns after the turn-on of the DMLs for the NGPON2 application. For CML applications, as discussed in Section 3.5, even more precise locking of the frequency within < 2 GHz on an optical filter is required within ~a few nanoseconds in time after the turn-on of the DMLs.

Low-frequency wavelength drift in DMLs can cause the other issue in fiber transmissions due to the dispersion which causes the timing skew at the receiver. This issue is depicted in Figure 3.16(a), which shows a single bit of 1 after a long string of 0 bits. If the FM response at lower frequency is not flat, then the frequency of the single bit of 1 can be different from the preceding 1 bits separated by the long string of 0 bits. A well-known origin for such FM non-flatness includes the thermal chirp; However, there are other reasons: for example, the spatial hole burning effect in DFB lasers [74]. In Figure 3.16(a), the 1 bit after the long string of 0 bits is delayed by as much as 50 ps after a transmission of 21 km fiber. Using the Equation (3.13), it can be found that this corresponds to 15 GHz of FM droop at the low frequency. Such timing skew causes an error floor in the BER curve at the receiver, as shown in Figure 3.16(b), where the strong intersymbol interference (ISI) skews the timing of the rising edge.

Thermal chirp behavior of DMLs is briefly described in Figure 3.17(a) in the time domain within ~10 μsec and Figure 3.17(b) as a function of the DC bias. The simple way to find the heat source is to use the energy conservation

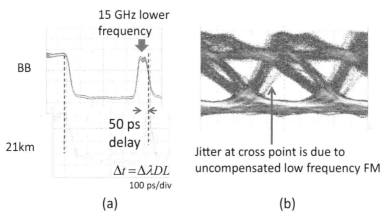

(a)   (b)

Figure 3.16 (a) Waveform of 1 bit after a long string of 0 bits in the back-to-back and after 21 km transmission. The frequency of 1 bit is 15 GHz lower than preceding 1 bits, resulting in 50 ps timing delay after 21 km transmission, and (b) the corresponding transmitted 10-Gb/s eye diagram showing the timing skew of 50 ps in the rising edge.

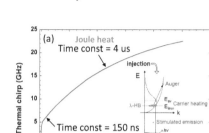

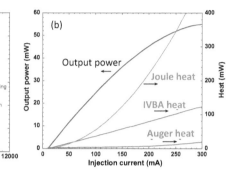

Figure 3.17 (a) Time-domain thermal chirp with three time constants. Inset shows the carrier energy relaxation process responsible for heat sources, (b) Light-Current curve on the left axis, and heat due to Joule heating, Auger, and IVBA processes on the right axis.

rule for the input power (*IV*) and the output power (*P*$_{out}$): namely, *heat* = *IV* – *P*$_{out}$. To determine more details about the heat sources and where they are located, the expression for *P*$_{out}$ in Equation (3.9) can be plugged into the rate equation; then the amount of heat related to each term in the rate equation can be calculated. Of those heat sources, the heating associated with IVBA, the spontaneous and stimulated emissions, and Auger recombination happens right in the core of the active region; therefore, the corresponding time constant can be fast (∼15 ns) because the effective thermal mass is small. The thermal time constant for Joule heating can be slower than this (∼150 ns) because of the greater distance between the optical mode and the location of the Joule heating. Heating of the whole chip itself and sub-mount creates a much slower thermal chirp (∼4 μsec) due to large thermal mass [83]. The breakdown of the heat sources is plotted in Figure 3.17(b) as a function of the bias. As the bias reaches closer to the roll-over point, the Joule heating dominates and the time constant becomes slow.

ii) Thermal chirp compensation scheme

Several schemes have been proposed to compensate for the thermal effect on the laser frequency. In Reference [84], the electrical contact for a DFB laser was split into two sections, where one of the sections was based high under CW and the other was turned on and off in the burst mode to reduce the thermal chirp. A transmission over 20 km was demonstrated at 10 Gb/s by lowering the ER to 5 dB. A short PRBS length of 7 was used.

Integrating a thermal heater near the active stripe was proposed and discussed. The time constant of the chirp caused by the integrated heater is in a range of ~10 μsec, even when the heater is integrated on top of the active laser stripe [85], and therefore the scheme is not suitable to compensate for the fast time constant. According to the concept of "counter heating", by turning off the heater in a complementary manner as the gain bias is turned on, it was possible to suppress the thermal drift [86, 87]; however, the compensation of the fast thermal component still seemed difficult with this approach alone.

The concept of compensating fast thermal chirp by adiabatic chirp was proposed in Reference [86]. After turning on the laser, the adiabatic chirp is larger than the thermal chirp over a certain period of time (~500 ns). For a 10-Gb/s NRZ signal with a PRBS of 2^{31}-1 pattern, the worst part of the thermal excursion continues over ~500 ns, and then the laser turns back to a cooling phase. Therefore, it is possible to compensate for the thermal chirp by gradually increasing the DC bias current as long as the thermal chirp is smaller than the adiabatic chirp (Figure 3.18). The temporal form of the bias current required for the compensation of the fast thermal chirp can be found after the Laplace transform of the thermal chirp waveform [86]:

$$I(t) = I_0 u(t) + I_0 \frac{C_T}{C_A - C_T} \left(1 - e^{-\frac{C_A - C_T}{C_A} \frac{t}{\tau}} \right) u(t), \qquad (3.18)$$

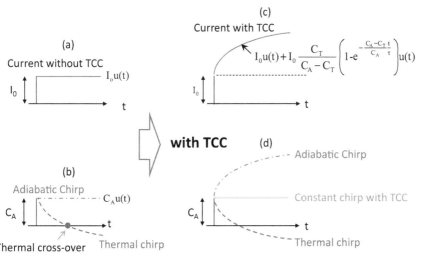

Figure 3.18 (a) Step function bias current to DML without thermal chirp compensation (TCC), (b) corresponding adiabatic chirp and thermal chirp, (c) bias current to DML with TCC, (d) compensated chirp with TCC (green).

where *u(t)* is a step function of the bias, C_A is the adiabatic chirp, and C_T is the thermal chirp. As long as $C_A > C_T$, the gradual increase in the bias current will suppress the thermal drift. At a thermal cross-over point ($C_A = C_T$), the required bias current goes up to infinity. This issue can be avoided if the counter heating scheme is combined in order to compensate for the low-speed thermal chirp (> 1 μsec) [86].

Designing DFB lasers with a long "thermal crossover" requires careful consideration of the spatial hole burning effect and the MQW structure design, and was extensively studied for coherent systems in the late 1980's when DMLs were used as an FM source. At that time, the non-flatness of the FM response at low frequency, known as the "FM dip issue", was addressed by the DFB chip design [88] or by the gain-lever effect [89] which achieved a flat FM response down to 100 kHz.

For tunable DBR lasers, complete compensation of thermal chirp is possible by taking advantages of a fast tuning speed (~1 ns) and a very high tuning efficiency (~12 nm). The thermal chirp is typically < 1 nm; therefore, an injection current of < 1 mA is sufficient to compensate for the thermal chirp. Figure 3.19(a) shows a simple thermal chirp compensation circuit (TCC), where a small modulation current (~1 mA$_{pp}$) was tapped from the modulated gain section and was directed to the DBR section. A simple low-pass filter

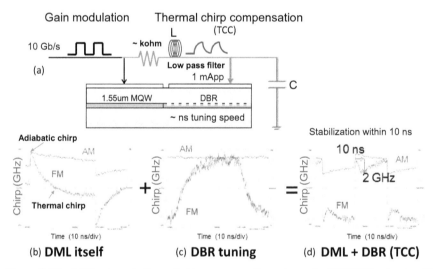

Figure 3.19 (a) Burst-mode thermal chirp compensation scheme for NGPON2 using tunable DBR laser, (b) AM and chirp waveforms of DML, (b) tuning response of DBR section itself, and (c) modulation of both DML and DBR tuning section used as TCC. (after Reference 21).

(LC circuit) was added to the DBR section to match the tuning speed of the DBR section with the thermal chirp time constant. Figure 3.19(b) shows the thermal chirp for the DBR laser itself and (c) for the tuning response of DBR section itself including a LC low-pass filter. The combined chirp waveform of the DBR laser at bit transitions is shown in Figure 3.19(d). By optimizing the time constants of the TCC circuit, the complete thermal compensation was possible from ∼10 ns down to millisecond regimes [21].

For DBR lasers, there is an additional fast frequency-chirp component originating from the DBR tuning section. In a ∼nanosecond time scale, one has to consider the photo absorption in the DBR tuning section at low bias conditions, which can add an additional blue-chirp behavior. At high bias conditions to the DBR section, the modulated light in the gain section depletes the carriers in the DBR section by stimulated emission to create a red chirping in a few nanosecond time scale [21].

3.5 Chirp Managed Laser (CML)

In the previous section, we reviewed the paths to extend the reach of DMLs by the material and cavity designs. It is essential to suppress the transient chirp and utilize the adiabatic chirp. For this purpose, slightly over-damped modulation response was realized by the cavity and MQW designs, followed by compensating for the roll-off with a simple LC circuit to make the S21 response flat. The transmission distances of 20 km – 40 km at 10 Gb/s have been achieved, while the ER has to be kept lower than ∼7 dB. This performance may be sufficient for 10-Gb/s PON applications; however, it is not applicable to metro > 80 km links. A CML was proposed in order to overcome this shortcoming of DMLs by increasing the ER through FM-AM conversion of the adiabatic chirp on the slope of an optical filter [24, 25]. A typical set up for CML is shown in Figure 3.20. Simply, the output of the DML is passed through an optical filter called an optical spectrum re-shaper (OSR) for FM discrimination. This is the basic concept of CML. In spite of the simple appearance, however, the interaction between the chirp of DMLs and the OSR produces a rich variety of effects such as FM-AM conversion, AM-FM conversion, the suppression of transient chirp, the DST effect, the ODB effect, minimum shift keying (MSK) format generation, the vestigial sideband (VSB) effect, the BW enhancement effect, and the 2nd order dispersion compensation effect for the FSK format, resembling the 3rd order dispersion compensation. This section looks in more details of those

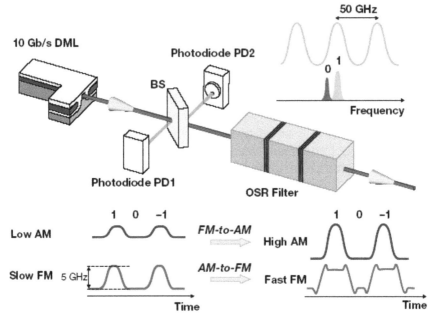

Figure 3.20   Schematic of Chirp Managed Laser (CML) with optical spectrum re-shaper (OSR) and wavelength locker (PD1 and PD2). (after Reference 24).

principles of CML which can extend the reach to 250 km at 10 Gb/s. The end of this section summarizes the experimental demonstrations of CMLs from 10 Gbit/s up to 56 Gbit/s.

3.5.1 Principles of CML

i) FM-AM conversion by optical spectral filtering

An enhancement of the ER by converting FM to AM through the slope of the OSR is one of the defining features of CML. As an example, the spectral filtering is applied to the DBR laser described in Figure 3.21(a). The OSR filter used was a 4-ch Gaussian-shaped AWG filter with a 100 GHz spacing [21], which could be used as a wavelength multiplexer for an optical line terminal (OLT) at the central office in an NGPON2 network. The 3-dB BW of the filter transmission spectrum was 40 GHz. The frequency of the 1 bits was aligned at the peak of the filter transmission. The frequency of the 0 bits was shifted by –27 GHz due to the adiabatic chirp where the transmission of the AWG filter was 5.5 dB lower from the peak. Therefore, the ER after

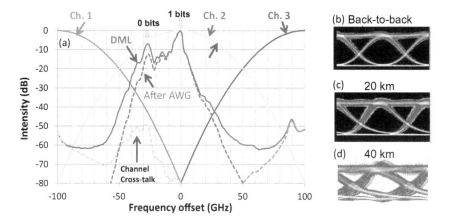

Figure 3.21 (a) Modulated spectrum of DBR laser (solid blue), 100 GHz-spaced AWG filter for NGPON2 wavelength multiplexing (light blue), DML spectrum after AWG (broken pink). The frequencies of 1 bits and 0 bits are indicated by arrows, 10-Gb/s eye diagrams for (b) back-to-back, after (c) 20 km, and (d) 40 km transmissions. (after Reference 21).

the OSR filter increased to 11 dB from 6.5 dB. The open eye diagrams in the BtB, after 20 km, and 40 km are shown in Figure 3.21(b)–(d), respectively. The BER performance was improved compared to that for the original DML without OSR effect over $+/-$ 12 GHz frequency range on the AWG filter locking position. However, such a large adiabatic chirp can, in turn, limit the transmission distance to < 40 km according to the DST condition described in Equation (3.13).

ii) Generation of Minimum Shift Keying signaling by using adiabatic chirp of DML

For further extending the reach, it is possible to narrow the spectral width by using a minimum shift keying (MSK) condition which is a special case of continuous-phase FSK where the frequency difference between the 1 and 0 bits are set to half the bit rate. The AM and FM waveforms corresponding to this condition are shown in Figure 3.22(a). The blue trace is the 10-Gbit/s AM waveform where the ER is set to around 2 dB. The corresponding chirp waveform is shown in a red trace. The adiabatic chirp is set to 5 GHz, which is half the bit rate. The corresponding phase shift is shown in Figure 3.22(b). The phase shift across the 0 bit can be calculated by integrating the adiabatic chirp over the bit period; namely, $2\pi \times 5$ GHz $\times 100$ ps $= \pi$. Therefore, the relative phase shift between the 1 bits before and after the 0 bit is π. In a time averaged spectrum, a pair of such π phase-shifted 1 bits

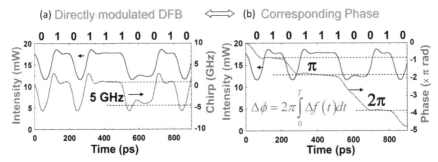

Figure 3.22   (a) AM and chirp waveforms of 10-Gb/s DML, and (b) corresponding phase shift on right axis. The integral of 5 GHz chirp (half bit rate) over 100 ps (bit period) corresponds to π shift. (after Reference 24).

cancels each other by the destructive interference. The time-averaged spectrum simulated for the un-chirped NRZ signal and the DML that satisfies MKS condition are shown in Figures 3.23(a) and (b), respectively. A commercial simulation software, VPIPhotonics, was used. The 20dB-down spectral width for the un-chirped NRZ is $\sim$2 $\times$ bit rate, excluding the carrier spikes spaced at 10 GHz. On the other hand, the 20dB-down spectral width for CML is $\sim$1 $\times$ bit rate, which is similar to the spectral efficiency of an

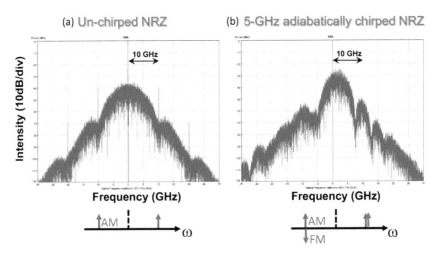

Figure 3.23   (a) Optical spectrum for un-chirped NRZ showing AM sideband carriers, and (b) NRZ with half the bit rate FM (5 GHz) satisfying the MSK condition. Spectrum narrows due to AM and FM sideband cancellation on the low frequency side.

ODB format. The spectral peaks corresponding to the 1 bits (+5 GHz) and the 0 bits (0 GHz) disappeared due the relative π phase shift between the bits. Contrastingly, the clear peaks for the 1 bits and the 0 bits can be seen in Figure 3.21(a), separated by the amount of the adiabatic chirp (~27 GHz) when the MSK condition is not satisfied. The carrier spikes also disappear for CML.

iii) AM-FM conversion effect for flat-chirp generation

When the modulated light from a DML passes through the OSR, both FM-AM conversion and AM-FM conversion occur. The reason for the AM-FM conversion is summarized in Figure 3.24. Here the slope of the OSR transmission on the edge is written as a linear function, $T(w) = a + b(w-w_0)$, where a is the transmission at the center of the OSR, and b is the slope of the OSR. According to Fourier theory, Fourier transformation of the slope of the filter, bw, is the time derivative of the signal intensity.

$$\omega \Leftrightarrow -i\frac{\partial}{\partial t}. \tag{3.19}$$

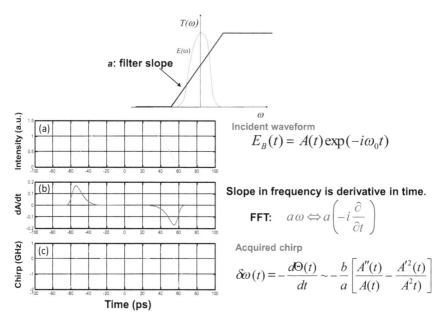

Incident waveform
$$E_B(t) = A(t)\exp(-i\omega_0 t)$$

Slope in frequency is derivative in time.

FFT: $\quad a\omega \Leftrightarrow a\left(-i\dfrac{\partial}{\partial t}\right)$

Acquired chirp
$$\delta\omega(t) = -\frac{d\Theta(t)}{dt} \sim -\frac{b}{a}\left[\frac{A''(t)}{A(t)} - \frac{A'^2(t)}{A^2 t)}\right]$$

Figure 3.24 AM-FM conversion effect on the slope of optical filter (top). And (a) for time domain AM waveform, (b) the time derivative of AM waveform, and (c) acquired chirp due to AM-FM conversion.

The acquired additional chirp due to the AM-FM conversion at the output of the OSR is given by:

$$\delta\omega \approx -\frac{b}{a}\left[\frac{A''(t)}{A(t)} - \left(\frac{A'(t)}{A(t)}\right)^2\right] \approx -\frac{d}{dt}\tan^{-1}\left(\frac{bA'(t)}{A(t)(a+b\omega(t))}\right), \quad (3.20)$$

where $A(t)$ is the amplitude envelope of the input electric field and $w(\tau)$ is the adiabatic chirp waveform. This means that the weighted derivatives of the input amplitude generates negative-going spike-like chirp, as shown in Figure 3.24(c), in the tails of the 1 bit. Also, the blue-shifted broader peaks at both the rise and fall transitions should be added to the original adiabatic chirp. The intensity and chirp waveforms before and after the OSR filter are shown in Figure 3.25. Very sharp negative-going chirp is introduced near the intensity minima of the 0 bits after the OSR, indicating that the phase shift is very abrupt. Also, the chirp across the 1 bits becomes flat-topped after the OSR filter. Measured intensity and phase waveforms after the OSR are shown in Figure 3.26 for a 01010 bit sequence with low and high ER conditions controlled by changing the slope of the OSR. A high-resolution complex optical spectrum analyzer (APEX AP2440A) was used for the measurement. The phase transition in the original DML is gradual because it is a result of an integration of the continuous FSK. For low ER condition, the phase transition in the middle of the 0 bit is still gradual. Contrary, for a high ER condition when the slope of OSR is high, an abrupt phase shift similar to ODB is generated in the middle of the 0 bits at the intensity minima. This is due to the AM-FM conversion effect.

iv) Vestigial sideband effect

In addition to the MSK condition, the spectrum of CML can be further narrowed by spectral filtering on the edge of the OSR. Therefore, there is a vestigial sideband effect (VSB) effect for CML. At the same time, Equation (3.14), which describes the condition for AM and FM sideband cancellation, requires a high ER when the FM is 5 GHz. Therefore, the ER enhancement through the OSR filer can reduce the spectral width based on the AM and FM sideband cancellation.

v) Similarity of CML to the Optical Duobinary format

An ODB modulation is known as a format that can extend the reach of a 10 Gbit/s signal over nearly three times the distance [23] that EMLs can bridge with a pre-chirping technique (~80 km). There is a similarity between

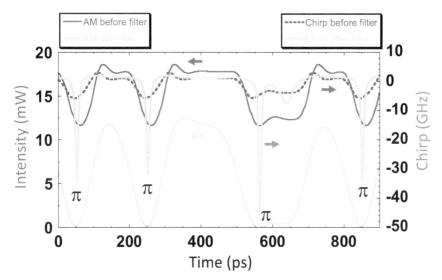

Figure 3.25   Intensity and chirp waveforms before and after OSR filter. AM-FM conversion effect creates negative-going spike chirp at the intensity minima in the 0 bits, corresponding to abrupt phase shift. The chirp across 1 bits becomes flat-topped after the OSR filter.

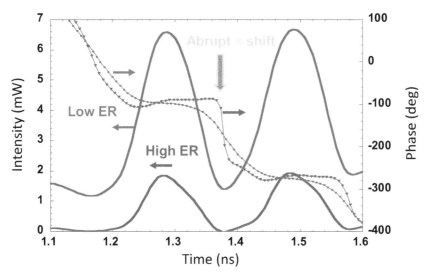

Figure 3.26   Measured chirp and intensity waveforms after OSR filter using APEX/ AP2441B chirp tester. For high ER obtained on the steep slope on the OSR, abrupt phase shift can be seen.

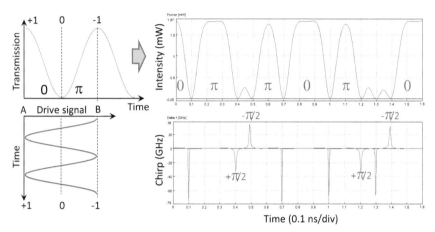

Figure 3.27 Principle of optical duobinary (ODB) signal generation based on Mach-Zehnder modulator. Abrupt π phase shift is introduced in the middle of 0 bits. Also, $\pi/2$ phase shift is introduced in the tails of isolated 1 bit.

ODB and CML. An ODB format can be generated using a Mach-Zehnder modulator (MZM), which is biased at a null point of the transfer function of the MZM, and then modulates between two transmission peaks on the higher and lower voltages of the null point (Figure 3.27). When the modulation voltage changes from the lower bias point A for one 1 bit to the higher bias point B for the other 1 bit, a π phase shift is introduced upon passing the null point for the 0 bits in between. To produce such a coding pattern, an encoder to map the NRZ binary signal onto the duobinary signal is required. The excellent dispersion tolerance of ODB is often explained by the π phase shift in the middle of 0 bit in the 101 bit meta-symbol pattern. As shown in Figure 3.28, after fiber transmission, the energies for the 1 bits spread due to the dispersion. For the standard NRZ format, the dispersed energies of the 1 bits constructively interfere in the middles of the 0 bit and close the eye. For ODB, thanks to the π phase shift, the dispersed energies destructively interfere, and keep the eye open. On the other hand, in CML, π phase shift can be *naturally* introduced between the bits simply by driving DMLs under the MSK condition of the continuous FSK.

vi) Similarity of CML to Dispersion Supported Transmission

Excellent dispersion tolerance of ODB can be explained by the π phase shift in the middle of 101 bit meta-symbol. How about other important meta-symbols like 010 bits, i.e., an isolated 1 bit? In fact, in an ODB format, $\pi/2$

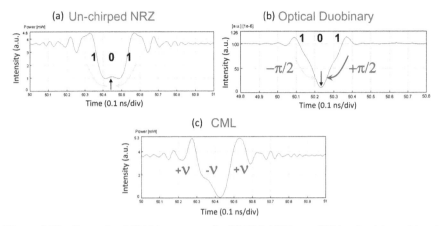

Figure 3.28 Transmitted 101 bit sequence for 10-Gb/s NRZ over 100 km for (a) un-chirped NRZ, (b) duobinary, and (c) CML formats. For ODB and CML, destructive interference of dispersed energies from 1 bits keeps the 0 bit open.

phase spikes are introduced in the tails of the isolated 1 bit which suppress the distortion of the 1 bit after ∼250 km of fiber transmission at 10 Gbit/s. This is also illustrated in Figure 3.27 for the ODB generated by the low-pass filter approach. The reason for the strong dispersion tolerance for an isolated 1 bit in the CML format is very different. CML is an ASK-FSK format and is more similar to the DST [26, 68] in the interpretation of the dispersion tolerance of the isolated 1 bit. As has already been explained in Section 3.4.1 in Figure 3.7(b), the frequency of the 1 bits is higher than that of the 0 bits and therefore travels faster in the anomalous dispersion of a fiber. Then, the waveform of the single bit of 1 leans forward, and creates ringing in the leading edge due to the interference between the 1 bit and the preceding 0 bit. The waveform of an isolated 1 bit after 200 km transmission is shown in Figure 3.29. It can be seen that such ringing can keep the eye open as long as the DST condition is satisfied, as shown in Equation (3.13). For example, for a 10-Gbit/s CML with a 4 GHz of adiabatic chirp, the predicted maximum transmission distance is 225 km, which is close to the experiment [24, 25]. If the MSK condition is the most important criterion for the reach extension, which is mainly chosen to mimic the ODB effect on the 101 bit sequence, 5 GHz of adiabatic chirp is supposed to be the optimum chirp. The fact that 4 GHz of adiabatic chirp extended reach more means that the DST condition, which controls the energy spread of the 010 bit sequence, seemed to be more important for CML.

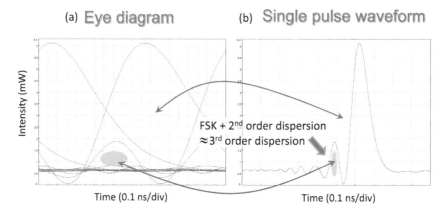

Figure 3.29   (a) Eye diagram of isolated single bit after 200 km transmission for 10-Gb/s CML, and (b) temporal waveform showing the ringing in the leading edge created by interference between 1 bit and preceding 0 bit due to 2nd order fiber dispersion. The interference keeps the eye open as long as DST condition is satisfied. This ringing in the leading edge can be compensated by 3rd order dispersion.

The difference between DST and CML is in the ER. The ER for DST is very low in the BtB; therefore, there is a large penalty in the receiver sensitivity. On the other hand, the ER for CML is > 10 dB and can also satisfy the AM/FM sideband cancellation conditions, which narrows the modulated spectrum.

The importance of flat-chirp for a single-pulse NRZ signal to improve the dispersion tolerance is illustrated in Figure 3.30. The pulse width of a single-bit NRZ signal can spread more quickly compared to un-chirped Gaussian pulse because of the faster rise-fall times. By adding the 5 GHz of adiabatic chirp with a top-hat chirp across the single-bit NRZ signal, as shown in Figure 3.25(b), the pulse broadening effect can be suppressed below 120 ps in FWHM after 200 km of the transmission. The broadening can be suppressed more for a pulse with a 10 dB ER rather than a 13 dB ER, indicating that the finite pulse energy in the precedent 0 bit interferes favorably and suppresses the broadening of the pulse.

vii) Effectiveness of 3rd-order dispersion compensation for CML

The ringing in the leading edge in Figure 3.29(b) is a result of an interaction of an ASK-FSK pulse with a 2nd-order dispersion of a fiber. The shape of ringing resembles that for the 3rd-order dispersion effect on the Gaussian pulse [90]. Indeed, it is possible to compensate for such ringing in the leading

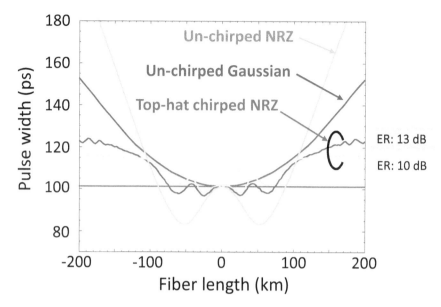

Figure 3.30 FWHM pulse width vs. fiber length for un-chirped NRZ, un-chirped Gaussian, top-hat-chirped CML with 10-dB and 13-dB ER.

edge by 3rd-order dispersion. The OSR used in CML has 3rd-order dispersion near the edge of the spectrum and can compensate for the fiber dispersion equivalent to nearly 40 km of a standard single mode fiber.

3.5.2 Experimental Demonstrations of CML

CML has been demonstrated at various bit rates from 10 Gb/s to 56 Gb/s for the NRZ format in C band. The implementation of CML is easier at higher bit rates because the required OSR filter BW can be wider and also the FM required for the MSK condition is higher so that the accuracy of locking on the OSR filter can be more relaxed. The commonly used type of an OSR filter is Gaussian shaped with a 3-dB BW of 1.0 ~1.5 × the bit rate. Multi-cavity etalon filters or a delayed-line interferometer (DLI) filter gives a shape close to a Gaussian filter, which performs well. The maximum reach for 10-Gb/s CML is around 200 km for an adiabatic chirp of ~5 GHz [24, 25], and a slightly lower chirp can extend the reach further to 250 km due to the DST condition as discussed in Section 3.4.1. Figure 3.31 shows the OSNR performance of 10-Gb/s CML as a function of a fiber dispersion. The plot also compares the case for a chirp-free NRZ format generated by a lithium

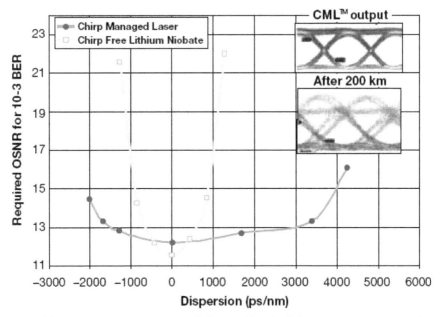

Figure 3.31   OSNR curves at 1e-3 BER for un-chirped NRZ generated by LN modulator compared with CML at 10 Gb/s. The insets are the eye diagrams for CML in the BtB and after 200 km. The center of the bathtub curve for CML is shifted toward the positive dispersion due to the dispersion compensation effect of the OSR.

niobate MZM. The transmission performance for the pure FSK format should be the same for both positive and negative dispersion as expected from Figure 3.7(b). In Figure 3.31, the center of the bathtub curve is slightly shifted toward the positive dispersion due to the dispersion compensation effect by the OSR filter on the edge. When there is a residual transient chirp component in CML, this favors the transmission in a negative dispersion, as is the same as a conventional DML.

The reach was further extended to 300 km by using electrical dispersion compensation (EDC) on the Rx side [91]. The maximum reach in CML is set by slow falling tails of consecutive 1 bits followed by a 0 bit since the 1 bits with higher frequency travel faster than 0 bits, leaving slow falling tails behind (Figure 3.29). By addressing this issue, 350 km reach was demonstrated by using a negative-going driving signal in the tails of 1 bits to reduce the fall time [92].

Full-C band 10-Gb/s CML was demonstrated using a high-speed modulated-grating Y-branch (MGY) laser tailored by minimizing the passive length. The schematic structure of the MGY laser is shown in

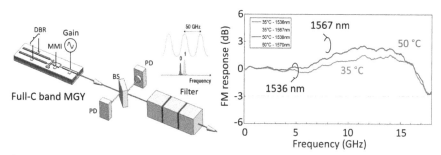

Figure 3.32 Schematic of full-C band 10-Gb/s CML based on tunable modulated-grating Y-branch (MGY) laser (left), and FM response of MGY laser at 35°C and 50°C operating at 1567 nm and 1536 nm wavelengths (right). (after Reference 93).

Figure 3.32 [93]. Since the passive DBR sections for Vernier tuning are located on one side through a 2×1 MMI phase section, instead of both sides of the gain section as in the super-structure grating (SSG) DBR lasers [94], the longitudinal confinement factor can be increased. This configuration improves the intrinsic speed of the MGY laser. As a result, a very wide FM modulation bandwidth of 16 GHz – 18 GHz was realized over the full C-band channels at 50°C. Full-C band 200 km transmission was demonstrated using a 2-cavity 50-GHz periodic etalon filter with a 3-dB BW of 7 GHz.

Monolithic CML based on a Si platform was demonstrated in Reference [95], where monitor PDs and a heater-tuned single-ring OSR filter were integrated as shown in Figure 3.33. An output of a DFB laser was coupled to a single-ring filter on a Si platform by a passive alignment process. An ER of 7.4 dB was obtained after the single-ring filter. A wide dispersion range of −1100 ps/nm to 5180 ps/nm or the maximum reach of 300 km was demonstrated at 10 Gb/s. In Reference [96], a compact 4×10 Gb/s CML on a Si Photonics platform was realized by integrating four thermally-tunable micro-ring resonators (MRR) as multiplexers, which were also used as OSRs, together with transparent in-line detectors, named ContactLess Integrated Photonic Probes (CLIPPs). The slope and 3-dB BW of the OSR filter was somewhat limited due to a wide FSR of 125 GHz for a single MRR, however, a transmission performance over 40 km was improved by passing through an MRR. In Reference [97], transmissions over 81 km at 17.5 Gb/s and 40 km at 20 Gb/s at 1.55 μm were demonstrated.

Using a 1.55-μm DBR laser described in Section 3.4.4, uncooled operation of 10-Gb/s CML was demonstrated for 100 km transmission at 87°C. An OSR filter effectively removed the transient chirp and enabled such high

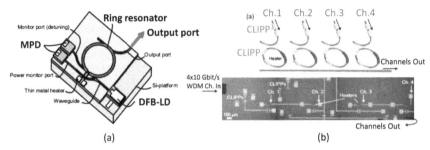

(a)   (b)

Figure 3.33   (a) Monolithic 10-Gb/s CML based on PLC ring filter (after Reference 95), and (b) based on four thermally-tunable micro-ring resonator (MRR) multiplexer integrated with transparent in-line detectors (CLIPP). (after Reference 96).

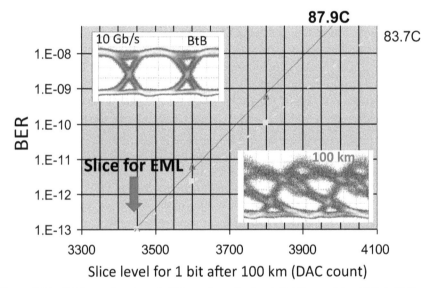

Figure 3.34   BER as a function of APD receiver slice level (DAC count) for 10Gb/s CML at 87.9 C after 100 km. Slice level for receiving EML is indicated by an arrow for comparison, indicating CML was interoperable with EML.

temperature operation. A combination of a slightly large adiabatic chirp of 8 GHz and a DLI filter with an FSR of 28 GHz improved the BtB eye opening as shown in the inset of Figure 3.34. An error-free operation was obtained by using the same slice level for the APD receiver as that for EMLs, indicating that CMLs can be interoperable with EMLs. The same DBR laser was also used at 25°C for 25.6 Gb/s CML for a transmission of 0–60 km, which was tunable over 12 nm. A combination of a 23-GHz 3-dB BW 3-cavity etalon filter and a 67-GHz FSR DLI filter were used together with an MSK adiabatic chirp condition of 12 GHz. The error-free BER performance and the eyes at BtB and 56 km are shown in Figure 3.35.

Unique implementation of CML was demonstrated where the adiabatic chirp was generated by integrating a phase modulator in the cavity of a 4-λ DBR laser array with a 100 GHz channel spacing [98] (Figure 3.36). The modulation BW in this case is determined by an RC frequency of the short

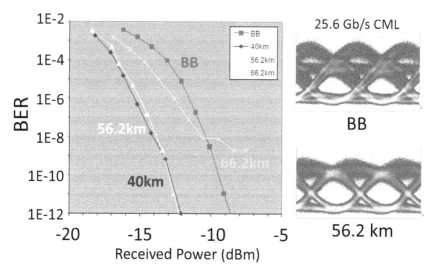

Figure 3.35 BER performance and eye diagrams of 25.6-Gb/s CML (12-nm tunable) transmitted over 0 km–66 km.

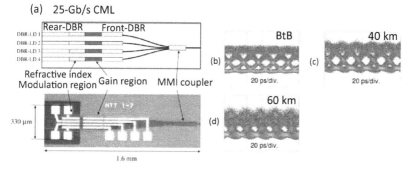

Figure 3.36 (a) Schematic diagram and photograph of in-cavity phase-modulated DBR laser array, and (b) 25-Gb/s NRZ signals in the back-to-back, (c) after 40-km transmission, and (d) after 60-km transmission. (after Reference 98).

phase modulator section, not by the intrinsic F_r. An 80 μm-length phase modulator section was integrated in the cavity of the DBR laser array, which employed a short gain section of 80 μm to enhance the FM efficiency (10 GHz/2 V_{pp}). A 40 km transmission at 25 Gb/s was demonstrated using a combination of a 50 GHz FSR DLI filter and a 13.5 GHz 3-dB BW etalon filter. The concept of the in-cavity phase modulation was also applied for a single-stripe SSG DBR laser, which can be modulated at 20 Gb/s and be tuned over 27 nm [99]. The gain and the phase sections were 350 μm and 200 μm, respectively. A transmission over 40 km was successfully demonstrated at 20-Gb/s using a combination of a 40-GHz FSR DLI filter and a 13.5-GHz 3-dB BW etalon filter. Further, 40-Gb/s CML transmission over 20 km was demonstrated using an in-cavity phase modulation of a DBR laser. In order to improve the FM efficiency, a DBR mirror was used only for the rear reflector (Figure 3.37(a)) for minimizing the passive length. The gain and phase modulator lengths were 180 μm and 90 μm, respectively. An FM of 10 GHz at a modulation voltage of 1.78 V_{pp} was obtained, which is half the MSK condition for 40 Gb/s CML. An ER of 4 dB was obtained after a combination of a 50 GHz FSR DLI filter and a 13.5 GHz 3-dB BW etalon filter [100].

A 40-Gb/s CML was also demonstrated based on a DFB laser with an F_r of 15 GHz [101]. CML utilizes FM response, which is faster than AM response. Therefore, CML enables applications to much faster bit rates than

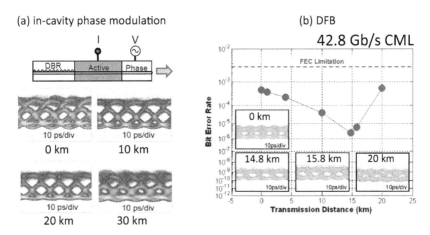

Figure 3.37 (a) 40-Gb/s signal transmission of CML based on in-cavity phase-modulated DBR laser over 0–30 km (after Reference 99), (b) 0–20 km transmission performance of 42.8-Gb/s CML based on DFB laser. (after Reference 100).

the BW of the AM response. A transmission over 20 km, which is a critical distance for PON applications, was demonstrated at a bit rate of 40 Gb/s. The BER performance and eye diagrams are shown in Figure 3.37(b). An adiabatic chirp of 11 GHz with 2.7 V_{pp} modulation was used, which increased the ER from 1.3 dB to 5 dB after an OSR filter with a 3-dB BW of 38 GHz.

A short-cavity two-section DBR laser with a gain length of 50 μm was made by cleaving the DBR laser described in Section 3.4.5 in order to increase the FM efficiency for 56 Gb/s CML [102] (Figure 3.38). The HR coating was applied on the rear facet at the end of gain section. The light was coupled *through* the DBR section. The peak-to-peak chirp was 10 GHz for a modulation of 1.6 V_{pp}. An F_r of 18 GHz was obtained. While 3-dB AM BW of the short-cavity DBR laser was limited to 20 GHz by RC roll-off, the FM response was extended to 33 GHz. A Finisar WaveShaper (model: 4000S) was used as an OSR filter, which is a programmable optical filter that supports arbitrary filter shapes in both transmission and dispersion. It was possible to improve the eye opening at 56-Gb/s and the ER from 2 dB to 8.1 dB by the FM-AM conversion; however, the phase of the WaveShaper was forced to compensate for the large timing delay between the AM and FM responses at a higher frequency than F_r (18 GHz). As a result, the natural Kramers-Kronig relation between the filter transmission and phase was lost.

The concept of utilization of both AM and FM from DMLs has been extended to generate various modulation formats, including RZ-AMI [103], DPSK [104], RZ-DPSK [105], and 3/4-RZ-DQPSK [106].

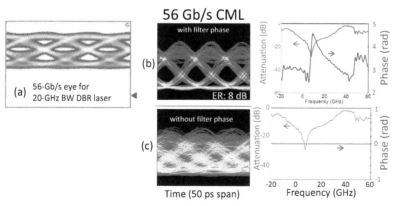

Figure 3.38 (a) 56-Gb/s eye diagram of 20-GHz BW short-cavity DBR laser at 1550 nm, (b) eye diagram after Finisar WaveShaper as an OSR filter with optimized filter phase, (c) without phase control. (after Reference 102)

In Reference [107], both AM and FM were used to improve the OSNR sensitivity of the DML-based coherent system by Maximum likelihood sequence estimation (MLSE). The local oscillator of a simplified coherent receiver without a digital signal processor (DSP) created a similar effect as CML with the VSB effect; however, executed in the electrical domain [108]. A link budget of 43 dB was obtained for a 100-km link, which is attractive for PON applications.

3.6 New Era of High-Speed DML Toward 100-GHz Bandwidth

In the early days of the development of high-speed DMLs, the progress was made mainly through focusing on the improvement of the differential gain by adopting strained MQW and new material systems, e.g., InGaAlAs/InP instead of InGaAsP/InP, which provided tighter confinement of electrons and achieves uniform hole injection. However, it seemed as though there were an invisible ceiling at around a BW of 30 GHz as shown in Figure 3.1. This is partly due to the limitation of material properties of MQWs and partly due to the limitation posed by a parasitic capacitance. In order to overcome these limitations, in this section we discuss three physical effects: 1. Detuned-loading effect, 2. Intra-cavity FM-AM conversion effect, and 3. PPR effect. The detuned-loading effect enhances the effective differential gain and therefore increases F_r by modulating the loss of the DBR mirror, which is in phase with the carrier density modulation [27, 109]. The intra-cavity FM-AM conversion effect also occurs on the flank of the DBR mirror and creates an effective HPF effect on the modulation response through FM-AM conversion, and therefore can counteract the RC roll-off limitation [28]. The PPR effect adds an additional peak in the modulation response at a frequency corresponding to the frequency difference between the main and side modes, which is often located at higher frequency than F_r peak, and therefore can extend the 3-dB BW of the S21 response beyond F_r limit [28–30]. The PPR effect does not improve the F_r by itself, or rather it can even degrade the F_r if the additional extended section to support the PPR effect, in turn, reduces the round-trip differential gain. Therefore, care must be taken in designing the cavity to utilize the PPR effect. For DBR lasers or DR lasers, all the above three effects can happen simultaneously when the lasing mode is located on the long wavelength flank of the DBR mirror. Indeed, the modulation BW around 60 GHz has been reported in spite of the limited RC cut-off frequency of 22 GHz [3]. In the following sections, we will discuss the details of those effects.

3.6.1 Detuned-Loading Effect

The detuned-loading effect on the DBR lasers is illustrated in Figure 3.39. This effect occurs on the flank of the Bragg reflector mirror of DBR lasers or DR lasers, where the chirp of DMLs, as the gain section is modulated, is translated into dynamic changes in the DBR mirror loss and the penetration depth. This can enhance the speed of DMLs beyond the limit of the material properties. The detuned-loading effect was first discussed by Vahala in terms of a quantum noise reduction [27]. It is important to note that such loss modulation is caused by the chirp, which is in phase with the carrier density, and is therefore in phase with the gain as well. Therefore, the DBR mirror loss modulation can be considered as a part of "gain modulation" in the imaginary part of the detuned-loading effect since those two are both in phase. On the other hand, for "Q-factor modulation" proposed in Reference [110], the cavity loss is modulated instead of the current. In this case, the photon density responds to the loss modulation, then, the carrier density (or the gain modulation) follows with a delay. Therefore, the loss modulation and the gain modulation are not in phase. As a result, there is no enhancement effect on the speed of DMLs. This is discussed in Section 3.6.3 in Figures 3.43 and 3.44. The detuned-loading effect can also affect the damping factor [71, 109] and the linewidth enhancement factor [72] as will be discussed later in this section.

Another important part of the detuned-loading effect is the dynamic modulation of the penetration depth into the DBR section or the cavity volume. Such dynamic changes in the cavity loss and the volume are not

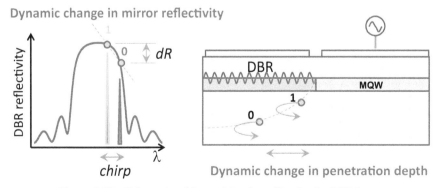

Figure 3.39 Schematic of detuned-loading effect for the DBR laser.

taken into account in a simple set of the rate equations in Equation (3.1). Numerically, these effects can be accurately modeled by solving dynamic coupled travelling wave equations [111, 112], based on the transfer-matrix method [113], on the transmission-line method [114], or on the power-matrix method [115]. It is still very useful to derive the analytical expressions for F_r and damping factor to build an intuitive picture of the detuned-loading effect.

In order to modify the rate equation to include the detuned-loading effect, it is important to note that the "photon density" in the rate equation, Equation (3.1), represents an average photon density in the "cavity"; however, it does not necessary reflect the useful "stored energy" that contributes to the high-speed operation of DMLs [116]. As an example, the simulation software, LaserMatrix, developed by Richard Schatz [71], was used to show the total energy and the stored reactive energy for a DFB laser with an HR coating on the back facet and AR coating on the front facet (Figure 3.40).

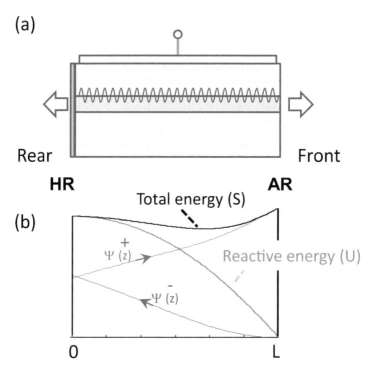

Figure 3.40   (a) Schematic of uniform-grating DFB laser coated with HR and AR on the facets, and (b) distribution of total energy (black), reactive energy (green), forward traveling wave (blue), and backward traveling wave (red).

For simplicity, uniform grating was assumed. Near the HR-coated rear facet, the forward and backward travelling waves have similar intensities and therefore form a very visible standing wave pattern, indicating that the light is well confined or stored. On the other hand, near the AR coated front facet, nearly 100% of the power is in the forward travelling wave, and there is very weak power for the backward travelling wave. Around this area of the DFB section, the dynamics of the carriers and photons are similar to those in SOAs, which is known for a low response speed. To find a ratio between the total photon energy to the "stored reactive energy", the following expressions can be used [71];

$$S = \int \left| \Psi_{(Z)}^+ \right|^2 + \left| \Psi_{(Z)}^- \right|^2 dz, \qquad (3.21(a))$$

$$U = \int 2\Psi_{(Z)}^+ \Psi_{(Z)}^- dz, \qquad (3.21(b))$$

$$V = S - |U|. \qquad (3.21(c))$$

where S is the total energy in the cavity and U is the reactive energy, which relates to the standing wave formed by forward and backward traveling waves. The traveling wave energy can be found by subtracting U from S. The integrations are executed over the whole cavity. Note the total wave function, $\Psi(z)$, can be a summation of the multiple modes if the laser cavity supports multiple modes. For each given Eigen mode, Φ, the ratio between S and U is given by:

$$\frac{S}{|U|} = \frac{\int \left| \Phi_{(Z)}^+ \right|^2 + \left| \Phi_{(Z)}^- \right|^2 dz}{\left| \int 2\Phi_{(Z)}^+ \Phi_{(Z)}^- dz \right|} \equiv \sqrt{K_z}, \qquad (3.22)$$

where K_z is the longitudinal Petermann's factor, also known as the excess spontaneous emission factor [71, 109]. The rate equation for the photons should be written in terms of the useful reactive energy; however, it can be rewritten in terms of the average photon density, S, using Equation (3.22) as:

$$\frac{dS}{dt} = \frac{d\sqrt{K_z}|U|}{dt} = \left[G(t) - \frac{1}{\tau_p(t)} + \frac{1}{2} \frac{dIn(K_z)}{dt} \right] \cdot S(t). \qquad (3.23)$$

Here, G is the gain, and τ_p is the time-dependent photon lifetime. Now, the third term on the right side of the equation is the correction term to the standard rate equation, representing the detuned-loading effect. The dynamic changes in the loss and the penetration depth into the DBR mirror depend on the instantaneous frequency of the lasing mode, which depends on the plasma effect of the free carriers. Therefore, the gain and the third terms are in phase since both follow the dynamics of the carriers. This can effectively enhance the differential gain and therefore the speed of DMLs. Strictly speaking, there is a finite delay in the built-up time of the mode, which can affect the damping of the relaxation oscillation [117].

Based on this modified rate equation including the detuned-loading effect, the expression for the F_r is given by [71]:

$$F_r = \frac{1}{2\pi}\sqrt{\frac{S}{\tau_p}v_g Re\left[\tilde{\Gamma}_z g_N\left(1 + i\alpha_H\right)\right]}, \qquad (3.24)$$

where v_g the group velocity of light, v_N the differential gain, and α_H the Henry's alpha parameter. $\tilde{\Gamma}_z$ is the complex longitudinal confinement factor in the axial direction, which is defined as:

$$\tilde{\Gamma} = \frac{L_a}{L_a + \tilde{\Gamma}_p} \qquad (3.25)$$

where L_a is the length of the active region, and $\tilde{\Gamma}_p$ is the complex effective length of the DBR section:

$$\tilde{L}_p = \frac{1}{2}iv_g dIn\left(R_p\right)/d\Omega, \qquad (3.26)$$

where the R_p is the reflectivity of the DBR mirror and Ω is the complex eigenvalue of the mode, corresponding to the Eigen mode $(\Phi^+, \Phi^-)^t$ in the travelling wave equation [109]. The real part of Ω is the detuning of the angular frequency for the mode from the reference frequency, and the imaginary part gives the decay rate of the mode. $\tilde{L}_p$ is complex valued, and the imaginary part is a measure for the slope of the DBR reflection spectrum. If the detuned-loading effect is ignored, $\tilde{L}_p$ reduces to a standard effective penetration depth of the DBR mirror defined at a 1/e decay point. Without detuned-loading effect, the differential gain is simply diluted due to the passive waveguide in the cavity, and therefore the speed of DMLs degrades. By including the detuned-loading effect, however, F_r can be faster when the slope of the DBR mirror is steep and the chirp of the DML is large. The simulation of S21 response performed by LaserMatrix [71] for the DR laser is shown in Figure 3.41(a). The output power was coupled through the DBR section located near the AR-coated front facet. An F_r peak of around 32 GHz was observed while that for DFB laser was 25 GHz under the same bias condition, which is close to the experimental results [3, 28].

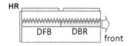

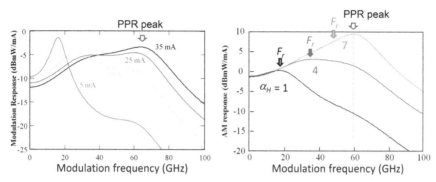

Figure 3.41 (a) Simulated S21 responses of DR laser, and (b) S21 responses for alpha parameters of 1, 4, and 7. The output power is coupled through AR-coated DBR section. RC roll-off was ignored.

The peak around 60 GHz is due to the PPR effect. Figure 3.41(b) shows the S21 responses where the linewidth enhancement factor was varied from 1 to 7. Since the effective differential gain increases with the chirp due to the detuned-loading effect, F_r can be significantly increased as the chirp parameter increases [118].

When the slope of DBR mirror becomes high, the effective damping factor is reduced, and the self-pulsation may result when the damping factor becomes negative [117, 119]. The reduction of the damping factor and self-pulsation for the detuned-loading situation is analogous to the case of saturable absorbers for the mode-locking or self-pulsation where the cavity loss is reduced with the increase of the photon density. The opposite is the gain compression which compresses the gain as the photon density increases, resulting in the damping of the relaxation oscillation (Equation (3.2)). The expression for the damping factor for detuned-loading is given by [71, 109]:

$$\gamma_d = \frac{1}{\tau_d} + \frac{S}{\tau_p}\frac{\partial}{\partial N}In[g\Gamma_z\sqrt{K_z}], \qquad (3.27)$$

where τ_d is the spontaneous carrier lifetime. The material property of the gain compression phenomenon, ε, is ignored in this expression. Since the chirp of DMLs is frequency up-chirp (blue chirp), when the lasing mode is located on the long wavelength side of the Bragg mirror, the penetration depth into the

DBR mirror is reduced with the current modulation. This increases the longitudinal confinement factor, hence $d\Gamma_z/dN > 0$ according to Equation (3.25). On the other hand, the increase of the Bragg mirror reflectivity increases the reactive photon energy, thus reduces the Petermann factor according to Equation (3.22). Therefore, $dK_z/dN < 0$ on the long wavelength side of the Bragg mirror. As a result, the damping factor is reduced on the long wavelength side of the Bragg reflector. In an extreme case, the damping factor can have a negative value; then pulsation may result.

The detuned-loading effect can also modify the linewidth enhancement factor. The expression is given below [72]:

$$\alpha_{eff} = \alpha \frac{\tau_L + \frac{d\phi_r}{d\omega} - \frac{1}{\alpha|r_m|}\frac{d|r_m|}{d\omega}}{\tau_L + \frac{d\phi_r}{d\omega} + \frac{\alpha}{|r_m|}\frac{d|r_m|}{d\omega}}, \qquad (3.28)$$

where τ_L is the cavity roundtrip time and α is the linewidth enhancement factor when the detuning from the Bragg reflector peak is zero. According to Equation (3.28), the detuned-loading effect predicts the smaller linewidth enhancement factor on the long wavelength side of DBR mirror when the slope of Bragg mirror $d|r_m|/d\omega > 0$. This is the case when the α parameter in the gain section determines the spectral linewidth. If the tuning current is injected into the DBR section, and the shot noise for the carrier recombination dominates the spectral linewidth, then the spectral linewidth can be broader on the long wavelength side of the DBR mirror. This effect is pointed out in the paper [120].

The consequence of a large effective alpha parameter on the short wavelength side of DBR mirror indicates that the roundtrip group delay can be negative. This can cause an instability of the mode, and the hysteresis at the mode hop condition of the DBR laser. The roundtrip delay for the DBR laser is given by;

$$\frac{d\varphi_{roundtrip}}{d\omega} = \frac{d}{d\omega}\left[2\frac{\omega}{c}n_a\left(N_{th}\left(\omega\right)\right)L_a + \arg\left(R_{DBR}\right)\right], \qquad (3.29)$$

where the refractive index of the active region n_a is given as a function of the threshold carrier density in the active region, which depends on the DBR mirror reflectivity at the lasing mode frequency [117]. This hysteresis behavior is a linear phenomenon and needs to be distinguished from the hysteresis of DBR mode hop [78] caused by the asymmetric cross-gain compression [75]. On the other hand, on the long wavelength side of the DBR mirror, the stability of the mode improves. When the lasing mode is pulled toward a longer wavelength by perturbations, for example, random noise or an optical feedback, the DBR mirror reflectivity is reduced, and then the threshold carrier density increases as a result. This pushes the lasing mode back to the shorter wavelength. Therefore, the mode on the long wavelength side of DBR mirror becomes more stable.

3.6.2 S21 High-Pass Filter Effect Due to In-Cavity FM-AM Conversion by the DBR Mirror

For a conventional DFB laser, the modulation responses for the light coupled from the front and rear facets should be identical. For DBR lasers or DR lasers, this is not the case because of the FM-AM conversion effect in the Bragg reflector. Since the FM response of DMLs is faster than the AM response, the high-speed performance of DMLs can be improved by FM-AM conversion. Also, FM-AM conversion can act as an HPF for an S21 response. This effect was first discussed in Reference [28] and is similar to the CML effect; however, this happens in the cavity of the DML through the slope of the DBR mirror. Figure 3.42 illustrates the in-cavity FM-AM conversion effect. When the modulation frequency is low, both AM and FM modulations

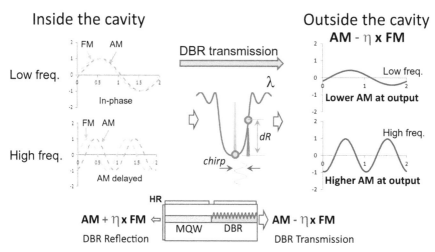

Figure 3.42   High-pass filter (HPF) effect in the modulation response of DML created by in-cavity FM-AM conversion for DBR laser coupled through DBR mirror transmission. At low (high)-frequency modulation, FM and AM responses are (not) in phase. The negative slope of DBR transmission reduces the AM response at low frequency while AM remains high at high frequency, creating HPF effect in S21 response.

can follow the current modulation with a small timing delay; therefore, the AM and FM modulations are in phase and the higher intensity in the AM modulation corresponds to the high frequency. In this case, the FM-AM conversion effect reduces the AM modulation depth after the transmission through the DBR mirror on the long wavelength side. On the other hand, for a higher modulation frequency than F_r, the phase of FM modulation is $\pi/2$ in advance to AM modulation as shown in Figure 3.42, and therefore the converted FM does not reduce the AM modulation depth as effectively as it does at the low frequency. Overall, the FM-AM conversion effect creates an effective HPF effect on the S21 response. This can enhance the modulation BW of DMLs at the expense of the lowered ER. The opposite is true for the DBR reflection side of the output.

It is known that the FM response for DMLs modulated by the injection current is faster than that for the AM response. On the other hand, for the case of "loss modulation", the AM response is faster than the FM response [121]. Figure 3.43(a) and (b) compare the AM and FM responses and the phase for the current and in-cavity loss modulated lasers, respectively. As shown, the AM (FM) response of the loss modulated laser is the same as the FM (AM) response of the current modulated laser. The 3-dB BW for the FM (AM) response is wider than that for the AM (FM) response for the current (loss) modulation. There is no enhancement effect for F_r for the pure loss modulation. In the detuned-loading situation, the internal loss is dynamically changing; however, this should be distinguished from the "loss modulation". Indeed, it was experimentally confirmed that the FM response for a DR laser with a strong detuned-loading effect is

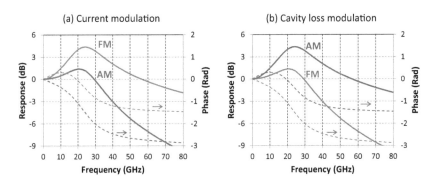

Figure 3.43 AM and FM responses and phase for (a) current modulated DML, and (b) in-cavity loss modulated laser.

faster than the AM response [28], as shown later in Figure 3.51(b), which is the same as the current modulation scheme. Therefore, the FM-AM conversion of the detuned-loaded DMLs can enhance the modulation BW. The importance of the relative delay between the current modulation and the in-cavity loss modulation in the hybrid modulation scheme was discussed by Mieda [122].

Figure 3.44 shows the AM and FM responses of the two-section DR laser for the outputs through the front DBR section and the rear gain section which is the reflection of the front DBR mirror. The simulation was performed by LaserMatrix [71]. The response peaks at around 30 GHz and 50 GHz are associated with the F_r and PPR, respectively. The AM and FM responses

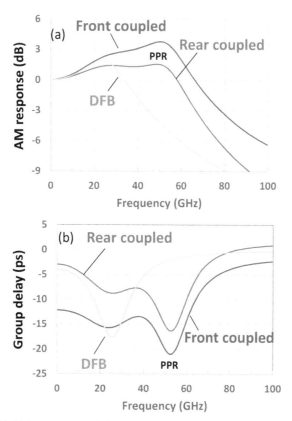

Figure 3.44 (a) AM responses, and (b) group delays for DFB laser (green), DR laser coupled through DBR transmission (front coupled), and DBR reflection (rear coupled). RC roll-off was ignored.

for a standard DFB laser are also shown for comparison. As experimentally confirmed, the FM response is faster than the AM response for the detuned-loaded DR laser. It can be seen that the AM response through the DBR section is faster than that for the output from the gain facet (DBR reflection) due to the FM-AM conversion effect. It is also interesting to note that the corresponding group delay ripples in Figure 3.44(b) are suppressed for the DR lasers below 40 GHz for the output through the DBR section. This should have an advantage in reducing the ISI effect in the eye diagrams under a large signal modulation.

The time domain large signal simulations including the detuned-loading effect and in-cavity FM-AM conversion effect are shown in Figure 3.45(a) for DBR reflection and (b) for DBR transmission. On the DBR transmission side, the ER is low compared to the DBR reflection side, as a consequence of the "HPF" effect since the low frequency component is suppressed. In the time domain waveform, the HPF effect can be seen as overshooting or undershooting at the bit transitions. What is not commonly seen for a conventional DMLs is the feature where the overshooting (undershooting) *before* the falling (rising) edge is observed for the output of the DBR trans-mission. This is because the chirp is in advance of the intensity modulation and the undershooting (overshooting) in the chirp at the falling (rising) edge is converted to the overshooting (undershooting) before the intensity starts to fall (rise). This results in the upward shift of the cross-point. Such shift in the cross-point was evidenced in the experiment in Reference [28].

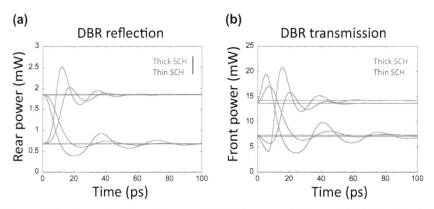

Figure 3.45 Large signal eye diagrams for DR laser coupled out of (a) DBR reflection, and (b) DBR transmission. Adiabatic chirp is higher for thick SCH (blue) compared to thin SCH (red).

3.6.3 Photon-Photon Resonance Effect

When a DML is modulated, the modulation sidebands broaden the spectrum around the main lasing mode. If the side mode of the laser cavity is present within the modulation spectrum, such sidebands can be coupled into the side mode and be resonantly amplified. This situation is depicted in Figure 3.46. This effect is called the photon-photon resonance (PPR) effect [29] and can enhance the modulation response at around a frequency corresponding to the frequency difference between the main and side modes [109].

This effect is different from a mode beating effect between the main and side modes, in which case two modes are simultaneously lasing in the CW operation, and the phase relation between the two modes could be random. The PPR effect is also different from a "mode locking" operation, where the FP modes are, again, all lasing under the CW operation and equally spaced at ω_1, ω_2, and ω_3. In this case, FWM generated at $2\omega_1-\omega_2$ can act as a seed to achieve a CW injection locking of the mode at a frequency ω_3. It takes some build up time for the modes to be injection locked. In the time domain picture, the phase locking among FP modes is achieved by a mode

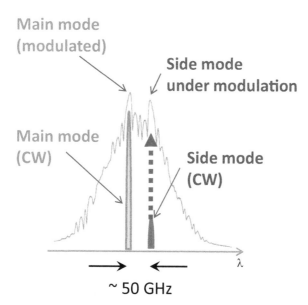

Figure 3.46 Simulated modulation spectrum of DR laser with PPR effect at 50 GHz. VPIcomponentMaker was used for simulation. Main and side modes under CW operation are shown schematically.

locker, for example, a saturable absorber, and needs many round trips to shape the noise into the pulse train [123, 124]. As a result, the response of the mode-locking is narrow-banded and, therefore, is not suitable for broadband communication.

On the other hand, the PPR effect is considered as a modulation sideband injection to the side mode, and amplification of the sidebands by the presence of the side mode. Therefore, the phase relation is naturally correlated. Even when the side mode is well suppressed in the CW spectrum, under the mod-ulation, the sideband injection can excite the side mode, and then improve the S21 response. This is somewhat similar to the case of the relaxation oscillation; in the CW optical spectrum, the peak associated with F_r is very weak (< 40 dB) or hardly be seen on an OSA spectrum. However, it is well known that the relaxation oscillation is a dominant factor for the dynamics of DMLs.

The expressions describing the AM modulation efficiency improvement by the PPR effect related to the high-frequency side mode are given by [117, 125];

$$\Delta S_{m,+\omega} = \nu_g \left|U_0\right| Re \left[\frac{\omega \left\langle \Phi_m \Phi_0^* \right\rangle_a \left\langle \Phi_m \mid \Phi_0 \right\rangle_a}{(\omega - \Omega_m)\,\Omega_m} (1 + i\alpha)\, \frac{\partial g}{\partial N} \Delta N_w e^{j\omega t} \right],$$

(3.30(a))

$$\left\langle \Phi_m \mid \Phi_0 \right\rangle_a = \int_{active} \Phi_m^+ \Phi_0^- + \Phi_0^+ \Phi_m^- dz,$$

(3.30(b))

$$\left\langle \Phi_m \mid \Phi_0^* \right\rangle_a = \int_{active} \Phi_m^+ \Phi_0^{+*} + \Phi_0^{-*} \Phi_m^- dz,$$

(3.30(c))

where U_0 is the stored energy for the main mode, Φ_0 and Φ_m are the longi-tudinal mode profiles of the main and side modes in the cavity, respectively, Ω_m is the frequency of the side mode relative to that of the main mode, α is the alpha parameter, $\partial g/\partial N$ is the differential gain, and ∂N_w is the carrier density modulation. The inner product in Equation (3.30(b)) is zero for the case of FP lasers due to the mode orthogonality; therefore, the PPR effect does not exist for FP lasers. In contrast, in multi-section DBR or DR lasers, the main and the side modes are not necessarily orthogonal; therefore, the PPR effect can enhance the modulation BW as the spatial overlap between the main and the side modes increases. For this purpose, DBR lasers [30, 126], or complex-cavity designs, including DR lasers [3, 127], two-section DFB lasers [128, 129], an active-feedback laser [130], a passive-feedback laser [131], and a coupled-cavity injection grating (CCIG) laser [132] were proposed. The expression in Equation (3.30) also shows that larger chirp for the gain section increases the PPR effect since it effectively broadens the spectrum. Therefore, once again, the large alpha parameter can increase the BW of DMLs in the framework of the PPR effect as well as the detuned-loading effect, as shown in Figure 3.41(b).

A PFL laser is well-known structure exploiting the PPR effect. It consists of a DFB section with an integrated phase section ($\sim$150 μm length), which provides an optical feedback from the HR coated facet reflection at the end of the phase section (Figure 3.47). The PFLs have been demonstrated both at 1.55 μm [45] and 1.3 μm [46]. A very strong complex-coupled grating is etched into InGaAsP MQW in order to realize a good mode stability under a strong optical feedback. A PPR frequency at 30 GHz was obtained, while F_r of the DFB section was only 12 GHz [46] (Figure 3.47(c)). By adjusting the phase condition for the feedback, a flat S21 response was achieved, as shown in Figure 3.47(b) for experiment and (c) for simulation. Understanding the external-cavity mode position and hysteresis behavior have been well established [133, 134]; however, the presence of the detuned-loading effect in PFL lasers may need further investigation. In Figure 3.47(b), the phase

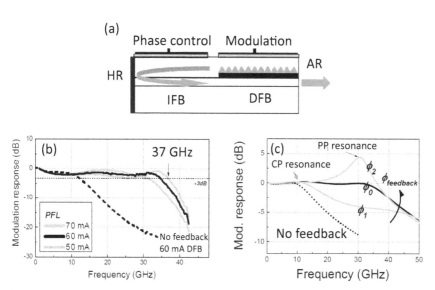

Figure 3.47 (a) schematic picture of passive-feedback laser (PFL), (b) measured S21 responses, and (c) simulated S21 responses for three different feedback phase conditions. (after Reference 46).

condition which gives the faster speed coincides with the low FM efficiency condition [46]. The trend seems to agree with the small α-parameter predicted by the detuned-loading effect in Equation (3.28).

3.6.4 Co-Existence of Photon-Photon Resonance and Detuned-Loading Effects

For DBR lasers or DR lasers, both detuned-loading and PPR effects can coexist. An example of the cavity mode positions on the two-section DBR laser exploiting both detuned-loading and PPR effects is shown in Figure 3.48 [125]. The main lasing mode is indicated by a red circle located on the long wavelength side of the DBR mirror; therefore, the detuned-loading effect is expected. The PPR mode is located at the bottom of the DBR mirror

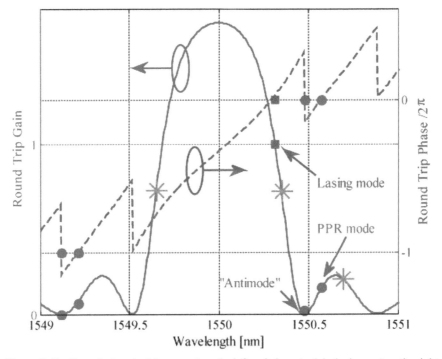

Figure 3.48   Round trip gain (blue curve) on the left and phase (red dashed curve) on the right at the DBR laser threshold. The squared red marker represents the lasing mode; the blue markers indicate non-lasing cavity modes. The green asterisks on the reflectivity curve represent the modes locations in the maximum detuned loading condition. (after Reference 125).

in the ripple, separated by 35 GHz from the main mode in this example. The output power was coupled from the gain facet with a 30% reflection. In this simulation study, it was found that the optimum length of the DBR section is 800 μm for a gain length of 120 μm. Similarly, Feiste and Morthier predicted by simulations that 55 GHz–70 GHz modulation BW is achievable using the optimum DBR length of 400 μm–700 μm for a gain length of 100 μm–120 μm, assuming a linewidth enhancement factor of 5 [109, 135]. Relatively long DBR length was reported to be an optimum in those simulations.

The PPR effect in a DBR laser was experimentally observed in Reference [30], where a 3-dB BW of 30 GHz was obtained. The light was coupled from the gain facet. The same group also showed a 26 GHz 3-dB BW by using a 100 μm–200 μm gain length and a 300-μm DBR length [136]. The PPR frequency was 18 GHz. Figure 3.49 shows the S21 responses comparing the lasing on the long and short wavelength sides of the DBR mirror. A clear benefit of the detuned-loading effect can be seen on the long wavelength side. In a more recent study, 27 GHz BW was achieved by a DR laser fabricated on an identical MQW structure used for both DFB and DBR sections [127]. The PPR effect enhanced the response at 20 GHz. The lengths of DFB and DBR sections were 220 μm and 380 μm, respectively.

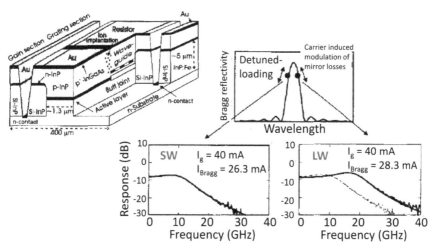

Figure 3.49 Schematic picture of a two section DBR laser with GSG contact configuration, (b) Small signal modulation response of a DBR laser lasing on the long (right) and short (left) wavelength sides of the Bragg reflection peak. (after Reference 136).

3.6.5 55-GHz Bandwidth Short-Cavity DR Laser and 56 Gbaud PAM4 Generation

Most of DBR lasers or DR lasers reported so far coupled the light through the facet on the gain section, or the reflection from the DBR section. In this case, the FM-AM conversion effect creates an effective LPF effect on the S21 response and therefore limits the modulation bandwidth (Figure 3.42). Matsui demonstrated 3-dB bandwidth of 55 GHz BW at 25°C and 50 GHz at 50°C for a DR laser by using a very short DFB length of 50 μm, integrated with a 200 μm-long DBR section (Figure 3.50) [3, 28]. The BH structure employed a PN blocking structure, resulting in a limited RC cutoff frequency of only 22 GHz. However, the RC limitation was overcome by coupling the light through 200 μm-length DBR section in order to exploit the HPF effect by the FM-AM conversion. The PPR effect was created between the DFB and DBR modes located on the long wavelength flank of the DBR mirror to utilize the detuned-loading effect. The short device length ensured the mode stability even when the lasing happens on the edge of the DBR mirror. As a result of the three combined effects, 55 GHz 3-dB BW was obtained at a 37 mA bias, together with a high F_r of 30 GHz and a PPR peak at 50 GHz, as shown in Figure 3.51(a). The plot also shows a subtracted S21 response between the biases at 36.2 mA and 30 mA in order to remove the extrinsic effect [137], which shows a smooth intrinsic response without a dip. The FM and AM responses were measured with an Advantest chirp tester (Q7606B) as shown in Figure 3.51(b). The FM bandwidth approaches to 65 GHz and is

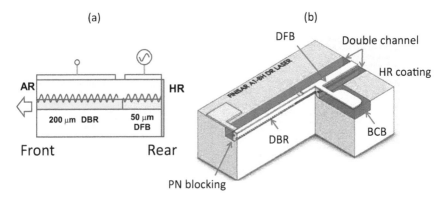

Figure 3.50 (a) Schematic side view of two-section DR laser where the light is coupled through DBR transmission, and (b) corresponding DML structure with PN blocking with double-channel stripe filled with BCB. (after Reference 3).

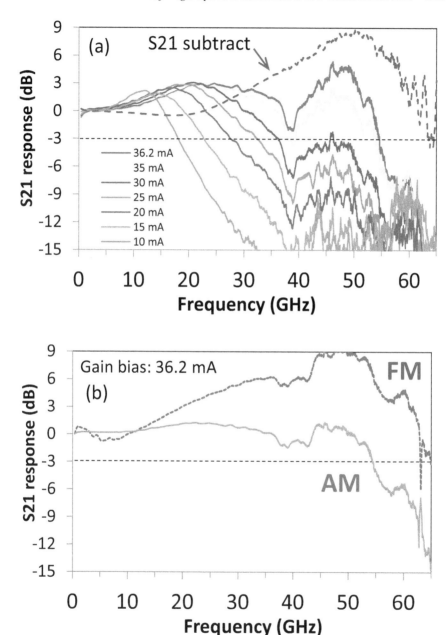

Figure 3.51 (a) S21 response of DR laser for a bias range from 10 mA to 36.2 mA. Subtracted S21 response between 36.2 mA and 30 mA is shown in red broken line, (b) AM and FM response at 36.2 mA bias.

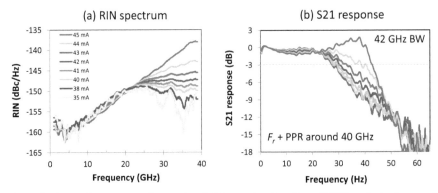

Figure 3.52 (a) RIN spectra for another DR laser showing F_r and PPR peaks both around 40 GHz at 45 mA, and (b) corresponding S21 response.

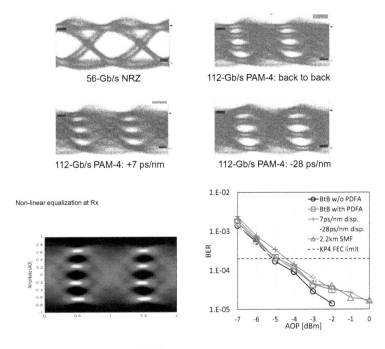

Figure 3.53 (a) 56-Gb/s NRZ, (b) 112-Gb/s PAM4 back-to-back, (c) transmission with 7 ps/nm fiber dispersion (~3 km at 1330 nm), and (d) –28 ps/nm fiber dispersion, (e) BER for BtB, positive (7 ps/nm) and negative dispersion (–28 ps/nm). The inset is the eye after non-linear equalization.

faster than the AM response, as expected for DMLs with the detuned-loading effect. A RIN spectrum for a DR laser chip with a slightly different design having a PPR peak at 38 GHz is shown in Figure 3.52(a), using a SYCATUS RIN tester (A0010A), and the corresponding S21 response in Figure 3.52(b). The RIN degrades when the frequencies of F_r and PPR are close and resonate each other (~40 GHz); however, it still remained below −140 dBc/Hz for a bias condition that achieved a flat S21 response (44 mA).

In Figure 3.53, eye diagrams for 56-Gbit/s NRZ, 56-Gbaud PAM4 in the BtB, over a fiber dispersion of +7 ps/nm, and −28 ps/nm are shown. No equalizer (EQ) was used on the Tx side. The ER was 4 dB. As predicted in Section 3.6.2, the PAM4 eye showed slightly closed eye height on the top compared to the eye at the bottom, due to the in-cavity FM-AM conversion effect. The BER curve is shown after non-linear EQ on the Rx side. A good eye opening and an average optical power of −4.5 dBm is achieved at a KP4 FEC level for a dispersion range equivalent to 2 km for CWDM and 10 km for LAN-WDM [3].

3.7 Conclusions

In this chapter, the importance of a chirp parameter enhancement in DMLs, stemming from, for example, the carrier transport effect in SCH layers, and resultant high FM response were discussed. This plays an important role not only in the reach extension by CML, but also for realizing a DML BW beyond 50 GHz through the detuned-loading effect, in-cavity FM-AM conversion, and PPR effects. In more recent result, a BW of 63 GHz was achieved as a result of the three combined effects.

A gain material or a cavity geometry which simultaneously and effectively realizes a large differential gain and large linewidth enhancement factor may further push the envelope of the BW toward 100 GHz. This may call for the realization of a strong optical confinement factor, for example, by using a high index contrast between the semiconductor and air, or a PhC cavity, while suppressing the thermal effect.

For extending the reach of DMLs, the emphasis has been on the fact that the dispersion limit of DMLs is *not* determined by the linewidth enhancement factor per se, but rather a suppression of the transient chirp is more important for both DML and CML. As more channels are added in CWDM systems for a 10-km distance or even longer for the data center applications, achieving both high BW and dispersion tolerance will be a crucial consideration for the future of DML technology. On the other hand, for a parallel systems, for example, based on multi-core fibers, the advantages of DMLs could be further exploited.

Acknowledgements

The author thanks Prof. Richard Schatz at KTH for very fruitful discussions and long-term support on his simulation software LaserMatrix.

References

[1] 100G Ethernet http://www.ieee802.org/3/ba/index.html

[2] 400G Ethernet http://www.ieee802.org/3/bs/timeline_3bs_0915.pdf

[3] Y. Matsui, T. Pham, W. A. Ling, R. Schatz, G. Carey, H. Daghighian, T. Sudo, C. Roxlo, "55-GHz bandwidth short-cavity distributed reflector laser and its application to 112-Gb/s PAM-4," Optical Fiber Communications Conference and Exhibition (OFC), PDP paper Th5B.4, 2016.

[4] Y. Arakawa, and A. Yariv, "Quantum well lasers - Gain, spectra, dynamics," IEEE Journal of Quantum Electron., vol. 22, pp. 1887–1899, 1986.

[5] E. Yablonovitch and E.O. Kane, "Band structure engineering of semiconductor lasers for optical communications," IEEE J. Lightwave Technol., vol. 6, pp. 1292–1299, 1988.

[6] Shun Lien Chuang, "Physics of Photonic Devices," Wiley, ISBN: 978-0-470-29319-5.

[7] E. Meland, R. Holmstrom, J. Schlafer, R.B. Lauer, and W. Powazinik, "Extremely high-frequency (24 GHz) InGaAsP diode lasers with excellent modulation efficiency," Electron. Lett., vol. 26, pp. 1827–1829, 1990.

[8] R. Olshansky, W. Powazinik, P. Hill, V. Lanzisera, and R.B. Lauer, "InGaAsP buried heterostructure laser with 22 GHz bandwidth and high modulation efficiency," Electron. Lett., vol. 23, pp. 839–841, 1987.

[9] J.E. Bowers, U. Koren, B.I. Miller, C. Soccolich, and W.Y. Jan, "High-speed, polyimide-based semi-insulating planar buried heterostructures," Electron. Lett., vol. 23, pp. 1263–1264, 1987

[10] M.C. Tatham, I.F. Lealman, C.P. Seltzer, L.D. Westbrook, and D.M. Cooper, "Resonance frequency, damping, and differential gain in

1.5 μm multiple quantum-well laser," IEEE J. Quantum Electron., vol. 28, pp. 408–414, 1992.

[11] I.F. Lealman, D.M. Cooper, S.D. Perrin, and M.J. Harlow, "Effect of Zn doping on differential gain and damping of 1.55 μm InGaAs/InGaAsP MWQ lasers," Electron. Lett., vol. 28, pp. 1032–1034, 1992.

[12] K. Uomi, M. Aoki, T. Tsuchiya, and A. Takai, "Dependence of high-speed properties on the number of quantum wells in 1.55 μm InGaAs-InGaAsP MQW lambda/4-shifted DFB lasers," IEEE J. Quantum Electron., vol. 29, pp. 355–360, 1993.

[13] G. N. Childs, S. Brand, and R. A. Abram, "Intervalence band absorption in semiconductor laser materials," Semiconductor Science and Technology, vol. 1, Number 2, pp. 116–120, 1986.

[14] L. Chiu, and A. Yariv, "Auger recombination in quantum-well InGaAsP heterostructure laser," IEEE J. Quantum Electron., vol. 18, pp. 1406–1409, 1982.

[15] H. Yasaka, K. Takahata, and M. Naganuma, "Measurement of gain saturation coefficients in strained-layer multiple quantum-well distributed feedback lasers," IEEE J. Quantum Electron., vol. 28, pp. 1294–1304, 1992.

[16] J. Shimizu, H. Yamada, S. Murata, A. Tomita, M. Kitamura, and A. Suzuki, "Optical-confinement-factor dependencies of the K factor, differential gain, and nonlinear gain coefficient for 1.55 μm InGaAs/InGaAsP MQW and strained-MQW lasers," IEEE Photon. Technol. Lett., vol. 3, pp. 773–776, 1991.

[17] P.A. Morton, R.A. Logan, T. Tanbun-Ek, P.F. Sciortino, A.M. Sergent, R.K. Montgomery, and B.T. Lee, "25 GHz bandwidth 1.55 μm GaInAsP p-doped strained multiquantum-well lasers," Electron. Lett., vol. 28, pp. 2156–2157, 1992.

[18] Y. Matsui, H. Murai, S. Arahira, S. Kutsuzawa, and Y. Ogawa, "30-GHz bandwidth 1.55-μm strain-compensated InGaAlAs-InGaAsP MQW laser," IEEE Photon. Technol. Lett., vol. 9, pp. 25–27, 1997.

[19] J.D. Ralston, S. Weisser, I. Esquivias, E.C. Larkins, J. Rosenzweig, P.J. Tasker, and J. Fleissner, "Control of differential gain, nonlinear gain and damping factor for high-speed application of GaAs-based MQW lasers," IEEE J. Quantum Electron., vol. 29, pp. 1648–1659, 1993.

[20] S. Weisser, E.C. Larkins, K. Czotscher, W. Benz, J. Daleiden, I. Esquivias, J. Fleissner, J.D. Ralston, B. Romero, R.E. Sah, A. Schonfelder, and J. Rosenzweig, "Damping-limited modulation bandwidths up to 40 GHz in undoped short-cavity $In_{0.35}/Ga_{0.65}/As$-GaAs multiple-quantum-well lasers ", IEEE Photon. Technol. Lett., vol. 8, pp. 608–610, 1996.

[21] Y. Matsui, W. Li, H. Roberts, H. Bulthuis, H. Deng, L. Lin, and C. Roxlo, "Transceiver for NG-PON2: Wavelength tunablity for burst mode TWDM and point-to-point WDM," Optical Fiber Communications Conference and Exhibition (OFC), Invited paper Tu2C.1, 2016.

[22] T. Sakamoto, T. Kawanishi, and M. Izutsu, "Optical minimum-shift keying with external modulation scheme," Optics Express, vol. 13, pp. 7741–7747, 2005.

[23] K. Yonenaga, and S. Kuwano, "Dispersion-tolerant optical transmission system using duobinary transmitter and binary receiver," IEEE J. Lightwave Technol., vol. 15, pp. 1530–1537, 1997.

[24] Y. Matsui, D. Mahgerefteh, Xueyan Zheng, C. Liao, Z.F. Fan, K. McCallion, and P. Tayebati, "Chirp-managed directly modulated laser (CML)," IEEE Photon. Technol. Lett., vol. 18, pp. 385–387, 2006

[25] D. Mahgerefteh, C. Liao, X. Zheng, Y. Matsui, B. Johnson, D. Walker, Z.F. Fan, K. McCallion, and P. Tayebati, "Error-free 250 km transmission in standard fibre using compact 10 Gbit/s chirp-managed directly modulated lasers (CML) at 1550 nm," Electron. Lett., vol. 41, pp. 543–544, 2005.

[26] B. Wedding, B. Franz, and B. Junginger, "10-Gb/s optical transmission up to 253 km via standard single-mode fiber using the method of dispersion-supported transmission," IEEE J. Lightwave Technol., vol. 12, pp. 1720–1727, 1994.

[27] K. Vahala, and A. Yariv, "Detuned loading in coupled cavity semiconductor lasers—effect on quantum noise and dynamics," Appl. Phys. Lett., vol. 45, p. 501, 1984.

[28] Y. Matsui, R. Schatz, T. Pham, W. A. Ling, G. Carey, H. M. Daghighian, D. Adams, T. Sudo, and C. Roxlo, "55 GHz Bandwidth Distributed Reflector Laser," IEEE J. Lightwave Technol., vol. 35, pp. 397–403, 2017 (Invited).

[29] A. A. Tager, and B. B. Elenkrig, "Stability Regimes and High-Frequency Modulation of Laser Diodes with Short External Cavity," IEEE J. Quantum Electron., vol. 29. pp. 2886–2890, 1993.

[30] O. Kjebon, R. Schatz, S. Lourdudoss, S. Nilsson, B. Stalnacke, and L. Backbom, "30 GHz direct modulation bandwidth in detuned loaded InGaAsP DBR lasers at 1.55 μm wavelength," Electron. Lett., vol. 33, pp. 488–489, 1997.

[31] Steven H. Strogatz, "Nonlinear Dynamics and Chaos: With Applications to Physics, Biology, Chemistry, and Engineering," Westview Press, ISBN-10: 0813349109.

[32] H. Yamazaki, A. Tomita, and M. Yamaguchi, "Evidence of nonuniform carrier distribution in multiple quantum well lasers," Appl. Phys. Lett. Vol. 71, pp. 767–769, 1997.

[33] I. Tomkos, I. Roudas, R. Hesse, N. Antoniades, A. Boskovic, and R. Vodhanel, "Extraction of laser rate equations parameters for representative simulations of metropolitan-area transmission systems and networks," Optics Communications, vol. 194, pp. 109–129, 2001.

[34] M.C. Tatham, C.P. Seltzer, S.D. Perrin, and D.M. Cooper, "Frequency response and differential gain in strained and unstrained InGaAs/InGaAsP quantum well lasers," Electron. Lett., vol. 27, pp. 1278–1280, 1991.

[35] M. Aoki, Uomi, T. Tsuchiya, M. Suzuki, and N. Chinone, "Enhanced relaxation oscillation frequency and reduced nonlinear K-factor in InGaAs/InGaAsP MQW l/4-shofted DFB lasers," Electron. Lett., vol. 26, pp. 1841–1843, 1990.

[36] K. Kikuchi, M. Kakui, C.E. Zah, and T.P. Lee, "Differential gain and linewidth enhancement factor of 1.5-μm multiple-quantum-well active layers with and without biaxially compressive strain," IEEE Photon. Technol. Lett., vol. 3, pp. 314–317, 1991.

[37] Y. Matsui, H. Murai, S. Arahira, Y. Ogawa, and A. Suzuki, "Novel design scheme for high-speed MQW lasers with enhanced differential gain and reduced carrier transport effect," IEEE J. Quantum Electron., vol. 34, pp. 2340–2349, 1998.

[38] Y. Matsui, H. Murai, S. Arahira, Y. Ogawa, and A. Suzuki, "Enhanced modulation bandwidth for strain-compensated InGaAlAs-InGaAsP MQW lasers," IEEE J. Quantum Electron., vol. 34, pp. 1970–1978, 1998.

[39] Y. Matsui, S. Kutsuzawa, S. Arahira, and Y. Ogawa, "Generation of wavelength tunable gain-switched pulses from FP MQW lasers with external injection seeding," IEEE Photon. Technol. Lett., vol. 9, pp. 1087–1089, 1997.

[40] Y. Matsui, S. Kutsuzawa, S. Arahira, Y. Ogawa, and A. Suzuki, "Bifurcation in 20-GHz gain-switched 1.55-μm MQW lasers and its control by CW injection seeding," IEEE J. Quantum Electron., vol. 34, pp. 1213–1223, 1998.

[41] Y. Matsui, M.D. Pelusi, and A. Suzuki, "Generation of 20-fs optical pulses from a gain-switched laser diode by a four-stage soliton compression technique," IEEE Photon. Technol. Lett., vol. 11, pp. 1217–1219, 1999.

[42] N. Akram, C. Silfvenius, O. Kjebon, and R. Schatz, "Design optimization of InGaAsP–InGaAlAs 1.55 μm strain-compensated MQW lasers for direct modulation applications," Semiconductor Science and Technology, vol. 19, pp. 418–421, 2004.

[43] N. Akram, O. Kjebon, S. Marcinkevicius, R. Schatz, J. Berggren, F. Olsson, and S. Lourdudoss, "The effect of barrier composition on the vertical carrier transport and lasing properties of 1.55-μm multiple quantum-well structures," IEEE J. Quantum Electron., vol. 42, pp. 713–724, 2006

[44] M. Möhrle, L. Morl, A. Sigmund, A. Suna, F. Reier, and H. Roehle, "InGaAsP/InGaAlAs 1.55 μm strain-compensated MQW BH lasers with 12.5 GHz cut-off frequency at 90°C," IEEE International Semiconductor Laser Conference, Conference Digest. Paper WB2, 2004.

[45] J. Kreissl, V. Vercesi, U. Troppenz, T. Gaertner, W. Wenisch, and M. Schell, "Up to 40 Gb/s Directly Modulated Laser Operating at Low Driving Current: Buried-Heterostructure Passive Feedback Laser (BH- PFL)," IEEE Photon. Technol. Lett., vol. 24, pp. 362–364, 2012.

[46] U. Troppenz, J. Kreissl, M. Möhrle, C. Bornholdt, W. Rehbein, B. Sartorius, I. Woods, and M. Schell, "40 Gbit/s directly modulated lasers: physics and application, "Proceedings of the SPIE, vol. 7953, id. 79530F, 2011.

[47] S. Matsuo, T. Fujii, K. Hasebe, K. Takeda, T. Sato, and T. Kakitsuka, "Directly Modulated DFB Laser on SiO2/Si Substrate for Datacenter Networks," IEEE J. Lightwave Technol., vol. 33, pp. 1217–1222, 2015.

[48] T. Kakitsuka, T. Fujii, K. Takeda, H. Nishi, T. Sato, K. Hasebe, T. Tsuchizawa, T. Yamamoto, K. Yamada, and S. Matsuo, "Lateral Current-injection Membrane Lasers Fabricated on a Silicon Substrate," NTT Technical Review, Vol. 14, No. 1, Jan. 2016.

[49] S. Matsuo, A. Shinya, T. Kakitsuka, K. Nozaki, T. Segawa, T. Sato, Y. Kawaguchi, and M. Notomi, "High-speed ultracompact buried

heterostructure photonic-crystal laser with 13 fJ of energy consumed per bit transmitted," Nature Photonics, 4, pp. 648–654, 2010.

[50] K. Takeda, T. Sato, A. Shinya, K. Nozaki, W. Kobayashi, H. Taniyama, M. Notomi, K. Hasebe, T. Kakitsuka, and S.Matsuo, "Few-fJ/bit data transmissions using directly modulated lambda-scale embedded active region photonic-crystal lasers," Nature Photonics 7, pp. 569–575, 2013.

[51] T. Kitatani, J. Kasai, Nakahara, K. Adachi, and M. Aoki, "High-performance GaInNAs Long-wavelength Lasers," in *Conf. Indium Phosphide Relat. Mater.* (IPRM'07), Techn. Digest, pp. 354–357, 2007.

[52] W. Kobayashi, T. Ito, T. Yamanaka, T. Fujisawa, Y. Shibata, T. Kurosaki, M. Kohtoku, T. Tadokoro, and H. Sanjoh, "50-Gb/s Direct modulation of a 1.3-μm InGaAlAs-Based DFB laser with a ridge waveguide structure," IEEE J. Sel. Top. Quantum Electron., vol. 19, Art. no. 1500908, 2013.

[53] K. Nakahara, T. Tsuchiya, S. Tanaka, T. Kitatani, K. Shinoda, T. Taniguchi, T. Kikawa, E. Nomoto, S. Fujisaki, M. Kudo, M. Sawada, T. Yuasa, and M. Mukaikubo, "115°C, 12.5-Gb/s direct modulation of 1.3-μm InGaAlAs-MQW RWG DFB laser with notch-free grating structure for datacom applications," Optical Fiber Communications Conference and Exhibition (OFC), Techn. Digest, paper PD-40, 2003.

[54] K. Nakahara, Y. Wakayama, T. Kitatani, T. Taniguchi, T. Fukamachi, Y. Sakuma, S. Tanaka, "Direct modulation at 56 and 50 Gb/s of 1.3 μm InGaAlAs ridge-shaped-BH DFB lasers," IEEE Photon. Technol. Lett., vol. 27, pp. 534–536, 2015.

[55] Y. Matsui, T. Pham, T. Sudo, G. Carey, and B. Young, "112-Gb/s WDM link using two directly modulated Al-MQW BH DFB lasers at 56 Gb/s," Optical Fiber Communications Conference and Exhibition (OFC), Techn. Digest, PDP paper Th5B.6, 2015.

[56] Y. Matsui, T. Pham, T. Sudo, G. Carey, B. Young, J. Xu, C. Cole, and C. Roxlo, "28-Gbaud PAM4 and 56-Gb/s NRZ Performance Comparison Using 1310-nm Al-BH DFB Laser," IEEE J. Lightwave Technol., vol. 34, pp. 2677–2683 (Invited), 2016.

[57] Y. Gao, J. C. Cartledge, A. S. Kashi, S. S.-H. Yam, and Y. Matsui, "Direct Modulation of a Laser Using 112-Gb/s 16-QAM Nyquist Subcarrier Modulation," IEEE Photon. Technol. Lett., vol. 29, pp. 35–38, 2017.

[58] Y. Gao, J. C. Cartledge, S. S.-H. Yam, A. Rezania, and Y. Matsui, "112 Gb/s PAM-4 Using a Directly Modulated Laser with Linear Pre-Compensation and Nonlinear Post-Compensation," 42nd European Conference on Optical Communication (ECOC), Paper M.2.C.2, 2016.

[59] M. Matsuda, A. Uetake, T. Simoyama, S. Okumura, K. Takabayashi, M. Ekawa, and T. Yamamoto, "1.3-μm-wavelength AlGaInAs multiple-quantum-well semi-insulating buried-heterostructure distributed-reflector laser arrays on semi-insulating InP substrate," IEEE J. Sel. Top. Quantum Electron., vol. 21, p. 1502307, 2015.

[60] W.F. Sharfin, J. Schlafer, W. Rideout, B. Elman, R.B. Lauer, J. LaCourse, and F.D. Crawford, "Anomalously high damping in strained InGaAs-GaAs single quantum well lasers," IEEE Photon. Technol. Lett., vol. 3, pp. 193–195, 1991.

[61] G. Agrawal, "Gain nonlinearities in semiconductor lasers: Theory and application to distributed feedback lasers," IEEE J. Quantum Electron., vol. 23, pp. 860- 868, 1987.

[62] W. Rideout, W.F. Sharfin, E.S. Koteles, M.O. Vassell, and B. Elman, "Well-barrier hole burning in quantum well lasers," IEEE Photon. Technol. Lett., vol. 3, pp. 784–786, 1991.

[63] N. Tessler, and G. Eistenstein, "On carrier injection and gain dynamics in quantum well lasers," IEEE J. Quantum Electron., vol. 29, pp. 1586–1595, 1993.

[64] A. Tatarczak, S. M. R. Motaghiannezam, C. Kocot, S. Hallstein, I. Lyubomirsky, D. Askarov, H. M. Daghighian, S. Nelson, I. Tafur Monroy, and J. A. Tatum, "Reach Extension and Capacity Enhancement of VCSEL-Based Transmission Over Single-Lane MMF Links," IEEE J. Lightwave Technol., vol. 35, pp. 565–571, 2017.

[65] D. M. Kuchta, T. N. Huynh, F. E. Doany, L. Schares, C. W. Baks, C. Neumeyr, A. Daly, B. Kögel, J. Rosskopf, and M. Ortsiefer, "Error-Free 56 Gb/s NRZ Modulation of a 1530-nm VCSEL Link," IEEE J. Lightwave Technol., vol. 34, pp. 3275–3282, 2016.

[66] D. M. Kuchta, A. V. Rylyakov, F. E. Doany, C. L. Schow, J. E. Proesel, C. W. Baks, P. Westbergh, J. S. Gustavsson, and A. Larsson, "A 71-Gb/s NRZ Modulated 850-nm VCSEL-Based Optical Link," IEEE Photon. Technol. Lett., vol. 27, pp. 577–580, 2015.

[67] M. Ahmed, A. Bakry, M. S. Alghamdi, H. Dalir, and F. Koyama, "Enhancing the modulation bandwidth of VCSELs to the

millimeter-waveband using strong transverse slow-light feedback," Optics Express, vol. 23, no. 12, pp. 15365–15371, 2015.

[68] D. Penninckx, M. Chbat, L. Pierre, and J.-P. Thiery, "The phase-shaped binary transmission (PSBT): a new technique to transmit far beyond the chromatic dispersion limit," IEEE Photon. Technol. Lett., vol. 9, pp. 259–261, 1997.

[69] J. Binder, and U. Kohn, "10 Gbit/s-dispersion optimized transmission at 1.55 μm wavelength on standard single mode fiber," IEEE Photon. Technol. Lett., vol. 6, pp. 558–560, 1994.

[70] D. Erasme, T. Anfray, M. E. Chaibi, K. Kechaou, J. Petit, G. Aubin, K. Merghem, C. Kazmierski, J.-G. Provost, P. Chanclou, and C. Aupetit-Berthelemot, "The Dual-Electroabsorption Modulated Laser, a Flexible Solution for Amplified and Dispersion Uncompensated Networks Over Standard Fiber," IEEE J. Lightwave Technol., vol. 32, pp. 3466–3476, 2014.

[71] M. Chacinski, and R. Schatz, "Impact of Losses in the Bragg Section on the Dynamics of Detuned Loaded DBR Lasers," IEEE J. Quantum Electron., vol. 46, pp. 1360–1367, 2010.

[72] K. Petermann, "Laser Diode Modulation and Noise," Kluwer Academic Publishers/Springer Netherlands, 1988.

[73] A. Mecozzi, "Cavity standing-wave and gain compression coefficient in semiconductor lasers," Optics Letters, vol. 19, pp. 640–642, 1994.

[74] R. Schatz, "Dynamics of spatial hole burning effects in DFB lasers," IEEE J. Quantum Electron., vol. 31, pp. 1981–1993, 1995.

[75] H. Kuwatsuka, H. Shoji, M. Matsuda, and H. Ishikawa, "Nondegenerate four-wave mixing in a long-cavity $\lambda/4$- shifted DFB laser using its lasing beam as pump beams," IEEE J. Quantum Electron., vol. 33, pp. 2002–2010, 1997.

[76] C. -Y.Tsai, C. -Y. Tsai, R.M. Spencer, Yu-Hwa Lo, and L.F. Eastman, "Nonlinear gain coefficients in semiconductor lasers: effects of carrier heating," IEEE J. Quantum Electronic, vol, 32, pp. 201–212, 1996.

[77] A. Mecozzi, S. Scotti, D'Ottavi, E. Iannone, and P. Spano, "Four-wave mixing in traveling-wave semiconductor amplifiers," IEEE J. Quantum Electron., vol. 31, pp. 689–699, 1995.

[78] H. Debregeas-Sillard, C. Fortin, A. Accard, O. Drisse, E. Derouin, F. Pommereau, and C. Kazmierski, "Nonlinear Effects Analysis in DBR Lasers: Applications to DBR-SOA and New Double Bragg DBR," IEEE J. Sel. Top. Quantum Electron., vol. 13, pp. 1142–1150, 2007.

[79] R.F.S. Ribeiro, J.R.F. da Rocha, A.V.T. Cartaxo, H.J.A. da Silva, B. Franz, and B. Wedding, "FM response of quantum-well lasers taking into account carrier transport effects," IEEE Photon. Technol. Lett., vol. 7, pp. 857–859, 1995.

[80] R. Nagarajan, and J. E. Bowers, "Effects of carrier transport on injection efficiency and wavelength chirping in quantum-well lasers," IEEE J. Quantum Electron., vol. 29, pp. 1601–1608, 1993.

[81] R. Nagarajan, R.P. Mirin, T.E. Reynolds, and J.E. Bowers, "Experimental evidence for hole transport limited intensity modulation response in quantum well lasers," Electron. Lett., vol. 29, pp. 1688–1690, 1993.

[82] R. Nagarajan, M. Ishikawa, T. Fukushima, R.S. Geels, and J.E. Bowers, "High speed quantum-well lasers and carrier transport effects," IEEE J. Quantum Electron., vol. 18, pp. 1990–2008, 1992.

[83] H. Shalom, A. Zadok, M. Tur, P. J. Legg, W. D. Cornwell, and I. Andonovic, "On the various time constants of wavelength changes of a DFB laser under direct modulation," IEEE J. Quantum Electron., vol. 34, pp. 1816–1822, 1998.

[84] V. Houtsma, D. v. Veen, S. Porto, N. Basavanhally, C. Bolle, and H. Schmuck, "Investigation of 100G (4 × 25G) NG-PON2 Upgrade using a Burst Mode Laser based on a Multi-Electrode Laser to enable 100 GHz Wavelength Grid," Optical Fiber Communications Conference and Exhibition (OFC), paper M1B.3, 2018.

[85] S. Y. Zou, R. W. Olson, B. Pezeshki, E. C. Vail, G. W. Yoffe, S. A. Rishton, M. A. Emanuel, and M. A. Sherback, "Narrowly Spaced DFB Array With Integrated Heaters for Rapid Tuning Applications," IEEE Photonics Technology Letters, vol. 16, pp. 1239–1241, 2004.

[86] B. Johnson, D. Mahgerefteh, K. McCallion, Z. F. Fan, D. Piede, and P. Tayebati, "Thermal chirp compensation systems for a chirp managed directly modulated laser (CML^{TM}) data link," Patent US7505694 B2, Nov 6, 2002.

[87] Hélène Debrégeas, Robert Borkowski, Rene Bonk, Romain Brenot, Jean-Guy Provost, Sophie Barbet, and Thomas Pfeiffer, "TWDM-PON Burst Mode Lasers With Reduced Thermal Frequency Shift," IEEE J. Lightwave Technol., vol. 36, pp. 128–134, 2018.

[88] P. Vankwikelberge, F. Buytaert, A. Franchois, R. Baets, P.I. Kuindersma, and C.W. Fredriksz, "Analysis of the carrier-induced FM response of DFB lasers: theoretical and experimental case studies," IEEE J. Quantum Electron., vol. 25, pp. 2239–2254, 1989.

[89] K. Y. Lau, "Frequency modulation and linewidth of gain-levered two-section single quantum well lasers," Appl. Phys. Lett. 57, p. 2068, 1990.

[90] G. Agrawal, "Nonlinear fiber optics," Academic Press, eBook ISBN: 9781483288031, 1995.

[91] X. Zheng, K. McCallion, D. Mahgerefteh, Y. Matsui, Z. F. Fan, J. Zhou, M. Deutsch, and Y. F. Chang, "Performance demonstration of 300-km dispersion uncompensated transmission using tunable chirp-managed laser and EDC integratable into small-form-factor MSAs," Proc. SPIE 7136, Optical Transmission, Switching, and Subsystems VI, 71361L, 2008.

[92] X. Zheng, S. Priyadarshi, D. Mahgerefteh, Y. Matsui, T. Nguyen, J. Zhou, M. Deutsch, V. Bu, K. McCallion, J. Zhang, and P. Kiely, "Transmission from 0–360 km (6120 ps/nm) at 10 Gb/s without optical or electrical dispersion compensation using digital pulse shaping of a chirp managed laser" Optical Fiber Communications Conference and Exhibition (OFC), paper OThE5, 2009.

[93] Y. Matsui, D. Mahgerefteh, X. Zheng, X. Ye, K. McCallion, H. Xu, M. Deutsch, R. Lewén, J. O. Wesström, R. Schatz, and P. J. Rigole, "Widely tuneable modulated grating Y- branch Chirp Managed Laser," European Conference on Optical Communication (ECOC), Post-deadline paper, 2010.

[94] A.J. Ward, D.J. Robbins, G. Busico, E. Barton, L. Ponnampalam, J.P. Duck, N.D. Whitbread, P.J. Williams, D.C.J. Reid, A.C. Carter, and M.J. Wale, "Widely tunable DS-DBR laser with monolithically integrated SOA: design and performance," IEEE J. Sel. Top. Quantum Electron., vol. 11, pp. 149–156, 2005.

[95] Y. Yokoyama, T. Hatanaka, N. Oku, H. Tanaka, I. Kobayashi, H. Yamazaki, and A. Suzuki, "10.709-Gb/s-300-km transmission of PLC-based chirp-managed laser packaged in pluggable transceiver," European Conference on Optical Communication (ECOC), paper We.1.C.4, 2010.

[96] S. Grillanda, R. Ji, F. Morichetti, M. Carminati, G. Ferrari, E. Guglielmi, N. Peserico, A. Annoni, A. Dedè, D. Nicolato, A. Vannucci, C. Klitis, B. Holmes, M.Sorel, S. Fu, J. Man, L. Zeng, M. Sampietro, and A. Melloni, "Wavelength Locking of Silicon Photonics Multiplexer for DML-Based WDM Transmitter," IEEE J. Lightwave Technol., vol. 35, pp. 607–614, 2017.

[97]   V. Cristofori, F. Da Ros, M. E. Chaibi, Y. Ding, L. Bramerie, A. Shen, A. Gallet, G.-H.Duan, L. K. Oxenløwe, and C. Peucheret, "Directly Modulated and ER Enhanced Hybrid III-V/SOI DFB Laser Operating up to 20 Gb/s for Extended Reach Applications in PONs," Optical Fiber Communications Conference and Exhibition (OFC), paper Tu3G.7, 2017.

[98]   S. Matsuo, T. Kakitsuka, T. Segawa, R. Sato, Y. Shibata, R. Takahashi, H. Oohashi, and H. Yasaka, "4 × 25 Gb/s Frequency-Modulated DBR Laser Array for 100-GbE 40-km Reach Application," IEEE Photon. Technol. Lett., vol. 20, pp. 1494–1496, 2008.

[99]   S. Matsuo, T. Kakitsuka, T. Segawa, N. Fujiwara, Y. Shibata, H. Oohashi, H. Yasaka, and H. Suzuki, "Extended transmission reach using optical filtering of frequency-modulated widely tunable SSG-DBR," IEEE Photonics Technology Letters, vol. 20, pp. 294–296, 2008.

[100] T. Kakitsuka, S. Matsuo, T. Segawa, Y. Shibata, Y. Kawaguchi, and R. Takahashi, "20-km transmission of 40-Gb/s signal using frequency modulated DBR laser," Optical Fiber Communications Conference and Exhibition (OFC), paper OThG4, 2017.

[101] J. Yu, Z. Jia, M. F. Huang, M. Haris, P. N. Ji, T. Wang, and G. K. Chang, "Applications of 40-Gb/s Chirp-Managed Laser in Access and Metro Networks," IEEE J. Lightwave Technol., vol. 27, pp. 253–265, 2009.

[102] A. S. Karar, J. C. Cartledge, Y. Matsui, I. Lyubomirsky, and D. Mahgerefteh, "Performance of a 56 Gbit/s directly modulated DBR laser with an optimized optical spectrum reshaper," European Conference on Optical Communication (ECOC), paper Tu. 3.6.4, 2014.

[103] X. Zheng, D. Mahgerefteh, Y. Matsui, X. Ye, V. Bu, K. McCallion, H. Xu, M. Deutsch, H. Ereifej, R. Lewén, J. O. Wesström, R. Schatz, and P. J. Rigole, "Generation of RZ-AMI using a widely tuneable modulated grating Y-branch chirp managed laser, "Optical Fiber Communication (OFC), paper OThE5, 2010

[104] A. S. Karar, J. C. Cartledge, and Y. Matsui, "Performance of a 10.709 Gb/s DPSK signal generated using a FIR filter and a DBR laser," 27th Biennial Symposium on Communications (QBSC), 2014.

[105] W. Jia, J. Xua, Z. Liu, K. H. Tse, and C. K. Chan, "Generation and Transmission of 10-Gb/s RZ-DPSK Signals Using a Directly Modulated Chirp-Managed Laser," IEEE Photon. Technol. Lett., vol. 23, pp. 173–175, 2011.

[106] W. Jia, Y. Matsui, D. Mahgerefteh, I. Lyubomirsky, and C. K. Chan, "Generation and Transmission of 10-Gbaud Optical 3/4-RZ-DQPSK Signals Using a Chirp-Managed DBR Laser," IEEE J. Lightwave Technol., vol. 30, pp. 3299–3305, 2012.

[107] D. Che, F. Yuan, and W. Shieh, "Maximum likelihood sequence estimation for optical complex direct modulation," Optics Express, vol. 25, pp. 8730–8738, 2017.

[108] R. Corsini, M. Artiglia, M. Presi, and E. Ciaramella, "10 Gb/s long-reach PON system based on directly modulated transmitters and simple polarization independent coherent receiver," Optics Express, vol. 25, pp. 17841–17846, 2017.

[109] U. Feiste, "Optimization of modulation bandwidth in DBR lasers with detuned Bragg reflectors," IEEE J. Quantum Electron., vol. 34, pp. 2371–2379, 1998.

[110] K. Iga, "Modulation limit of semiconductor lasers by some parametric modulation schemes," Institute of Electronics and Communication Engineers of Japan, Transactions, Section E, vol. E68, pp. 417–420, 1985.

[111] D. D. Marcenac and J. E. Caroll, "Distinction between multimoded and singlemoded self-pulsations in DFB lasers." Electronics Letters, vol. 30, pp. 1137–1138, 1994.

[112] P. Vankwikelberge, G. Morthier, and R. Baets, "CLADISS-A longitudinal multimode model for the analysis of the static, dynamic, and stochastic behavior of diode lasers with distributed feedback," IEEE Journal of Quantum Electronics., vol. 26, pp. 1901–1909, 1990.

[113] M. G. Davis and R. F. O'Dowd, "A new large-signal dynamic model for multielectrode DFB lasers based on the transfer matrix method," IEEE Photonics Technology Letters, vol. 4, pp. 838–840, 1992.

[114] A. Lowery, A. Keating, and C. P. Murtonen, "Modeling the static and dynamic behavior of quarter-wave shifted DFB lasers," IEEE J. Quantum Electron., vol. 28, pp. 1874–1883, 1992

[115] L. M. Zhang and J. E. Carroll, "Large-signal dynamic model of the DFB laser," IEEE Journal of Quantum Electronics, vol. 28, pp. 604–611, 1992.

[116] U. Bandelow, R. Schatz, and H.-J. Wunsche, "Correct single-mode photon rate equation for multisection lasers," IEEE Photon. Technol. Lett., vol. 8, pp. 614–616, 1996.

[117] R. Schatz, "Ultra-High-Speed Optical-Cavity-Enhanced DMLs", Optical Fiber Communications Conference and Exhibition (OFC), Th3B.1 (Tutorial Talk), 2018.

[118] Y. Matsui, R. Schatz, G. Carey, T. Sudo, and C. Roxlo, "Direct modulation laser technology toward 50-GHz bandwidth," IEEE International Semiconductor Laser Conference (ISLC) 2016, WA1 (Invited), 2016.

[119] R. Schatz, O. Kjebon, S. Lourdudoss, S. Nilsson, and B. Stalnacke, "Enhanced modulation bandwidth and self- pulsations in detuned loaded InGaAsP DBR-lasers," IEEE International Semiconductor Laser Conference (ISLC), pp. 93–94, 1996.

[120] J.-M. Verdiell, U. Koren, and T.L. Koch, "Linewidth and alpha-factor of detuned-loaded DBR lasers," IEEE Photon. Technol. Lett., vol. 4, pp. 302–305, 1992.

[121] Larry A. Coldren, and Scott W. Corzine, "Diode Lasers and Photonic Integrated Circuits," Chapter 5, pp. 199–200, Wiley, ISBN-10: 0470484128.

[122] Shigeru Mieda, Nobuhide Yokota, Ryuto Isshiki, Wataru Kobayashi, and Hiroshi Yasaka, "Frequency response control of semiconductor laser by using hybrid modulation scheme," Optics Express, vol. 24, pp. 25824–25831, 2016.

[123] H.A. Haus, "Mode-locking of lasers," IEEE J. Sel. Top. Quantum Electron., vol. 6, pp. 1173- 1185, 2000.

[124] S. Arahira, Y. Matsui, and Y. Ogawa, "Mode-locking at very high repetition rates more than terahertz in passively mode-locked distributed-Bragg-reflector laser diodes," IEEE J. Quantum Electron., vol. 32, pp. 1211–1224, 1996.

[125] A. Laakso, and M. Dumitrescu, "Modified rate equation model including the photon–photon resonance," Opt Quant Electron, vol. 42, pp. 785–791, 2011

[126] Paolo Bardella, and Ivo Montrosset, "A New Design Procedure for DBR Lasers Exploiting the Photon–Photon Resonance to Achieve Extended Modulation Bandwidth," IEEE Journal of Selected Topics in Quantum Electronics, vol. 19, pp. 1502408–1502408, 2013.

[127] Yuanfeng Mao, Zhengliang Ren, Lu Guo, Hao Wang, Ruikang Zhang, Yongguang Huang, Dan Lu, Qiang Kan, Chen Ji, and Wei Wang, "Modulation Bandwidth Enhancement in Distributed Reflector Laser Based on Identical Active Layer Approach," IEEE Photonics Journal, vol. 10, Article Sequence Number 1502308, 2018.

[128] Hans Wenzel, Uwe Bandelow, H.-J. Wunsche, and Joachim Rehberg, "Mechanisms of fast pulsations in two section DFB lasers," IEEE J. Quantum Electron., vol. 32, pp. 69–78, 1996.

[129] M. Shahin, K. Ma, A. Abbasi, G. Roelkens, and G. Morthier, "45 Gb/s Direct Modulation of Two-Section InP-on-Si DFB Laser Diodes," IEEE Photon. Technol. Lett., vol. 30, pp. 685–687, 2018.

[130] O. Brox, S. Bauer, M. Radziunas, M. Wolfrum, J. Sieber, J. Kreissl, B. Sartorius, and H.-J. Wünsche, "High-frequency pulsations in DFB lasers with amplified feedback," IEEE J. Quantum Electron., vol. 39, pp. 1381–1387, 2003.

[131] M. Radziunas, A. Glitzky, U. Bandelow, M. Wolfrum, U. Troppenz, J. Kreissl, and Wolfgang Rehbein, "Improving the Modulation Bandwidth in Semiconductor Lasers by Passive Feedback," IEEE J. Sel. Top. Quantum Electron., vol. 13, pp. 136–142, 2007.

[132] L. Bach, W. Kaiser, J.P. Reithmaier, A. Forchel, T.W. Berg, and B. Tromborg, "Enhanced direct-modulated bandwidth of 37 GHz by a multi-section laser with a coupled-cavity-injection-grating design," Electron. Lett., vol. 39, pp. 1592–1593, 2003.

[133] André Loose, "Multistability due to Delayed Feedback and Synchronization of Quasiperiodic Oscillations Studied With Semiconductor Lasers," Dissertation, Faculty of Mathematics and Natural Sciences, Humboldt University of Berlin, 2011.

[134] R. Lang, and K. Kobayashi, "External optical feedback effects on semiconductor injection laser properties," IEEE J. Quantum Electron., vol. 16, pp. 347–355, 1980.

[135] G. Morthier, R. Schatz, and O. Kjebon, "Extended modulation bandwidth of DBR and external cavity lasers by utilizing a cavity resonance for equalization," IEEE J. Quantum Electron., vol. 36, pp. 1468–1475, 2000.

[136] O. Kjebon, R. Schatz, C. Carlsson, and N. Akram, "Experimental evaluation of detuned loading effects on distortion in edge emitting DBR lasers," Microwave Photonics, 2002. International Topical Meeting on, 2002.

[137] P.A. Morton, T. Tanbun-Ek, R.A. Logan, A.M. Sergent, P.F. Sciortino, D.L. Coblentz, "Frequency response subtraction for simple measurement of intrinsic laser dynamic properties," IEEE Photon. Technol. Lett., vol. 4, pp. 133–136, 1992.

4

PAM4 Modulation Using Electro-absorption Modulated Lasers

Trevor Chan and Winston Way

Neophotonics, California, USA

4.1 Introduction

1.3 μm EMLs based on MQWs can be applied to optical links which require high speeds (e.g., 56 Gbaud) or long transmission distances (e.g., $\geq$ 10 km). An EML operates by sending a CW DFB laser light into an electro-absorption modulator based on the QCSE [1], which is characterized by a quadratic relationship between the energy shift and the applied voltage. As a result, in an electro-absorption modulator, the application of an electric field perpendicular to quantum wells shifts the absorption band-edge to a longer wavelength. So while the modulator may appear transparent at a specific wavelength, once the electric field is applied, the wavelength will be absorbed by the material since the band-gap has decreased. The response to a modulating electrical field is considerably fast and linear up to a saturation region, and has been tested well with PAM4 and DMT modulations [2].

EMLs, due to the fact that they are externally modulated, have the advantage over DMLs because they have minimum frequency chirp under modulation. They also can achieve a much higher extinction ratio than that of DMLs. Comparing to silicon photonic modulators, EMLs do not have to deal with the challenging optical input/output coupling, whose economical advantage is of particular use in the case of multiple wavelengths (e.g., 4 or 8 wavelengths). However, conventional EMLs do have their weak points in that a TEC is needed due to the temperature balance between a laser and an electro-absorption modulator, and the TEC increases both cost and power consumption. In the last couple of years, we have seen increasing efforts in the industry to make un-cooled EMLs without hermetic sealing. The issue with un-cooled EMLs is that GaInAsP that is typically used has weak electron confinement. The consequence for this weak confinement is carrier overflow at high temperatures which significantly penalizes the laser output power and the modulator extinction ratio by reducing the QCSE. Other materials, such as InGaAlAs and InGaNAs have been shown to be better candidates for uncooled operation [3]. The issue that remains with these materials is the detuning that happens between the lasing wavelength (0.1 nm/°C) and the quantum well bandgap (0.4–0.5 nm/°C) [4]. This detuning can be mitigated by separate design of the laser and quantum well where they are eventually coupled by a butt-joint structure [5]. The potential success of this effort could make EML a strong candidate for future intra-data center applications.

PAM4, on the other hand, is a paradigm shift from NRZ OOK for the purpose of doubling a transmission link capacity without increasing the optical transceiver bandwidth. As a result of using PAM4 modulation, 28 and 56 Gbaud transmission capacity can be achieved by using EMLs with a bandwidth of $\sim$20 GHz and $\sim$30 GHz, respectively, as will be described in later sections.

In this chapter, we will review our experience in using EMLs in various transmission distances (2, 10, 40, and 100 km) and baud rates (28 and 56 Gbauds) over the last 4 years. We will also review our pioneering work in experimentally proving that the system penalty due to MPI can be small given that a proper optical reflectance specification is imposed on optical connectors, transmitter, and receiver. The chapter is organized as follows: Section 4.2 will provide insight into PAM4 signalling and compensation methods that enable an EML to operate with this signal format. Section 4.3 details our work with PAM4 at 28 GBaud which served as a base for IEEE802.3bs 400 GBASE-LR8 links. Section 4.4 explores PAM4 at 56 GBaud for short reach links consisting of 100 Gbps/wavelength. Section 4.5 extends these links out to 40 km by using APD and gain-clamped SOA. Section 4.6 shows the technical feasibility of using EML and PAM4 for 100 km links. Section 4.7 details our experimental study of MPI with PAM4 modulation. Finally, Section 4.8 is the summary.

4.2 General PAM4 Optical Transceiver and Link Considerations

4.2.1 PAM4 Signal and Optical Link Characteristics

A PAM4 signal is typically obtained by combining two pseudo random signals with 3 dB amplitude difference. The larger amplitude signal corresponds to the MSB and the smaller one corresponds to the LSB. The MSB and LSB were added together to achieve the four amplitude levels (corresponding to 10, 11, 01, and 00) which becomes the PAM4 signal with levels (+3, +1, −1, −3). Generally speaking, PRBS15 is not long enough to test conditions such as baseline wandering. Instead, PRBS31 signals which have sections that are heavily weighted towards the same bit should be used. However, in exploratory studies where PAM4 PHYs are not available for real time error counting, entire waveforms are captured and processed offline. Unfortunately, the PRBS31 signal is far too large to fit within the memory of currently available oscilloscope and DAC/ADC evaluation boards. So, we are limited to using pattern lengths close to PRBS15. As an alternative, the sections of the PRBS31 signals with low frequency components can be isolated and spliced with sections of higher frequency components in suitable proportions. This has been done in a signal known as SSPRQ (Short Stress Pattern Random – Quaternary) which has been widely adopted as a benchmark test signal [6].

For an electrical PAM4 signal, by assuming all PAM4 levels are equally likely, all inner eyes have the same amplitude, and each PAM4 level has an additive Gaussian noise, we can obtain the relationship between BER and SNR according to Equation (4.1) [7]:

$$BER = \frac{3}{8} erfc \left(\frac{\sqrt{SNR}}{\sqrt{10}} \right) \tag{4.1}$$

where SNR is defined as the ratio of average signal power to noise where the average signal power is $(\sqrt{5} \times d/2)^2$ (d is the symbol spacing, i.e., the amplitude between adjacent PAM4 levels) and σ is the standard deviation of the Gaussian noise distribution. This equation is plotted in Figure 4.1. We have measured a PAM4 PHY ASIC's output SNR of 26.3 dB which registered no bit errors due to the low BER probability (less than 1e-20). Note that in Equation (4.1) we have assumed that the data is gray coded so that BER = (1/2)SER, where SER represents "symbol error rate".

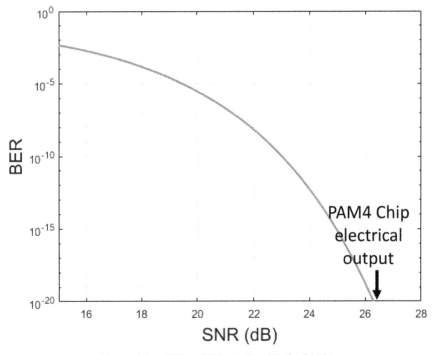

Figure 4.1   BER vs. SNR relationship for PAM4.

For a received optical PAM4 signal, we can have the spacing between adjacent PAM4 levels given by

$$d = \frac{2}{3} \frac{ER - 1}{ER + 1} ROP \tag{4.2}$$

where ER is the extinction ratio of the PAM4 signal and ROP is the received optical power. By substituting $\sqrt{SNR} = \sqrt{5} \times d/2$ and Equation (4.2) into Equation (4.1), we obtain [8]:

$$BER = \frac{3}{8} erfc \left(\frac{\left(\frac{ER-1}{ER+1} \right) ROP}{3\sqrt{2}\sigma} \right) \tag{4.3}$$

where σ is the combination of thermal noise, shot noise, and RIN noise as shown in Equation (4.4):

$$\sigma^2 = \left[\sqrt{\langle i_{th}^2 \rangle} \right]^2 \triangle f + 2qI_i \triangle f + RINI_i^2 \triangle f \tag{4.4}$$

In Equation (4.4), $\sqrt{\langle i_{th}^2 \rangle}$, Δf, q, I_i and RIN are the thermal noise spectral density, receiver bandwidth, elementary charge, average detected photocurrent and average relative intensity noise. Note that $I_i = ROP \cdot \rho$, where ρ is the photo-detector responsivity.

For 28 and 56 Gbaud PAM4 systems using state-of-the-art optical and electronic components, one can expect BER values shown in Figure 4.2. Note that the typical inter-symbol interference-induced error floor is not included in this simple model. In the 28 GBaud case, typical parameters are 15 pA/$\sqrt{\text{Hz}}$ thermal noise, 0.8 A/W responsivity, 20 GHz bandwidth, < -132 dB/Hz RIN and 6.3 dB extinction. In the 56 GBaud case, bandwidth, noise and responsivity performances are harder to achieve for the faster components. In this case, typical system values are 15 pA/$\sqrt{\text{Hz}}$ thermal noise, 0.7 A/W responsivity, 30 GHz bandwidth, < -136 dB/Hz RIN and

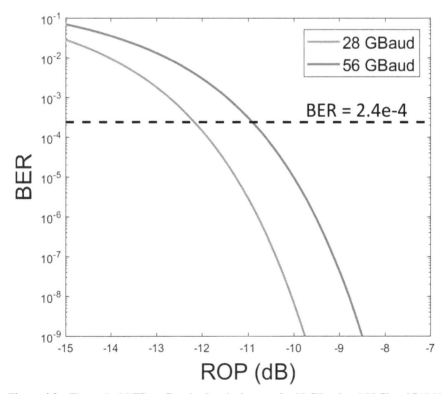

Figure 4.2 Theoretical BER vs. Received optical power for 28 GBaud and 56 Gbaud PAM4 transmission using state-of-the-art components.

5.6 dB extinction. As can be observed in Figure 4.2, with a pre-FEC BER threshold of 2.4e-4 (see Section 4.2.3), a receiver sensitivity of −12.2 and −10.9 dBm can be obtained for 28 and 56 Gbaud, respectively. For the extinction ratio numbers we used, these correspond to an outer OMA of −11.3 and −10.3 dBm for 28 and 56 Gbaud, respectively.

Furthermore, for any two adjacent PAM4 symbols, there are 16 possible level transitions and only 4 are the same level transition. The other 12 transitions are non-adjacent level transitions. Consequently, the three inner PAM4 eyes opening is mainly limited by the inter-symbol interference (ISI) caused by the non-adjacent level transitions. This implies that the bandwidth of a PAM4 optical transceiver should be slightly higher than what is offered by an OOK/NRZ transceiver for the same baud rate, although this depends on the strength of the transmitter and receiver linear equalizers.

4.2.2 EML Biasing and Nonlinear Equalization

Due to the nonlinear response of an EML at its modulation extremities, the driving range must be kept in the linear region of the response, as shown in Figure 4.3a). However, a higher driving amplitude helps achieve a higher extinction ratio, leading to more separation of the PAM4 levels and fewer bit errors (see Equation (4.3)). Therefore, there is a trade-off between a high extinction ratio and nonlinearity. For our EML, the best driving range typically results in a 5 to 7 dB extinction ratio. Naturally, a bias voltage must also be optimized such that the optical modulation is placed in the middle of the linear response.

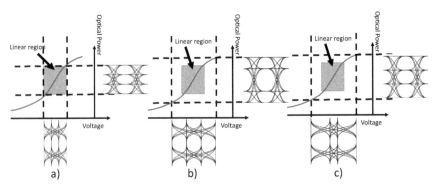

Figure 4.3 Electrical to optical relationship of an EML. With PAM4 signals, the signal can be a) limited to the linear region, b) have increased amplitude, thereby inheriting nonlinearities of the transfer curve or c) pre-compressed so the output signal has equal eye openings.

The modulation range can be extended even further if nonlinear correction is involved in the signal generation. On a first order level, this can be accomplished by increasing the least significant bit amplitude. This adjustment is equivalent to increasing the opening of the upper and lower eyes relative to the middle eye, as illustrated in Figure 4.3(b) and (c). This adjustment is usually a feature in most PAM4 signal generators and PAM4 PHY ASIC. In the experiment described here, the LSB amplitude is increased relative to the MSB. Although this allows for a larger OMAs and extinction ratios (>7 dB), the driving amplitude still needs to be restricted as the noise on the middle levels also increase. Also, the outer levels experience saturation and deviate from a normal Gaussian distribution. This saturation penalizes the ability of linear correction. Therefore, the ideal extinction ratios are typically found to be 5 to 7 dB. In this range, a slight amount of nonlinearity can be observed and managed by decreasing the inner OMA.

The nonlinear correction by adjusting the outer eye amplitudes corrects for the instantaneous nonlinearities. Theoretically, the nonlinear correction can be accomplished by using a Volterra series [9]. pre-emphasis which, in addition to correcting the instantaneous nonlinearities, corrects for the memory-associated nonlinearities. Since both the high-speed modulator and driver introduce nonlinearities with memory, a Volterra series correction would effectively improve the performance. However, due to the multiplicative number of higher order coefficients in the Volterra series, this process become computationally intensive. In practical systems, it is sufficient to correct the nonlinear response only by outer eye amplitude adjustment.

4.2.3 Forward Error Correction (FEC) and Data Rates for PAM4 Links

For optical links up to 10 km, IEEE802.3bj KP4 FEC based on Reed Solomon coding (544, 514) has been used by 200G and 400G in IEEE802.3bs. The pre-FEC threshold of KP4 is 2.4e-4 [10].

Both 200G and 400G physical coding sublayer (PCS) are based on a 64B/66B code. The 64B/66B code is then transcoded to 256B/257B encoding to reduce the overhead and make room for FEC. The 256B/257B encoded data is then FEC encoded by a Reed-Solomon encoder RS(544, 514) before being transmitted. Therefore, for a 50 Gbaud Ethernet data, the net baud rate becomes 50 × (257/256) × (544/514) = 53.125 Gbaud. Similarly, for a 25 Gbaud Ethernet data, the net baud rate becomes 26.5625 Gbaud. Therefore, 26.5625 and 53.125 Gbaud are the two KP4 FEC-based Ethernet data rates used for intra-data center applications.

For inter-data center or inter-central office applications, stronger, propri-etary FECs have been implemented in PAM4 PHY ASIC, which cause the baud rate to increase slightly above the KP4 FEC-based Ethernet data rates mentioned above. Another baud rate for inter-data center applications is based on the Optical Transport Network (OTN) G.709 [11] framing. This signal comes from or goes to a host board which the pluggable optical transceiver is plugged into. The OTN G.709 hierarchy defines mapping 100G Ethernet to an optical channel transport unit OTU4 (111.809 Gb/s), which uses optical channel transport lanes OTL4.4 to interface with an optical transceiver mod-ule with four line-side wavelengths or fibers, each carrying 27.952 Gb/s. To double the transmission capacity per transceiver, the baud rate is doubled to 55.904 Gbaud. All of our experiments have been carried out based on 28 or 56 Gbaud to ensure that the highest baud rate case is covered.

Table 4.1 shows all standards and MSAs that are using PAM4 modula-tions at different wavelengths, distances, and baud rates. It can be observed that EML can find the most applications (except for PSM4) in the table, including cooled EML for ≥ 10 km and uncooled EMLs for ≤ 2 km.

4.2.4 Sampling Rate and Analog Bandwidth

Baud rate sampling has been used in today's low-power, 28 and 56 Gbaud PAM4 chips. In these PAM4 chips, each ADC is preceded by an anti-aliasing analog filter (AAF), whose optimum bandwidth is compromised between minimum aliasing and ISI [15]. To have a better control of the clock recovery timing, Muller and Muller timing recovery circuit which takes its input directly from an ADC output is used.

As far as the required analog bandwidth to carry a PAM4 signal, a min-imum of 50% of the baud rate must be used according to Nyquist Theorem. For example, a 28 GBaud PAM4 signal must be transmitted with at least 14 GHz bandwidth (the so called "Nyquist frequency"), which results from a sum of all the frequency limitations of a system, including DAC, ADC, driver amplifier, EML, and receiver.

In reality, however, it is often necessary to have more than the Nyquist frequency from all the components. To understand the implications of more bandwidth, one can imagine a signal with infinite bandwidth. With infinite bandwidth, symbol switching is done instantaneously and the three eyes of the PAM4 signal becomes three rectangles. As the bandwidth decreases down to the Nyquist frequency, the eyes will remain open (as dictated by the Nyquist Theorem), but they become narrow, especially due to the

Table 4.1 IEEE standards or MSA related to PAM4 modulation

Optical Device	Wavelength	Standard or MSA	Number of λ's & Fibers	Link Distance	Baud Rate (Gbaud)	Application
EML SiPho	1.3 µm CWDM	400 G-FR [12]	4 λ, 2 fibers	< 2 km	53.125	Intra-DC
EML	1.3 µm CWDM	400 G-LR [12]	4 λ, 2 fibers	< 10 km	53.125	Inter-DC Inter-CO
SiPho	∼1.31 µm	400 GBase-DR4 [13]	1 λ, 8 fibers	< 500m	53.125	Intra-DC
EML DML	1.3 µm LAN-WDM	400 GBase-FR8 [13]	8 λ, 2 fibers	< 2 km	26.5625	Intra-DC
EML	1.3 µm LAN-WDM	400 GBase-LR8 [13]	8 λ, 2 fibers	< 10 km	26.5625	Inter-DC Inter-CO
SiPho	∼1.31 µm	200 GBase-DR4 [13]	1 λ, 8 fibers	< 500m	26.5625	Intra-DC
EML DML	1.3 µm LAN-WDM	200 GBase-FR4 [13]	4 λ, 2 fibers	< 2 km	26.5625	Intra-DC
EML	1.3 µm LAN-WDM	200 GBase-LR4 [13]	4 λ, 2 fibers	< 10 km	26.5625	Inter-DC Inter-CO
EML SiPho	∼1.31 µm	100 G-FR [12]	1 λ, 2 fibers	< 2 km	53.125	Intra-DC
EML	∼1.31 µm	100 G-LR [12]	1 λ, 2 fibers	< 10 km	53.125	Inter-DC Inter-CO
EML SiPho	1.55 µm DWDM (non-standard)	100 G-ER/ZR [14]	48 λ, 2 fibers	< 40 ∼80 km	28.125	Inter-DC Inter-CO

non-adjacent level transitions which were mentioned in Section 4.2.1. This is shown in Figure 4.4 where a) a high bandwidth PAM4 signal is wide open compared to b) a PAM4 signal with bandwidth close to the Nyquist frequency. In reality, the signal would not be sampled at the exact eye opening and despite the Nyquist theorem, timing errors and jitter would result in bit errors. Thus, it is still necessary to transmit with more than the Nyquist frequency as higher frequencies allow the eyes to widen, making it less sensitive to errors in the sampling time. As far as how much more bandwidth is needed than the Nyquist bandwidth depends on the strength of the transmitter and receiver equalizers.

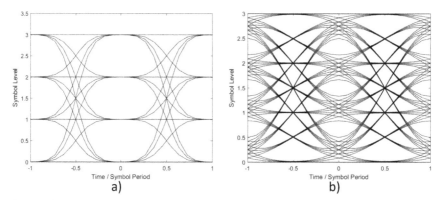

Figure 4.4   a) A PAM4 signal with bandwidth in excess (2x) of the Nyquist frequency and b) a PAM4 signal bandwidth near the Nyquist frequency showing nearly closed eyes.

4.2.5 FFE and DFE Equalization

While in the linear regime, FFE and DFE are usually employed to exploit as much usable system bandwidth as possible. FFE is a time domain correction where each symbol is a sum of the symbols value added to its neighbour's values weighted in an optimum fashion. In this case, the taps are referred to as "T-spaced" taps, which correspond to baud rate sampling mentioned in Section 4.2.4. In some cases, the signal can be at twice the baud rate, so that "T/2-spaced" taps are used. T/2-spaced taps correct twice the bandwidth of the T-spaced taps while T-spaced taps only correct frequencies up to the Nyquist frequency. However, T-spaced taps are more common as it is often sufficient (according to Nyquist Theory), much easier to implement, and saves power consumption. The taps consist of pre-cursors and post-cursors. Pre-cursor taps are weights from symbols that occur later while post-cursor taps are from symbols occurring earlier in time. Only 1 or 2 precursor taps are usually helpful as there is little power leakage to earlier symbols. More post-cursor taps are usually required.

The location of FFE equalization can be flexible. It can be applied for equalization in the transmitter to create a pre-distorted signal, or it may be applied after signal detection to equalize the signal before symbol decisions are made. If it is applied for predistortion, one side effect is that the signal will exhibit overshoot and be modulated at a greater amplitude than the symbol levels. To preserve this equalization and avoid the effects of nonlinear saturation regions, the symbol levels become compressed and the OMA is ultimately reduced, as illustrated in Figure 4.5.

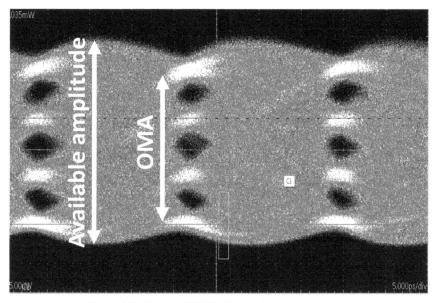

Figure 4.5 Reduced OMA after applying pre-emphasis.

As FFE takes place in the time domain, a Fourier transform of the equalizer taps reveals the frequency correction which takes place. In frequency domain, this is equivalent to amplifying high frequencies that have been overly attenuated. One price of this type of equalization is that noise will also be amplified in the high frequency region.

Conversely, DFE applies only to the receiver once a symbol decision is made. Here, the detected symbols from earlier times are weighted in the decision slicer level. Since a decision must first be made, only post-cursor taps are possible here. Noise amplification has been avoided here since discrete symbol levels are used as feedback which do not contain any noise fluctuations; however, error propagation from misread symbols may result from this.

4.3 28 Gbaud PAM4 Transmission [16, 17]

This experiment was carried out in 2015, and was the first attempt to test the technical feasibility of existing EMLs, originally designed for 28 Gb/s and 40 Gb/s NRZ-OOK data, for 28 Gbaud PAM4 transmission. The purpose of the study was to understand the system performance limitations of EMLs

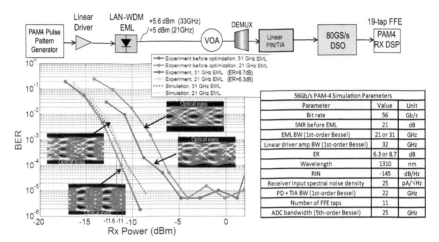

Figure 4.6   Basic setup of the 28 Gbaud PAM4 link. Experimental and simulated results of two different bandwidth EMLs. Sensitivity and error floor improvement is gained using higher bandwidth EMLs with proper level equalization.

such as transmitter linearity, bandwidth, and extinction ratio, and also the effect of receiver thermal noise and bandwidth.

The experimental setup is shown in Figure 4.6. We used two 1310 nm EMLs with 21 GHz and 31 GHz 3 dB bandwidths, respectively. A driver amplifier with a 3 dB bandwidth of 32 GHz was used to achieve extinction ratios of 6.3 and 8.7 dB for the two EMLs, respectively. The faster EML enables a higher extinction ratio. This was followed by a VOA to simulate link loss. The signal was then filtered by a LAN-WDM demultiplexer and measured by a linear PIN and TIA ROSA. The PIN had 0.5 A/W responsivity, <20 pA/$\sqrt{\text{Hz}}$ spectral noise density. The TIA had a 22 GHz 3 dB bandwidth. The differential output of the TIA was recorded using a 33 GHz, 80 GSample/second digital storage oscilloscope (DSO) whose waveforms were recorded for offline processing.

At a BER of 2.4e-4, a receiver sensitivity of -11 and -11.6 dBm (equivalent to -14.9 and -14.6 dBm inner OMA, respectively) were obtained for the two EMLs, respectively. The difference is not large because the sensitivity is primarily limited by the responsivity and spectral noise density in the receiver.

In a similar study, a sensitivity down to -18.3 dBm inner OMA (-13.3 dBm ROP) was achieved using a receiver with 0.85 A/W responsivity and 15 pA/$\sqrt{\text{Hz}}$ spectral noise density [18]. This is similar to results obtained using a DML which exhibited −18.2 dBm inner OMA (−13.3 dBm ROP) sensitivity using a 0.8 A/W PIN [19]. Comparing our results in Figure 4.6 to these experiments, we can see the importance of requiring a receiver with a high receiver responsivity (> 0.8 A/W) and a low thermal noise (< 15 pA/$\sqrt{Hz}$). The reason why the receiver sensitivity of [18, 19] is ~1 dB better than what we obtained in Figure 4.2 is probably due to the smaller receiver bandwidths used in these experiments.

Note that due to the processing time, error rates below 1e-6 were observed to be errorless and are not shown. Therefore, the difference in error floors are not observed here but they are primarily influenced by the signal bandwidth. A clearer picture of this will be seen in Section 4.4 where the noise floor is measured in a more bandwidth limited system. Also shown are the simulation results by using the parameters listed in the table on the right side of the figure. The dashed lines on the graph show good agreement between the measured and simulated results.

In the offline processing, the signal was resampled to 1 sample per symbol with proper clock recovery. LMS optimization was used to find the 19 tap FFE coefficients to equalize the transmission bandwidth. From the symbol level distribution, the slicer decision levels were selected and the BER was directly measured by error counting.

In Figure 4.6, the effect of nonlinearities is also shown when nonlinear correction was not used. The nonlinearities can easily be seen from the two right-hand side eye diagrams where the outer eyes have become closed. This results in optical power penalties over 3 dB. Under this nonlinear condition, both EMLs exhibited noise floors of 5e-6 and was limited by the nonlinear response.

4.4 56 Gbaud PAM4 Transmission over 2 km Experiment

This experiment was carried out in 2015, and it was the first single-wavelength 56 Gbaud PAM4 experiment with a reasonably good optical performance. The experimental setup is shown in Figure 4.7(a). This experiment uses a 50 GHz linear driver to drive a 31 GHz bandwidth EML (the same EML used in the previous section) and obtain an extinction ratio of 6.1 dB. The system is bandwidth limited since its bandwidth is just 10% larger than the required Nyquist frequency of 28 GHz. Therefore to mitigate this bandwidth limit, a 19 tap FFE pre-emphasis was employed at the transmitter, and the corrected electrical and optical eyes are shown in Figure 4.7 which would otherwise be closed. While the FFE has opened the eyes, there are a

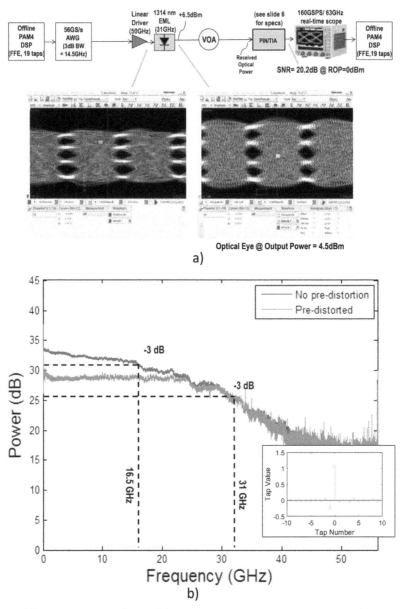

a)

b)

Figure 4.7 (a) Basic setup for a 56 Gbaud PAM4 system. The pre-distorted electrical and optical eye diagrams show all three eyes are open. (b) The frequency response of the combination of DAC, driver, and EML without (blue) and with (green) pre-distortion (FFE taps shown in the inset). Generally speaking EML's bandwidth still needs further improvement, and it is the main limiting factor for today's 53 Gbaud EML's TDECQ.

few trade-offs that are observed here, as has been mentioned in Section 4.2.5. First, these eyes now show overshoot regions between symbols. This overshoot occupies the total available optical amplitude and effectively reduces the extinction ratio (and OMA) between the top and bottom symbol levels. As with the 28 GBaud case, a balance must again be achieved to maximize the OMA while minimizing the nonlinearities. With the addition of the overshoot, the optimization now must consider the overshoot region which is effectively restricting the OMA.

Additionally, since the FFE effectively enhances the higher frequencies with respect to the lower, high frequency noise also becomes amplified and compounds the problem. This phenomenon can be clearly observed in Figure 4.7(b) where the electrical spectrum is the frequency response of the combination of DAC, driver, and EML with and without FFE pre-distortion. The pre-distortion also comes with the cost of additional processing operations required to optimize the taps and condition the signal.

The BER versus ROP results of the 56 Gbaud PAM4 transmission are shown in Figure 4.8 using two different receivers labelled as receiver A and receiver B, whose key parameters are listed in the figure. In this example, receiver A has a larger bandwidth than receiver B, but worse responsivity and input spectral noise density. The theoretical BER, based on Equation (4.3), is also shown in this plot for 56 GBaud. Equation (4.3) used to calculate this curve does not factor in the effects of either ISI penalties or FFE correction; therefore, the theoretical curve produces no noise floor. With ISI playing a role in the measured data, the obtained noise floor was 5e-6 using receiver A which is half as much as receiver B. The noise floor difference indicates that the noise floor is improved with a higher system bandwidth.

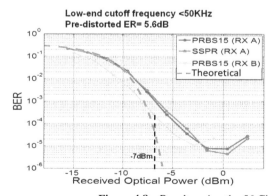

	RX A	RX B
Responsivity (A/W)	0.4	0.7
3dB BW (GHz)	40	30
Spectral noise density (pA/√Hz)	40	35

Figure 4.8   Results using the 56 Gbaud PAM4 link.

This is due to the excess bandwidth over the Nyquist frequency, as discussed in Section 4.2.4.

Conversely, a sensitivity of −7 dBm (corresponding to −6.2 dBm outer OMA) was demonstrated using receiver B which had 0.7 A/W responsivity. In contrast, receiver A has 0.4 A/W responsivity which results in about 2 dB worse sensitivity – consistent with the measured difference in responsivities. Additionally, both receivers had high spectral noise density which was also limiting the results. Consequently, it is critical to develop a receiver with a high responsivity ≥ 0.7A/W, a low spectral noise density =15pA/$\sqrt{\text{Hz}}$, and a high 3 dB bandwidth ≥ 35 GHz, in order to obtain a receiver sensitivity < -10 dBm ROP and an error floor of $\sim$1e-6.

In 2017, we have used an optical receiver with a 3 dB bandwidth of 30 GHz, and a much improved responsivity of 0.6–0.75 A/W and input spectral noise density of 15 pA/$\sqrt{\text{Hz}}$ to achieve a receiver sensitivity of OMA < -10 dBm (or ROP of $\sim$10.8 dBm). The BER versus OMAouter is shown in Figure 4.9.

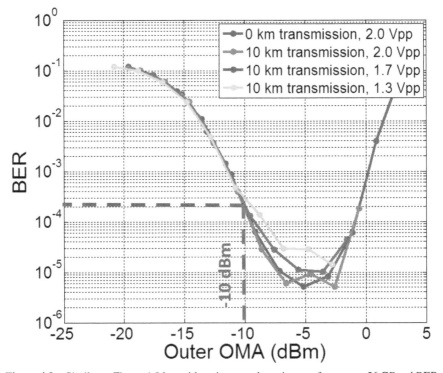

Figure 4.9   Similar to Figure 4.8 but with an improved receiver performance. 56 GBaud BER vs. Outer OMA when the signal is transmitted through 0 and 10 km of fiber.

Although the test setup in 5(a) does not include a 2 km transmission single-mode fiber, we have proven the technical feasibility in a separate experiment, which shows the negligible fiber dispersion penalty when a 56 Gbaud PAM4 signal is sent through a 2 km SMF at a CWDM wavelength of 1264.5 or 1337.5 nm [17].

4.5 40 km PAM4 Transmission

In order to extend the transmission distance to 40 km based on PAM4 modulation, we have proposed two solutions in 2015: 1) APD, and 2) using Gain- Clamped SOA [20].

4.5.1 Avalanche Photodiode (APD)

APDs are knowns for the high sensitivity achieved by its inherent optical gain. This optical gain is created through impact ionization or the Avalanche effect. Although the APDs are known to be non-linear [21], they have been fabricated for high-linearity in order to handle the higher order modulation formats [22].

When using an APD in the receiver of a PAM4 link, the BER vs. ROP was measured and is shown in Figure 4.10. Owing to the higher responsivity of the APD, a sensitivity of −17 dBm received optical power (corresponding to −15.5 dBm outer OMA) was achieved in the optical back-to-back case, with a dynamic range of 4.5 dB. This result was similarly achieved at NTT [22]. The dynamic range was achieved without using any adaptive gain control of the APD and is still limited by the non-linearity of the APD. After 40 km transmission, the sensitivity improved to −18 dBm. This 1 dB improvement was caused by the slightly negative dispersion associated with the 1308 nm EML optical wavelength in a standard single mode fiber. It has been shown that upper power nonlinearity-limit can be increased to > -10 dBm with adaptive bias voltage control [22].

4.5.2 Gain Clamped Semiconductor Optical Amplifier (GC-SOA)

Optical amplification can also be achieved by way of a Semiconductor Optical Amplifier, which has been used in 100 GBase-ER [23], for example. The SOA is essentially a gain region which is under a population inversion from the injection of an electrical current. As a photon passes through this gain region, stimulated emission occurs which leads to an optical gain. Unwanted side effects of the SOA include spontaneous emission and non-radiative recombination.

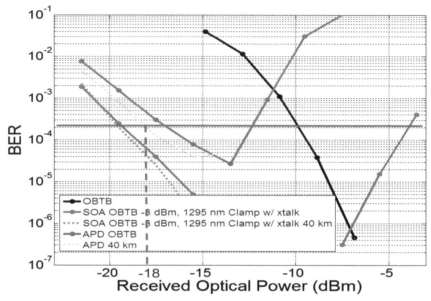

Figure 4.10   28 Gbaud PAM4 transmission using an APD or a GC-SOA for a 1308 nm signal wavelength.

Naturally, the SOA will have a maximum output power that is governed by the number of carriers possible in the population inversion. The carrier lifetime of an SOA is usually finite, meaning that if too many photons are injected into the gain region, the SOA gain will eventually become saturated. As the SOA approaches its maximum output power limit, its gain will decreases to the point that it will start to absorb optical energy. This decrease in gain at high input powers is known as gain compression. Also, since there is a recharge time for the carriers, the decrease in gain will be present for the next period of photons and lead to ISI.

Having established this relationship between gain and signal power, we can consider the effect from having multiple wavelength channel flowing through the SOA. Since the SOA replenishes the carriers at a rate (approximately 300 ps) comparable to 25 and 50 Gbaud rates considered here, the gain starts to mirror the symbols in time. Other wavelength channels will experience this gain fluctuation, meaning that they will show a copy of the original channel. This is known was cross gain modulation.

The above problems have been mitigated by intentionally bringing the SOA out of saturation by clamping its gain to a constant value. This can be accomplished by various means of light injection into the gain region [24–27] which effectively keeps the carriers away from a complete population inversion. In the 40 km PAM4 transmission experiment presented in this section, a gain clamp was introduced by simultaneously transmitting an out of band CW signal to stimulate carriers out of the conductance band. This creates an equilibrium of excitation and relaxation transitions where there is now a reservoir of carriers in the valence band constantly replenishing the conductance band. This results in constant gain versus input power, thereby solving the SOAs limitations described above.

An experimental setup shown in Figure 4.11 based on 28 Gbaud PAM4 signals and EML was used to demonstrate the effectiveness of a GC-SOA. Two distinct PAM4 modulated signals were wavelength multiplexed and combined for 112 Gbps transmission. Both 0 km and 40 km transmission distances were tested to show the effect of fiber dispersion. After the transmission loss, an out of band CW optical signal was introduced into the fiber to act as a gain clamp in the SOA. All signals were polarization controlled to align with the maximum gain mode of the SOA. Signal detection and bit recovery was done in the same way mentioned in the prior section.

The results of the gain clamped transmission through 0 km fiber are shown in Figure 4.12. A single wavelength transmission is shown by the blue curve at 1308 nm. Here, the sensitivity has improved down to -19.8 dBm due to the pre-amplification of the SOA. At higher powers, many bit errors

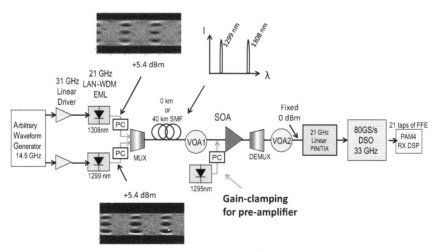

Figure 4.11 Experimental Setup for two 28 Gbaud PAM4 wavelengths passing a GC-SOA after 40 km SMF.

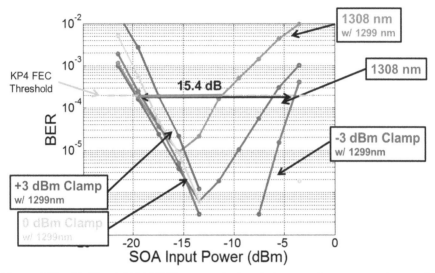

Figure 4.12   28 Gbaud PAM4 BER vs. received optical power for systems using a gain clamped SOA and PIN as the receiver.

are created by SOA saturation, limiting the signal to a noise floor of 7e-7. The green curve shows the bit errors with the addition of a 1299 nm channel where crosstalk adds to the impairment. This increases the noise floor to 9e-6 while decreasing the dynamic range by 5 dB.

The effect of a −3, 0 and +3 dBm gain clamp is shown in the red, teal and purple curves respectively. Significant improvement can be seen even at −3 dBm with a noise floor below the lowest detectable BER of 1e-7. This was also accompanied by an increase in dynamic range from 8.2 dB to 15.4 dB. As the gain clamp increases in power, the effect of cross talk and saturation cannot be observed in the graph as there was no detected bit errors at high powers. However, with higher power gain clamps, there is fewer carriers in the conductance band resulting in less optical gain. This trade-off is seen in the reduction of sensitivity which has increased by as much as 2.5 dB with the +3 dBm gain clamp.

A comparison of the gain clamped SOA with multiplexed channels and the APD is shown in Figure 4.10 for the 1308 nm EML. This shows that the gain clamped SOA has a much better sensitivity and dynamic range, owing to the superior linearity earned by the gain clamp. The effect of 40 km SMF dispersion is also shown. As mentioned previously, negative dispersion leads to pulse compression, improving the sensitivity in both cases.

The dynamics of the semiconductor optical amplifier allows it to be used for higher baud rates PAM4 signals, which is another advantage over APD. By using the 31 GHz EML, 50 GHz linear driver and 13 GHz AWG, a 56 GBaud PAM4 signal on 1314 EML was transmitted through the gain clamped SOA, as shown in Figure 4.13. Similar to the 28 Gbaud case, the gain clamp is introduced into the transmission line via a wavelength multiplexer before it is injected into the SOA. The 40 GHz PIN/TIA receiver was used in conjunction with a 63 GHz DSO and offline DSP to recover the symbols.

The results of the transmission can be seen in Figure 4.14. Recall in Figure 4.8 that without the SOA, a sensitivity of −7 dBm was obtained when RX B was used. The blue curve in Figure 4.14, when the SOA without gain clamping was used, shows a back to back sensitivity improvement of 6 dB with a BER noise floor of 7e-5. The performance improves with the introduction of a −3 dBm gain clamp. This brings the noise floor down below 2e-5 and increases the dynamic range to 13.2 dB. With or without the gain clamp, the 40 km of fiber lead to a 2.8 dB dispersion penalty. This dispersion penalty is due to using a wavelength with a slight positive fiber dispersion (contrary to the wavelengths in Figure 4.10 which experience negative fiber dispersions), leading to pulse broadening and ISI.

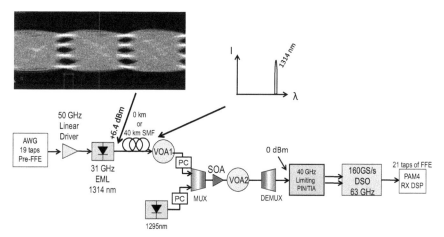

Figure 4.13 Experimental setup using an GC-SOA to transmit a 56 Gbaud PAM4 signal over 40 km SMF.

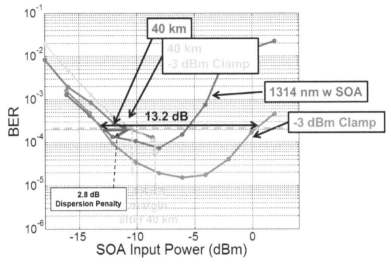

Figure 4.14 56 Gbaud PAM4 BER vs. power received into the SOA using different gain clamp power levels. Back-to-back and after 40 km SMF transmissions are shown. With a −3 dBm gain clamp, a 1.5 dB margin after 40 km SMF transmission can be obtained.

4.6 100 km PAM4 Transmission

Unlike all previous sections which discussed the applications of 1.3 μm EMLs, this section will explore the use of 1.55 μm EMLs for inter-datacenter and metro links with a transmission distance up to 100 km. In this effort, PAM4 and DMT [28, 29] have gained a lot of traction due to their high spectra efficiency while having lower cost and power consumption afforded by direct detection. Both PAM4 and DMT have explored the use of 10G non-tunable optical transmitters to carry 28 Gbaud PAM4 signals with the help of pre-emphasis and FEC. While this review focuses on the use of EMLs, 100 km PAM4 transmission can also be achieved using DMLs [30], InP MZMs [31], LiNbO3 MZMs [32], and silicon photonic MZM [33].

4.6.1 Experimental Setup

The test setup for this 100 km transmission is shown in Figure 4.15. In this scenario, a 28 Gbaud PAM4 ASIC was used together with a linear driver to drive the EML. Up to 96 such C-band wavelengths can be multiplexed with a DWDM MUX and boosted by an EDFA. The launch power here needs to be limited as to avoid optical fiber nonlinearities whose effect will also be

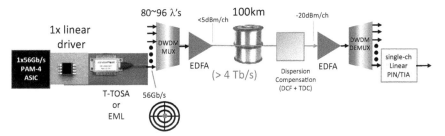

Figure 4.15 System for 100 km transmission using a 10 G, C-band EML.

explored later in this section. Unlike O-band, the C-band wavelengths carry a significant amount of dispersion which must be handled by a section of TDC and an optional DCF. The removal of these optical dispersion compensators can be expected by using innovative and practical modulation techniques soon. In Figure 4.15, the dispersion compensation block is composed of a fixed-length DCF and a TDC. This dispersion compensation block lies before a pre-amplifier in order to maintain low power into the DCF to avoid non-linear effects. The channels are demultiplexed by a DWDM DEMUX and received by a linear PIN/TIA. Finally, FFE was applied to the received signal using 3 pre-cursors and 6 post-cursors.

4.6.2 Single Channel Characteristics

By scanning the TDC, the sensitivity of the required dispersion compensation is shown in Figure 4.16(a). An order of magnitude loss in BER occurs from a dispersion offset of about $+/- 100$ ps/nm. Thus, the signal is very sensitive to chromatic dispersion and required fine dispersion adjustment.

The system's OSNR limit is shown in Figure 4.16(b). For the KP4 FEC limit of 2.4e-4, the OSNR sensitivity was 30 dB at a received optical power of -6.5 dBm.

4.6.3 Effect of Fiber Nonlinearities

The effect of 2 adjacent neighbor channels with 50 GHz spacing is shown in Figure 4.17. The neighbor channels consist of two 28 Gbaud PAM4 modulated wavelengths spaced 50 GHz apart from the measured center channel near 1550 nm. Their spectra is shown in the inset of Figure 4.17. As the launch power increases, the effect of cross-phase modulation induced intensity fluctuation becomes apparent above $+5$ dBm launch power.

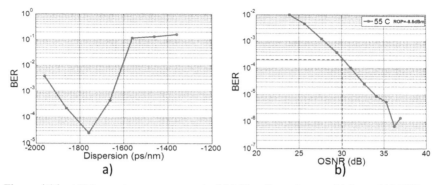

Figure 4.16   100 km system measurement of (a) fiber dispersion sensitivity and b) BER vs. OSNR.

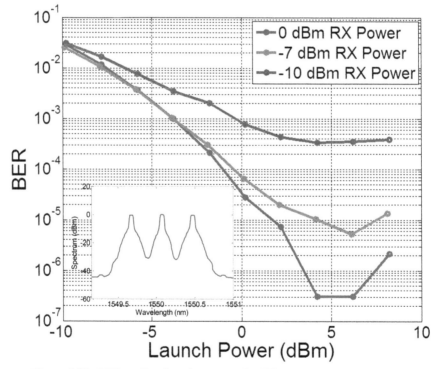

Figure 4.17   BER vs. fiber launch power under different received optical power.

This is shown at 0, −7 and −10 dBm received powers where signal integrity improved at higher received powers. Each of these cases consistently show that the launch power should be limited to under +5 dBm in order to avoid the negative effects of fiber nonlinearity. For a 100 km link, and assuming 0.25 dB/km fiber loss, this gives a per channel pre-amplifier (with a noise figure of 6 dB) input power of −20 dBm which corresponds to an OSNR of 32 dB. According to Figure 4.16(b), this gives a BER of 5e-5 which is within the KP4 FEC limit for error free transmission.

4.7 Multipath Interference [34, 35]

Multipath interference (MPI) have been studied since 1989 [36–38] for 10 Gb/s OOK/NRZ signals. The underlying physics is that one or more doubly-reflected signals (see Figure 4.18) which interfere with the direct-path signal could cause an interferometric effect which converts laser phase noise to intensity noise, thereby increasing the system RIN noise (see Equation (4.4)) significantly. Since PAM4 signal is much more sensitive to intensity noise than OOK-NRZ due to its reduced inner eye amplitudes, MPI was thought to be detrimental to PAM4 systems. However, thanks to the advancement in forward-error-correction, while the MPI effect on 10 Gb/s systems were studied with respect to a BER of 1e-9 or lower, PAM4 today can tolerate a pre-FEC BER of 2.4e-4 in a single-mode fiber transmission link.

Each doubly-reflected signal in Figure 4.18 experiences a different polarization rotation and phase change. Polarization rotations are described by the Jones vectors and randomized according to these vectors. It interferes

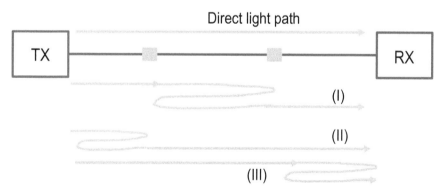

Figure 4.18 A diagram illustrating multipath interferences due to doubly reflected light paths: (I) between connectors; (II) between a connector and TX reflectance; and (III) between a connector and RX reflectance.

with the original signal according to their corresponding dot product. The phase changes are associated with the path lengths. On the wavelength scale, these phase changes are random due to temperature drifts and changes in the apparent transmission fiber length. Additionally, the signal has a randomized amplitude between 0 and the maximum amplitude of the PAM4 signal.

Multipath interference can be examined using an upper-bound approach and using a time-domain mixing approach [40]. The upper-bound approach is the worst case with zero linewidth signal where phase, polarization and amplitude are all in alignment at the maximum PAM4 amplitude and represented with one signal that is mixed with the original. The time-domain mixing approach considers the explicit sum of all reflections which have varying phases, amplitudes and polarizations. The experimental and theoretical embodiments of these two approaches will be explored here.

MPI for PAM4 links have been studied in IEEE802.3bs intensively. For example, by using the upper bound-approach, Bhatt [40] had found that the triple link with 35 dB return loss connectors was sufficient with PAM4 with a calculated MPI penalty of 0.34 dB. Farhood [41] took the upper-bound approach and improved it with a more statistical representation of MPI amplitude. The main difference here was that Bhatt had assumed maximum signal amplitude in the MPI while Farhood considered signals with a random optical amplitude. In doing so, an even lower MPI penalty of 0.26 dB was found for the same triple link study. The time domain mixing approach was also studied by King [42] using a statistical model where TX and RX reflections were also included. A link to a spreadsheet of this analysis can be found in his contribution so that the reader can also explore this model [43]. In his study, he randomized MPI intensity and phase while keeping polarizations aligned, and used the best cases falling within 99.9999% probability to determine the MPI penalty. For 35 dB return loss connectors (at fiber, transmitter and receiver facets), this penalty was under 0.3 dB. When the transmitter and receiver had lower return loss at 26 dB, the penalty increases to 0.9 dB.

We have carried out two 28 Gbaud PAM4 MPI experiments, to be described below, one belongs to the Upper Bound, and the other belongs to the time-domain mixing approach. Table 4.2 summarizes the difference in the statistical assumptions of polarization, phase, and amplitude in various studies.

Table 4.2 Statistical Assumptions of Different Parameters in MPI Studies

Reference	Random Polarization	Random Phase	Random 4-level Amplitude
[40]	No	No	No*
[41]	No	No	Yes
[42]	No	Yes	Yes
Our experiment#1 (Figure 4.20)	No	No	No**
Our experiment#2 (Figure 4.23)	Yes	Yes	Yes

*The received signal and interference are both at the peak PAM4 amplitude.
**The received signal and interference are both at one of the four PAM4 amplitudes.

4.7.1 Experimental Demonstration of the Upper Bound MPI Scenario

To experimentally find the upper bound of multipath interference, the setup shown in Figure 4.19 was used. This is based off a 4-lane 28 Gbaud PAM4 configuration using a quad EML TOSA which includes a wavelength multiplexer.

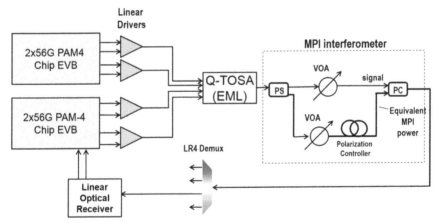

Figure 4.19 MPI interferometer used to introduce multipath interference into the signal. The bottom arm in the interferometer is the path for MPI whose polarization is aligned to the signal for maximum interference.

In this experiment, a 31 GHz driver was used to amplify the signal to around 2.0 Vpp. The signal was then biased around -0.6 V to center the modulation within the linear region of the EML response. This was done for each channel of a quad, LR4 TOSA which consisted of 4 EMLs and a multiplexer. The single channel output power was ~ 1.6 dBm. The back reflection at the EML output was measured to be -41.9 dB which is low enough to be negligible compared to the artificially-generated MPI levels in this study.

The multiplexed signal is divided into two paths, i.e., the main and interferer paths, to simulate the interferometer effect. The interferer path is decorrelated from the main channel by a couple meters of extra fiber length. A polarization controller was used to align the polarization of the two paths for the worst-case interference. Since this setup uses only one interferer path, this can be thought of as equivalent to a multipath case where every path is added in perfect synchronized phase, amplitude, and polarization. Therefore, this represents the upper bound of multipath interference. Note that although the amplitudes of the signal and interference are aligned, they are aligned at any of the PAM4 levels. This is unlike the case of [39] where the signal and interference are aligned at the peak PAM4 amplitude.

The mixed signal was then filtered by a demultiplexer and measured by a linear PIN and TIA ROSA. To adequately detect the signal, a 22 GHz TIA was selected with 17 pA/$\sqrt{(\text{Hz})}$. With received powers down to -10 dBm, the PIN with 0.45 A/W responsivity was used. At the PIN output, -47.3 dB optical back reflection was measured which is negligible compared to the induced MPI levels in this study. The differential output of the TIA was recorded using a 33 GHz, 80 GSample/second DSO whose waveforms were recorded for offline processing.

From Figure 4.20, the MPI penalty can be seen by comparing the inner OMA power at the KP4 FEC limit of 2.4e-4. For example, the difference between the case of no MPI and -34 dB MPI was 1 dB. In the triple link channel, -34 dB MPI is equivalent to -26 dB reflectance on the transmitter and receiver, and -35 dB reflectance at patch cable connectors where the polarization and phase from each reflection is aligned, and the PAM4 amplitudes are aligned. Although this penalty is large, one must keep in mind that this is the worst case scenario where the polarization, phase and amplitude of each path are all aligned. In reality, this has almost no chance of ever occurring. As a result, a guideline for MPI cannot be based on the worst case scenario, but rather should be based on a more statistical analysis. For this, different Monte Carlo simulations can be employed, as shown in Table 4.2.

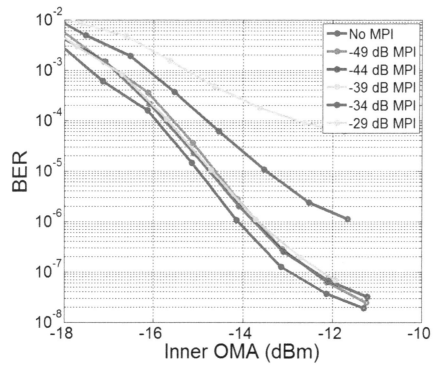

Figure 4.20 BER vs. inner OMA for various MPI powers.

In IEEE802.3bs, [42] which used aligned polarization and random phase and amplitude, and additional 3 dB link attenuation, an MPI penalty of 0.5 dB was adopted. The added 3 dB attenuation reduce the MPI penalty from the previously mentioned 0.9 dB to 0.5 dB.

4.7.2 Time-Domain Mixing Monte Carlo Simulation

The Monte Carlo simulation of MPI involves the addition of random phase, amplitude and polarization for each possible double pass fiber path permutation. Although paths exist consisting of even number of reflections greater than two, it can be assumed that the power level in these paths is negligible compared to two reflection paths. The MPI is then described by Equation (4.5).

$$\overrightarrow{E} = Ae^{-i\theta}\hat{u} + \sum_{m}\sum_{n} r_m r_n A_{mn} e^{-i\theta_{mn}} \hat{u}_{mn} \qquad (4.5)$$

This equation describes the resulting field (E). The main signal is described by the first term on the right hand side of the equation with amplitude A, phase θ and polarization u. The second term is the summation of the double reflection paths caused by connectors m and n. Connector m has a reflectance, r_m, and results in a field with randomized amplitude, phase and polarization. For PAM4, the amplitude is constrained between the top and bottom PAM4 levels. As shown in Table 4.2, this is different from King's simulations as this now assumes that polarization is also randomized.

A result derived from a Monte Carlo simulation is shown in Figure 4.21. This was done for the condition of -26 dB connector reflectance and -26 dB reflectance at the transmitter and receiver ports. The histogram shows the distribution of crosstalk power that has resulted from samples with random phase, amplitude and polarization. For PAM4, this power would be added to the PAM4 signal (by adding the equivalent MPI electric filed) and create a period of burst errors for the time that the given MPI state is present. Each sample in the Monte Carlo simulation represents one MPI state. This histogram contains 40000 samples where the distribution averages to

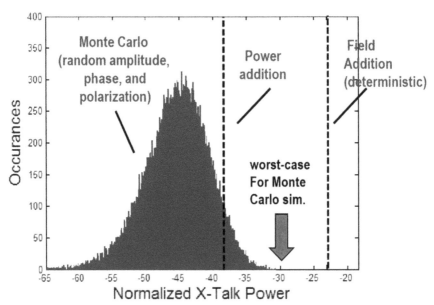

Figure 4.21   Monte Carlo histogram of MPI crosstalk powers from randomized polarization, phase, and symbol levels from each connector reflection in a triple link system. Sample size = 40,000.

around -45 dB. 40000 samples were simulated to allow for reasonable simulation time. In the next section, we will show that 40000 samples represents signal transmission times over thousands of years. After 40000 samples, the worst observed case was below -30 dB. The theoretically worse case (when the amplitude is the greatest with phase and polarization alignment) is shown at -23 dB. This was obtained through perfect constructive interference of all interfering paths by adding the reflected signal electric fields (at the maximum PAM4 symbol level) in perfect coherent alignment (phase and polarization are matched). The large 7 dB difference between the worst observed case and the worst theoretical case supports the fact that perfect multipath alignment is extremely unlikely to occur and a statistical approach is more useful to determine the MPI characteristics. In this case, -30 dB cross-talk power is a more accurate representation of the MPI and is observed in 0.0025% of all MPI states. By using different sample sizes in a Monte Carlo simulation and comparing with the experimental results in the next section, we can estimate how long in real time it takes to achieve this low probability state. In our experiments, BER will be observed over an 8 hour period within which MPI will cause burst errors. The worst measured instantaneous BER within these 8 hours will be considered as the representative number. We then find the Monte Carlo sample size which also gave the same BER as this worst case. This sample size will represent an 8 hour period of time. From there, we can use the simulated sample sizes to extrapolate how the MPI will behave over longer periods of time. For example, in Figure 4.23, the light blue curve was the worst obtained BER in the measurement and is closely overlapping with the simulated yellow curve that corresponds to 500 Monte Carlo samples. The curve with 40,000 Monte Carlo samples has 80 times more samples so we can estimate that if the experiment was taken 80 times longer, we should see a worst case BER matching with the Monte Carlo simulation of 40,000 samples.

4.7.3 MPI Experiment with Multiple Connectors

A lab experiment was carried out to reflect the Monte Carlo simulation in a real world trial. The experiment is based on a triple link setup [39]. A triple link consists of three link sections connected through patch cords and equipment cords at the ends, see Figure 4.22(a). Schematically, this is represented by 6 fiber to fiber connections and 2 fiber to port connections, as shown in our experimental setup (Figure 4.22(b)).

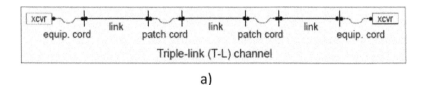

a)

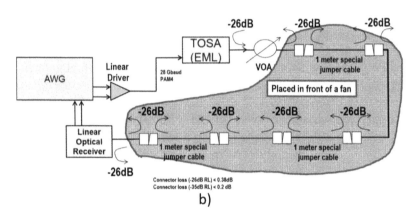

b)

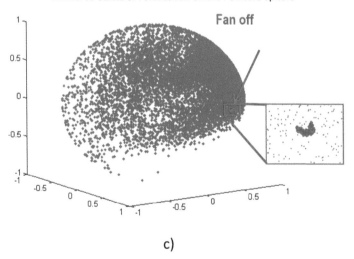

c)

Figure 4.22   Experimental setup used to create randomized MPI in a PAM4, a) triple link channel [39], b) the experimental setup based on a triple link model where the connector reflectance was purposely polished to be −26 dB, and c) the Poincare sphere showing significant polarization variation by shaking the fiber over 8 hours. When the fan was off, the SOP was spread in the narrow red arc range.

In this experiment, multipath interference was generated using connectors which were purposely polished to have around −26 dB reflectivity. A 28 Gbaud PAM4 signal was transmitted in this triple link channel arrangement. As mentioned earlier, the EML and PIN had −41.9 and −47.3 dB reflectivity, respectively. To create a real world scenario, additional −26 dB reflectivity connectors were used at the EML and PIN to imitate a higher system input and output return loss. The set of patch cables were hung up in front of a strong fan to induce random polarization fluctuation. This motion causes polarization and phase changes to the double reflection paths. Random amplitudes are inherent from the PAM4 symbol modulation. These three random parameters combined were the basis for the Monte Carlo simulation.

Using a polarimeter, the polarization was measured in order to verify sufficient randomness in this test. The polarimeter showed an adequate distribution of samples enveloping the Poincare sphere, as shown in Figure 4.22(c). Using the 50% state of polarization decorrelation time as a comparison metric [44], it was estimated that the polarization state change in our experimental setup is 36,000 times faster than a fiber located in a stable intra-data center environment. The factor of 36,000 comes from comparing the average 50% polarization decorrelation times for fiber when the fan is on versus when the fan is off. When the fan was on, the average decorrelation time was 0.8 s whereas when the fan was off, the state of polarization remained constant for longer than 8 hours.

The BER vs. inner OMA results from this experiment are shown in Figure 4.23. The BER at each measured inner OMA point in the graph was measured for 8 hours and get the worst result out of the distribution. With an acceleration factor of 36,000 times, this is equivalent to 26 years of static operation. An OBTB static case is shown here to contrast with the triple link channel. It can be seen that the link with and without back reflecting jumpers are identical with respect to their average BER. The worst measured BER during the 8 hour run times are also plotted which increase as the MPI comes more into alignment. As expected, the presence of the jumpers produced a higher worst case BER because MPI is now present from multiple connectors. To extrapolate the data, we draw a comparison to the Monte Carlo simulation samples, as shown by the dashed lines in Figure 4.23. One sample of the Monte Carlo simulations consists of 2^{15} randomized PAM4 symbols which experiences one state of MPI interference. This assumes that the MPI will be constant for this period, generating a corresponding BER. Although the MPI power is likely not held over this many bits, this will give us a conservative estimate for the BER by overestimating the burst error. Different Monte Carlo

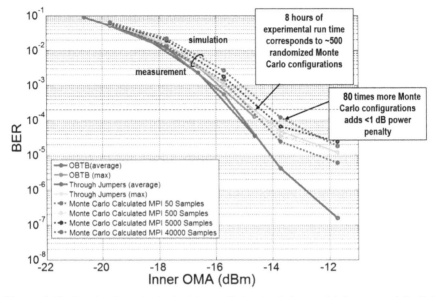

Figure 4.23   Results from both the Monte Carlo simulation and laboratory triple link experiment using connectors with −26 dB return loss. The experimental sample size can be compared with the simulations to extrapolate the long term MPI penalties.

sample set sizes are also shown in Figure 4.23 where more samples lead to an expected increase in the BER. From the light blue curve (measured worst case in the 8-hour experiment) and the dotted-light green curve, we can see that a sample size of 500 matches with the 8 hours of experimental run time, or 26 years of normal operation. From the dotted blue curve, it is shown that by the increase in Monte Carlo sample sizes to 40,000 samples, an extra 1 dB penalty is incurred when comparing to the dotted yellow curve (500 samples). Since this experiment was equivalent to 26 years, this implies that a 1 dB penalty in BER (dotted-light green versus dotted-blue curve) occurs in a time frame 80 times longer (= sample size ratio (40,000/500)), and therefore will statically occur in 2080 years.

The same experiment and simulations were carried out using jumpers with −35 dB connector reflections, with results shown in Figure 4.24. In contrast to the case of six −26 dB connectors, this represents a much safer requirement as the MPI power becomes less significant. In terms of both measured average and worst-case results, we see indistinguishable difference in OMA at BER=2e-4 with or without the multiple jumpers. In terms of Monte Carlo results, we see that the OMA penalty at BER=2e-4 increases

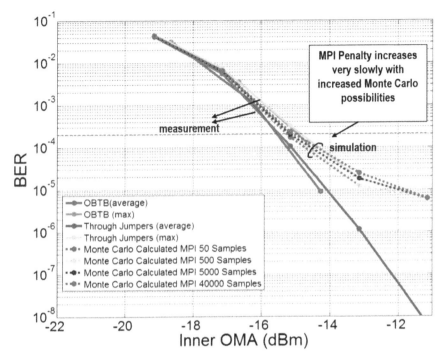

Figure 4.24   Results from both the Monte Carlo simulation and laboratory triple link experiment using connectors with −35 dB reflectivity. Large MPI sample sizes do not show much increase in power penalty due to the low MPI power from the connectors.

very slowly with additional samples. At 40,000 randomized samples (the dotted blue curve, corresponding to 2080 years), the penalty is virtually zero compared to the short term case consisting of 50 samples (the dotted-purple curve). Consequently, IEEE802.3bs has set the connector reflectance to −35 dB for the triple-link case with 6 connectors.

4.8 Summary

We have reviewed our work in using various high-speed EMLs to transport 28 and 56 Gbaud PAM4 signals over 2, 10, 40, and 100 km standard single-mode fibers. These work can serve as a base for future improvement in system performance by further improving (a) EML bandwidth, linearity and RIN; (b) receiver bandwidth and thermal spectral noise density, and (c) power-efficient transmitter and receiver linear and nonlinear equalization.

We believe that in the near future, EMLs can find many important and novel applications. These include: 28 and 56 Gbaud PAM4 EMLs for 5G front- and back-haul transmissions; $53 \sim 56$ Gbaud PAM4 EMLs for 10 km transmission with sufficient link margin; and for potentially cost-effective access [45] and regional [46] transmission systems.

References

[1] D. A. Miller, D. S. Chemla, T. C. Damen, A. C. Gossard, W. Wiegmann, T. H. Wood, C. A. Burrus, 'Band-edge electroabsorption in quantum well structures: The quantum-confined Stark effect', Physical Review Letters, 53(22), 2173, 1984.

[2] T. Chan, I. C. Lu, J. Chen, W. I. Way, '400-Gb/s transmission over 10-km SSMF using discrete multitone and 1.3-μm EMLs', IEEE Photonics Technology Letters, 26(16), 1657–1660, 2014.

[3] K. Koyama, J. I. Hashimoto, T. Ishizuka, Y. Tsuji, T. Yamada, C. Fukuda, Y. Onishi, K. Fujii, T. Katsuyama, 'GaInNAs electroabsorption modulated laser' IEEE 19th International Conference on In Indium Phosphide & Related Materials, 2007.

[4] M. R. Gokhale, P. V. Studenkov, Ueng- J. McHale, J. K. Thomson, J. Yao, J. van Saders, 'Uncooled, 10 Gb/s 1310 nm electroabsorption modulated laser', Optical Fiber Communication Conference, Atlanta, GA, 2003.

[5] W. Kobayashi, M. Arai, T. Yamanaka, N. Fujiwara, T. Fujisawa, T. Tadokoro, K. Tsuzuki, Y. Kondo, F. Kano, 'Design and fabrication of 10-/40-Gb/s, uncooled electroabsorption modulator integrated DFB laser with butt-joint structure', Journal of Lightwave Technology, 28(1), 164–171, 2010.

[6] http://www.ieee802.org/3/bs/public/adhoc/smf/16_04_19/anslow_01_04 16_smf.pdf

[7] https://www.xilinx.com/publications/events/designcon/2017/112gbps-serial-transmission-over-copperpam4-vs-pam8-slides.pdf

[8] http://www.ieee802.org/3/100GNGOPTX/public/mar12/plenary/trem blay_01_0312_NG100GOPTX.pdf

[9] N. Stojanovic, F. Karinou, Z. Qiang, & C. Prodaniuc, 'Volterra and Wiener equalizers for short-reach 100G PAM-4 Applications', Journal of Lightwave Technology, 35(21), 4583-4594, 2017.

[10] http://www.ieee802.org/3/bs/public/14_11/parthasarathy_3bs_01a_1114. pdf

[11] http://www.itu.int/itu-t/recommendations/index.aspx?ser=G

[12] http://100glambda.com/specifications

[13] http://www.ieee802.org/3/bs/index.html IEEE802.3bs^{TM}/D3.5

[14] N. Eiselt, et al., 'Evaluation of Real-Time 8×56.25 Gb/s (400G) PAM-4 for Inter-Data Center Application Over 80 km of SSMF at 1550 nm', J. Lightwave Tech., 35(4), 955–962, 2017.

[15] S. Bhoja, 'PAM4 signaling for intra-data center and data center to data center connectivity (DCI)', Optical Fiber Communication Conference, Los Angeles, CA, 2015.

[16] http://www.ieee802.org/3/bs/public/14_05/way_3bs_01a_0514.pdf

[17] http://www.ieee802.org/3/bs/public/15_01/way_3bs_01a_0115.pdf

[18] http://www.ieee802.org/3/bs/public/15_03/stassar_3bs_01a_0315.pdf

[19] R. Motaghiannezam, T. Pham, A. Chen, T. Du, C. Kocot, J. Xu, & B. Huebner, '52 Gbps PAM4 receiver sensitivity study for 400 GBase-LR8 system using directly modulated laser', Optics express, 24(7), 7374–7380, 2016.

[20] T. Chan and W. I. Way, '112 Gb/s PAM4 transmission over 40 km SSMF using 1.3 μm gain-clamped semiconductor optical amplifier', Optical Fiber Communications Conference, Los Angeles, CA, 2015.

[21] T. Kagawa, Y. Kawamura, and H. Iwamura, 'Saturation of multiplication factor in InGaAsP/InAlAs superlattice avalanche photodiodes', *Appl. Phys. Lett.* 62(10), 1122, 1993.

[22] M. Nada, S. Kanazawa, H. Yamazaki, Y. Nakanishi, W. Kobayashi, Y. Doi, T. Ohyama, T. Ohno, K. Takahata, T. Hashimoto, and H. Matsuzaki, 'High-linearity avalanche photodiode for 40-km transmission with 28-Gbaud PAM4', Optical Fiber Communication Conference, Los Angeles, CA, 2015.

[23] IEEE802.3ba-D21

[24] K.-P. Ho and C. Lin, 'Distortion and crosstalk reduction in semiconductor laser amplifier for WDM systems', IEEE LEOS, 1996.

[25] Y. Sun, A. K. Srivastava, S. Banerjee, J. W. Sulhoff, R. Pan, K. Kantor, R. M. Jopson, and A. R. Chraplyvy, 'Error-free transmission of 32×2.5 Gb/s DWDM channels over 125 km of AllWave^{TM} fiber using cascaded in-line semiconductor optical amplifiers', Optical Amplifiers and their Applications, 30, 1999.

[26] S. Verspurten, G. Morthier and R. Baets, 'Experimental and numerical small-signal analysis of two types of gain-clamped semiconductor optical amplifiers', IEEE J. Quantum Electron 42(3), 302, 2006.

[27] M. Yoshino and K. Inoue, 'Improvement of saturation output power in a semiconductor laser amplifier through pumping light injection', IEEE Photon. Tech. Lett. 8(1), 58, 1996.

[28] T. Takahara, T. Tanaka, M. Nishihara, L. Li, Z. Tao, and J. C. Rasmussen, '100-Gb/s (2 × 50-Gb/s) transmission over 80-km using 10-Gb/s class DML', Proc. OECC, 417, 2012.

[29] A. Dochhan, H. Grieser, M. Eiselt, and J.-P. Elbers, 'Flexible bandwidth 448 Gb/s DMT transmission for next generation data center interconnects', European Conference on Optical Communications, Anaheim, CA, 2014.

[30] S. Zhou, X. Li, L. Yi, Q. Yang, and S. Fu, 'Transmission of 2 × 56 Gb/s PAM-4 signal over 100 km SSMF using 18 GHz DMLs', Opt. Lett. 41(8), 1805–1808, 2016.

[31] S. Yin, T. Chan, and W. I. Way, '100-km DWDM Transmission of 56-Gb/s PAM4 per lambda via Tunable Laser and 10-Gb/s InP MZM', IEEE Photonics Technology Letters, 27(24), 2531–2534, 2015.

[32] N. Eiselt, J. L. Wei, H. Griesser, A. Dochhan, M. Eiselt, J.-P. Elbers1, J. J. Vegas Olmos, and I. T. Monroy, 'First Real-Time 400G PAM-4 Demonstration for Inter-Data Center Transmission over 100 km of SSMF at 1550 nm', Optical Fiber Communications Conference, Anaheim, CA, 2016.

[33] M. Filer, S. Searcy, Y. Fu, R. Nagarajan, and S. Tibuleac, 'Demonstration and performance analysis of 4 Tb/s DWDM metro-DCI system with 100G PAM4 QSFP28 modules', Optical Fiber Communications Conference, Los Angeles, CA, 2017.

[34] http://www.ieee802.org/3/bs/public/15_09/way_3bs_01a_0915.pdf

[35] http://www.ieee802.org/3/bs/public/16_01/way_3bs_01a_0116.pdf

[36] J. L. Gimlett, N. K. Cheung, 'Effects of phase-to-intensity noise conversion by multiple reflections on gigabit-per-second DFB laser transmission systems', Journal of Lightwave Technology, 7(6), 888–895, 1989.

[37] C. R. Fludger, M. Mazzini, T. Kupfer, & M. Traverso, 'Experimental measurements of the impact of multi-path interference on PAM signals', Optical Fiber Communication Conference, San Francisco, CA, 2014.

[38] P. V. Mena, E. Ghillino, A. Ghiasi, B. Welch, M. S. Khaliq, & D. Richards, '100-Gb/s PAM4 link modeling incorporating MPI', Optical Interconnects Conference (OI), 2015.

[39] http://www.ieee802.org/3/bs/public/14_05/kolesar_3bs_01_0514.pdf

[40] http://www.ieee802.org/3/100GNGOPTX/public/may12/bhatt_01_0512_optx.pdf

[41] http://www.ieee802.org/3/bm/public/nov12/farhood_01_1112_optx.pdf

[42] http://www.ieee802.org/3/bs/public/adhoc/smf/16_01_07/king_01a_0116_smf.pdf

[43] http://www.ieee802.org/3/bs/public/adhoc/smf/16_01_07/king_02_0116_smf.7z

[44] D. Waddy, P. Lu, L. Chen, and X. Bao, 'The measurement of fast state of polarization changes in aerial fiber', Optical Fiber Communication Conference, Anaheim, CA, 2001.

[45] B. Schrenk, and F. Karinou, "World's first TO-can coherent transceiver", Optical Fiber Communication Conference, San Diego, CA, 2018.

[46] C. Xie, "Transmission of 128-Gb/s PDM-4PAM generated with electroabsoption modulators over 960-km standard single-mode fiber", Optical Fiber Communication Conference, San Francisco, CA, 2014.

5

Optical Fiber for Datacenter Connectivity

Yi Sun and John Kamino

OFS Fitel LLC, Norcross, GA, USA

5.1 Introduction

Modern optical fiber communications originated from a paper published in 1966 by Dr. Charles Kao [1–3]. Kao proposed using a thin strand of glass to transmit an optical signal and outlined all basic properties of a single-mode optical fiber, in addition to the associated manufacturing processes. The conclusion of his paper stated the following: *"Theoretical and experimental studies indicate that a fiber of glassy material constructed in a cladded structure with a core diameter of about lambda zero and overall diameter of about 100 lambda zero represents a possible practical waveguide with important potential as a new form of communication medium."* Kao's prediction accurately described what would come to be known as optical fiber. When Kao published his paper, the attenuation of contemporary bulk glass was an enormous 1000 dB/km. That attenuation was reduced to the level of tens of dB/km applicable to telecommunication [4], but modern commercially available optical fiber has much less attenuation, as low as 0.15 dB/km for commercial submarine fiber [5]. The lowest reported attenuation is 0.1419 dB/km in [6]. The world as it is today, from homes to cities, through continents and oceans, is covered by an interwoven network of optical fiber telecommunication cables. The deployed optical fibers and cables from a single vendor, OFS Fitel, LLC (formerly Lucent Technologies), is enough to make 628 round-trips between the Earth and the Moon. This intricate web of connectivity forms the physical foundation of the World Wide Web, the three eras of which are defined by Steven Case [7]. The first era covers the construction of the internet between 1985 and 2000. The second era involves the creation of new internet-dependent services between 2000 and 2015 and also encompasses the foundation of global cloud infrastructure, including datacenters and interconnected networks. The third and current era involves creating an internet of everything (IoT), establishing a cloud-integrated network (CIN), and exploiting Augmented Intelligence, the ability to transform data into action based on big data analysis [7]. Some experts argue that the first and second eras of the internet form the beginning of a third industrial revolution, and that advancements in technology and economic processes, more radical than any previously experienced, lie ahead [7]. The current backbone fiber communication paradigm, based on single-mode fiber and the optical fiber amplification, is already within a factor of two from maximum usage of the current telecommunications spectrum [7]. Global networks are experiencing increased demand for high-quality optical fibers while next-generation data traffic is quickly surpassing that of traditional telecom across the internet backbone.

A large portion of traffic through optical networks is composed of data transmitted within (intra-) and between (inter-) datacenters. Figure 5.1 illustrates a global optical network with emphasis on datacenter data transmission. This ever-increasing amount of information demands higher speed optics and greater capacity in intra-datacenter, metro, long haul, and submarine optical networks. "Cisco Visual Networking Index (VNI): Forecast and Methodology, 2016–2021" (published in 6/6/2017) predicts a 24% compound annual growth rate (CAGR) of global IP traffic between 2016 and 2021, from 1.2 to 3.3 ZB/year. By the end of the 2021, Cisco predicts that the number

Figure 5.1 Illustration of a global optical network formed by datacenters and inter-datacenter backbone optical cables.

of IP devices will be $3\times$ the global population, and monthly IP traffic will reach 35 Gbits/month.

Traffic within and between datacenters will grow at even higher rates. Per [8], global datacenter traffic will reach 15.3 ZB/year in 2020, up from 4.7 ZB in 2015, with the CAGR of 27%. The traffic within datacenters, from datacenter to user, and from datacenter to datacenter comprises 77%, 14%, and 9%, respectively, of all global data traffic [8]. Hyper-scale datacenters accounted for 34% of total traffic within datacenters in 2015 and will make up 53% of total internal datacenter traffic by 2020 [9]. By 2020, hyper-scale datacenters will account for 47% of all installed datacenter servers [9].

The Ethernet Alliance has developed a roadmap for speeds, driven by increasing traffic demand from both traditional and cloud carriers [10]. Inside the datacenter, enterprise server I/O speed is currently 10G, while cloud services are migrating to 25G, and will increase to 50G around 2019 and to 100G thereafter. In general, enterprise datacenters migrate to higher speeds at a slightly slower rate than the hyper-scale/cloud market. The speed of switch ports, meanwhile, is generally four times that of server I/O. 100G datacenter switches have already been deployed in 2016 in hyper-scale, but 40G remains common in enterprise.

As speeds have increased from 10 Gbps, optical fiber has dominated as the medium of choice in datacenters, except for the shortest server connections. While copper was widely deployed for 1 Gbps links and some short 10 Gbps links, single-mode and multimode fiber deployment now dwarfs its formal rival. Multimode fiber is typically used in short-reach applications inside the datacenter, while single-mode fiber is used in longer link distances, sometimes found in hyper-scale applications and inter-datacenter links.

Both 40G and 100G IEEE 802.3 standards adopted parallel multi-mode optics. Some proprietary multimode solutions use wavelength division multiplexing to reduce the number of fibers required for 40G and 100G transmission. For longer links ($\geq$300/500 m) encountered in hyper-scale datacenters, proprietary 40 and 100G single-mode solutions have been developed that utilize either parallel fibers or lower cost WDM technology. 200G and 400G standards are currently under development, and several multi-source agreements (MSA) and proprietary solutions have been announced.

Server speeds and switch port speeds are also relevant when discussing intra-datacenter transmissions. For inter-datacenters, 100G client optic speeds are currently being deployed and 400G speeds will soon be needed for long-haul transport. 1.6 Tbps speeds are on the roadmap after 400G [10].

400G-DR4, FR8, and LR8 cover maximum reaches of 500 m, 2 km, and 10 km, respectively. IEEE 802.3 has created a CFI to consider 400G options that would cover a >10 km reach, including PAM4 direct detection and leveraging 400G-ZR coherent solution created by OIF [11].

The size of datacenters can vary from several servers and a top-of-rack switch to hyper-scale, measuring hundreds of thousands of square feet. These datacenters can be in one building or several buildings located in one or several campuses in a metro area less than 80 km apart. Cloud providers have warehouse-scale datacenters located across the globe where the datacenter interconnect distance can far exceed 80 km, often necessitating long-haul territorial and submarine networks [12]. Within datacenters, 90% of links are shorter than 100 m and involve traditional copper, multimode optical fiber, and single-mode optical fiber communication mediums. For high-speed links, copper is capable of reaching up to only tens of meters. For 10 Gbps speeds, multimode optical fiber supports reaches up to 550 m. Single-mode fiber is preferred for reaches longer than 500 m. The reach supported by IEEE 802.3 40/100G-SR4 is reduced as a trade-off for a relaxed transceiver specification, and the standard defined link length is 150 m for 40G and 100 m for 100G. However vendors have developed proprietary transceivers to extend the reach comparable with 10G-SR link distances (300–400 m). The speed, modulation format, fiber and cable types, the counts of fiber, and wavelengths used in existing and proposed IEEE and proprietary Ethernet applications from 25 to 400 Gbps are summarized in Table 5.1.

The rest of this chapter is structured as follows: Section 5.2 reviews fiber and cable types for datacenter connectivity, including MMF and single-mode fiber (SMF) recommended by IEEE 802.3 standards and novel fibers in research; Section 5.3 covers the fundamentals of waveguide design, mode structure and the physical mechanism of time response in SMF and MMF; Section 5.4 focuses on three types of multimode optical fibers (OM3, OM4, OM5) for high-speed short-reach interconnects. Fiber characterizations on differential modal delay (DMD) and bandwidth (BW) are explained in detail; in Section 5.5, system evaluation methodology is reviewed and the VCSEL-MMF is validated as a low-cost short-reach interconnect solution through several system experiments conducted at speeds from 10 to 400 Gbps; Section 5.6 summarizes the requirements for single-mode fiber and cabling in IEEE 802.3 standards, concluding with Section 5.7.

Table 5.1 Speed, modulation format, baud rate, fiber type, cable type, fiber, and wavelengths specified for existing and proposed IEEE and proprietary Ethernet applications from 25 to 400 Gbps

Speed	Applications	Format	Baud Rate	Fiber Type	Cable Type	# of fiber	# of wavelength	
25Gbps	25G BASE-SR	NRZ	25.8G	MMF	Duplex	2	1	IEEE 802.3by
40Gbps	40G BASE-SR4	NRZ	10.3G	MMF	MPO parallel	8	1	IEEE 802.3-2015
	40G SWDM4	NRZ	10.3G	MMF	Duplex	2	4	proprietary
	40G BASE-SR-BiDi	NRZ	20G	MMF	Duplex	2	2	proprietary
50Gbps	50G BASE-SR	PAM4	26.6G	MMF	Duplex	2	1	IEEE802.3cd task force
100Gbps	100G BASE-SR10	NRZ	10.3G	MMF	MPO parallel	20	1	IEEE 802.3-2015
	100G BASE-SR2	PAM4	26.6G	MMF	parallel	4	1	IEEE802.3cd task force
	100G BASE-SR4	NRZ	25.8G	MMF	MPO parallel	8	1	IEEE 802.3-2015
	100G SWDM4	NRZ	25.8G	MMF	Duplex	2	4	SWDM MSA
	100G PAM4 BiDi	PAM4	25.8G	MMF	Duplex	2	2	proprietary
200Gbps	200G BASE-SR4	PAM4	26.6G	MMF	MPO parallel	8	1	IEEE802.3cd task force
	200G BASE-SR1.4	PAM4	26.6G	MMF	Duplex	2	4	Pre-IEEE 802.3 CFI
400Gbps	400G BASE-SR16	NRZ	26.6G	MMF	MPO parallel	32	1	IEEE802.3bs task force
	400G BASE-SR4.2	PAM4	26.6G	MMF	MPO parallel	8	2	Pre-IEEE 802.3 CFI
25Gbps	25G BASE-FR	NRZ	25.8G	SMF	Duplex	2	1	IEEE802.3cc task force
	25G BASE-LR	NRZ	25.8G	SMF	Duplex	2	1	IEEE802.3cc task force
40Gbps	40G PSM4	NRZ	10.3G	SMF	MPO parallel	8	1	proprietary
	40G BASE-LR4	NRZ	10.3 G	SMF	Duplex	2	4	IEEE 802.3-2015
	40G BASE-ER4	NRZ	10.3 G	SMF	Duplex	2	4	IEEE 802.3-2015
	40G BASE-FR	NRZ	41.25 G	SMF	Duplex	2	1	IEEE 802.3-2015
50Gbps	50G BASE-FR	PAM4	26.6G	SMF	Duplex	2	1	IEEE802.3cd task force
	50G BASE-LR	PAM4	26.6G	SMF	Duplex	2	1	IEEE802.3cd task force
100Gbps	100G BASE-DR	PAM4	53.1G	SMF	Duplex	2	1	IEEE802.3cd task force
	100G PSM4	NRZ	25.8G	SMF	MPO parallel	8	1	100G PSM4 MSA
	CWDM4/CLR4	NRZ	25.8G	SMF	Duplex	2	4	CWDM4 MSA
	100G BASE-LR4	NRZ	25.8G	SMF	Duplex	2	4	IEEE 802.3-2015
	100G BASE-ER4	NRZ	25.8G	SMF	Duplex	2	4	IEEE 802.3-2015
200Gbps	200G BASE-DR4	PAM4	26.6G	SMF	PO parallel (PSM)	8	1	IEEE802.3bs task force
	200G BASE-FR4	PAM4	26.6G	SMF	Duplex	2	4	IEEE802.3bs task force
	200G BASE-LR4	PAM4	26.6G	SMF	Duplex	2	4	IEEE802.3bs task force
400Gbps	400G BASE-DR4	PAM4	53.1G	SMF	PO parallel (PSM)	8	1	IEEE802.3bs task force
	400G BASE-FR8	PAM4	26.6G	SMF	Duplex	2	8	IEEE802.3bs task force
	400G BASE-LR8	PAM4	26.6G	SMF	Duplex	2	8	IEEE802.3bs task force

5.2 Fiber Type for Datacenters

The waveguide at the core of an optical fiber can be designed to support a single propagating mode or a multiplicity of propagating modes. For longer reach links, single-mode fiber has the distinct advantage that chromatic dispersion is the primary source of inter-symbol interference (ISI). However, with a core diameter approximately 8 microns, it requires sub-micron alignment tolerances and clean, high-quality connectors to achieve low loss mating. Multimode fiber introduces the dispersion between modes as a source of ISI, but with a core size of 50 μm, the tolerance required for connections is much larger, and leads to significant cost savings in transceivers and connectors and higher tolerance to dirty connectors. Only with the advent of hyper-scale DCs has single-mode fiber become necessary, when connections beyond 200 or 300 m were needed at speeds beyond 10G.

5.2.1 Multimode Fiber Types for Datacenters

For distances less than 500 m, MMF is the preferred choice due to its low system cost. There are two general MMF fiber types defined by core size: 50 μm and 62.5 μm. 62.5 μm fiber is designated as OM1 in the ISO/IEC 11801 standard, the most common reference for multimode fiber types. Developed in the mid 1980s, 62.5 μm fiber was designed for use with LED sources, and isn't well suited to today's laser based applications. OM1 fiber only supports 33 m reaches at 10G, and is not recognized as an option for faster 40G and 100G speeds.

OM2, OM3, OM4, and OM5 fibers (also defined in ISO/IEC 11801) all have a 50 μm diameter core but have different bandwidths or information carrying capacities. OM2 fiber, like OM1, is nearly obsolete and no longer recommended for new installations. OM1 and OM2 fiber are available, but typically only used in legacy and application-specific areas. For speeds 10 Gbps and greater, OM3, OM4, and OM5 fiber should be used. OM3 was the first 50 μm laser-optimized multimode fiber, designed for 1 and 10+ Gbps applications operating at 850 nm. OM4 fiber has similar design parameters to OM3, but has higher 850 nm bandwidth and can support longer link distances. OM5 fiber has only recently been developed. It has the same 850 nm bandwidth as OM4 fiber, but is also designed for operation over a range of wavelengths from 850 to 953 nm. This allows the fiber to support WDM-based applications that are now available in the market. Table 5.2 displays each fiber type and its corresponding designation in various industry standards.

Table 5.2 MMF types in industry standards

Fiber Type	ISO/IEC 11801 (draft)	IEC 60793-2-10	TIA-568.3-D	TIA/EIA 492AAAX	ITU-T
62.5/125	OM1	Alb	TIA 492AAAA (0M1)	492AAAA	—
50/125	OM2	Ala.lb	TIA 492AAAA (OM2)	492AAAB	G.651.1
50/125	OM3	Ala.2b	TIA 492AAAC (0M3)	492AAAC	—
50/125	OM4	Ala.3b	TIA 492AAAD (0M4)	492AAAD	—
50/125	OM5	Ala.4b	TIA 492AAAE (0M5)	492AAAE	—

Table 5.3 40 and 100 Gbps multimode applications and their reach over OM3, OM4, and OM5 MMFs

Speed (Gb/s)	Application	# fibers	Reach (m) OM3	OM4	OM5
40	40GBASE-SR4*	8	100	150	150
40	40GBASE-eSR4	8	300	400	400
40	40G-BiDi	2	100	150	150/200
40	40G-SWDM4	2	240	350	440
100	100GBASE-SR4*	8	70	100	100
100	100GBASE-eSR4	8	200	300	300
100	100G-BiDi#	2	70	100	150
100	100G-SWDM4*	2	70	100	150

*-IEEE Standard
#-Announced

Table 5.3 summarizes available or announced 40 and 100 Gbps multimode applications as well as their reach over OM3, OM4, and OM5 MMFs. Note that 40/100GBASE-SR4 are designated in IEEE 802.3 standards, and both use eight parallel fibers for transmitting/receiving data to achieve 40G and 100G speeds. The other applications listed are proprietary solutions to either extend the reach (40GBASE-eSR4, 100GBASE-eSR4) or reduce fiber counts (40G-BiDi, 40G-SWDM4, 100G-BiDi, and 100G-SWDM4).

OM5 wideband fiber was first introduced in 2015, and was standardized in TIA in 2016. Standardization was completed at the end of 2017 in the international ISO/IEC arena. Multiple structured cabling vendors have begun to offer OM5-based solutions to their customers. Traditionally, it has taken 3–5 years for the enterprise structured cabling market to begin large-scale adoption of a new fiber type. For example, the first laser-optimized 50 μm multimode fiber solution that eventually became OM3 went to market in 1998. Standardization of OM3 took place in 2003. Shipments of OM3 did not overtake shipments of OM1 fiber until 2012 in North America, and 2013 worldwide (CRU International, February 2017). Similarly, enhanced OM3+ fibers with extended link distance support began shipping as early as 2003, but didn't become standardized as OM4 fiber until 2010. In 2017, CRU expects that OM3 shipments will continue to exceed OM4 volume (CRU International, February 2017). OM5 adoption is expected to be more rapid, as SWDM-based transceivers that take advantage of wideband capabilities become more widely available and are quickly adopted, but may not achieve dominance in the immediate future. However, OM5 wideband fiber provides the best support for next-generation SWDM and PAM4 solutions, while providing backward compatibility to existing OM3 and OM4 applications.

5.2.2 Single-mode Fiber Types for Datacenters

For distances greater than 500 m, single-mode fiber is required. IEEE 802.3 Ethernet 40G and 100G standards recommend three IEC 60793-2-50 defined SMFs for short-reach interconnect applications defined as B1.1, B1.3, and B6.a. Table 5.4 describes the recommended SMF fiber types, all of which are undispersion-shifted single-mode fibers, and lists its designation in industry standards. For each of these fibers, the zero dispersion wavelength is around 1310 nm. The main difference between B1.1 and B1.3 is the attenuation specification. B1.3 fiber has an attenuation specification not only at 1310 nm and 1550 nm, but also at 1385 nm. B1.1 fiber allows a high attenuation at 1385 nm, caused by OH contamination. Low attenuation across the wavelength range from 1260 to 1625 nm allows B1.3 fiber to cover the full SMF wavelength spectrum window.

Table 5.4   The designation of SMF in various industry standards

| | Industry Standards | | | | |
| | IEC | | ISO/IEC | | |
Fiber Type	60793-2-50	ITU-T	11801	TIA/EIA	Comments
Standard SMF	B1.1	G.652.A or B	OS1	492CAAA	
Low Water Peak SMF	B1.3	G.652.C or D	OS2	492CAAB	
Bending Insensitive SMF	B6.a	G.657.A1	OS2		G.652.D Compliant "Bend-Insensitive" Single-Mode Fiber
	B6.a	G.657.A2	OS2		
	B6.a	G.657.B2	Non-compliant		G.652.D Compatible "Bend-Insensitive" Single-Mode Fiber
	B6.a	G.657.B3			

IEC 60793-2-50 B6.a is a bend-insensitive SMF. This allows smaller diameter loops with lower attenuation than B1 fibers. The IEEE 802.3 requirements of SMFs and a more detailed explanation of ITU-T G.652.D and G.657.A and B fibers are given in Section 5.6.

The SMF fibers in Table 5.3 support reaches from 500 m to 2 km sometimes encountered in hyper-scale intra-datacenter applications and from 2 to 30+ km for inter-datacenter transmission at 40 and 100 Gbps speeds. However, with the surging requirements of cloud computing and the growing need to connect multiple warehouse-scale datacenters within a metropolitan area, the requisite fiber length can exceed 30 km. For these demands, dispersion-shifted fiber (ITU-T G.655 and G.656) is considered for metro DCI in order to take advantage of its smaller chromatic dispersion in C-band for CWDM and DWDM [13]. Web 2.0 companies, including Facebook and Google, also build their own long-haul networks to connect datacenters distributed globally over thousands of kilometers [14, 15]. Private long-haul network capabilities are added to connect remote datacenters and transport information from these areas to population centers [12, 14, 15]. These long-distance networks can leverage advanced SMF with ultra-low loss and large effective area (ITU-T G.654) to lower long-haul system cost, eliminating the need for signal regeneration [16–18]. Only SMFs recommended in IEEE 802.3 standards and ITU-T G.652 and G.657 fibers are discussed in Chapter 5. Readers interested in understanding dispersion-shifted fiber, large effective area fibers, and ultra-low-loss fibers should refer to ITU-T G.654, G.655, and G.656 standards documents and fiber vendor information [19, 20].

5.2.3 Optical Cabling for Datacenters

Common cable types for intra-datacenter links are duplex LC and MPO/MTP® pre-terminated cables [21]. Duplex cables consisting of two SMF or MMF fibers terminated with LC connectors are the "building blocks" of 10 Gbps transmission. Optical fiber cables have migrated from Duplex to Parallel links for short-reach 40/100 Gbps. Industry standard MPO options include 8, 12, and 24 fiber counts, all using the same external ferrule dimensions and connector housing. 12-fiber MPO connectors have been typically used for modular fiber optic cabling within datacenters, but both 8-fiber and 24-fiber solutions are becoming more popular.

MPO pre-terminated cables are used to minimize installation time in datacenters. To support 10 Gpbs duplex LC applications, breakout modules, or cable fanouts are installed. Modules are typically mounted in an enclosure and provide a transition from the MPO pre-terminated cable to a duplex LC cable assembly. Cable fanouts provide the same functionality, but without the enclosure.

MPO pre-terminated cables are ideally suited for parallel fiber-based solutions. Both industry standard and proprietary 40/100G parallel solutions typically use 8-fiber links. 8-fiber MPO pre-connectorized cables provide the most straightforward link, but also create installation difficulties because of the large number of connectors that need to be routed through the cable trays. In many cases, network designers use 24-f MPO pre-connectorized cable and a breakout module that provides a transition from a 24-f MPO to 3 8-f MPO connections. This reduces the connector count on the pre-terminated cable by 3x.

Migration from 100 to 400G using 400G-SR16 PMD requires 32 fibers, and is not attractive due to the increased complexity of cable management. It is further hampered by the need for a new 32-fiber MPO connector that is not backward compatible to existing 8, 12, or 24-f MPOs. 400G-SR4.2 has been proposed as an 8-fiber solution for 400G link, with each fiber carrying 100 Gbps over two wavelengths using PAM4 signaling. In this new nomenclature, the 4 in SR4.2 refers to the number of fiber pairs, and the two references the number of wavelengths. This solution allows the use of the existing MPO pre-terminated cabling infrastructure supporting 400G transmission.

For inter-datacenter transmission, high density optical cable is an alternate solution to wavelength division multiplexing (WDM). In some cases, it is simply more cost-effective to run more fibers. Fiber counts for this application have grown at a significant rate, and many thousands of fibers in a single cable are being discussed.

As fiber counts increase, cable diameter becomes an important issue and duct size becomes a limiting factor. Management of the fiber for ease of splicing also becomes more of an issue. A number of developments have improved capabilities in this area. Bending insensitive fiber (ITU G.657.A1 and A2) can help mitigate loss due to fiber bends, and 200 μm-coated fiber enables the creation of smaller cables. Standard polymer-coated optical fibers are 250 μm in diameter but recently, multiple fiber vendors have begun to offer 200 μm-coated fibers. The glass diameter remains constant at 125 μm.

Rollable ribbons, designed with intermittent bonds between fibers so that they can be rolled into smaller packages than flat ribbons, allow higher density fiber cables due to their smaller diameter and weight. These ribbons also facilitate handling for splicing purposes. Ultra-high fiber count cables can have as many as 1728 fibers in a $1\frac{1}{4}''$ duct, while $\geq$3456 fiber cables are possible [22].

5.2.4 Multicore (MCF) and Few-Mode Fiber (FMF) for SDM

Recently, multicore fiber (MCF), few-mode fiber (FMF), and spatial-division multiplexing (SDM) have attracted research interest as they may serve as an effective path to increase capacity beyond the Tbps threshold. They have potential to reduce the footprint of short-reach optical interconnects and overcome the non-linear Shannon limit "capacity crunch" in long-haul networks. As the name would suggest, multicore fiber contains several cores inside a single strand of fiber. These cores can be single-mode, few-mode, or multimode. The highest core count and core density for single-mode MCF are 37 cores in a cladding diameter of 248 μm and 12 cores in a 125 μm cladding diameter [23, 24]. Relevant to intra-datacenter applications, recent papers have reported 70 Gbps transmission over 550 m and 120 Gbps over 100 m in single-strand multicore MMF containing seven 26 μm multimode cores [25, 26]. Although multi-core fibers offer great promise for high-density connections, some practical issues will need to be addressed to make multicore fiber a cost-effective solution. Further work on a high-performance connector and/or a low-attenuation splicing method will be needed before widespread deployment can occur. The total cost of fabricating multi-core fiber and aligning it to matching laser and receiver arrays in a transceiver will need to prove-in at lower system cost than the alternatives.

Few-mode fiber (FMF) has the highest spatial density in all SDM schemes and has both step-index and graded-index variants. Graded-index FMF is a version of graded-index MMF with a lower core diameter and delta than the multimode fibers discussed earlier. Two-mode and four-mode step-index and graded-index FMFs are commercially available [27]. Step-index FMF has less coupling between different spatial modes and is suitable for MIMO-less applications while graded-index FMF can be designed to achieve low-loss, low differential group delay (DGD) and low coupling loss for two LP modes [28]. However, when there are more than four LP modes, some modes are degenerate modes and form a principal mode group. Those spatial modes inside the same principal mode group are strongly coupled with a similar group delay. A good review of the design principles for graded-index FMF are given [29, 30]. A rectangular core is proposed to break the degeneracy of the degenerate high-order LP modes for intra-datacenter applications [31]. Commercial 50 μm core MMF can be a good benchmark to estimate how many spatial modes can be practically supported for SDM in a graded-index design. At 1550 nm, a 50 μm core MMF supports nine well-guided principal mode groups, comprising approximately 25 LP modes with about 2/3 of those being two-fold spatially degenerate. Recently, mode-multiplexed transmission using selected three and six spatial modes of a conventional multimode fiber has been demonstrated [32]. At this time, it is not obvious that the link cost of few-mode fiber transmission will be lower than that of the alternatives for future higher speed, high-density datacenter applications.

5.3 Waveguide Design, Modal Structure, and Time Response of SMF and MMF for Datacenters

5.3.1 Fundamentals of Waveguide Design and Mode Structures of SMF and MMF

Transmission optical fiber is a hair-thin cylindrical waveguide formed from core, cladding, and coating layers. The core and cladding are either pure silica or silica doped with materials which raise or lower the refractive index to form a weakly guiding waveguide. The core is typically doped with Germanium (Ge) to increase the index of refraction, although other dopants can be used. Fiber designs using a pure silica core fiber with a depressed cladding region are also possible. The cladding is typically a pure silica layer extending to a 125 μm diameter. Other clad diameters are used for specific applications. The cladding can have a depressed refractive index trench surrounding the core, created using Fluorine as the dopant. This depressed index area can improve the bending performance of the fiber. Some optical fibers also have a thin ring layer with a higher index of refraction adjacent to the trench layer, up-doped with Ge to facilitate cutoff and effective area control. The coating layer consists of primary and secondary coating layers, with an outer diameter of 240–250 μm. Figures 5.2a and b show the side view and cross section of an optical fiber.

The index of the cladding is given as n_2 and the index of the core at radius $r = 0$ is given as n_1. The index of a circularly symmetric optical fiber, *n(r)*, at radial position r is given by

$$n(r) = \begin{cases} n_2 + (n_1 - n_2)\left(1 - \left(\frac{r}{a}\right)^\alpha\right) & r \leq a \\ n_2 & r > a \end{cases} \tag{5.1}$$

where a is the fiber radius and α is the index profile-shape parameter. Relative refractive index difference Δ is defined as $\frac{n_1 - n_2}{n_1}$. 50 μm diameter core

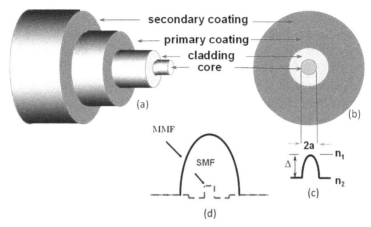

Figure 5.2 Physical structure and refractive index profiles of optical fibers. (a) Side view of an optical fiber; (b) cross section view of an optical fiber; (c) refractive index profile of a graded-index fiber; (d) comparison between the index profiles of MMF and SMF.

multimode optical fiber for modern datacom applications has a core radius of $a = 25$ μm and $\Delta = 1\%$, while early OM1 MMF design had a core diameter of 62.5 μm and $\Delta = 2\%$. The index profile shape parameter is approximately 2.1 for high-speed datacom MMF, sometimes referred to as graded-index MMF (GI-MMF). A standard SMF has a core radius less than 5 μm and a Δ between 0.3 and 0.4%. The index shape parameter approaches infinity for a step-index SMF. Figure 5.2d compares the core refractive index profile of MMF and an SMF with depressed cladding region surrounding the core.

The refractive index difference between the core and cladding causes total internal reflection at the boundary of the waveguide (core-cladding interface), forcing the optical signal to propagate along the longitudinal axis of the optical fiber. Intuitively, the wave guiding inside the optical fiber can be visualized in a ray picture while the physically accurate description of the optical fiber is presented in the concept of spatial modes. It can be shown that the effective index n_{eff}, β/k, of a guided mode must satisfy the inequality $n_2 < n_{eff} = \beta/k < n_1$, where β is the propagation constant of the mode and $k = 2\pi/\lambda$ is the propagation constant of a plane wave in free space. When n_{eff} is larger than the cladding index, the transverse field in the cladding is evanescent, resulting in a confined mode or guided mode in the waveguide structure. When n_{eff} is less than the cladding index, the transverse field in the cladding is oscillatory, resulting in radiative modes.

The mathematical treatment of the symmetric cylindrical waveguide starts by solving the Maxwell equations with appropriate boundary conditions. The wave equation can be greatly simplified by using the weak guiding approximation $n_1 \approx n_2$, yielding analytic solutions for the field power distribution. Unlike the plane wave in a slab waveguide, both meridional rays passing only through the longitudinal direction and skew rays exhibiting a spiral-like path down the core exist in a cylindrical waveguide. Meridional rays have pure transverse electric (TE) and magnetic (TM) field configurations. Skew rays are "hybrid" modes with longitudinal components of both E and H (designated as EH or HE). The eigenmodes are classified by the angular (or azimuthal) mode number q and radial mode number m., i.e., TE_{qm}, TM_{qm}, EH_{qm}, and HE_{qm}. Rather than EH, HE, TE, or TM nomenclature, linearly polarized (LP) modes are generally used in the description of optical fiber modes. Linear polarized (LP_{lm}) modes are the superposition of TE_{qm}, TM_{qm}, EH_{qm}, and HE_{qm} in a degenerate set. The resulting LP_{lm} is linearly polarized in the transverse field which conveniently represents how linearly polarized light is coupled to and will be maintained in an ideal optical fiber. The radial mode number m in LP mode notation remains the same. The azimuthal mode number l is related to q by $l = 1$ (TE_{om}, TM_{om}), $q+1$ (EH_{qm}), or q-1(HE_{qm}). The LP_{0m} modes, those without azimuthal variation of spatial power distribution and including the fundamental LP_{01} mode, are two degenerate HE_{1m} modes. The LP_{1m} modes have two spatial and two polarization configurations, resulting from the superposition of TE_{0m}, TM_{0m}, and HE_{2m} transverse modes. The 2 degenerate guiding modes in SMF are LP_{01} modes in orthogonal polarizations. GI-MMF with a 50 μm core diameter, on the other hand, has over 100 LP modes. The LP modes are classified into 18–19 principle mode groups shown in the following paragraphs. Readers interested in exploring additional details of waveguide theory would benefit by consulting John Buck's book [33] and C. K. Kao's landmark paper [1]. In practice, the field solutions and propagation constants of LP modes for real fiber designs are usually calculated numerically using finite element methods [34, 35]. The vectorial 2D finite-difference method has been further developed to accurately handle fiber parameters in specialty fiber and bending loss in bending-insensitive fiber [36, 37].

The normalized frequency parameter, often called the V-number and sometimes referred to as the waveguide strength, is an important parameter to estimate the number of guided modes. V-number is defined as

$$V = k \, a \left(n_1^2 - n_2^2 \right)^{1/2} \approx k \, n_1 a \sqrt{2\Delta}. \tag{5.2}$$

V-number embodies the fiber waveguide structural parameters (n_1, a, Δ) and frequency (k). It is used to determine cutoff condition, propagation constant, and information on power confinement. A larger V indicates more guided modes. When a, n_{core}, and Δ are chosen to make $0 \leq V < 2.405$, only two degenerate fundamental modes with perpendicular linear polarizations can propagate. Such fiber is defined as single-mode fiber. When V~6.5, the fiber supports five LP modes (LP01, LP11, LP02, LP12, and LP21). The V number is 27.1 for 50 μm GI-MMF at 850 nm.

The normalized propagation constant is a useful parameter to evaluate a given LP mode and is defined as $b_{lm} = \frac{n_{eff}-n_2}{n_1-n_2}$. b_{lm} varies between zero and one, and approaches zero at the cutoff wavelength and unity far above the cutoff. Figure 5.3 shows the normalized propagation constant of modes LP_{01} to LP_{21} versus V-number for a step-indexed fiber. The V-number is generally chosen close to the cutoff for the LP_{11} mode so that a strong waveguide for the fundamental mode LP_{01} gives better mode confinement and lower susceptibility to bending loss than would otherwise be possible. Meanwhile, LP_{11} must be sufficiently lossy at the shortest operating wavelength so that multiple-path-interference (MPI) between a strongly guided fundamental mode and a weak higher order mode is small enough to be neglected. The b_{lm} of all 850 nm guided LP modes for a 50/125 μm GI-MMF is represented by red stars on Figure 5.3, from lower order LP modes at the top to the highest LP mode at the bottom. The normalized propagation constants of the ~100 LP

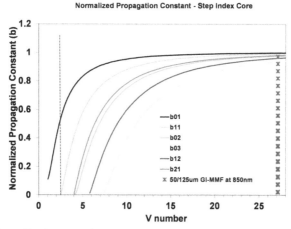

Figure 5.3 Normalized propagation constant b_{lm} versus V-number for a step-indexed fiber (colored lines) and for 50/125 μm GI-MMF at 850 nm.

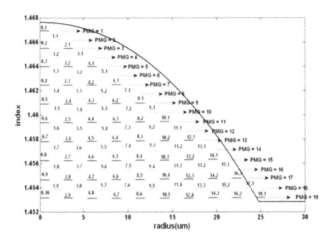

Figure 5.4   Effective index of LP modes at 850 nm for 19 PMG of 50/125 μm GI-MMF [38].

modes are clustered into 19 values, indicating 19 principal mode groups. The normalized propagation constant of the fundamental mode is near unity, while that of the highest principle mode group (#19) is close to zero, indicating that the modes in PMG #19 are lossy.

Figure 5.4 illustrates the effective index of all LP modes for a 50/125 μm, 1% delta GI-MMF at 850 nm. The mode number (l,m) is marked above the bar of effective index for each mode. The modes with near equivalent effective index are near-degenerate and clustered into one principal mode group (PMG). PMG1 has only the fundamental mode LP_{01}, PMG2 has only the first higher order mode LP_{11}, and PMG19 has 10 LP modes (0,10), (2,9), (4,8), (6,7), (8,6), (10,5), (12,4), (14,3), (16,2), and (18,1). Figure 5.5 shows the radial power distributions of the ten LP modes in PMG19. The lines for $LP_{18,1}$ (in black) and $LP_{0,10}$ (in red) are thickened to compare the radial peaks/nodes of the two modes with largest and smallest radial mode numbers. Different modes in the MMF travel at different speeds (group delay). The index profile of a GI-MMF is designed to minimize this speed difference between different modes, which will be explained in the next section.

5.3.2 Fundamentals of the Time Response of Optical Fiber

The time response of optical fiber is related to the information capacity of the fiber. An optical pulse transmitted into the optical fiber will spread in the time domain as it travels down the fiber, leading to bit errors at the

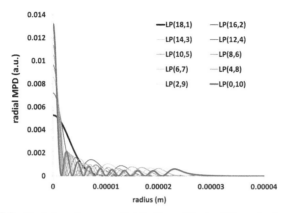

Figure 5.5 Radial mode power distribution of the ten LP modes in PMG19.

receiver. The pulse spread of an optical signal in optical fiber is referred to as dispersion. It is best to start from the concept of refractive index to understand the concept of fiber dispersion. The speed of light transmitting in a dielectric material is slower than light propagating in vacuum. The photons are coupled to the electrons in the dielectric material causing the light to travel more slowly. The ratio of the speed of light in vacuum to the dielectric material is the refractive index n. The refractive index is a function of wavelength and is different for different dopants within materials. Following the three-term Sellmeier equation, [39] describes how the refractive index is related to wavelength. Sellmeier coefficients $A_{i=1,2,3}$ and $a_{i=1,2,3}$ listed in Table 5.5 are the coupling strength and resonant wavelengths for pure silica, silica doped with 13.5 mole% Germanium, and silica doped with 1 mole% Fluorine (F), respectively. The refractive index decreases as wavelength increases. Silica doped with 13.5 mole% Ge has a higher refractive index than pure silica, and pure silica has a higher refractive index than 1 mole% F doped silica at

Table 5.5 Sellmeier coefficients for silica, Germanium-doped silica, and Fluorine-doped silica [39]

Sellmeier Coefficient	Al	al	A2	a2	A3	a3
Pure Silica	0.69680	0.06907	0.40820	0.11570	0.89080	9.90100
Ge-doped Silica 13.5 mole%	0.71104	0.06427	0.45189	0.12941	0.70405	9.42548
F-doped Silica 1 mole%	0.69110	0.06840	0.40790	0.11620	0.89750	9.89600

all wavelengths (shown in Figure 5.6). Note that the core of 50/125 μm GI-MMF is doped with about 0–10 mole% Ge from the edge to the center of the core while the core of SMF is doped with about 3–4 mole% Ge. An F-doped trench is often used to control bend loss in both SMF and MMF, and is doped with between 0.2 to 4 mole% F. A 1 mole% F-doped trench is common for SMF.

$$n^2 - 1 = \sum_{j=1}^{M} \frac{\lambda^2 B_j}{\lambda^2 - \lambda_j^2} \tag{5.3}$$

The group delay per unit length τ is given as

$$\tau = \frac{1}{c} \frac{d\beta}{dk_0} \tag{5.4}$$

where $\beta = n_{eff} * k_0$, k_0 is the free-space wave-number, and c is the speed of light in a vacuum. τ can be rewritten as a function of wavelength λ and refractive index n (Equation 5.5). The group delay of light in pure silica is faster than that of Ge-doped silica and slower than F-doped silica shown in Figure 5.7, consistent with the refractive index in Figure 5.6. The group delay is a function of wavelength, and the first-order derivative, or slope, of the group delay is the chromatic dispersion.

$$\tau = \frac{1}{c} \left(n - \lambda \frac{dn}{d\lambda} \right) \tag{5.5}$$

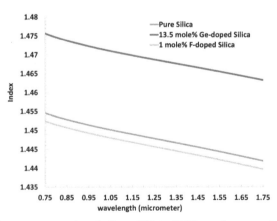

Figure 5.6   Index versus wavelength from 750 to 1750 nm for pure silica, 13.5 mole% Ge-doped Silica, and 1 mole% F-doped Silica.

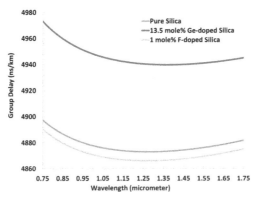

Figure 5.7 Group delay versus wavelength from 750 to 1750 nm for pure silica, 13.5 mole% Ge-doped Silica, and 1 mole% F-doped Silica.

The chromatic dispersion of the fiber is given by the first order wavelength derivative of the group delay. Since lasers used in transmitting optical signals always have a spectral width, chromatic dispersion (CD), together with parameters derived from it including the zero dispersion wavelength (λ_0) and zero dispersion slopes (S_0), are measured in fiber qualification tests and used to specify optical fibers' time response versus wavelength. For a bulk material, the chromatic dispersion is solely dependent on the refractive index of that dielectric material with a certain chemical composition and, therefore, is equivalent to the so-called material dispersion (Figure 5.8). The material dispersion of pure silica is -84, -106, and -83 ps/(nm*m) at 850 nm for the pure, Ge-doped, and F-doped silica examples shown in Table 5.5. The zero dispersion wavelengths are 1275, 1371, and 1271 nm, respectively.

$$\frac{d\tau}{d\lambda} = -\frac{\lambda}{c}\frac{d^2n}{d\lambda^2}. \tag{5.6}$$

The refractive index and, consequently, group delay of dielectric materials with different components (e.g., the pure, Ge-doped, and F-doped silica in Figures 5.6 and 5.7) are different, and so, for a spatial mode of the optical fiber with power distribution in both Ge-doped core and silica cladding (and/or F-doped trench), its group delay and chromatic dispersion are the composite group delay and chromatic dispersion determined by the waveguide structure in addition to the material composition, resulting in a curve laying in between those for pure, Ge-doped, and F-doped silica.

For MMF, the situation is even more complicated. Since there are plural guided LP modes and each mode has its own spatial power distribution (as

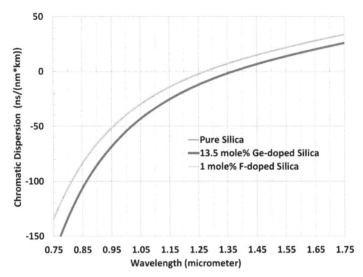

Figure 5.8 Chromatic dispersion versus wavelength for pure silica, 13.5 mole% Ge-doped silica, and 1 mole% F-doped silica.

the ten LP modes of PMG19 shown in Figure 5.5), each LP mode sees its own unique material composition and waveguide structure resulting in different group delay and chromatic dispersion. Figure 5.9 shows several examples of LP modes for a 50/125 µm and $\alpha = 2.16$ GI-MMF with 1% delta of Ge-doped core. The chromatic dispersion for a GI-MMF in fiber measurement is an averaged CD resulting from an LED source illuminating equivalent power for all modes. The design profile parameters for the GI-MMF example are chosen to minimize the group delay of all modes at 850 nm (as shown in Figure 5.10). However, the group delay difference increases as wavelength deviates more from 850 nm. Small manufacturing deviation from the optimized ideal index profile also differentiates the group delay between different modes. This group delay difference is the modal dispersion for MMF (Note that modal dispersion has the unit of ps/m or ns/km, different from the unit of ns/(nm.km) for chromatic dispersion). The time response and bandwidth of GI-MMF is a combined contribution of chromatic dispersion and modal dispersion. This is different from SMF, where modal dispersion is not a consideration above the cutoff wavelength.

Since the invention of GI-MMF, great efforts have been made to reduce its modal group delay. To further illustrate why such work is needed, the simulated modal delays of all LP modes at 850 nm for the example GI-MMF

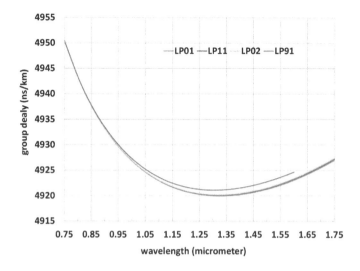

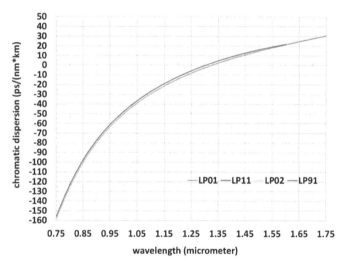

Figure 5.9 Group delay and chromatic dispersion of LP01, LP11, LP02, and LP91 modes for GI-MMF optimized at 850 nm.

mentioned above is plotted in Figure 5.10. α is optimized to reduce modal delay at 850 nm, and the modal delays are nearly equal for LP modes below PMG15. Above that, modes with high modal delay are very lossy due to a strong coupling to the cladding caused by their near-equal effective index. Further, in practice, the modal delay is not as high as shown in Figure 5.10

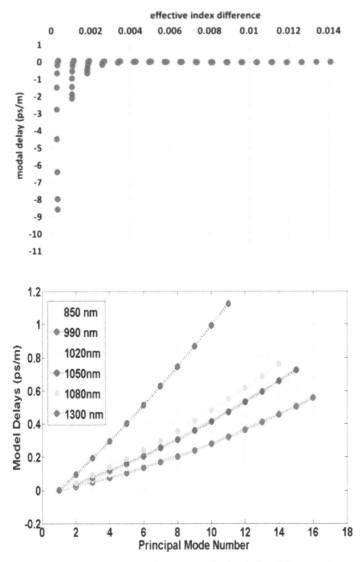

Figure 5.10   Top: delays of all LP modes versus effective index difference. Bottom: Modal delay of PMG versus principal mode number from 850 to 1300 nm for GI-MMF [38].

for MMF with pure silica cladding. After tens of meters, each principal modal group moves with an averaged modal delay. For MMF with a down-doped trench to optimize bending loss, controlling higher order modes is more challenging and will be revisited in a later section. The modal delay of the example GI-MMF is minimized at 850 nm. Moving away from 850 nm to longer wavelengths, the modal delay deviates from the optimum benchmark, growing larger than 1 ps/m at 1300 nm. To expand the capacity of MMF links, careful optimization of conventional MMF and novel design with co-dopants have been developed in recent years. These will be covered in the Wideband MMF section.

5.4 Multimode Optical Fiber for High-Speed Short-Reach Interconnect

5.4.1 Laser-optimized MMF (OM3 and OM4)

5.4.1.1 What is Laser-optimized MMF?

Most theoretical foundations regarding profile design of GI-MMF were established by the late 1980s [38, 40–49]. The concepts related to GI-MMF waveguide design which had been explored and/or well established by that time were: a parabolic-shaped core profile [40, 41], optimum alpha for wavelength dispersion consideration [42], trimming core edge to reduce higher order mode modal delay deviation [47–49], and multi-alpha, multi-components to expand the wavelength ranges for high bandwidth [45, 46]. In the following 15 years, most of the advances in commercial MMF for Datacom usage were due to improvement in processing feedback from experimenting with the ideal alpha shape to achieve higher bandwidth around 850 nm and better process control. Common multimode fiber types include FDDI, OM1, OM2, OM3, and OM4, defined by TIA and ISO. As time progressed, 850 nm/1300 nm modal bandwidths improved from 160/500, 200/500 (OM1), and 500/500 MHz.km (OM2) to 2000/500 (OM3) and 4700/500 (OM4) MHz.km as the supported data rate evolved from 10 Mbps to 40 and 100 Gbps. OM1 usually has a 62.5 μm core and OM2 typically has a 50 μm core, but these fibers can have different core sizes. OM3 and OM4 have a core diameter of 50 μm. OM3 and OM4 specifications define a multimode fiber that optimizes coupling to Vertical Cavity Surface Emitting Lasers (VCSELs), which were developed for low-cost, short-reach optical interconnects at speeds from 1 to >10 Gbps [50, 51]. These standards first defined laser-optimized MMF, and introduced the concept of effective modal bandwidth (EMB), often referred to as "laser bandwidth."

5.4.1.2 Differential Modal Delay (DMD)

Even in an OM3/OM4 GI-MMF with delta $\sim$1% and an alpha optimized to reduce the modal delay difference between all the guided modes at 850 nm, there is still a small spread in intrinsic modal delays, as shown by the yellow dots in Figure 5.10. In addition to the theoretical modal delay spread, manufacturing deviations from an ideal profile (alpha-shift, center dips/peaks, kinks, and core-cladding deviations) will increase modal delays between different principal modal groups. Fibers with smaller deviations have less pulse spreading and higher bandwidth.

These deviations are measured using Differential Mode Delay (DMD). This technique uses a short laser pulse launched into a special SMF with a cut-off of <850 nm, which is scanned across the face of the MMF. DMD is defined as the difference between the 25% threshold level of the slowest pulse-trailing edge and the fastest pulse-leading edge between specified radial positions. High-resolution DMD (HRDMD) measurement uses a $\sim$4 ps near transform-limited pulse from an SMF to scan the face of the MMF, ideally in one micron step. The temporal response at each radial position from 0 to 25 μm is then plotted as the dark curves shown in Figure 5.11, yielding a high-resolution picture of the modal delays excited at the cross section of the MMF.

For OM3 fiber, six DMD "templates" with inner (5–18 μm) and outer (0–23 μm) delay pairs (masks) and a sliding mask of 7 μm span are specified in TIA-492AAAC and IEC 60793-2-10 [52] to limit the effect of various profile deviations. The DMD masks for OM3 MMF are shown on the left of

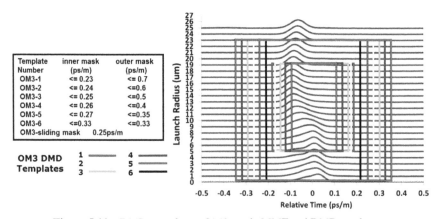

Figure 5.11 DMD scan for an OM3-grade MMF and DMD mask sets.

Figure 5.11. The value of the inner mask and outer mask is used to balance the modal dispersion caused by the medium-order mode groups and higher or lower order mode groups. The sliding mask is developed to limit the DMD variation across the range of offsets, typically caused by the slope of alpha profile variation and kinks.

OM4 MMF has a tighter DMD specification than that of OM3 to enable support of longer link distances. The first, fourth, and sixth inner and outer DMD mask specifications for OM3 are divided by a factor of 2.35, resulting in three DMD inner/outer mask pairs for OM4. The sliding mask for OM4 MMF is also constrained to less than 0.11 ps/m. Figure 5.12 shows the DMD trace for an OM4 MMF. The scaling of one DMD mask is indicated as the red and pink cursors. An MMF with higher bandwidth than that of OM3 was first proposed in a TIA-492AAAC informative annex, and, later, OM4 was specified in TIA-492AAAD standard and ISO60793-2-10 [53].

DMD is not only a specified fiber parameter, but also provides necessary processing feedback for precise refractive index profile control in manufacturing, as proposed by Petermann [54]. Centroid delays, $T(r)$, are defined as the average of measured DMD scan traces at each radial position r. $T(r)$ is related to the radial position by a polynomial series given as

$$T\left(r\right) = \sum_{i=1,2,\ldots,n} p_i * \left(\frac{r}{a}\right)^{2i} \tag{5.7}$$

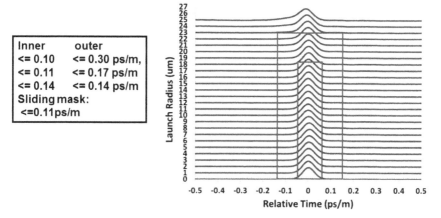

Figure 5.12 OM4 DMD traces (blue) and DMD masks. Measured DMD traces at 850 nm for an OM4 fiber. The 0–18 μm, 0–23 μm, and sliding mask widths are 0.043, 0.049, and 0.037 ps/m, respectively. The red line is the OM4 mask at 0.1 ps/m and the purple one is 0.3 ps/m.

where the coefficients p_i for $i = 1, 2, \ldots, n$ are extracted by fitting the equation for the centroid delays [48]. The refractive index corrections are given by Petermann.

$$\delta n\,(r) = -2p_1(r/a)^2 ln\left(r^2/a^2\right) - \sum_{i \geq 2} p_i \frac{i+1}{i-1}\left\{(r/a)^{2i} - (r/a)^2\right\} \quad (5.8)$$

Development of HRDMD testing technique is essential to provide sufficient feedback to ensure viable mass production of OM4 MMF. Dips/peaks in the index profile at the core center-impact lower order principal mode groups and cause a split pulse as shown from $r = 0$ to 5 µm at the bottom of DMD scan traces in Figure 5.11. A kink causes a sudden shift of the time traces at the radial positions of the DMD scan (e.g., at $r = 10$ and 20 µm in Figure 5.11). Defects near the core/cladding interfaces impact higher order modes and cause multiple/split pulses for DMD at a larger radial position. These issues are more significant for bending optimized MMF which will be discussed in the next section. The modal delay deviations due to the center defects and kinks can lead to significant pulse spreading, and often limit system performance, especially in legacy OM1 and OM2 fibers.

An alpha shift can cause a smooth slope either tilted to the left or right on the DMD scan traces. This could either benefit or degrade the system performance depending on wavelength. This will also be discussed in the WBMMF section.

5.4.1.3 Bandwidth of MMF Links

5.4.1.3.1 *Overfilled modal bandwidth*
The Overfilled Launch Bandwidth (OFL BW) of MMF is specified in TIA FOTP 204 [55] at 850 and 1300 nm. This measurement procedure uses LEDs to illuminate the core of the MMF. LEDs distribute optical power uniformly throughout all modes of the fiber core. The OFL BW requirements for OM3 and OM4 are 1500 MHz.km and 3500 MHz.km at 850 nm, respectively. At 1300 nm, both OM3 and OM4 have OFL-BW requirements of 500 MHz.km, the same as OM1 and OM2. OFL BW is reliable in specifying information-carrying capacity of MMF links with LED sources, but it cannot accurately predict high-speed ($>$1Gbps) link performance where lasers are used.

5.4.1.3.2 *Effective modal bandwidth*
10/40/100GBASE-SR MMF links use 850 nm VCSELs. Effective modal bandwidth was developed to capture the combined effect of VCSEL modal power distribution and DMD of OM3/OM4. Compliance of 850 nm VCSELs

to the 802.3ae standard is assured by measuring the encircled flux (EF) of the source according to TIA/EIA FOTP 203. The EF specification is specifically intended to minimize the impact of residual center dips or core-clad deviations. The integrated power within a 4.5 µm radius must be less than 30%, while at least 86% of the power must be contained within a 19 µm radius. This has the effect of placing most optical power away from the area where MMF is most likely to experience non-ideality (Figure 5.13). Figure 5.13a shows the EF specification (within the gray area) and the measured EF of four off-shelf 10GBASE-SR transmitters. Ten weighting functions (Figure 5.13d) were chosen in an attempt to represent worst-case VCSEL mode power distributions, in order to calculate EMB [56]. These ten VCSEL MPD weighting functions are convolved with measured fiber DMD traces (Figure 5.13c) resulting in ten impulse responses. The ten impulse responses are inverse Fourier-transformed into the frequency domain. The minimum of the ten -3 dB bandwidth is defined as the minimum calculated effective modal bandwidth (EMBc). The EMB of a fiber is calculated by EMB=minimum EMBc * 1.13 [50]. OM3 has an EMB requirement of ≥ 2000 MHz.km, while OM4 has a minimum EMB of ≥ 4700 MHz.km, scaled up by a factor of 2.35 from OM3.

5.4.1.3.3 *Chromatic bandwidth*
Chromatic bandwidth is determined by the chromatic dispersion of the fiber and the spectral content of the laser launched into the fiber. Since SMF only propagates a single mode, it only has chromatic bandwidth. MMF has both modal and chromatic bandwidth. Since chromatic bandwidth is associated with the laser source, it is not a fiber specification. But for short-reach links using MMF and VCSEL/LED sources, chromatic bandwidth combined with modal bandwidth is included for composite exit impulse response at the output of the fiber to calculate the inter-symbol-interference penalty [57]. Figure 5.14 shows the modal bandwidth of an ideal alpha optimized at 850 nm and the chromatic bandwidth using a RMS = 0.65 nm VCSEL source. The blue curve is the total effective bandwidth. The chromatic bandwidth increases with wavelength. The total effective bandwidth is closer to the chromatic bandwidth at wavelengths less than 870 nm and closer to the modal bandwidth approaching 990 nm.

5.4.1.3.4 *True bandwidth of a short-reach interconnect system channel*
EMB is an emulated fiber modal bandwidth using 10 VCSEL weighting functions and fiber differential modal delay. For each individual

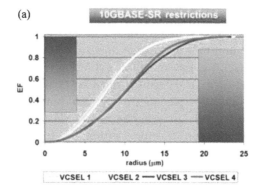

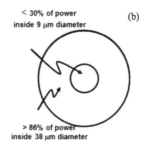

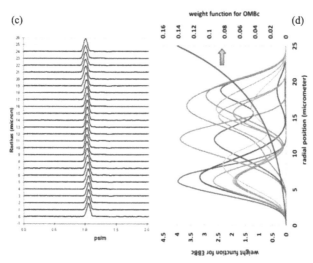

Figure 5.13 (a) EF specification of IEEE 802.3ae compliant 850 nm VCSELs and (d) 10 weighting functions for EMBc calculations. (b) Cross-section view of EF flux radius and power ratio; (c) DMD scan of a OM4 fiber.

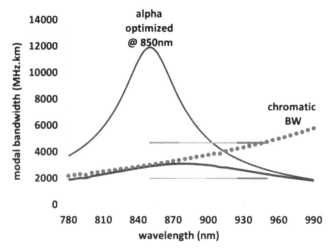

Figure 5.14 Modal BW and chromatic BW of ideal alpha 50/125 μm MMF. Blue is the composite bandwidth. The horizontal bars are OM3 and OM4 EMB specification limit.

VCSEL-MMF link, the true channel bandwidth is determined by the power distribution of that particular VCSEL, the time response of the MMF including both modal and chromatic dispersion factors, how the factors are coupled, and the bandwidth of the receiver. The true bandwidth of a particular VCSEL-MMF link can be measured directly by using a Vector Network Analyzer (VNA). This is not a standardized measurement method, but it is helpful to directly associate the VCSEL-MMF link bandwidth to its system performance for better understanding of the various penalties. Figures 5.15 shows a test setup for direct-link bandwidth measurement. In this experiment, the VNA has 40G bandwidth, and the photo-diode detector has 30G bandwidth. The VCSEL is driven at 8 mA, with a center wavelength of 853 nm and RMS spectral width of 0.35 nm. MMF is cut at 100, 200, 300, 400, 500, and 600 m, and has an 850 nm EMB of 5700 MHz.km. The CD BW is 5501 MHz.km, and the total bandwidth calculated using the EMB and CD BW for the MMF from 100 and 600 m are plotted at right as the blue diamonds in Figure 5.15. The directly-measured bandwidth of the MMF from 100 to 600 m using the setup on the left of Figure 5.15 is plotted with red circles. The measured fiber EMB (in red) is slightly less than the calculated fiber EMB (in blue) indicating the mode power distribution of the VCSEL is likely close to the MPD of the emulation VCSEL.

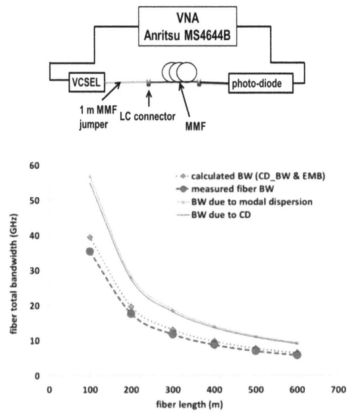

Figure 5.15   Left: Test setup for direct measurement of VCSEL-MMF bandwidth; Right: an example of measured fiber bandwidth versus calculated bandwidth from EMB and chromatic bandwidth (CD BW).

5.4.2 Bend-optimized OM3/OM4 and Overfilled Effective Modal Bandwidth

In 2009, bend-insensitive or bend-optimized OM3/OM4 was introduced to the market and standardized by TIA. Bend-optimized MMF is fully compliant with conventional OM3/OM4 standards with the added benefit of improved macrobend performance.

Bend-insensitive or bend-optimized OM3/OM4 uses a 50 μm graded-index profile similar to the conventional OM3/OM4 but with an optical trench surrounding the core. The trench helps to better confine the guided modes but also allows "leaky modes" to propagate over a much longer distance before they dissipate than standard OM3/OM4. Figure 5.16 highlights the guided higher order modes (PMG16 to PMG 19 shaded in blue) that remain confined and guided by the trench when the index profile is tilted by bending. More optical power is preserved under bent-condition in the BO-MMF link by the larger output spot size shown in Figure 5.16 left. Table 5.6 compares the macrobend loss for standard and bend-insensitive MMF [59].

Another difference regarding modal propagation between standard MMF and BI-MMF is the enhancement of the "leaky modes" (in red). In standard MMF, leaky modes, modes with an effective index right below the cladding index normally dissipate within several meters. In BI-MMF, the trench helps to trap the leaky modes so that they can travel a few hundred meters without bending. The stronger guided higher order modes and existence of leaky modes impact the characterization and measurement of the BI-MMF. A new modal bandwidth called OMBc (Calculated Overfilled Modal Bandwidth) is specified in FOTP-204 [55] to control the modal delay of higher order modes. OMBc is calculated based on measured DMD and a weighting function, similar to EMBc. However, the weighting function resembles an overfilled power distribution which grows larger as the radius increases (shown in red

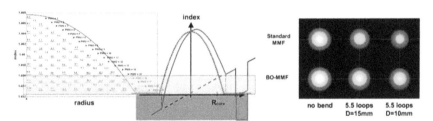

Figure 5.16 Left: Index profile and effective indices of all LP modes in a 50/125 μm core GI-MMF; Middle: tilted refractive index profile (blue); Right: mode power distribution at the output of a standard MMF and BO-MMF w/ and w/o bends [58].

Table 5.6 Comparison of macrobend loss for standard and bend-insensitive MMF

	Macrobend Loss		
	100 turns @ r = 37.5mm	2 turns @ r = 15mm	2 turns @ r = 7.5mm
Standard 50/125 MMF	0.5 dB (850 & 1300 nm)	1 dB (850 & 1300 nm)	Not specified
BI 50/125 MMF	0.5 dB (850 & 1300 nm)	0.1 dB (850 nm) 0.3 dB (1300 nm)	0.2 dB (850 nm) 0.5 dB (1300 nm)

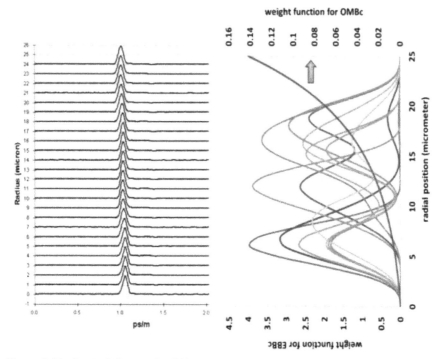

Figure 5.17   DMD (left) and 10 VCSEL mode power weighting functions (right) for EMBc and the mode power weighting function for OMBc (in red).

in Figure 5.17). The minimum OMBc is defined as $\geq$3500 MHz.km for OM4 grade BI-MMF and $\geq$1500 MHz-km for OM3 fiber.

5.4.3 Wideband MMF (OM5)

The profile shape parameter of OM3/OM4, both conventional and bend-resistant, is optimized to support operation at 850 nm. As networks migrate to 400 and 800G, a new MMF type has been introduced to support wavelength division multiplexing (WDM), reducing fiber counts, and enabling to an eight-fiber solution for these speeds, as adopted by 40/100GBASE-SR4. Wideband MMF has standardized in TIA-492AAAE [60] and IEC 60793-2-10 Edition 6.0. It is also recognized in ANSI/TIA-568.3-D structured cabling standard as a new MMF fiber type, and will be included as OM5 fiber in the latest revision of ISO/IEC 11801.

OM5 MMF is fully backward compatible and meets the same geometry (core size, NA, core-cladding eccentricity) and 850/1300nm optical requirements (DMD, EMB, OMB, OFL) as OM4 MMF. In addition, the optical specifications of OM5 were extended to cover a range from 850 to 953 nm. EMB requirements for OM5 fiber are shown in Table 5.7 and in red in Figure 5.18. The EMB at 840 and 953 nm shown in Table 5.7 are informative. The dashed black curve is the EMB of an OM4 MMF that failed the TIA WBMMF specification at wavelengths longer than 850 nm. The blue dots and curve are the total effective bandwidth calculated using the EMB of TIA-492AAAE WBMMF specification and the chromatic bandwidth using an RMS = 0.65 nm laser (in green). High chromatic bandwidth at longer wavelength ensures OM4-equivalent performance from 840 to 953 nm despite the lower EMB of OM5 at longer wavelengths.

The TIA-492AAAE EMB specification can be met by changing the index profile of OM4 fiber slightly to shift the peak modal bandwidth to a

Table 5.7   EMB versus wavelength of WBMMF

Wavelength, nm	EMB, MHz*km
840	*3840*
850	**4700**
930	*2565*
953	**2470**

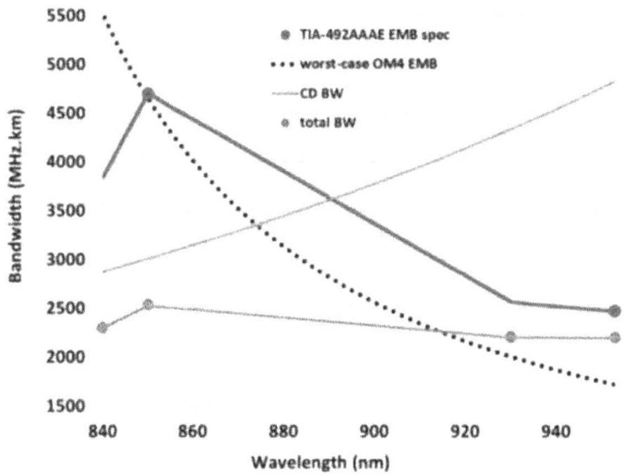

Figure 5.18   [61, 62]. TIA-492AAAE EMB specification versus wavelength for WBMMF (in red) and worst case OM4 (in black dot). Bandwidth due to chromatic dispersion (in green) and total bandwidth (CD BW + EMB) for WBMMF (in blue).

longer wavelength, but there is a negative impact at shorter wavelengths. In addition, such a design would not have sufficient bandwidth beyond 953 nm for possible future long-wavelength transmission schemes. A newly designed NG-WBMMF employing multiple dopants in the core broadens the wavelength range over which high EMBc can be obtained [63]. In theory, such a design can exceed the minimum EMB of OM4 fiber at all wavelengths [63]. The design methodology also allows consideration of realistic manufacturing tolerances. The blue circles in Figure 5.19 represent the EMB of a NG-WBMMF with a much wider bandwidth profile sample. Though the longer wavelength optimized OM4 passes the TIA EMB spec at 953 nm, the new design has twice the bandwidth. The higher EMB can offer extra margin and also allows operation at longer wavelengths, such as 980 nm. Meanwhile, the longer wavelength-optimized OM4 fails to meet EMB required for a 100-m reach at 970 nm, when using the same simulation approach used to estimate EMB from 850 to 953 nm.

The new design uses multiple dopants and allows OM4 equivalent EMB in a window from up to 200 nm wide. Figure 5.20 shows the EMB of a NG-WBMMF that is capable of supporting eight wavelengths, each carrying 50 Gbps, and operating from 850 to 1066 nm. The blue dashed line provides interpolated values from the actual measured data points. The EMBc from 850 to 980 nm is greater than 4700 MHz*km (the OM4 EMB specification at

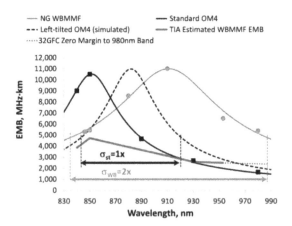

Figure 5.19   EMB versus wavelength for standard OM4 (solid black curve), left-titled OM4 (dashed black curve), and NG-WBMMF (solid blue dots). TIA estimated WBMMF EMB and extension to 980 nm using 32GFC zero-margin assessment represented in red [61].

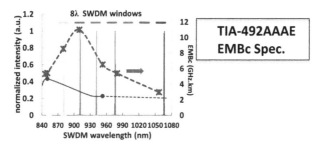

Figure 5.20 EMB versus wavelength of a NG-WBMMF supporting eight lambda SWDM window [64].

850 nm). As shown in Figure 5.20, the EMB at 1066 nm is 2750 MHz*km, better than the TIA-492AAAE specification at 953 nm. The green dashed line in Figure 5.20 represents a possible extrapolation of the TIA-492AAAE standard (using the same methodology found in [60], Appendix F). This would indicate a minimum EMBc of 2251 MHz*km at 1066 nm for acceptable link performance. The EMBc of the NG-WBMMF used in this study substantially exceeded that value.

5.5 High-Speed VCSEL-MMF Short-Reach Optical Interconnect System

Though MMF optical cable costs more than SMF, it can tolerate far less precise alignment, allowing lower cost coupling and use of low power consumption VCSELs. This results in a lower total link cost for multimode short-reach optical interconnects. The typical distance of optical interconnects in datacenters is less than 100 m, with the possible exception of hyper-scale datacenters. Even in hyper-scale installations, interconnects between server and switch make up the bulk of links, and are rapidly migrating to optical from copper links. These links are typically far shorter than 100 m. For high-performance computing (HPC), the interconnect distance is usually less than 100 m. IBM reported that the largest P775 system located in Endicott NY had >250,000 links, with 28 m as the longest cable length (Kuchta, 20 March 2016). OM3 supports 300 m and OM4 supports 400 m reach using 10GBASE-SR standards based transceivers. The IEEE standards based reach for 40/100GBASE-SR4 is reduced to 150 and 100 m respectively, due to relaxed transceiver specifications, but proprietary 40/100G eSR4 transceivers extend the reach of OM3 and OM4 to coverage similar to 10GBASE-SR. 40/100G SR4 PMDs use an 8-parallel-fiber MPO cable.

The first 400 Gbps PMD uses parallel 10 Gbps lanes at 850 nm, requiring a 32 fiber MMF cable which is unappealing due to the complexity of cable management. With the introduction of OM5 MMF in 2016, multimode fiber designed to support multiple wavelength transmission from 850 to 953 nm supports 40/100 Gbps duplex solution and provides a migration path to a 400/800 Gbps eight-fiber MPO solution, while maintaining backward compatibility with all OM4 specifications. Higher order modulation formats such as PAM4 further reduce the fiber counts of VCSEL-MMF links, allowing 50G-1λ, 100G-2λ, and 200G-4λ duplex links, and 400G-2λ transmission over eight-fiber MPO cables. In this section, the various factors contributing to the penalties of MMF-VCSEL links are reviewed, followed by an example using the IEEE MMF-VCSEL system link model to determine the required OM5 EMB for a 100 m 32G FC link at various wavelengths. Finally, a series of system experiments demonstrating the feasibility of low-cost VCSEL-MMF transmission up to 400 Gbps are reviewed.

5.5.1 System Evaluation Methodology

The VCSEL-MMF channel consists of two optical transceivers with a VCSEL based transmitter and a PIN-diode receiver, linked together with optical fiber/cables, patch panels, and jumpers. There are several ways to assess system performance: (1) using a pulse-pattern generator (PPG) and error detector (ED); (2) using an optical network tester (ONT); (3) using CMOS-based IC chipset evaluation kit to generate and detect bit patterns; (4) using a network traffic generator and commercial switches. An optical link is considered to operate error-free if data can be transmitted with errors occurring less frequently than a specified limit. For the 10GBASE-SR, 40GBASE-SR4, and 100GBASE-SR10 IEEE 802.3 standards, that limit is specified as 10^{-12}. For the standards based on 25 Gbps serial speed, receiver equalization and feed-forward error correction (FEC) are adopted. The pre-FEC error ratio is at a value recoverable by the FEC coding used: 2.4×10^{-4} for Reed Solomon ($N = 544$, $K = 514$, $m = 10$, $T = 15$) (KP4) and 5×10^{-5} for Reed Solomon *(528,514)* (KR4). With FEC enabled, the error ratio is expected to be less than 1×10^{-15}.

The non-stressed receiver sensitivity is the minimum received power at which the target BER is achieved without fiber/connectors in the link, denoted as Back to Back (BtB). The sensitivity depends on the thermal noise of the receiver circuit and the shape of the received waveform. The received waveform is determined by the bandwidth of the transmitter and receiver electronics while the difference between the transmitter (Tx) output power and the non-stressed receiver (Rx) sensitivity is the available power budget of the link. In standards development, the power budget between channel insertion loss (CHIL) and penalties induced by ISI and signal-borne noise are often allocated (Figure 5.21). The channel insertion loss includes attenuation of both cabled fiber and connectors and bend-induced loss. ISI can be inherent in the transmit waveform or incurred through dispersion, both chromatic and modal, in the fiber. Signal-borne noise sources include relative intensity noise (RIN), mode-partition noise (MPN), and modal noise (MN). The stability of the link performance is assessed by the margin, defined as the subtraction of various power penalties and CHIL from the power budget. CHIL is tested using a power meter. The receiver sensitivity versus BER can be varied by sweeping the received power using a variable optical attenuator, resulting in plots often called "waterfall curves." The sum of the power penalties is extracted from the difference of the BER waterfall curves between BtB and the fiber under test. Note that standards define the worst case for large statistics, i.e., the non-stressed receiver sensitivity is the maximum allowed minimum receiver sensitivity, and the transmitter power is minimum allowed optical power. In practice, a specific fiber and transceiver forms a unique link depending on the transmitter spectral content, modal power distribution (MPD), fiber modal and chromatic dispersion, and how the specific transmitter is coupled to the specific fiber. Figure 5.22 shows an example of measured BER waterfall curves using an off-the-shelf 10GBASE-SR transceiver. The black curve is the BtB curve, essentially a measurement of the receiver sensitivity of the specific waveform produced by the Tx in this

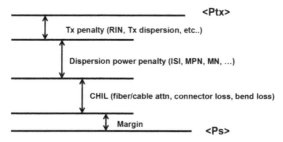

Figure 5.21 Power budget allocation in VCSEL-MMF system. Ptx is the launch power of the transmitter. Ps is the non-stressed receiver sensitivity.

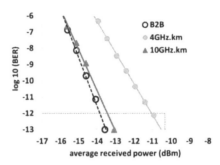

Figure 5.22   Waterfall curve showing bit error rate as a function of added attenuation for a low EMB (4 GHz.km) and high EMB (10 GHz.km) multimode fiber over 550 m compared to the back-to-back system test. Path penalty measurements have error bars of ±0.2 dB [65].

example. It captures the thermal noise and ISI penalties associated with the transceiver. The green curve represents tests with a 550 m OM4 of EMB >10 GHz.km, while the red curve represents tests with a 550-m enhanced OM3 with an EMB of about 4 GHz.km. These two curves capture ISI from fiber modal and chromatic dispersion and the impact of MPN, RIN, and MN. The CHIL is normalized out in the measurement. The figure indicates that the high bandwidth OM4 yields a power penalty of 0.5 dB, and the other fiber yields a 3 dB power penalty. The link with the lower bandwidth MMF has 2.5 dB less margin. To achieve the same BER, the lower bandwidth link will need a 2.5 dB higher output power on the Tx or less allocation to connector loss.

The system transmission experiment is limited to one or several specific MMF-VCSELs. In addition, the dispersion power penalty extracted from BER waterfall curves is a summation of various penalties from multiple sources including inter-symbol-interference (ISI), jitter, relative intensity noise (RIN), and modal partition noise (MPN). In order to generalize the conclusions of the transmission experiment with specific MMFs and transceivers, one could model the differences between worst-case transceivers and those used in the experiment.

The IEEE Ethernet spreadsheet model [66] is a good tool to understand the changes in penalties as transceiver and MMF properties are varied. This model is intended to be pessimistic and cover the worst case. One use of the model is to figure out the required EMB to reach 100 m link distances at wavelengths longer than 850nm. To estimate the EMB required at longer wavelengths to match 32 Gbps fiber channel performance at 850 nm, the IEEE excel spreadsheet for 100GBASE-SR4 is used. The RMS spectral width is 0.6 nm for all wavelengths and the total power budget is 8.2 dB, with

1.5 dB assigned to connector loss. All parameters are held constant, with the exception of EMB, which is adjusted to achieve zero margin in a 100 m link, at each wavelength. Figure 5.23 shows the penalties due to ISI, jitter, and RIN are dominant at a 100 m reach and do not change significantly versus wavelength. Penalties due to modal partition noise and attenuation decrease as wavelength increases. The effective modal bandwidths required to obtain the zero link margin are $\geq$4700 MHz*km at 850 nm, $\geq$3300 MHz*km at 875 nm, $\geq$2900 MHz*km at 900 nm, $\geq$2700 MHz*km at 925 nm, and $\geq$2550 MHz*km at 950 nm. Note that TIA spec uses the spreadsheet modeling for 32G FC 100 m transmission, where the EMB value ends up slightly lower than the values estimated using 100GBASE-SR4 parameters.

5.5.2 High-Speed VCSEL-MMF System Transmission Validation

5.5.2.1 10GBASE-SR transmission over OM3 and OM4 MMF

Around 2002, several fiber and transceiver companies, under the auspices of the TIA and its FO-2.2.1 working group, studied how the MMF and VCSEL work jointly for a 10-Gbps link transmission. A round-robin test including 12–300 m fibers from three manufacturers and 21 10-Gbps VCSEL samples (bootstrapped to 1000 source) were carried for ISI (inter-symbol-interference) and EMB analysis. The results helped to set the DMD/EMB specification for OM3 MMF and Encircled Flux (EF) for 10 Gbps VCSELs [51]. It is challenging for experiments to probe all the fiber-VCSEL parameter spaces, and modeling of MMF-VCSEL links was established with consideration of the complex coupling of MMF modes and VCSEL modes [52].

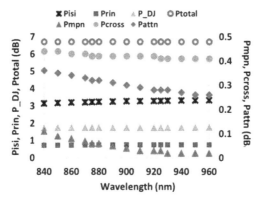

Figure 5.23 System penalties at a 100-m reach versus wavelength per IEEE excel spreadsheet modeling for 100GBASE-SR4 [67].

Table 5.8   Characterization of 10GBASE-SR transceivers used in system tests

	VCSEL (1)	VCSEL (2)	VCSEL (3)	VCSEL (4)
Spectral width (nm)	0.207	0.234	0.327	0.384
EF flux @ 4.5 μm	21.01%	18.35%	10.58%	10.16%
EF flux @ 19 μm	99.65%	99.86%	96.78%	97.72%

The multimode-link model takes into account the interactions of the laser, the transmitter optical subassembly, and the fiber, as well as effects of connections and the receiver preamplifier. Several years of experiment and modeling resulted in the standardization of OM3 fiber and 10GBASE-SR over 300 m of OM3 fiber [51]. Later, experiments on enhanced OM3 fiber led to the standardization of OM4 supporting reach beyond 300 m. The system experiment which was instrumental in determining the OM4 specification is summarized in the following paragraphs. Later, it also recognized that the interaction between multiple transverse modes of VCSEL and LP modes of MMF is more complicated than what was considered in theoretical modeling of [51]. It turns out that the modal and chromatic dispersion interaction between VCSEL and MMF benefit OM3/OM4 with left-tilted DMD slope to extend the transmission distance [68].

IEEE P802.3ae 10GBASE-SR recommends 300 m transmission over OM3 MMF, but some enterprise applications require coverage over distances longer than 300 m. 10GBASE-SR system transmission experiments helped to improve understanding of enhanced OM3, later standardized by TIA-492AAAD, and commonly referred to as OM4 fiber. In 2008, 10 Gbps system transmission tests of 550 m links of enhanced OM3 fiber from several vendors were conducted using four VCSELs with a selected range of RMS spectral widths and encircled fluxes(Table 5.8) [65]. VCSEL 1 and 2 had smaller RMS spectral widths and tighter encircled fluxes than those of 3 and 4. Statistically, the dispersion power penalties of the MMFs when using VCSEL 3 and 4 (Figure 5.24 in red) are higher than those when using VCSEL 1 and 2 (Figure 5.24 in green). BER waterfall curves in Figure 5.25 show the range of penalties measured in 30 combinations of 15 fibers from five vendors and two transceivers (VCSEL 2 and VCSEL 3). The order of the dispersion power penalties with VCSEL 2 and 3 are different for different fibers showing the combination effect of VCSEL mode power distribution and fiber DMD/EMB. For VCSEL 3 with larger RMS spectral width, some fibers fall out of the compliance limit. Ethernet standards were later amended to call out 400 m reach by 10GBASE-SR over OM4 fiber.

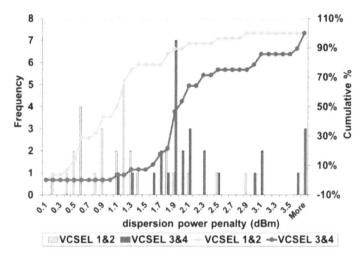

Figure 5.24 Statistical distribution of dispersion power penalty of 550 m enhanced MMFs using VCSEL 1 and 2 and VCSEL 3 and 4 [65].

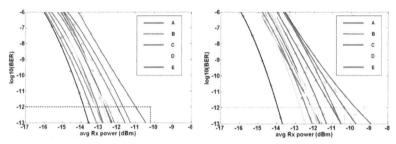

Figure 5.25 BER waterfall curves using VCSEL 2 (left) and VCSEL 3 (right). Fibers from five vendors A, B, C, D, and E are tested [65, 58].

5.5.2.2 40GBASE-eSR4 and 100G eSR4 extended reach demonstration over OM4 MMF

The IEEE 802.3 40GBASE-SR4 and 100G-SR10 standards were built from the optical specifications of 10GBASE-SR applied to parallel cabling with MPO connectors. Link distance support over OM3/OM4 MMF is reduced to 100/150 m in order to relax transceiver specifications and provide the most cost-effective short-reach solutions. To support longer 10GBASE-SR link distances, transceiver makers introduced premium 40GBASE-eSR4 transceivers with tighter specifications. This provided end users with a choice – most users only have links shorter than 100 m, and could use the standards based product, while those with longer link distances could use the premium transceivers. VCSEL capable of 25 Gbps per lane was also developed to reduce the fiber counts of 100G-SR10 link from 20 to 8, same as 40GBASE-SR4 [69, 70]. Premium 100G-SR4 transceivers were released from several vendors to extend the reach of 100G-SR4 transmission. The extended reach of OM4 using premium 40G/100G transceivers at both room and elevated temperatures has been validated in the experiment described below. The OM4 MPO cables used in the experiment conducted at room temperature included OM4 fibers with marginal DMD and EMB. Four fibers had a MW18 of almost 0.14 ps/m, 2 two more fell into the range from 0.14 to 0.16 ps/m, and ∼30% of the fibers had EMBc below 5000 MHz.km, as shown in Figure 5.26a [71].

In the first part of the experiment, 10G RF signals were used to simultaneously drive 2 40GBASE-eSR4 transceivers. The VCSEL's RMS spectral width versus center wavelength of the 2 transceivers is shown in Figure 5.26(c). 125, 150, and 400 m OM4 MPO cables were concatenated using MPO jumpers. 4 bidirectional 10 Gbps optical signals were transmitted over the >675-m OM4 link. Bit error ratio waterfall curves were measured at room temperature and dispersion power penalties (DPP) at BER = 10^{-12} were extracted. Figures 5.26(b) and (c) shows the BER waterfall curves over 675 m OM4 cable links for upper and down-direction transmission. There is significant margin for all links [71].

To test the reliability over a temperature range from 25° to 65°C, a 40GBASE-eSR4 transceiver was placed in a controlled environmental chamber and operated over 550 m of OM4 cable with an EMBc of 5171 MHz.km. Error-free transmission was obtained at all temperatures (Figure 5.27). The receiver sensitivities varied by less than 0.5 dB, and the maximum dispersion power penalty at BER = 10^{-12} was 4.06 dB at 55° C.

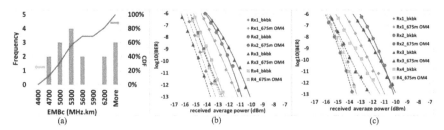

Figure 5.26 (a) EMB of 16 out of 24 active fibers in the OM4 MPO cables under test; (b) BER waterfall curves of B2B and 675 m OM4 for uplink transmission; (c) BER waterfall curves of B2B and 675 m OM4 for downlink transmission [71].

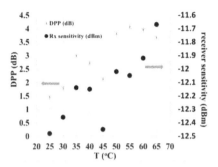

Figure 5.27 40GBASE-eSR4 transmission: receiver sensitivity and dispersion power penalty (DPP) from 25° to 65°C [72].

In the second part of the experiment, a 100GBASE-SR4 CFP transceiver module was tested using some of the same OM4 cables. The 100 and 150 m lengths of OM4 cable were connected using MPO jumpers and MPO-12LC fanouts to form a 224 and 324 m loop-back optical link between the four 25.78 Gbps transmitters and receivers. The optical eyes remained open after 324 m for all four channels (Figure 5.28). The BER was measured to be $9.8*10^{-9}$, $3.2*10^{-11}$, $3.8*10^{-11}$, and $2.2*10^{-7}$ for the 324 m link, well below the pre-FEC BER threshold. The pre-FEC BER for the 224 m link was five to six orders of magnitude below the threshold.

One 12F out of the 24F MPO is tested using two 100G-eSR4 transceivers. The optical link includes the 400 m MMF cable and several jumpers with five MPO or LC connections excluding the interface to the transceiver and a variable optical attenuator. The details of the results was published in

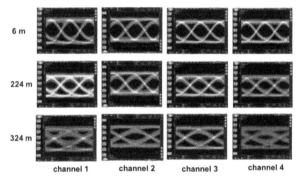

Figure 5.28 Optical eyes of 100GBASE-SR4 signals at the B2B (6 m), after 224 and 324 m of OM4 MPO cable [71].

IWCS 2017 [73]. We summarize the main discovery of that paper. The eight VCSELs in those two transceivers all have RMS spectral width of less than 0.45 nm and EF compliant to IEEE 802.3 100G-SR4 standard. Full Ethernet traffic with PRBS31 pattern from a 100-G optical network tester was used to generate and evaluate the 100G signals. Forward error correction (FEC) was disabled to evaluate BER waterfall curves. Figures 5.29(a) and (b) show the BER waterfall curves of 100G eSR4 transceiver #1 and #2, over eight fibers of the MPO12 F cable each. Dashed lines are BtB curves. The BER is at least two orders less than the FEC threshold 5×10^{-5} in all cases. The maximum dispersion power penalty at 5×10^{-5} is 5.66 dB and the smallest margin is 4.98 dB among the 16 VCSEL/fiber combinations. Error-free is achieved after FEC is turned on.

5.5.2.3 40/100 Gbps SWDM over OM5 MMF

Short wavelength-division multiplexing (SWDM) multiplexes the optical signals from VCSELs of multiple wavelengths and transmits them on a single MMF. Today's VCSELs are limited to 25 Gbaud/s, and while advances continue to be made in raising speeds to 50 Gbaud, network demands are pushing component vendors to look at new and different ways to increase the capacity of multimode fiber. Multiple participants in the datacom industry have researched SWDM in recent years as a means of reducing the number of multimode fibers in a link and providing a low-cost 400+ Gbps solution [74].

40G SWDM over 300 m OM3 with receiver equalization was reported as early as 2006 at OFC [75]. The VCSELs operated at 980, 1020, 1050, and

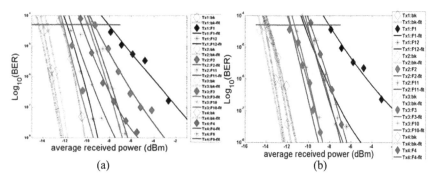

Figure 5.29   BER waterfall curves of 100G-eSR4 transmission over BtB and 400 m corner case OM4 cable. (a) 100GBASE-eSR4 transceiver module #1; (b) 100GBASE-eSR4 transceiver module #2 [73].

1080 nm. The optical eyes after 300 m OM3 are measured before and after an electronic dispersion compensation circuit from Scintera (Figure 5.30b). The eyes are wide open after EDC. The impulse response before and after 300 m fiber are measured and emulated with DFE with 8 (T/2 space) FF Taps and 3 FB Taps for dispersion power penalties assessment (Figure 5.30c). Work in [75] demonstrated the feasibility of transmitting 40 Gbps over 300 m of OM3 fibers using CWDM with 990 nm, 1020 nm, 1050 nm, and 1080 nm VCSELs and a commercial EDC circuit, showing a low-cost path to 40 Gbps links using CWDM and EDC. Though it was not under commercial development then, we predicted that CWDM and EDC could offer low-cost optics and generalization to even higher capacity. Today, commercialized SWDM transceivers have either two operation wavelengths (around 855 and 907 nm) or four wavelengths (around 850, 880, 910, or 940 nm) to allow operation over legacy OM3/OM4 MMF. In this section, several 40G/100G SWDM transmission experiments are reviewed.

The first commercialized WDM MMF transceiver, the Cisco 40G-SR-BiDi transceiver, is widely deployed in datacenters. A 40G-SR-BD transceiver uses the QSFP form factor and has 4 lanes of 10 Gbps electrical signals before the transmitter and after the receiver, similar to a 40G-SR4

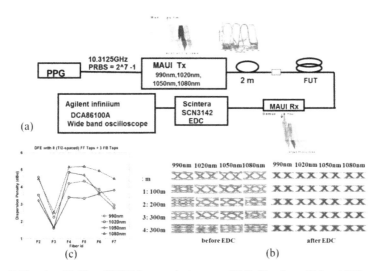

Figure 5.30 4 × 10 Gbps CWDM transmission over OM3 fiber from 990 to 1080 nm using electronic dispersion compensation (EDC) at the receiving side. (a) experimental setup; (b) optical eyes after OM3 fibers before and after EDC; (c) calculated dispersion penalties using measured impulse response after 300 m OM3 fiber with DFE of 8 FF taps and 3 FB taps [75].

transceiver. 40GBASE-SR4 converts the four electrical signals into four 10G parallel 850 nm optical signals and transmits them on four fibers in a 12F MPO bundle. The four 10G optical signals are detected by receivers and converted into four 10G parallel electrical signals. 40GBASE-SR-BD transceivers multiplex the four 10G electrical signals into two 20 Gbps lanes, one driving an 850 nm VCSEL while the other drives a 900 nm VCSEL. Two 20G 850 nm and 20G 900 nm optical signals transmit bidirectionally on a duplex LC-based fiber link, making this WDM solution compatible with existing 10G infrastructure. The reach over OM4 is specified to be 150 m.

One 40-G-SR-BD transceiver was tested over two 200-m OM4 MMFs [71]. The center wavelength and RMS spectral wavelength were 853.80 and 906.92 nm, and 0.2909 nm and 0.4463 nm, respectively. Both fibers were bend-optimized OM4 MMFs. One had an 850-nm EMBc of 6538 MHz.km and a 900-nm EMBc of 6565 MHz.km while the other had an 850-nm EMBc of 6014 MHz.km and a 900-nm EMBc of 2503 MHz.km. The MW18/MW23 of the two fibers were similar at 850 nm, but at 900 nm, one was 0.113/0.122ps/m and the other was 0.249/0.302 ps/m. The dispersion power penalties are small (0.69 and 1.07 dB) for both MMFs at 850 nm. At 900 nm, DPP is 0.99 and 3.13 dB, consistent with the difference on DMD and BW of the two MMFs at 900 nm. At OFC 2016, two 40G-SR-BD transceivers were plugged into two 40-G ports of an Ethernet switch and operated over a 300-m OM5 duplex cable. The demonstration ran three days without any interruption [76–78].

In another experiment [77, 78], four streams of 10G RF signals from a pattern generator were used to drive a 40-G-SWDM4 transceiver transmitting 40 Gbps optical signals. The SWDM4 transceivers used 4 10-G VCSELs operating at 850, 880, 910, and 940 nm, multiplexed onto a single fiber. The 40-G SWDM4 transceivers had RMS spectral width of less than 0.45 nm and were EF compliant to the IEEE 802.3ba standard. Two OM5 NG-WBMMF samples, one with EMB of 5500 MHz.km and the other with EMB of 3600 MHz.km at 950 nm, were used in the transmission experiment. Both achieved error-free transmission at room temperature with maximum DPP of 3.7 dB for the fiber with 3600 MHz.km EMB at 950 nm. The 40-G SWDM4 transceiver was placed into an environmental chamber and the temperature increased from 31°C to 62°C. BER waterfall curves were tested at 850 and 940 nm over the 500 m OM5 NG-WBMMF cables. Error free transmission was achieved at all temperatures at both wavelengths. The average received optical power (AOP) variation was less than 1 dB.

A comparison test between two OM5 and two OM4 samples was performed to assess the performance difference at longer wavelengths, using 40-G-SWDM4 transceivers [77, 78]. The 950-nm EMB was 5500, 3600, 3060, and 1610 MHz for the four samples. The two OM5 NG-WBMMFs were 500 m, and OM4 fiber with EMB of 3060 MHz-km at 950 nm EMB was also 500 m. The OM4 fiber with 953 EMB of 1610 MHz-km was 250 m. The BER waterfall curves are plotted in Figure 5.31 while the extracted DPP and fiber parameters are summarized in Table 5.9 The DPP correlates well with the EMB and the OM5 fiber shows clear performance advantages over standard OM4 MMF at longer wavelengths.

In another experiment, three Finisar 100-G-SWDM4 transceiver modules were tested using duplex LC terminated OM5 cables. The RMS spectral width of all 12 VCSELs in the three transceivers was less than 0.6 nm. For the OM5 F1, the optical eyes gradually narrowed as the link length was increased from 100 to 400 m, indicating increasing ISI. However, the eyes remained open at all wavelengths through 400 m (Figure 5.32). The BER after 100 and 200 m is barely increased from that of the B2B measurements. At 300 m, the

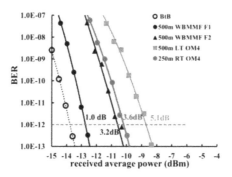

Figure 5.31 940 nm BER waterfall curves of 40-G SWDM4 transmissions over BtB, 500 m of NG-WBMMF F1, 500 m of NG-WBMMF F2, 500 m of OM4 F1 passing the TIA 492-AAAE spec, and 250 m of OM4 F2 failing the TIA 492-AAAE spec [77, 78].

Table 5.9 950 nm EMB and 940 nm DPP at BER = 10^{-12} for two NG-WBMMF and two OM4 MMFs [77, 78]

	Length (m)	950 nm EMB (MHz.km)	Pass/Fail TIA-492AAAE	940 nm DPP (dB)
OM5 F1	500	5500	Pass	1.0
OM5 F2	500	3600	Pass	3.2
OM4 F1	500	3060	Pass	5.1
OM4 F2	250	1610	Fail	3.6

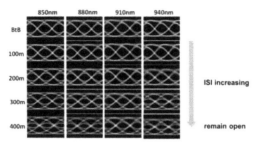

Figure 5.32   Optical eyes of the BtB link and after 100, 200, 300, and 400 m of OM5 F1 from 850 to 940 nm using a 100G SWDM4^{TM} transceiver [77, 78].

pre-FEC BER was five orders of magnitude better than 5×10^{-5}. At 400 m, the maximum DPP at BER = 5×10^{-5} was 3.4 dB from 850 to 940 nm. Using OM5 F2, a pre-FEC BER below 5×10^{-5} was achieved at 300 m. The 100 and 150 m BER was only slightly increased from the B2B results at all four wavelengths from 850 to 940 nm.

40G/100G SWDM4 over extended link lengths of duplex OM5 cable was validated using a 40G/100G commercial switch. Live video traffic from a content server was transmitted over a 500 m OM5 cable using 40G SWDM4 modules, then over a 300 m OM5 duplex cable using 100G SWDM4 modules. The signal was received by a second server and displayed on its monitor. The live demo ran three days at OFC 2016 without any interruption [76–78].

5.5.2.4 High-Speed PAM4 SWDM transmission over OM5 MMF

A high modulation format, PAM4, can double the bits per symbol and reduce fiber count by a factor of two. PAM4 is written into IEEE 802.3cd draft for 50GBASE-SR, 100GBASE-SR2, and 200GBASE-SR4 PMDs, operating at 50 Gbps/fiber. Supported link distances are 70 m using OM3 and 100 m over OM4 and OM5. 100GBASE-SR2 and 200GBASE-SR4 uses four and eight parallel fibers. A proprietary 100GBASE-PAM4-BiDi solution using two wavelengths for bidirection transmission over a duplex MMF cable was announced at OFC 2017. While similar to 40G-SR-BiDi, instead of 20 Gbps NRZ signaling over two wavelengths, 100GBASE-PAM4-BiDi transceivers use 50 Gbps PAM4 signals at each wavelength. 100G BiDi offers a path to 400 Gbps using eight parallel fibers. Combinations of 50 Gbps PAM4 and to eight SWDM wavelengths can provide a duplex fiber solution supporting 200 and 400 Gbps. Feasibility experiments have shown an upgrade path for duplex VCSEL-MMF links to 200 Gbps and even 1 Tbps using PAM4/SWDM over OM5 fiber.

The feasibility of transmitting an aggregated 206.25 Gbps PAM4 signal over 150 m of OM5 fiber using SWDM TOSAs from 850 to 940 nm has been demonstrated [79]. A KeySight Arbitrary Waveform Generator (AWG M8195A) was used to generate a 51.56-Gbps PAM4 electrical eye. The PAM4 electrical signals drove one of the four SWDM TOSAs in sequence. The TOSAs were originally developed for 25 Gbps NRZ operation. The optical PAM4 eyes propagated through 150 m of OM5 fiber and were collected by a 30 GHz bandwidth OE converter. The optical eyes were post-processed with a linear equalizer and extracted for bathtub curve analysis. The OM5 EMB used in the transmission experiment is shown in Figure 5.33(b). The Optical PAM4 eyes after 150 m of OM5 with (pink) and without (blue) linear equalization are shown in Figure 5.33(c). The red line inside the PAM4 eyes is the BER contour at 5×10^{-5}. Linear equalization improved the eye height and width as indicated by the wider opening of the bathtub curve in Figure 5.33(d). The jitter added by the fiber is less than 0.07 UI. This experiment demonstrated the feasibility of transmitting >200 Gbps aggregated speed over a single OM5 fiber, at a reach comparable to what OM4 covers at 40G/100G SR4. It suggested that transmitter pre-emphasis

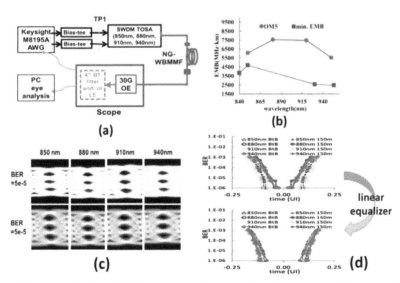

Figure 5.33 (a) SWDM-PAM4 experimental setup; (b) EMB versus wavelength for OM5 fiber sample and TIA OM5 spec; (c) optical PAM4 eyes after 150 m of OM5; (d) bath-tub curve w/ and w/o 150 m OM5 w/ and w/o linear equalizer [79].

and receiver equalization are necessary for 50 Gbps PAM4 using VCSEL based MMF links.

In following experiments, Inphi 28 nm CMOS-based 40G/100G PAM4 chipsets were used with real-time DSP to repeat the transmission test. BERs below the pre-FEC KP4 threshold was achieved at 53 Gbps from 850 to 940 nm over 100, 200, and 300-m OM5 fiber using the SWDM TOSAs and a wideband ROSA with linear TIA. 53 Gbps PAM4 transmission over 300-m OM5 fiber was also successfully achieved using a 980 nm TOSA developed for a 25 Gbps NRZ system (Figure 5.34 [80]). The PAM4 chips were designed to be integrated into SWDM transceiver modules including the QSFP28 and CFP8 form factors.

The experiments discussed in this section demonstrate that the approach using SWDM optics and PAM4 integrated circuits for >200 Gbps transmission over a single OM5 may be practical and become commercially viable in the near future.

53 Gbps PAM4 tests were extended to cover eight SWDM wavelength windows using a TOSA centered at 1066 nm and the aforementioned wideband ROSA with linear TIA. Figure 5.35 shows DPP (a) and total bandwidth (b) versus wavelength for 100, 200, and 300 m of OM5. The expected correlations between link bandwidth, DPP, link length, and transmission wavelength were realized. This was the first experiment to demonstrate the potential of a single OM5 fiber to support 400 Gbps transmission using $8\lambda \times 53$ Gbps PAM4 with current technology, using 25 Gbps NRZ optics and a PAM4 IC CMOS chipset [64].

A practical 50 Gbps -2λ PAM4 transceiver provides a duplex MMF solution at 100 Gbps. It expected to have the similar cost as the current two lanes 100GBASE-SR2 solution, however it offers a compelling route to a future 400 Gbps PMD based on four fibers per direction. OM4 supports 100 m and OM5 supports 150 m 100G-BiDi-PAM4 transmission. The baseline proposal for the 100 Gbps duplex MMF using two wavelengths and PAM4

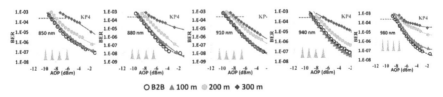

Figure 5.34   BER waterfall curves of 53 Gbps transmission over NG-WBMMF at 100 to 300 m from 850 to 980 nm [80].

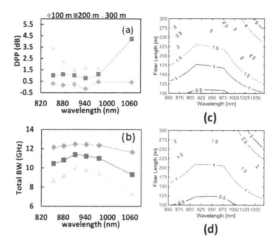

Figure 5.35   DPP (a) and total bandwidth (b) versus wavelength. DPP contour with infinite taps of linear equalizer (c) and infinite taps of DFE (d) [64].

was discussed by [81]. We demonstrated 400 m over selected OM5 sample using an alpha version 100G-PAM4-BiDi transceiver from FIT [73].

5.6 Datacom Transmission over Single-Mode Optical Fiber

Hyper-scale datacenters may have optical links up to 500 m inside one building and require up to 2-km links between switches in different buildings on the same campus. This has led to the adoption of SMF for these longer lengths at 100 Gbps and beyond. From campus to campus and datacenter to datacenter in the same metro region (DCI), the distance can range from 10 km up to 30 or 40 km. IEEE 802.3 standards call out fiber optic cables containing IEC60793-2-50 type B1.1 (dispersion unshifted single-mode), type B1.3 (low-water-peak single-mode), or type B6_a (bend-insensitive) fibers. IEC60793-2-50 type B1.1, B1.3, and B6_a fiber optic cables correspond to G.652.A and B, G.652.C and D, or G.657.A fiber and cables in ITU-T standards. In the following sections, fiber attributes of SMFs are first described, the fiber and cable requirements in IEEE 802.3 are reviewed for intra- and inter-datacenter high-speed SMF interconnect, and ITU-T G.652 and G.657 fibers recommended by ITU-T standards and their applications to Datacom connectivity are briefly described. The application of a parallel SMF infrastructure in a hyper-scale datacenter architecture is described.

The specified attributes of SMF include:

(a) Chromatic dispersion: range of zero dispersion wavelengths and zero dispersion wavelength slope.
(b) Mode field diameter (MFD).
(c) Fiber and cable attenuation.
(d) Macrobending loss.
(e) Polarization mode dispersion (PMD).
(f) Fiber and Cable cutoff.

Section 5.3.2 describes some fundamentals of chromatic dispersion. Readers interested in more in-depth understanding of fiber attributes are referred to [33, 39].

The above table summarizes the maximum fiber length and the requirement for fiber specifications in several IEEE 802.3 standards for 50 and 100 Gbps bit rates. The maximum fiber length in the above table covers the intra-datacenter reach from 500 m (100G-DR) to 2 km (50G-FR and 100G-CWDM4) and inter-datacenter reach up to 10 km (50G-LR, 100G-LR4, 100G-4WDM10) and from 30 km to 40 km (100G-ER4).

The transmitter wavelength range for each standard is summarized at the bottom of the table. The nominal fiber specification wavelength of operation is around 1310 nm with zero dispersion wavelength from 1300 to 1324 nm and maximum dispersion slope S_0 as 0.093 ps/(nm^2.km). The maximum (positive) and minimum (negative) allowed chromatic dispersion and maximum differential group delay (DGD) accumulated over the maximum allowed fiber/cable length at the transmitter operational wavelength ranges are summarized in the table as well.

IEC 60793-2-50 B1.1 fiber is standard SMF with attenuation up to 1 dB/km defined at both 1310 and 1550 nm but not specified at 1385 nm. IEC 60793-2-50 B1.3 fiber is a low-water-peak SMF with attenuation value up to 0.4 dB/km at 1310, 1385, and 1550 nm. IEC 60793-2-50 B6.a is bending insensitive fiber. B1.1 fiber correlates to ITU-T G.652.A or B, IEC 60793-2-50 B1.3 correlates to ITU-T G.652.C or D, and IEC 60793-2-50 B6.a correlates to ITU-T G.657, ITU-T G.657. {Yi, does that one ISO/IEC code B6.a map to all ITU G.657 categories?} A1 and A2 are compliant to G.652.D. Since most datacenter single-mode fibers operate around the 1310 nm range, the nominal mode field diameter is specified at 1310 nm. It is 8.6–9.5 μm for G.652.A, B, C fibers and 8.6–9.2 μm for G.652.D and G.657.A and B fibers.

Table 5.10   IEEE 802.3 recommended fiber/cable properties [82]

Standard	50G-FR	50G-LR	100G-DR	100G-LR4	100G-ER4	100G-ER4	100G-CWDM4	100G-4WDM10	100G-PSM4
Max length (km)	2	10	0.5	10	30	40	2	10	0.5
Nominal fiber specification wavelength (nm)					1310 nm				
Zero dispersion wavelength (λ_0) (nm)					1300 nm to 1324 nm				
Dispersion slope (max) S_0 (ps/(nm^2.km))					0.093				
Positive dispersion (max) (ps/nm)	3.2	16	0.8	9.5	28	36	6.7	33.5	1.2
Negative dispersion (min) (ps/nm)	−3.7	−18.6	−0.93	−28.5	−85	−114	−11.9	−59.5	−1.4
DGD_max (ps)	3	8	2.24	8	10.3	10.3	3	8	2.24
Tx_wavelength range (nm)	1304.5 nm to 1317.5 nm		1304.5 nm to 1317.5 nm	1294.53 to 1296.59 nm, 1299.02 to 1301.09nm, 1303.54 to 1305.63 nm, 1308.09 to 1310.19nm			1264.4 to 1277.5nm, 1284.5 to 1297.5nm, 1304.5 to 1317.5nm, 1324.5 to 1337.5nm	1264.4 to 1277.5nm, 1284.5 to 1297.5nm, 1304.5 to 1317.5nm, 1324.5 to 1337.5nm	1295 nm to 1325 nm
# of fibers	2	2	2	2	2	2	2	2	8
# of wavelength	1	1	1	4	4	4	4	4	1

ITU-T G.652 describes a single-mode optical fiber and cable which has zero-dispersion wavelength around 1310 nm, generally referred to as the dispersion un-shifted fiber. The maximum cable cutoff wavelength (λ_{cc}) of G.652 fiber is 1260 nm. It is optimized to be used in the 1310 nm range (O-band) due to its low chromatic dispersion around 1310 nm, but it can also be used at 1550 nm region (C-band). There are four types of ITU-T G.652 SMF fiber and cables. G.652.B and G.652.D have tighter polarization mode dispersion (PMD) than G.652.A and G.652.C. G.652.A and B do not have attenuation specified around 1383 nm, while G.652.C and G.652.D have attenuation specified around 1383 nm. The latter two are referred as "low water peak". Low attenuation around 1383 nm allows the "low-water-peak" SMF to support the full spectrum of wavelengths from 1260 to 1625 nm for system transmission. The maximum macrobend loss of G.652 fibers is 0.1 dB/100 turns at a bending radius of 30 mm. Today, G.652.A, B and C are more or less obsolete everywhere outside of Japan. ITU-T G.652.D corresponds to OS2 in ISO/IEC 11801 and B.1.3 in IEC 60793-2-50.

ITU-T G.657 describes two categories of SMF with improved bending loss performances: G.657.A and G.657.B. ITU-T.G.657 is equivalent to B6.a in IEC 60793-2-50 recommendations. It was initially developed for access network applications including inside buildings at the end of networks where limited space and easy operation were important considerations. ITU-T G.657.A is compliant with G.652.D and can be considered as a sub-category of G.652.D with tightened macrobend loss restrictions. G.657.B is not necessarily compliant with ITU-T G.652.D in terms of the chromatic dispersion coefficients and PMD, but it is compatible with G.657.A in its support of all system applications. Both categories A and B have two sub-categories: A1 and A2 and B2 and B3. A1 fibers are appropriate for a minimum bend radius of 10 mm, A2 and B2 are appropriate for a minimum bend radius of 7.5 mm, and B3 are appropriate for a minimum bend radius of 5.0 mm. Figure 5.36 shows the macrobend losses of the four G.657 fibers. The cable cutoff of G.657 fibers is less than 1260 nm.

Two basic categories of transceivers have been developed for single-mode fibers. One is commonly referred to as PSM4 and operates over four pairs of single-mode fibers, with either an MPO connector or four pairs of pigtailed duplex LC connectors. Transceivers known by the monikers LR4 or CWDM4 operate over a single pair of fibers, terminated with duplex LC connectors. At the time of this writing, 100G QSFP28 modules are popular datacenter form factors, and CWDM4, PSM4, and LR4 modules are shipping

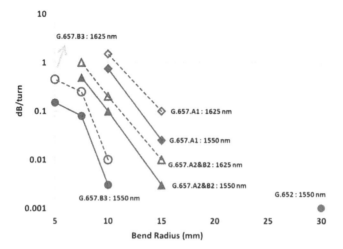

Figure 5.36 Macrobend loss of G.657.A and G.652 fibers in ITU-T standards.

in high volumes, all operating 25 Gbps NRZ lanes near the zero-dispersion wavelength of single-mode fiber near 1310 nm.

LR4 modules utilize four wavelengths with a spacing closer to DWDM than CWDM, and they have been defined by IEEE P802.3 standards without FEC (802.3ba) and with FEC (802.3bm) for 10 km reach, with significant allocation for loss in the link. The choice of wavelength grid was designed to minimize chromatic dispersion in the link, but reduce cost relative to DWDM. CWDM4 modules use wider 20 nm wavelength spacing and are not standardized, and these were originally aimed at a 2-km reach specification especially for datacenter applications. "Lite" versions of both transceiver types relax specs for shorter reach requirements. PSM4 transceivers are seen to have cost advantages when implemented in silicon photonics, where, for example, one higher power laser can be split four ways into a photonic integrated circuit containing four modulators. The corresponding CWDM4 transceiver requires four separate lasers as well as mux/demux and more complex packaging. Although the parallel single-mode infrastructure costs more to install initially than a duplex single-mode infrastructure, the PSM4 transceivers for 40 and 100 Gbps have generally been lower cost than the duplex transceivers, outweighing the higher cabling cost. It seems likely that PSM4 optics will always enter the market at lower initial cost than the duplex versions, rendering the parallel single-mode cabling architecture very cost-effective for future upgrades.

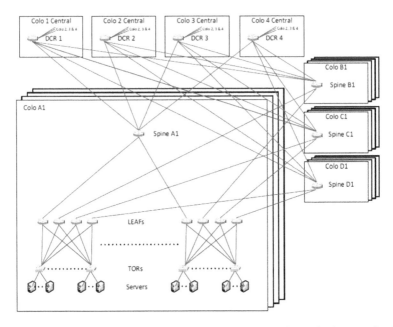

Figure 5.37 Representative datacenter architecture from Microsoft in contribution issenhuth_400_01_0713.pdf at the IEEE P802.3 Plenary Meeting in Geneva, July 2013. The server-TOR connections are direct-attach copper cabling up to 3 m, while the TOR-LEAF connections are active optical cables up to 20 m, and the LEAF-SPINE connections are MPO-terminated parallel single-mode fiber cabling with PSM4 transceivers up to 400 m. The shorter reach connections under 20 m are intended to be replaced as server and switch speeds advance, while the parallel single-mode infrastructure should be upgradeable through multiple generations.

While many companies that build hyper-scale datacenters keep their designs proprietary, Microsoft and Facebook have been vocal about transceiver and cable strategies. Facebook has expressed a strong public interest in duplex single-mode fiber cabling, whereas Microsoft has advocated PSM4 over parallel single-mode cabling. Figure 5.37 diagrams the cabling of the generation of hyper-scale datacenters that Microsoft was planning for the 2015 time frame.

5.7 Conclusions

VCSEL-MMF links have historically been the low-cost solution for short-reach Datacom optical interconnects up to 500 m. The developments of OM5, PAM4 and SWDM provide a upgrade path for multimode solutions to support low-cost 100, 200 and 400 Gbps, with proprietary solutions available for extended reach support beyond 100/150 m. At 500 m and beyond, single-mode fiber is the medium of choice. Selecting the proper medium type for links in between these distances can depend on a variety of factors, including the number of longer links and the migration path to higher speeds.

Hyper-scale cloud datacenters, with their massive size and enormous buying power, have typically adopted single-mode fiber for a large portion of their links, especially for those that are longer than 70/100 m. Server-switch links, and in some cases, TOR to Leaf switch links, are areas where multimode can provide a cost advantage if the link distance is suitable. In many cases, these cables are not reused, so the ability to support next-generation applications is unimportant. Because of this, OM3 is often the multimode medium of choice, rather than OM4 or OM5, which can support higher speeds over longer link distances.

These customers typically install leading-edge equipment and are among the first to adopt new technology. Because data management is part of their core competency, and their large size provides tremendous buying power, hyper-scale datacenters are not strictly tied to traditional architectures and standards. They are more willing to adopt proprietary technologies and unique architectures if they provide sufficient advantages.

As hyper-scale datacenters deploy small edge datacenters to provide faster response/access to their customers, multimode fiber again becomes a popular option because of its cost advantage. This is also the case in service providers' central offices as they migrate to a more data-centric architecture. Both of these markets provide growing demand for short-reach optical interconnects.

Even very large enterprise datacenters are smaller than hyper-scale/cloud centers, and remain a key multimode market. While cloud-based datacenters have been growing at a rapid rate and taking an ever-increasing part of the market, enterprise datacenters are still growing at high single-digit rates [8]. These datacenters are more traditional, using structured cabling and standards-based architectures. Cabling is often reused in these datacenters as they migrate to higher speeds, so the ability to support future technology generations is important. Many of these customers were the first to adopt OM3 and OM4 fiber, and are among the first to begin the adoption of OM5 fiber.

Multimode still has many advantages in short-reach applications that have been enhanced by the introduction of WDM technology. How-ever, single-mode solutions have become more prevalent in hyper-scale cloud-based datacenters that have much longer reach requirements. For those links, and for inter-datacenter links less than 30 or 40 km, low-water-peak, bend-insensitive SMF is the best solution. Dispersion-shifted fiber and ultra-low-loss large-effective-area fibers are options to optimize metro-DCI and long-haul system connecting remotely located hyper-scale datacenters. High density fiber counts cables enabled by rollable ribbon is an attractive alternative to CWDM/DWDM for inter-datacenter connections in reach of <20 km. For ultimate capacity increase for both intra- and inter-datacenters, spatial division multiplexing enabled by multicore and few-mode fiber has been intensively explored in research and development in recent years.

Acknowledgement

We acknowledge Robert Lingle Jr. of OFS for critical reading and significant contribution to Section 5.6. We acknowledge SPIE for reprint Figure 5.4 [38].

References

[1] Kao, K. C., and Hockham, G. A. (1966). "Dielectric-fibre surface waveguides for optical frequencies," *in Proceedings of the Institution of Electrical Engineers*, 113, 1151–1158.

[2] Kao, K. C. (2009). Sand from centuries past: Send future voices fast. *Nobel Lecture* [Accessed: December 8, 2009].

[3] Royal Swedish Academy of Sciences (2009). *Scientific Background on the Nobel Prize in Physics 2009: Two Revolutionary Optical Technologies.* Available at: http://www.nobelprize.org/nobel_prizes/physics/laure ates/2009/advanced-physicsprize2009.pdf

[4] Kapron, F. P., Keck, D. B., and Maurer, R. D. (1970). Radiation losses in glass optical waveguides. *Appl. Phys. Lett.* 17, 423–425.

[5] OFS Fitel (2017). http://fiber-optic-catalog.ofsoptics.com/Asset/TeraWa ve-Scuba-Ocean-Fibers-fiber-168-web.pdf

[6] Tamura, Y., Sakuma, H., Morita, K., Suzuki, M., Yamamoto, Y., Shimada, K., et al. (2017). "Lowest-ever 0.1419-dB/km loss optical fiber," *in Proceedings of the Optical Fiber Communication Conference (OFC)*, Th5D.1.

[7] Weldon, M. K. (2016). *The Future X Network – A Bell Labs Perspective.* New York, NY: CRC Press.

[8] Cisco (2016). *Cisco Global Cloud Index: Forecast and Methodology, 2015–2020.* [Accessed: November, 2016].

[9] Kamino, J. (2017). *What Is Going On With Data Center Standards and Technology?* Vancouver: BICSI Canadian Conference and Exhibition.

[10] Ethernet Alliance (2016). *The 2016 Ethernet Roadmap*. Available at: https://ethernetalliance.org/roadmap/

[11] D'Ambrosia, J. (2017). "CFI Consensus – Beyond 10 km Optical PHYs," *in Proceedings of the IEEE 802 July 2017 Plenary*, Berlin. Available at: http://www.ieee802.org/3/cfi/0717_1/CFI_01_0717.pdf.

[12] Xie, C. J. (2017). "Datacenter optical interconnects: Requirements and challenges," *in Proceedings of the Optical Interconnect Conference*, Santa Fe, NM.

[13] Filer, M., Searcy, S., Fu, Y., Nagarajan, R., and Tibuleac, S. (2017). "Demonstration and performance analysis of 4 Tb/s DWDM metro-DCI system with 100G PAM4 QSFP28 modules," *in Proceedings of the OFC/NFOEC, W4D.4*.

[14] Gill, V., and Lam, C. F. (2010). *Building Very Large-Scale Computing Infrastructure and Related Global Network Infrastructure*, Yokohama. Available at: http://www.suboptic.org/wp-content/uploads/2014/10/Vijay_Cedric_Preso-1-.pdf

[15] Facebook (2015). *Facebook builds world's longest optical network*. Available at: https://www.telecomasia.net/blog/content/facebook-builds-worlds-longest-optical-network

[16] Lingle Jr., R., McCurdy, A., and Balemarthy, K. (2017). *Benefits of TeraWave^{TM} ULL Optical Fiber for Improving Capacity, Reach, and Economics with Coherent Transport*, OFS Whitepaper. Available at: http://www.ofsoptics.com/request-wp.php?ID=TeraWave-ULL-White-Paper

[17] Peckham, D., Klein, A., Borel, P. I., Jensen, R., Levring, O., Carlson, K., et al. (2016). "Optimization of large area, low loss fiber designs for C+L band transmission," *in Proceedings of the Optical Fiber Communications Conference and Exhibition (OFC)*.

[18] Zhu, B., Zhang, J., Yu, J., Peckham, D., Lingle, R., Yan, M. F., Zhu, B. Y., et al. (2016). "34.6 Tb/s (173 × 256 Gb/s) Single-band transmission over 2400 km fiber using complementary Raman/EDFA," *in Proceedings of the Optical Fiber Communications Conference and Exhibition (OFC)*.

[19] ITU-T Standards (2008). "Series G: Transmission Systems and Media, Digital Systems and Networks, Transmission media and optical systems characteristics – Optical fibre cables": G.654 "Characteristics of a cut-off shifted single-mode optical fibre and cable"; G.655 standards

"Characteristics of a non-zero dispersion-shifted single-mode optical fibre and cable"; G.656 standards, "Characteristics of a fibre and cable with non-zero dispersion for wideband optical transport."

[20] OFS Fitel (2016). "Single-mode optical fiber selection guide – Terrestrial applications [OEM Fiber]," *OFS White Paper*. Available at: http://fiber-optic-catalog.ofsoptics.com/asset/Single-Mode-Optical-Fiber-Selection-Guide-fap-164-web.pdf

[21] Yang, F. (2016). *The Journey to Data Centers in the Future, EA ETF 2016 "The Road to Ethernet 2026"*, Santa Clara.

[22] OFS Fitel (2016). *More fibers, less space – advances in fiber/cable density*. Available at: http://iseexpo.com/wp-content/uploads/2016/10/OFS_More-fibers-less-space-%E2%80%93-advances-in-fiber-and-cable-density-9-16.pdf

[23] Sasaki, Y., Takenaga, K., Aikawa, K., Miyamoto, Y., and Morioka, T. (2017). "Single-mode 37-core fiber with a cladding diameter of 248 μm," in *Proceedings of the Optical Fiber Communications Conference and Exhibition (OFC)*.

[24] Sakamoto, T., Aozasa, S., Mori, T., Wada, M., Yamamoto, T., Sagae, S., et al. (2017). "Randomly-coupled single-mode 12-core fiber with highest core density," in *Proceedings of the Optical Fiber Communications Conference and Exhibition (OFC)*.

[25] Zhu, B. Y., Taunay, T. F., Yan, M. F., Fishteyn, M., Oulundsen, G. E., and Vaidya, D. S. (2010). 7×10 Gb/s multicore multimode fiber transmissions for parallel optical data links. *Photon. Technol. Lett.* 22, 1647–1649.

[26] Lee, B. G., Kuchta, D. M., Doany, F. E., Schow, C. L., Pepeljugoski, P., Baks, C., et al. (2012). End-to-end multicore multimode fiber optic link operating up to 120 Gb/s. *J. Lightwave Technol.* 30, 886–892.

[27] OFS Fitel (2017). *FMF Website*. Available at: http://fiber-optic-catalog.ofsoptics.com/viewitems/few-mode-optical-fiber-series/few-mode-optical-fiber-series1

[28] Grüner-Nielsen, L., Sun, Y., Nicholson, J. W., Jakobsen, D., Lingle Jr., R., and Pálsdóttir, B. (2012). "Few mode transmission fiber with low DGD, low mode coupling and low loss," in *Proceedings of the OFC/NFOEC Postdeadline Papers*.

[29] Sun, Y., Lingle, R., McCurdy, A., Peckham, D., Jensen, R., and Gruner-Nielsen, L. (2013). "Few mode fiber for mode-division multiplexing," in *Proceedings of the Photonics Society Summer Topical Meeting Series*.

[30] Peckham, D. W., Sun, Y., McCurdy, A., Lingle Jr., R. (2013). "Few-mode fiber technology for spatial multiplexing," in *Optical Fiber Telecommunication VI*", eds I. Kaminow, T. Li, and A. E. Willner (Amsterdam: Elsevier Inc.).

[31] Marom, D., Dangui, V., Ip, E., Butler, D., Tomkos, I. (2017). *OFC 2017 SDM Workshop Presentation.*

[32] Ryf, R., Fontaine, N. K., Chen, H., Guan, B., Huang, B., Esmaeelpour, M., et al. (2015). Mode-multiplexed transmission over conventional graded-index multimode fibers. *Optics Express* 23, 235–246.

[33] Buck, J. A. (1995). *Fundamentals of Optical Fibers.* Hoboken, NJ: A Wiley-Interscience Publication.

[34] Kawano, K., and Kitoh, T. (2001). *Introduction to Optical Waveguide Analysis.* New York, NY: John Wiley and Sons.

[35] Lenahan, T. A. (1983). Calculation of modes in an optical fiber using the finite element method and EISPACK. *Bell Sys. Tech. J.* 62, 2663–2694.

[36] Dangui, V., Digonnet, M. J., and Kino, G. S. (2006). A fast and accurate numerical tool to model the modal properties of photonic-bandgap fibers. *Optics Express* 14, 2979–2993.

[37] Guo, S., Wu, F., and Albin, S. (2004). Loss and dispersion analysis of microstructured fibers by finite-difference method. *Optics Express* 12, 3341–3352.

[38] Sun, Y., Lingle, R., Oulundsen, G., McCurdy, A. H., Vaidya, D. S., Mazzarese, D., et al. (2008). "Advanced multimode fiber for high speed short reach interconnect," *in Proceedings of the Asia Pacific Optical Communication Conference.*

[39] Lingle Jr, R., Peckham, D. W., McCurdy, A., and Kim, J. (2007). "Light-guiding fundamentals and fiber design," in *Optical Fiber Telecommunication VI A*, eds I. Kaminow, T. Li, and A. E. Willner (Amsterdam: Elsevier Inc.).

[40] Miller, S. E. (1965). Light propagation in generalized lens-like media. *Bell Syst. Tech. J.* 44, 2017–2064.

[41] Gloge, D., and Marcatilli, E. A. J. (1973). *Bell Syst. Technol. J.* 52, 1562; "Graded-index optical fiber," *US patent 3823997.*

[42] Olshansky, R., and Keck, D. R. (1976). Pulse broadening in graded-index optical fibers. *Appl. Opt.* 15, 483–491.

[43] Olshansky, R., and Keck, D. (1976). Pulse broadening in graded-index optical fibers. *Appl. Optics* 15, 483.

[44] Olshansky, R. (1976). Pulse broadening caused by deviations from the optimal index profile. *Appl. Optics* 15, 782–788.

[45] Olshansky, R. (1979). Multiple-alpha index profiles. *Appl. Optics* 18, 683–689.

[46] Kaminow, I. P., and Presby, H. M. (1977). Profile synthesis in multi-component glass optical fibers. *Appl. Optics* 16, 108–111; Gloge, D. C., Kaminow, I. P., Presby, H. M. "Graded-index fiber for multimode optical communication", *US patent 4025156.*

[47] Okamoto, K., and Okoshi, T. (1976). "Analysis of wave propagation in optical fibers having core with alpha-power refractive-index distribution and uniform cladding", *in Proceedings of the IEEE Trans. Microw. Theory Tech.*, MTT-24, 416–421.

[48] Okamoto, K., and Okoshi, T. (1977). "Computer-aided synthesis of the optimum refractive-index profile for a multimode fiber," *in Proceedings of the IEEE Trans. Microw. Theory Tech.*, 213–221.

[49] Stolz, B., and Yevick, D. (1983). Correcting multimode fiber profiles with differential mode delay. *J. Opt. Commun.* 4, 139–147.

[50] Pepeljugoski, P., Golowich, S. E., Ritger, A. J., Kolesar, P., and Risteski, A. (2003). Modeling and simulation of next-generation multimode fiber links. *J. Lightwave Technol.* 21, 1242–1255.

[51] Pepeljugoski, P., Hackert, M. J., Abbott, J. S., Swanson, S. E., Golowich, S. E., Ritger, A. J., et al. (2003). Development of system specification for laser-optimized 50-μm multimode fiber for multigabit short-wavelength LANs. *J. Lightwave Technol.* 21, 1256–1275.

[52] TIA-492AAAC (2003). "Detailed specification for 850 nm laser-optimized, 50μm core diameter/125μm cladding diameter class 1A graded-index multimode optical fibers," *in Proceedings of the Telecommunications Industry Association*; IEC/CEI 60793-2-10 (2007). 3rd Edition: "Optical fibers – part 2–10: product specifications – sectional specification for category A1: multimode fibers," *in Proceedings of the Telecommunications Industry Association.*

[53] TIA-492AAAD (2009). "Detailed Specification for 850 nm laser-optimized, 50-μm core diameter/125-μm cladding diameter class La Graded-Index multimode optical fibers suitable for manufacturing OM4 cabled optical fiber", *in Proceedings of the Telecommunications Industry Association.*

[54] Petermann K. (1978). Simple relationship between differential mode delay in optical fibers and the deviation from optimum profile. *Electron. Lett.* 14, 793–794.

[55] ANSI/TIA-455-204-A, and FOTP-204 (2013). "Measurement of bandwidth on multimode fiber," *in Proceedings of the Telecommunications Industry Association.*

[56] TIA/EIA-455-220, and FOTP-220 (2001). "Differential mode delay measurement of multimode fiber in the time domain," *in Proceedings of the Telecommunications Industry Association.*

[57] Cunningham, D., and Dawe, P. (2002). "Review of the 10 Gigabit Ethernet link model," *Agilent Technologies, ONIDS 2002* White Paper.

[58] Oulundsen, G., Yan, M., Kim, J., Sun, Y., Jiang, X., Weimann, P., et al. (2009). "Consideration in multimode fibers optimized for robust installation," *in Proceedings of the 58th International Wire and Cable Symposium (IWCS/IICIT).*

[59] Mazzarese, D. (2012). "LaserWave® *FLEX* Bend-Optimized Fiber: Innovative Design, Proven Performance," *OFS White Paper.*

[60] "TIA-492AAAE (2016). *Detail Specification for 50-μm Core Diameter/ 125-μm Cladding Diameter Class 1a Graded-Index Multimode Optical Fibre with Laser-Optimized Bandwidth Characteristics Specified for Wavelength Division Multipexing.*"[Accessed: June, 2016].

[61] Shubochkin, R., Braganza, D., Sun, Y., Balemarthy, K., Lingle, Jr., R., Kamino, J., et al. (2017). Next generation widband multimode fiber for high-speed SWDM datacom links," *in Proceeding of the IWCS-UL China Symposium*, Shanghai.

[62] "MMF Link Model Spreadsheet for 32 GFC" (2012). Available at: http://www.t11.org/t11/docreg.nsf/udocs/201237600

[63] Balemarthy, K., Shubochkin, R., and Sun, Y. (2016). "Next-generation wideband multimode fibers for data centers", *in Proceeding of the SPIE977504.*

[64] Sun, Y., Lingle, R., Chang, F., McCurdy, A. H., Balemarthy, K., Shubochkin, R., et al. (2017). SWDM PAM4 transmission from 850 to 1066 nm over NG-WBMMF using 100G PAM4 IC chipset with real-time DSP. *J. Lightwave Technol.* 35, 3149–3158.

[65] Oulundsen III, G., Sun, Y., Vaidya, D. S., Lingle Jr, R., Irujo, T., and Mazzarese, D. (2008). "Important performance characteristics of enhanced OM3 fibers for 10 Gb/s operation," *in Proceedings of the 57th International Wire & Cable Symposium (IWCS)*, 327–334.

[66] IEEE (2014). *IEEE Excel spreadsheet modelling: "Example MMF Link Model 130503.xlsx.* Available at: http://ieee802.org/3/bm/public/may13/ ExampleMMF%20LinkModel%20%20130503.xlsx.

[67] Sun, Y., Shubochkin, R., and Zhu, B. Y. (2015). "New development in optical fibers for datacenter applications," *in SPIE Proceedings, eds B. B. Dingel, K. Tsukamoto, Broadband Access Communication Technologies IX*, 9387.

[68] Castro, J. M., Pimpinella, R., Kose, B., and Lane, B. (2012). Investigation of the interaction of modal and chromatic dispersion in VCSEL-MMF channels. *J. Lightwave Technol.* 30, 2532–2541.

[69] Ji, C., Wang, J. Y., Södertröm, D., and Giovane, L. (2009). "High data rate 850 nm oxide VCSEL for 20 Gbit/s application and beyond," *in Proceedings of 2009 SPIE, Asia Communications and Photonics Conference*, 7631, 763119.

[70] Giovane, L. M., Wang, J., Murty, M. R., Harren, A. L., Chang, H. H., Wang, C., et al. (2016). "Volume manufacturable high speed 850 nm VCSEL for 100 G Ethernet and Beyond," in *Proceedings of the Optical Fiber Communications Conference and Exhibition (OFC)*, Tu3D.5.

[71] Sun, Y., Lingle Jr., R., Kamino, J., and Knight, D. (2015). "Extended reach of 40 G and 100 G transmission on OM4 MMF," *in Proceedings of the 64th IWCS Conference, International Wire and Cable Symposium (IWCS15)*.

[72] Sun, Y., Lingle Jr., R., King, T., Kamino, J., Shubochkin, R., and Knight, D. (2014). *in Proceedings of the 63rd IWCS Conference, International Wire and Cable Symposium (IWCS14)*.

[73] Sun, Y., Kamino, J., Shubochkin, R., Swartz, A., Lingle Jr., R., and Braganza, D. (2017). "High speed short reach optical interconnect over OM4 and OM5 multimode optical fiber," *to be published in Proceedings of the 66th IWCS Conference, International Wire and Cable Symposium (IWCS 2017)*.

[74] SWDM MSA. Available at: http://www.swdm.org/msa/

[75] Sun, Y., Ali, M. E., Balemarthy, K., Lingle, R. L., Ralph, S. E., and Lemoff, B. E. (2006). "10 Gb/s transmission over 300 m OM3 fiber from 990-1080 nm CWDM with electronic dispersion compensation," *in Proceedings of the Optical Fiber Communication Conference, 2006 and the 2006 National Fiber Optic Engineers Conference*, OTuE2.

[76] OFS Press Release (2016). *OFS Demonstrates Short-wave Wavelength Division Multiplexing over Wideband Multimode Fiber.* Available at: http://www.cablinginstall.com/articles/2016/03/ofs-wideband-multimode-fiber-demo-ofc.html.

[77] Sun, Y., Scott, K., Shubochkin, R., McCurdy, A., Kamino, J., Swartz, A., et al. (2016). "High speed short wavelength division multiplexing

transmission over next generation wideband multimode optical fiber," *in Proceeding of the IWCS-UL China Symposium*, Shanghai.

[78] Sun, Y., Scott, K., Shubochkin, R., McCurdy, A., Kamino, J., Swartz, A., et al. (2016). "High speed short wavelength division multiplexing transmission over next generation wideband multimode optical fiber," *in Proceedings of the 65th International Wire and Cable Symposium (IWCS) Conference*, 476–484.

[79] Sun, Y., Lingle, R., Shubochkin, R., Balemarthy, K., Braganza, D., Gray, T., et al. (2016). "51.56 Gb/s SWDM PAM4 transmission over next generation wideband multimode optical fiber," *in Proceedings of the Optical Fiber Communications Conference and Exhibition (OFC)*.

[80] Sun, Y., Lingle, R., Shubochkin, R., McCurdy, A. H., Balemarthy, K., Braganza, D., et al. (2017). SWDM PAM4 transmission over next generation wide-band multimode optical fiber. *J. Lightwave Technol.* 35, 690–697.

[81] Ingham, J. (2016). "Updated baseline proposal for the 100 Gb/s MMF objective using two-wavelength PAM4 transmission," *in Proceedings of the IEEE P802.3cd.* Available at: http://www.ieee802.org/3/cd/public/Sept16/ingham_3cd_01a_0916.pdf

[82] IEEE Standard for Ethernet (2015). *IEEE Std. 802.3.* Available at: http://standards.ieee.org/getieee802/download/802.3-2015.zip

[83] ITU-T Standards (2008). Series G: Transmission Systems and Media, Digital Systems and Networks, Transmission media and optical systems characteristics – Optical fibre cables; G.652 "Characteristics of a single-mode optical fibre and cable"; G.657 "Characteristics of a bending-loss insensitive single-mode optical fibre and cable."

6

PAM4 Signaling and its Applications

Frank Chang

Inphi Corp, California, USA

6.1 Introduction

Over the last few decades, the use of optical communication networks has exploded. Today internet and mobile applications demand a huge amount of bandwidth for transferring video, photo, music, and other multimedia files. For example, social networks like Facebook or Tencent, process more than 500 TB of data daily. With such increasing trends on data transfer, it's obvious that the demand for bandwidth will continue to rise rapidly in the data center and server sector, which drives the higher speed connectivity dramatically.

Just recently after years' long industrial debate, the IEEE 400 GbE P802.3bs and 802.3cd Task Forces [1] have adopted optical four level Pulse Amplitude Modulation or PAM4 signaling as the only viable technology standard of data center interconnect for 2 km and above distances as well as for low end of 500m and below. Current 100G data center solutions use either four fibers or four wavelengths at 25 Gbps per lane/wavelength. The optics usually take up main part of the whole optical module cost. The optical PAM4 modulation reduces optics counts by doubling the bits per symbol at the same baud rate, and transfers the complexity into CMOS electronics with PAM4 encoding, real-time DSP and FEC technologies, so bandwidth improvement can be achieved at a lower cost as compared to existing NRZ solution.

This chapter highlights PAM4 transceiver structures that achieves 100 Gbps Dual Channel transmission over electrical and optical interconnects used within world's largest Datacenters. The transceiver chip, built in 28 nm or 16 nm CMOS processes, employs a PAM4 transmitter with 3 (or more) tap FFE, a PAM4 receiver with CTLE, ADC and DSP based equalization and FEC logic.

The PAM4 applications and measured lab results are presented showing 100G PAM4 transmission over optical fibers linking the top of rack (TOR), aggregation and core switches typically found within layer-3 core switches and/or edge routers for cloud data centers. The architectural roadmap to 200G and 400G Ethernet will also be discussed in details. At the end of the chapter, we will give a brief discussion on the future trends beyond 400G.

6.2 A Brief History

About a dozen years ago there were two pioneered PAM4 SerDes designs out there, by Rambus and Accelerant, respectively, targeting multi-Gb/s rates (5–10 Gbps) applications [2–4]. The primary focus was on backplane and coaxial cables with 3 dB bandwidth in the range of 1.2 GHz limited by cable skin-effect loss and the process technology. This design differs in its use of a receiver equalizer in combination with a transmitter filter to compensate for the cable characteristics.

The subsequent researches on high-speed (>10 Gb/s), short-range (<100m) serial links over electrical backplanes or optical fibers have revealed the design trends for next generation, e.g., the chip-to-chip, chip-to-module and board-to-board communication, moving toward 20 Gb/s, and 100 Gb/s Ethernet [5, 6]. The chip designs operated PAM4 signaling using commercially available 90 nm CMOS technology.

Figure 6.1 summaries the evolution in PAM4 electrical transceivers. The strong industry race to commercialize PAM4 signaling has been driven by the major demand on 200G and 400G Ethernet for cloud data centers. The 28 nm

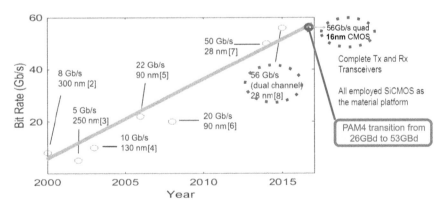

Figure 6.1 Evolution in PAM4 electrical transceivers.

and 16 nm CMOS has nowadays become the mainstream [7, 8] for chip design. It targets the 56 Gb/s data rates with transition to 112 Gb/s per lane or wavelength.

The chip implementation using the state-of-the-art 7 nm (or sub-10 nm) CMOS downsizing is expected to appear from the middle of 2019 with primary goal for the further reduction in power consumption and die size.

6.3 PAM4 IC Implementation Challenges

Before the real chip implementation becomes available, prior measurements of PAM4 optical transmission at 25Gbaud and higher have heavily relied on offline post-processing techniques with the aid of expensive instruments, and some high BER floors around $10^{-5} \sim 10^{-6}$ have been expected [9–12] due to limited memory sizes of captured waveforms.

To implement the PAM4 architecture over transition from NRZ, the key integrated circuit (IC) building blocks for an optical transceiver module, as shown in Figure 6.2, include the PAM4 DSP engine incorporating real time DSP processing and FEC as well as high linearity, low power and high bandwidth transimpedance amplifiers (TIA) and drivers.

For optics front, depending on link distances, the transmitter type can be Vertical Cavity Surface Emitting Lasers (VCSELs) for multimode technology ranging up to ~100 meters, and Directly-modulated DFB lasers (DMLs) or Electro-absorption modulated lasers (EMLs) ranging to 10 or even 40 kilometers. PIN and Avalanche photodiodes (APD) receivers are

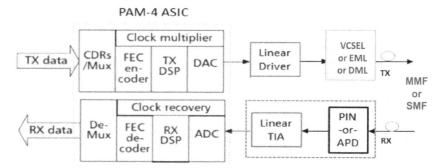

Figure 6.2 Typical optical PAM4 transmitter and receiver IC building blocks (EML: Electro-absorption modulated lasers, DML: Directly-modulated DFB lasers, VCSEL: Vertical Cavity Surface Emitting Lasers).

both designed for high speed data rates, and are widely used in laser-based fiber optic systems to convert optical data into electrical form. They are high-sensitivity, high-speed semiconductor light photodiodes. The main advantage of the APD is that it has a greater level of sensitivity compared to PIN.

The multi-purpose PAM4 PHY chip with real time DSP capability is critical device for generating the optical PAM4 signals. A block diagram of the DAC based PAM4 PHY chip is also shown in Figure 6.2. The transmit path starts with a CAUI-4 4 × 25.78125 Gbps CEI-28G-VSR compliant NRZ Receiver (RX) to interface to the host ASIC. The Ethernet traffic originating from the host is FEC encoded and transmitted out of the chip as two 25.78125 or 26.5625 GBaud PAM4 data streams. The PAM4 transmitter leverages transmit (TX) DSP, and DAC which maps the input MSB and LSB (Most & Least Significant Bits) bit streams into the PAM4 symbols. The FEC encoding function can be bypassed if this function is implemented in the host ASIC.

The PAM4 receive path starts with ADC+DSP based receiver. The details are outlined in Section 6.3.2. The output data stream from the DSP is passed on to the FEC decoder block, which if enabled, recovers the original Ethernet stream. The 100G data is transmitted back to the host through a CAUI-4 4 × 25.78125 Gbps CEI-28G-VSR compliant NRZ TX. In addition the chip implements both MDIO and I2C management interfaces to program registers for device configuration and diagnostic features.

6.3.1 PAM4 Transmit Architectures

A high-level block diagram of the TX data path is illustrated in Figure 6.3. Each TX lane takes the raw bit stream from two host RX lanes when the FEC is bypassed, or a post-FEC encoded 50G data stream and outputs an electrical PAM4 signal. The transceiver chip has two such lanes to give an aggregate 100G line throughput.

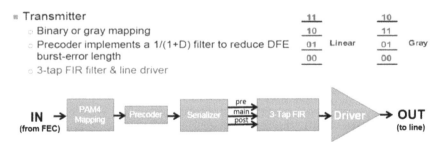

Figure 6.3   PAM4 Transmit Path Implementation Block Diagram.

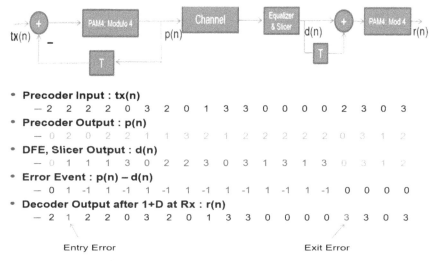

- **Precoder Input : tx(n)**
  − 2 2 2 2 0 3 2 0 1 3 3 0 0 0 0 2 3 0 3
- **Precoder Output : p(n)**
  − 0 2 0 2 2 1 1 3 2 1 2 2 2 2 2 0 3 1 2
- **DFE, Slicer Output : d(n)**
  − 0 1 1 1 3 0 2 2 3 0 3 1 3 1 3 0 3 1 2
- **Error Event : p(n) − d(n)**
  − 0 1 -1 1 -1 1 -1 1 -1 1 -1 1 -1 1 -1 0 0 0 0
- **Decoder Output after 1+D at Rx : r(n)**
  − 2 1 2 2 0 3 2 0 1 3 3 0 0 0 0 3 3 0 3

Entry Error   Exit Error

Figure 6.4 PAM4 TX Precoder Example ([13]).

The first stage in the TX data path is the mapping block to select between binary (linear) or gray mapping for the PAM4 symbols. The gray mapping function alters the normal binary mapping of symbol bits to voltage levels as shown in the inset of Figure 6.3.

The mapped output is then sent to the precoder block which implements a simple $1/(1+D)$ filter on the data stream. When the transmitter is partnered with a link receiver implementing a 1-tap DFE, the precoder reduces the DFE burst error which typical happens 2 errors per error event – one at the entry and one at the exit. Figure 6.4 illustrates an example of this precoder implementation.

A serializer block is then followed which generates individual pre-cursor, main-cursor and post-cursor serialized PAM4 symbol stream to the 3-tap FIR filter. The FIR filter has independent control on the PAM4 MSB and LSB paths. Overall, the filter coefficients can range from 0 to –0.25 for the pre-cursor and from 0 to –0.5 for the post-cursor.

The FIR output is fed into an output driver which provides swing levels up to ∼1.4Vppd with independent control on the MSB and LSB paths. The MSB to LSB ratio can also be altered for providing pre-distortion on the PAM4 eye, which is useful in applications where the PAM transmitter interfaces with optical drivers.

Figure 6.5(a) presents the electrical PAM4 eye directly from the PAM4 TX DAC output, and (b) subsequent PAM4 optical eye diagram from standard

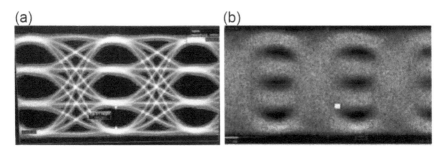

Figure 6.5   a) 25Gbuad PAM4 TX electrical eye; b) subsequent PAM4 optical eye diagram from standard 100 GbE-LR4 based EML with output power of +2 dBm at 1299.8.nm.

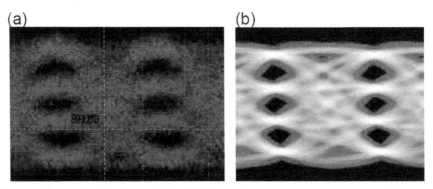

Figure 6.6   a) 1λ 42.5 Gbps PAM4 VCSEL optical eye; and b) subsequent PAM4 eye processed by linear equalizers.

100 GbE-LR4 based EML with output power of +2 dBm at 1299.8 nm (Courtesy of NeoPhotonics).

PAM4 chip can be implemented for $1\lambda \times 42.5$ and $2\lambda \times 53.125$ Gbps systems running at KP4 RS(544,514) FEC rates of 21.25 GBaud and 26.5625 GBaud, respectively. As another PAM4 TX example, Figure 6.6(a) presents the electrical PAM4 eye directly from VCSEL, and (b) subsequent PAM4 optical eye diagram processed by linear equalizers.

6.3.2 PAM4 Receive Architectures

The block diagram of the ADC+DSP Receiver is shown in Figure 6.7 as described elsewhere [8, 14, 15]. The two 50Gb/s PAM4 RX lanes give an aggregate 100G line throughput. Each RX signal terminates with 50 Ohms at the Continuous-time Linear Equalizer (CTLE) block. The CTLE

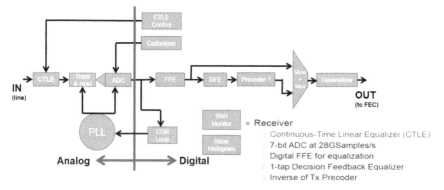

Figure 6.7 PAM4 Receive Path Implementation Block Diagram.

provides a programmable high-frequency boost ranging from 0 dB to 8 dB. It also has a gain range >12 dB in 0.1 dB steps to scale the input signal voltage to match the ADC input swing requirements. The CTLE drives the Track-and-Hold (T/H) and a 7-bit time-interleaved ADC which can operate up to 28 GSa/s.

For the clocking scheme, the PLL block takes timing recovery controls from the DSP to track the phase variations of the incoming data. The clocking for the T/H and ADC is derived from a PLL operating off an external reference clock. The clock recovery scheme can be made truly reference-less by taking advantage of the reference-less host VSR link. The recovered clock is filtered prior to ADC sampling. This allows the reference to be powered down.

The DSP core employs a set of parallel FFEs for channel equalization. The parallel factor was chosen to be a multiple of the number of sub-ADC channels to optimize power consumption. Then a multiple of the number of sub-channels in the time interleaved ADC allows for any bandwidth mismatch between the different AFE paths to be compensated by independent adaptation of the FFE slices.

The DSP employs an adaptive PAM4 Decision Feedback Equalizer (DFE). The feedback taps are limited to 1 tap to reduce the impact of error propagation. Baud rate clock recovery is used which is based on the well-known Mueller-Muller timing recovery scheme [16] but taking inputs directly at the ADC output, thus eliminating interaction problems with FFE-DFE adaptation while providing a low latency clock recovery path. The inverse of the TX precoder is implemented after the DFE.

Generally speaking, the DSP engine has the challenging tasks to combat various distortion and noise effects [12] from e.g., bandwidth limiting, intersymbol interference, chromatic dispersion and nonlinear behaviour etc. For receive DSP, the FFE and DFE are considered as two equalizer options for 25GBaud PAM4. Depends on various actual implementation, it normally consist of 10–21 taps of FFE and 1–4 taps of DFE. The front ADC and FFE may operate at 1–2 samples per symbol. The FEC decoder allows the PAM4 decoding at pre-FEC BER values, e.g., better than $\sim 10^{-5}$, then the decoded PAM4 digital signal is further enhanced, using standard FEC codes such as KR4 or KP4 FEC [1], to yield "error free" signal, i.e., better than 10^{-15} BER.

For the DSP to be self-adaptive at real time, the tap weights, for both the FFE and DFE, are updated automatically by an adaptive algorithm, and usually implemented using well established least-mean-squared (LMS) algorithms. This is accomplished by applying a close-loop feedback mechanism together with a means for estimating signal quality, e.g., error signal. When the adaptive algorithm controls the weights of the FFE and DFE, it converges to a minimum BER state, when is typically coincident with maximum eye opening.

Another important element of implementing PAM4 is the linear O/E front-end and automatic gain control (AGC) due to the analog nature of the ADC and equalization process. The linearity of the O/E front-end minimizes any further distortion on top of the impaired signal. As the input power can vary widely, a high amount of dynamic range is required. The AGC is normally integrated into the TIA and support an output signal in the range of 10 to $\sim$400 mVpp/side, and holds the output swing steady regardless of input signal strength over the entire dynamic range.

For link diagnostics and training, DSP bring-up is handled by a combination of Finite State Machines implemented on an on-chip micro-controller. There is a Signal-to-Noise Ratio (SNR) monitor block. The SNR is derived from calculating the mean square of the error signal which is simply the delta between the slicer input and the ideal PAM4 expected levels.

Both slicer histogram and eye diagram can be captured in real-time without interrupting traffic and represents true slicer margin at the sampling phase. Figure 6.8 depicts a typical output of slicer histogram block by giving more details of the distribution of samples at the final slicer around the 4 signal amplitude levels. This is accomplished by counting the number of hits at the slicer within a finite number of discrete bins (effectively determining a certain voltage resolution). The x-axis is address of the bins indicating the PAM4

Eye Diagrams

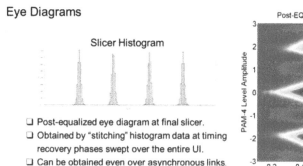

Slicer Histogram

❏ Post-equalized eye diagram at final slicer.
❏ Obtained by "stitching" histogram data at timing recovery phases swept over the entire UI.
❏ Can be obtained even over asynchronous links.

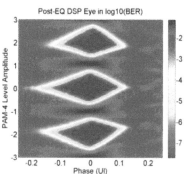

Figure 6.8 PAM4 On-chip Real-time DSP Processing for Slicer Output Histogram (left) and Post-equalized PAM4 Eye Diagram (right).

levels. The y-axis is simply a count of the number of hits within each bin. As shown in Figure 6.8, a more conventional post-equalized eye diagram at the final slicer can be obtained by stitching the slicer histogram data at timing recovery phases swept over the entire UI.

To conclude, the current PAM4 chips have been fabricated in 28 nm or 16 nm CMOS by the industry and become commercially available off the shelf from open market. Next three subsection will focus on the application of such PAM4 chips under various optical link environments.

6.4 PAM4 SMF Performance

In this subsection, we present the availability of industry first 100G PAM4 IC chipset operating error free under KP4 FEC threshold at ~25GBaud covering various scenarios from over 2 km of single mode fiber: 1) single lambda (1λ) 40Gb/s; 2) dual lambda (2λ) 100 Gbps with either PIN or APD devices. Using the realized PAM4 IC platform, we experimentally demonstrate up to 40km transmission is successfully achievable for 100 Gb/s with reference to the state of art high sensitivity APD optical receiver [17, 18].

6.4.1 Experimental Setups

The experimental PAM4 setups for $1\lambda \times 41.25$ and $2\lambda \times 51.5625$ Gb/s systems used in this work are depicted in Figure 6.9 (a) and (b) running at KR-FEC rates of 20.625GBaud and 25.78125GBaud, respectively. The multi-purpose PAM4 PHY chip is the heart of the setup. In both experiments, the embedded PRBS generator and checker inside the chip facilitate the true real time BER tests of the end-to-end optical PAM4 link. If it's not specified, we used PRBS $2^{31} - 1$ throughout these tests.

In the $1\lambda \times 41.25$Gb/s PAM4 test, one 41.25 Gb/s PAM4 signal was generated differentially at 500 mVpp/side from PAM4 PHY TX output. A 1299 nm DML with integrated linear Shunt driver was used as optical source [19] (Courtesy of SEDI). One 10 km SMF spool (with an optical loss of 5.6 dB) was inserted before VOA (variable optical attenuator) for varying the input power to the optical receiver. For lower cost considera-tion, a 10G LRM ROSA of 8 GHz 3 dB bandwidth (Courtesy of Oplink) was chosen for optical-to-electrical conversion in front of PAM4 PHY receiver (RX).

The $2\lambda \times 51.5625$Gb/s experimental setup is similar to that of 1λ PAM4 (Figure 6.9), but with two wavelengths each running simultaneously at 51.5625 Gb/s. The linear driver (IN3214SZ) was fed with differential input and generated single-ended output for driving EML. The two wavelengths at 1299 nm and 1304 nm from commercial quad EML TOSA module (Courtesy of NeoPhotonics, Avago) were then combined via a LAN-WDM multiplexer and attenuated through VOA, which controlled the input power to the ROSA. Several SMF spools of various distances of 2, 10, 25, 40 km (with optical losses of 1.8, 5.6, 8.7, 13.6 dB, respectively) are then inserted for transmission link test. We investigated two kinds of different commercial ROSAs with either PIN (IN3250TA) or APD (IN2860TA). The optical power of each wavelength was controlled and calibrated by VOA and wavemeter before it feeds into the ROSA and was demultiplexed. The insertion losses of

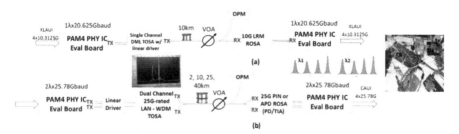

Figure 6.9   PAM4 test setup at (a) 1λ 40Gb/s DML, (b) 2λ 100Gb/s EMLs, with RX DSP recovered histograms shown in inset.

LAN-WDM multiplexer, demultiplexer, and VOA are 1.6, 1.5, and 2.6 dB, respectively.

6.4.2 1λ 40G 10 km Transmission

The measured BER curves versus (vs.) receiver (RX) optical power at 20.625GBaud for back-to-back (B2B) and 10 km are presented in Figure 6.10. The 1299 nm DML is biased at DC current of ∼80mA with optical output power of + 4.6 dBm. The extinction ratio (ER) was around 4.8 dB. The power penalty at pre-FEC BER threshold 2.4×10^{-4} for 10 km is negligible as compared to B2B case, while the BER floors at BER values of ∼1 order of magnitude higher. This could be understandable, as there exist large bandwidth gap required for the operating baud rate from the 10GLRM ROSA. When the FEC is enabled, the link ran error free with 10 km at sensitivity below –9.5dBm.

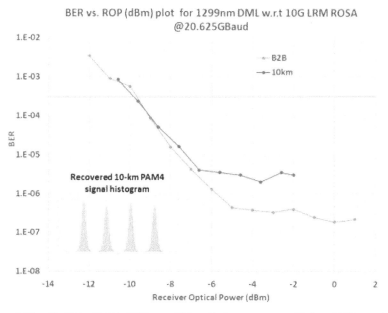

Figure 6.10 40 Gb/s PAM4 BER vs. RX optical power over 10 km SMF running at 20.625GBaud.

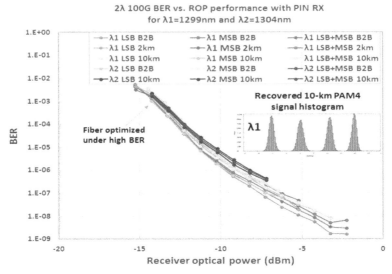

Figure 6.11 $2\lambda \times 50G$ PAM4 transmission results with PIN RX.

6.4.3 2λ 100G 10 km and 40 km Transmissions

Figure 6.11 shows the BER curves with PIN receiver for LSB, MSB and combined (LSB+MSB) bit streams after various distances and with B2B. Both EMLs for $\lambda 1$ and $\lambda 2$ are biased at 80 mA and maintained at 42°C. The extinction ratios were around 6.8 dB, and 6.5 dB, respectively. A slightly worse performance in MSB was observed due to high error probability on the symbol. The RX sensitivity was around –12.9 dBm, while power penalty for 10 km fell within 0.5 dB than B2B at 2.4×10^{-4} pre-FEC threshold.

The experimental results of $2\lambda \times 51.5625Gb/s$ over 40 km are shown in Figure 6.12 for $\lambda 1$ and $\lambda 2$ respectively. Similarly, the BER for LSB, MSB and combined LSB+MSB are monitored for both wavelengths. Both EMLs for $\lambda 1$ and $\lambda 2$ are biased at 100 mA and maintained at 42°C for slightly higher power of over +2 dBm. The extinction ratio are around 7.2 dB, and 7.0 dB, respectively after the negative biasing voltage to the EMLs was optimized. The APD voltage was biased at $\sim$25 Vdc. APD shows more than $\sim$6.5 dB better sensitivity in B2B than PIN. Around $\sim$1 dB penalty was observed between B2B and 40 km. The PAM4 signal histograms for 40 km were shown in inset indicating both $\lambda 1$ and $\lambda 2$ running robustly. When the FEC is enabled, the link ran error free for 40 km with margin.

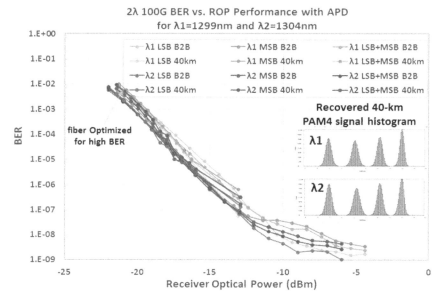

Figure 6.12 $2\lambda \times 51.5625$Gb/s PAM4 BER vs. RX optical power plots with APD for 40 km SMF transmission at $\lambda 1 = 1299$ nm and $\lambda 2 = 1304$ nm.

6.4.4 Technical Options for 200/400G Over SMF

By leveraging newly developed PAM4 IC chipset, the link performance with real-time DSP in miniaturized silicon format were extensively studied above for distance of standard single mode fiber from 2 km and above, and it showed error free transmission with great margin under KR4 FEC threshold at 1310 nm wavelengths for 10 km, up to 40 km distance. The results demonstrated the 40 km optical transmission with the availability of complete IC platform for enabling small form factor modules such as CFP4 and QSFP28 for 100 Gb/s [20].

Moving beyond 100 Gb/s, one can easily use the scale-out method to achieve 200/400G by increasing the number of lanes with lower lane speeds, such as $4/8 \times 50$G. Optical lanes can be either fiber lanes or wavelength lanes, as shown in Table 6.1 with technology choices. To reduce the requirements on the bandwidths of electronic and optical components, IEEE suggested using PAM4 for both electrical and optical signals when lane speeds are larger than 25G. PAM4 has a higher requirements than NRZ modulation format on the transmitter power, SNR and relative intensity noise (RIN). Higher-gain FEC are suggested by IEEE for 400G with the so-called KP4, which has a pre-FEC threshold at BER of 2.4×10^{-4}.

Table 6.1 200G/400G options over SMF

Standard	Designation	Modulation	Baudrate (GBaud)	Reach (km)	#Lanes or #λ	TOSA/ROSA	Sources of Optics Specs
50GbE	50GBASE-FR	PAM4	26.5625	2	1	DML/PIN	IEEE 802.3cd
	50GBASE-LR	PAM4	26.5625	10		DML,EML/PIN	
	50GBASE-ER	PAM4	26.5625	30-40?		EML/APD	IEEE B10K SG
100GbE	100GBASE-LR4	NRZ	25.78125	10	4	DML/PIN	IEEE 802.3ba
	100GBASE-ER4	NRZ	25.78125	30-40[a]		EML/APD	
	100GBASE-DR	PAM4	53.125	0.5	1	DML, Sipho/PIN	IEEE 802.3cd
	100G-FR	PAM4	53.125	2	1	DML, Sipho/PIN	100G Lambda MSA
	100G-LR	PAM4	53.125	10	1	EML/PIN	
200GbE	200GBASE-DR4	PAM4	26.5625	0.5	4	DML, Sipho/PIN	IEEE 802.3bs
	200GBASE-FR4	PAM4	26.5625	2		DML/PIN	
	200GBASE-LR4	PAM4	26.5625	10		EML/PIN	
	200GBASE-ER4	PAM4	26.5625	30-40?		EML/APD	IEEE B10K SG
400GbE	400GBASE-DR4	PAM4	53.125	0.5	4	DML, Sipho/PIN	IEEE 802.3bs
	400GBASE-FR8	PAM4	26.5625	2	8	DML/PIN	
	400GBASE-LR8	PAM4	26.5625	10		EML/PIN	
	400GBASE-ER8	PAM4*	26.5625	30-40?	8	EML/APD	IEEE B10K SG
	400G-FR4	PAM4	53.125	2	4	EML /PIN or APD	100G Lambda MSA

IEEE has specified three 400G SMF transceiver technologies so far, which are DR4, FR8, and LR8, and is now working toward to explore 4th ER8 option. DR4, FR8, and LR8 are specified for reach up to 500m, 2 km, and 10 km, respectively and ER8 could be 30–40 km over standard SMF. DR4 is equivalent to PSM4, which uses four parallel SMF lanes with the lane speed of 100G. FR8 and LR8 use eight wavelength lanes with the lane speed of 50G. An additional four wavelengths shorter than LAN WDM are assigned to FR8 and LR8, and the wavelength assignments are the same for FR8 and LR8. The wavelength spacing in the upper and lower four lanes is 800 GHz, same as LAN WDM, but the spacing between the upper four wavelengths and lower four wavelengths is 1.6 THz. Temperature control may have to be used due to the narrow wavelength range requirement.

For datacenter operators, it is desirable to have four optical lanes for 400G single-mode solutions, either four fiber lanes (DR4) or wavelength lanes (FR4), which will count on single lambda 100G technologies as described in the following Section 6.8. To meet the power and cost requirements, a larger channel spacing than that of FR8 may be needed so that TECs are not required for lasers. But with the lane speed at 100G, chromatic dispersion induced penalties may become an issue at larger channel spacing, and some stronger equalization techniques may need to be used.

6.5 PAM4 MMF Performance

VCSEL-based optical links operating at 850 nm over multimode fiber (MMF) still represent the vast majority (over 75%) of fiber deployments within data centers for low cost, low power consumption and short reach connectivity. Current 40 and 100G upgrades known as SR4 use four laser/detector pairs and 4 parallel fibers for transmitting and 4 parallel fibers for receiving at 10G or 25 Gbps per fiber. For most cases, optics usually take up main part of optical module cost. The optical PAM4 modulation dramatically reduces optics counts by doubling the bits per symbol at the same baud rate. So far PAM4 signaling is the only viable non-NRZ signaling technology of choice with MMF for data center interconnects standards by various IEEE802.3 groups [1] such as the 802.3cd 50/100/200 Gb/s Ethernet Task Force.

To allow transmission over duplex fibers with 10 Gbps deployments, create building blocks for future 400 and 800 Gbps parallel modules, and maintain longer and extended reaches achievable at higher transmission rates, the combination of PAM4 signaling with short wavelength division multiplexing (SWDM) [22] and novel wideband MMFs (WBMMF) [23] are highly desirable.

Prior MMF measurements of PAM4 transmission at ~25 GBaud or higher have either heavily relied on the offline post-processing techniques with the aid of expensive instruments [24–28] or to a much shorter reach on OM3/OM4 at 850 and 880 nm using PAM4 chip with real time BER function [29]. The longest reach on MMF was 300 m at 45 Gbps per wavelength using PAM4 chip [30]. To implement the PAM4 architecture with VCSEL lasers over transition from NRZ, the key IC building blocks, as shown in Figure 6.2, include PAM4 DSP engine incorporating real-time DSP processing and FEC as well as high linearity, low power and high speed transimpedance amplifiers (TIA) and drivers suitable for MMF environments.

It's also useful to leverage recent advance in silicon photonics with the introduction of the Ge/Si APD (avalanche photodiode) [31–32] operating at the range of ~25G and above to address improved optical receiver sensitivity. A better receiver sensitivity potentially achievable with the low cost Ge/Si APD is useful for upgrades to higher data rates using current VCSEL and PAM4 CMOS technology, especially with the loss of receiver sensitivity inherent to PAM4 or insertion of optical MUX/DeMUX into the link for SWDM.

6.5.1 Experimental Setups

The multi-purpose, 2-channel PAM4 PHY chip with real-time DSP capability is critical for generating the optical PAM4 signals [33, 34]. A block diagram of the DAC (digital-to-analog converter) based PAM4 PHY chip is shown in Figure 6.2 and was described in details in Section 6.3. PAM4 transmitters leverage TX DSP and DAC which maps the input MSB and LSB (most & least significant bits) streams into PAM4 symbols. Figure 6.6 (a) in the prior Section 6.3.1 presents the optical PAM4 eye directly from VCSEL, and (b) subsequent PAM4 equalized eye diagram processed by linear equalizers.

The experimental PAM4 setups for $1\lambda \times 42.5$ and $2\lambda \times 53.125$ Gbps systems used in this work are depicted in Figure 6.13 (a) and (b) running at KP4 RS(544,514) FEC rates of 21.25 GBaud and 26.5625 GBaud, respectively. The multi-purpose 100G PAM4 PHY chip is the heart of the setup. In both experiments, the embedded PRBS generator and checker inside the chip facilitate the true real-time BER tests of the end-to-end optical PAM4 link. We used the worst-case PRBS $2^{31} - 1$ pattern throughout this work to emulate the real data traffic.

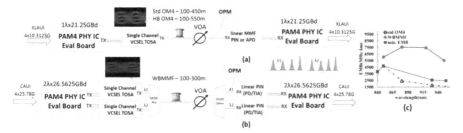

Figure 6.13 PAM4 setup at (a) 1λ 40/50 Gbps VCSEL, (b) 2λ (or more) 100/200 Gbps VCSELs, with 850 nm TX optical eye diagram and RX DSP recovered histograms shown in inset; (c) shows the EMB vs. wavelength for conventional OM4 and WBMMF.

The 2λ × 53.125 Gbps experimental setup is similar to that of 1λ PAM4 (Figure 6.13), but with two of the four SWDM4 wavelengths (850 and 880 nm, or 910 and 940 nm) each running simultaneously at 53.125 Gbps. Two wavelengths are optically multiplexed after TOSAs and de-multiplexed before ROSA using two out of four ports of a packaged NETWORK Cube Multimode CWDM-MUX-4 Module from Huber+Suhner.

For fiber spools, we used two OM4 samples including an off-the-shelf OM4 fiber passing specification of minimum effective modal bandwidth (EMB) of 4.7 GHz* km (shown as green square in Figure 6.13(c)) and an OM4 with higher EMB of 8.8 GHz*km at 850 nm. A next generation WBMMF (NG-WBMMF) with EMB over 6 GHz*km (shown in red in Figure 6.13(c)) was used for 2λ × 53.125 Gbps experiments.

At the receive side, we investigated the novel Ge/Si APD ROSAs with same linear TIA (IN2860TA) as for the reference PIN for potential low cost. The high speed TIA incorporates AGC function to maintain high linearity.

6.5.2 1λ 40G Transmission Over 550m OM4

The system performance was quantified with measurements of BER versus (vs.) average received (RX) optical power (AOP). The results with PIN and APD over various distances are illustrated in Figures 6.14 and 6.15 for the two OM4 MMFs. The maximum practical distance over the off-the-shelf OM4 can be achieved to be over 400m for one order of magnitude lower BER than KP4 FEC limit at 2.4×10^{-4}, while such distance is extended to 550m for the high bandwidth OM4. There is error floor for 450m or 500m distance due to modal dispersion. The APD shows –15.9 dBm sensitivity at B2B, which is ~4.8 dB better over PIN. The improvement is done by optimizing electronic equalizers embedded with the PAM4 chip. The power penalties at pre-FEC

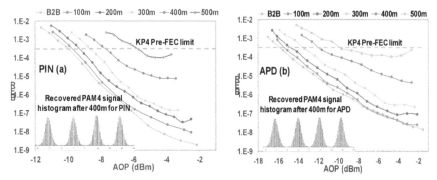

Figure 6.14 BER vs. RX optical power plots with PIN (a) and APD (b) for standard **OM4** transmission for VCSEL at 850 nm.

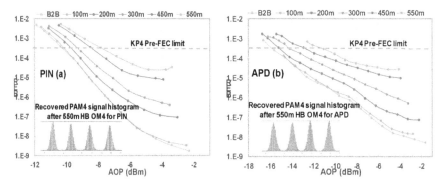

Figure 6.15 BER vs. RX power plots with PIN (a) and APD (b) for high bandwidth **OM4** transmission for VCSEL at 850 nm.

BER threshold increase dramatically with distance, while for the 100m case it is almost negligible ($<$0.2 dB) as compared to B2B case. The link was tested to run error free for the PIN case above sensitivity when the FEC is enabled with full 100G Ethernet traffic.

6.5.3 2λ 100/200 Gbps 300m Transmission

Figure 6.16 depicts the BER vs. AOP curves running at 53.125 Gbps for B2B, 200 and 300m at the SWDM4 wavelengths of 850, 880, 910, and 940 nm, respectively. The link budget (difference between TX optical power and B2B receiver sensitivity), total penalties including attenuation, dispersion power penalty (DPP) and attenuation at BER $= 2.4 \times 10^{-4}$ for 200 and 300m of NG-WBMMF at SWDM4 wavelengths are summarized in Table 6.2. The

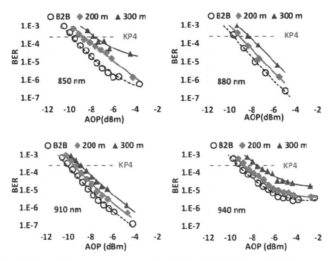

Figure 6.16 BER vs. RX optical power with PIN for WBMMF at SWDM wavelengths from 850 nm to 940 nm.

Table 6.2 Link budget, total penalties, DPP and attenuation for 200 and 300 m at four SWDM wavelengths

λ_c (nm)	Link budget (dB)	Total penalty (dB)		DPP (dB)		Attenuation (dB)		
		200 m	300 m	200 m	300 m	200 m	300 m	MUX / DeMUX
853	7.9	3.5	4.7	1.1	2.1	2.3	2.6	1.8
881	7.1	3.4	4.4	0.6	1.4	2.9	3.0	2.3
911	7.7	2.6	3.3	0.3	0.9	2.3	2.4	1.8
941	8.8	3.2	4.4	1.1	2.3	2.1	2.2	1.8

maximum DPP is 2.3 dB and attenuation is 3.0 dB where 2.3 dB is attributed to each MUX/DeMUX pair. The total penalty is 4.7 dB at maximum for 300m distance, allowing 2.7 to 4.4 dB margin from 850 to 940 nm.

6.5.4 Technical Options for 200/400G Over VCSEL/MMF

In this subsection, by leveraging the newly developed PAM4 IC chipset, link performance with real-time DSP in miniaturized silicon format was extensively studied for standard OM4 and WBMMF for upgrades to 40/50 and 100/200 Gbps. The experimental data in this wavelength region showed the PAM4 transmission for record reach of 550 m with the Ge/Si APD and record aggregated rate of 212.5 Gbps over WBMMF with real-time DSP processing. The results demonstrate 550 m transmission at 850 nm for OM4 and 300m across 4 SWDM wavelengths over WBMMF with MUX/DeMUX included. This availability of complete IC platform enables small form factor 100G modules such as QSFP28 for the wide MMF deployments covering ∼100m MMF distances in cloud data centers.

Going beyond 100G, Table 6.3 lists the 200G/400G technical options over VCSEL/MMF [35]. IEEE has specified 400G MMF transceiver technologies so far, based on SR16. SR16 uses 16 MMF lanes with a lane speed of 25G NRZ and can reach up to 100 m over OM4 fiber. It requires 32-fiber multi-parallel optic (MPO) cables and connectors.

By leveraging PAM4 signaling, the 400G throughput rates with fewer fiber lanes can be achievable by SR8 using 8 × 50G PAM4 signals or SR4.2 by introducing one additional SWDM wavelength [35]. As fiber cost is a significant part for the cost of multimode solutions, it is preferred not to use more than four fiber lanes in real implementation. Before the maturity of single lambda 100G VCSELs, one can further reduce the fiber lanes of multimode solutions with either SWDM technique [36], which may transmit two VCSEL wavelengths in one fiber, or bidirectional technique, where two optical lanes counter-propagate in one fiber [37].

Currently IEEE P802.3cm 400Gb/s over multimode fiber Task force [35] is actively exploring to standardize both 400G SR8 and SR4.2 multimode solutions targeting two specific distinct application scenarios: SR8 is cost optimized for greenfield new install/brownfield upgrade offering breakout capability to support 50G/100G/200G connection, while SR4.2 leverages to maximize the large deployment base of 4 pairs cabling.

Table 6.3   200/400G technical options over VCSEL/MMF

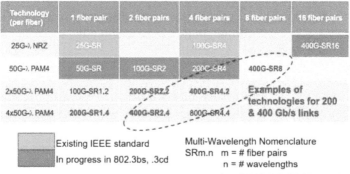

6.6 PAM4 for OSNR-limited Systems at 1550 nm

For inter-data center connections, there is demand to transport 100G or
400G data rates over longer distances up to 80 km or 100 km [38, 39],
which is beyond what has been currently ratified for 100/400 GbE. For such
applications, optically amplified WDM systems are typically specified, using
the C-band window at 1550 nm. It can be cost-effective to scale up from
the short reach IM/DD based intra-data center optics. The PAM4 challenge
resides in the OSNR performance and the chromatic dispersion tolerance.

From literature survey, it has been shown that 56 Gb/s IM/DD DMT on
8 channels can successfully bridge 240 km of SMF [40]. Single channel
112 Gb/s PAM4 transmission was demonstrated over 80 km SMF [12, 41, 42]
and a 100 km DWDM transmission of 3 × 56 Gb/s PAM4 via tunable laser
and 10 Gb/s InP MZM was shown [23], both using a dispersion compensating
module [44] in the transmission setup. However, all these demonstrations utilized offline DSP processing. The real-time transmission of 400G (8 × 50G)
PAM4 DWDM signals [38, 45–47] for data center interconnects up to 100
km SSMF is described in this subsection.

6.6.1 Experimental Setups

Figure 6.17 shows the experimental setup for the real-time generation and
decoding of two 25.78125GBaud PAM4 signals [38, 45]. At the transmit
side, the PHY consisted of a forward error correction (FEC) block, a PRBS
generator, a DSP unit, to map the bits into PAM4 symbols. Additionally,
the TX DSP provided a programmable 3-tap equalizer and a level shifting
function to compensate for limited bandwidth and the nonlinear behavior
of the modulator, respectively. The non-pre-equalized eye has a measured
SNR of over 27 dB. At the receive side, the PHY consisted of an integrated
ADC and an adaptive FFE-DFE equalizer with adaptive thresholds prior to
the PRBS bit error checker and the FEC decoder.

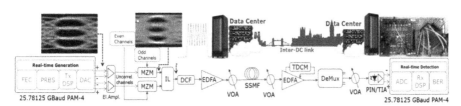

Figure 6.17   Experimental setup for PAM4 DWDM transmission at 1550 nm.
Courtesy: ADVA Optical Networks [33].

The two generated electrical PAM4 signals were amplified with a linear 35 GHz amplifier before driving two 27 GHz LiNbO3 Mach-Zehnder modulators (MZM). When transmitting 400G, each MZM modulated a group of four 100 GHz spaced wavelengths, with a relative offset of 50 GHz between the two groups. Both modulators were biased at quadrature point and driven at full Vπ, resulting in an extinction of approx. 12 dB. After the MZMs an inline power meter was used to determine the operating point of the MZM, before a 50 GHz interleaver combined the two modulated four-channel groups leading to an eight-channel, 50 GHz spaced DWDM signal.

For the fiber link, a conventional 80 km or 100 km SSMF link, both pre-compensated with a dispersion compensating fiber (DCF) for 80 km was used for the experiment, representing typical transmission distances for data center interconnects. The input power into the DCF was set to –8 dBm per channel. Furthermore, the DCF ensures sufficient decorrelation of the jointly modulated channels before the WDM signal is launched into the SSMF. An Erbium-doped fiber amplifier (EDFA), followed by VOA was used to set the launch power level into the fiber. After transmission, a VOA allowed the adjustment of the OSNR for single channel performance evaluation.

For DWDM transmission, the OSNR was set to the maximum value of approx. 34 dB for 80 km and approx. 31.5 dB for 100 km transmission. As the transmission link exhibited a loss of approx. 18 dB and 22 dB, for 80 km and 100 km, respectively, the second EDFA was needed to ensure sufficient power into the receiver. In case of 100 km transmission, the tunable dispersion compensation module (TDCM) placed inside the second stage of this EDFA, was used to compensate the last 20 km. The DCF was not replaced, emulating a realistic system scenario. Finally, the channels were separated with a demultiplexer of approx. 39 GHz optical bandwidth, and a VOA controlled the power into a 23 GHz photodiode with PIN/TIA. In case of optical back-to-back (B2B) transmission, only one EDFA was used in the transmission setup and amplified spontaneous emission (ASE) noise was added by means of a 3 dB coupler.

6.6.2 OSNR and Dispersion Performance

The following bit-error rate (BER) results were achieved with the dedicated PRBS generators and checkers within the PHY circuit. A PRBS $2^{31} - 1$ pattern was chosen, and for single-wavelength transmission the optical carrier frequency was set to 194.05 THz. For DWDM-transmission the wavelengths on the ITU grid between 194 THz and 194.35 THz were used.

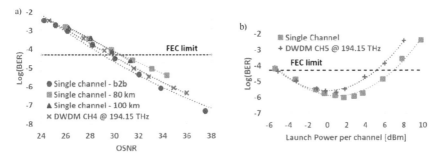

Figure 6.18 BER vs. OSNR plot for single channel DWDM transmission (a); BER vs. launch power into the fiber (b).

Figure 6.18(a) shows the measured BER vs. OSNR performance of a single PAM4 channel for optical B2B, for 80 km and for 100 km transmission as well as the B2B performance of channel 4 of the DWDM system. Additionally, at BER of 5.2×10^{-5} the threshold for the KR4 FEC is shown, which guarantees a post-FEC error rate below 10^{-12}. A 64B/66B to 256B/257B transcoder allows to apply the KR4 FEC RS(528,514) in-band, without increasing the gross rate of the signal [8]. At this FEC threshold an OSNR penalty of approximate 1.5 dB between optical B2B and the transmission over 80 km and 100 km SSMF is observable; associated with fiber nonlinearities, as the launch power into the fiber was set to 2 dBm. No performance difference between 80 km and 100 km is noted. Furthermore, this graph reveals a negligible impact of linear crosstalk, as nearly no performance differences between single channel results and DWDM results for optical B2B is shown.

In Figure 6.18(b), the BER vs. launch power per channel into 80 km SSMF is displayed for DWDM transmission (+−marker) and for single wavelength transmission (squared-marker). At lower input power, the link is OSNR limited, while at higher input power nonlinearities cause significant distortions. Nonlinear cross phase modulation (XPM) distorts the DWDM transmission at higher input power and thus, a different optimum launch power as well as a different minimum BER between DWDM transmission and single channel transmission is observed. The optimum launch power for a single channel is about +2 dBm while for 8 channels it is approximate +1 dBm.

Figure 6.19 shows an example of the impact of residual dispersion on DWDM PAM4 transmission. Here, the residual dispersion was introduced with the TDCM while transmitting over the DCF compensated 80 km link.

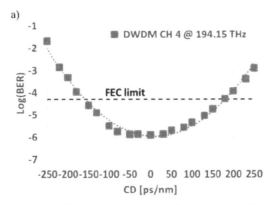

Figure 6.19 Tolerance to residual dispersion with reference to LiNbO3 MZM Transmitter.

The TDCM has an operating bandwidth of up to 40 GHz and can be used for full C-band chromatic dispersion compensation. A residual dispersion up to ± 170 ps/nm was measured to be tolerable for the given FEC-limit.

The experimental results described in this subsection clearly demonstrates an interesting option for a low-cost 100G solution for inter-data center connections under practical OSNR and fiber dispersion conditions. With the aid of WDM wavelengths, the system concept allows > 4 Tb/s links [21, 38, 47] for next generation of data center interconnects as shown in Figure 6.20.

A detailed discussion of this implementation by utilizing the combination of silicon photonics and PAM4 technology was introduced subsequently in Chapter 10.

6.7 PAM4 Compliance Tests

Ethernet networks for datacom typically assume that the transmitter, channel, and receiver will be produced from different vendors. To allow module

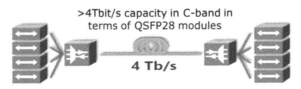

Figure 6.20 Application scenario for PAM4 based transceiver modules for data center interconnects.

Courtesy: Microsoft.

interoperability towards large volume production, the main components of the systems are individually specified such that even in the worst case scenario the link budget will be closed and the system bit-error-ratio (BER) will be achieved.

The use of PAM4 signaling leads to some substantial and important changes in how optical transmitters and receivers are tested.

TDECQ (transmitter dispersion and eye closure quaternary) has been developed [48, 49] for PAM4 Optical Transmitters as a metric that is similar to traditional transmitter dispersion penalty (TDP) measurements. It is intended to determine the additional power that is required for the transmitter being tested to achieve the symbol error ratio (SER) that is achievable when using an ideal reference transmitter. Two important differences exist between TDP and TDECQ. Rather than use a physical reference transmitter, a virtual reference transmitter is mathematically created based on the measured OMA of the transmitter under test. Also, equalization will be employed to overcome the significant impairments of the overall system. In addition to a virtual reference transmitter, a virtual equalizer is used in the TDECQ test to emulate operation of the actual communications system.

Similar to transmitter performance testing, a digital reference equalizer is required to compute various signal metrics during stress signal calibration for receiver stress testing.

Because of the significant sensitivity penalty resulting from the shift from NRZ to PAM4, the optical transceiver is not expected to operate error-free under the stress conditions defined by the standards or during typical use, while FEC is typically performed outside the transceiver module.

For purposes of system conformance, IEEE802.3 defines the PMD sublayer at the points described shown in Figure 6.21 for either parallel fiber or WDM architecture. The optical transmit signal is defined at the output end of cabling (TP2), between 2 m and 5 m in length for all transmitter measurements and tests. The optical receive signal is defined at the output of the fiber optic cabling (TP3) at the MDI (Medium dependent interface) for all receiver measurements and tests

Table 6.4 listed the comparison of current TX TDECQ & Receive (RX) SRS specifications defined by IEEE 802.3bs. The max TDECQ penalty is the critical pass/fail specs on optical TX at TP2. And Optical Stressed Receiver Sensitivity (SRS) is critical pass/fail specs on optical RX at TP3.

The details of PAM4 TDECQ and SRS testing methodologies are described in details in the following subsections.

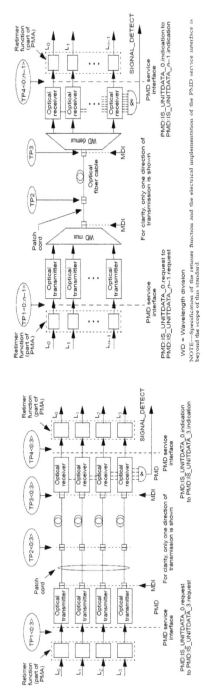

Figure 6.21 Block diagram for optical module transmit/receive paths for a) parallel fiber architecture; b) WDM architecture with duplex fiber.

Table 6.4 Comparison of current TX TDECQ & RX SRS specifications.

Parameter	Units	200GBASE-DR4	200GBASE-FR4	200GBASE-LR4	400GBASE-FR8	400GBASE-LR8	50GBASE-FR 50GBASE-LR	100GBASE-DR	400GBASE-DR4	50GBASE-SR 100GBASE-SR2 200GBASE-SR4
Baudrate	GBaud	26.5625	26.5625	26.5625	26.5625	26.5625	26.5625	53.125	53.125	26.5625
Reference Rx bandwidth	GHz	13.28125	13.28125	13.28125	13.28125	13.28125	13.28125	26.5625	26.5625	11.2
Reference Rx equalizer	Taps, Spacing	5, T	5, T	5, T	5, T	5, T	5, T	5, T	5, T	5, T
TDECQ (max)	**dB**	**3.4**	**3.3**	**3.4**	**3.1**	**3.3**	**3.2** / **3.4**	**3.4**	**3.4**	**4.9**
SRS	**dBm**	**-4.1**	**-3.6**	**-5.2**	**-3.1**	**-4.7**	**-5.1** / **-6.4**	**-1.9**	**-1.9**	**-3**
Dispersion (max)	ps/nm	0.8	6.7	9.5	1.9	9.5	3.2 / 16	0.8	0.8	–
Dispersion (min)	ps/nm	-0.93	-11.9	-28.4	-10.2	-50.8	-3.7 / -18.6	-0.93	-0.93	–
$\lvert\Delta T\text{disp} / T\text{symbl}\rvert$ (dispersion max)	%	0.3%	2.7%	3.8%	0.8%	3.8%	1.3% / 6.4%	1.3%	1.3%	–
$\lvert\Delta T\text{disp} / T\text{symbl}\rvert$ (dispersion min)	%	0.4%	4.7%	11.3%	4.1%	20.2%	1.5% / 7.4%	1.5%	1.5%	–
Draft	–	P802.3cd D2.1	P802.3bs D3.3	P802.3bs D3.3	P802.3bs D3.3	P802.3bs D3.3	P802.3cd D2.1	P802.3cd D2.1	P802.3bs D3.3	P802.3cd D2.1
Clause	–	121	122	122	122	122	139	140	124	138

Note: T_{disp} is approximate spread due to chromatic dispersion for transform-limited signals at the baud rate.. T_{symb} is symbol time.

6.7.1 Transmitter Dispersion Eye Closure for PAM4 (TDECQ)

Typically optical transmitter specifications include parameters like optical modulation amplitude, extinction ratio, and eye-masks [50]. In tranditional NRZ systems, yet only a transmission dispersion penalty (TDP) measurement provides a direct assessment of interoperability, due in large part to the inability of other metrics to observe low-probability statistical performance required for system level BER requirements in the 10^{-12} range. For TDP measurements a reference transmitter is connected to a BER tester (BERT) through an optical attenuator and receiver. The transmitter under test replaces the reference transmitter and attenuation is adjusted to achieve the BER of the reference transmitter and a power penalty metric is obtained. A BER based metric can provide an accurate assessment of transmitter system level performance to very low probabilities. However, the TDP test is considered relatively expensive, complicated and slow to implement.

Besides as Ethernet transmission rates increase, typical link BER levels of 10^{-12} become difficult and costly to achieve in manufacturing environments. Also the receivers could tolerate a few hits inside the mask and still deliver strong, error-free performance by using equalization.

In the 802.3bm-2015 specification, TDEC measurements replace both the mask test and TDP test. The inherent reason is that FEC and equalization techniques allow higher link BER levels while still achieving very low system level BERs. TDEC measurement is performed on an optical eye diagram, such as is captured by an oscilloscope. The use of an oscilloscope based test provides a more consistent and convenient test set up than a typical TDP test bed. Like traditional TDP measurements, TDEC is effectively a power penalty measurement which compares the closure of the eye to the nominal optical modulation amplitude. This approach has been adopted because it provides a comprehensive measurement result that accounts for the transmitter waveform shape and the role of inter-symbol interference (ISI) jitter in closing the eye vertically.

In TDEC measurements for NRZ signaling, four vertical histograms are captured from the eye diagram, centered at 0.4 UI and 0.6 UI, above and below average optical power. This is illustrative as an example by the four boxes in Figure 6.22. The choice of time window positions takes into account the receiver's timing. These histograms are then accumulated into a cumulative density function (CDF) to obtain TDEC in terms of decibels (dB). The output is equivalent to the BER contour measurement of the past and gives a good indication of what happens to the eye as the BER changes.

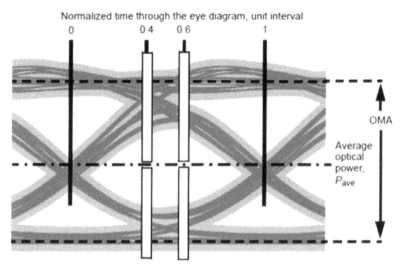

Figure 6.22 TDEC measurement for NRZ signaling as described in the IEEE 802.3bm-2015 focusing on vertical eye closure.

The TDEC measurement technique was developed for NRZ signals and can be extended to PAM4. Thus TDECQ add the "Q" (quaternary) to the TDEC as a measurement result. As it extends the principle of the TDEC measurement to PAM4 signals, TDECQ requires significant reconstruction of the TDEC mathematics to account for the four possible signal levels and 2 bit symbols [49].

The TDECQ measurement on PAM4 signaling is an extension of the TDEC measurement approach used on NRZ signaling. Like TDEC, TDECQ is a transmitter waveform-shape penalty measurement. Figure 6.23 shows a simplified measurement set up for TDECQ for PAM4 signaling. It starts with the DUT transmitter, and then the optical channel, the splitter, test fiber, the O/E-to-clock recovery, and an oscilloscope. For the most part, this a traditional setup with the addition of a reference equalizer to open the eye. Without the reference equalizer, the eye could be completely closed due to channel impairments.

TDECQ is a measure of the optical transmitter's vertical eye closure (via closure with adding noise) when transmitted through a worst case optical channel, as measured through an optical to electrical converter (O/E) with a bandwidth equivalent to a reference receiver, and equalized with the reference

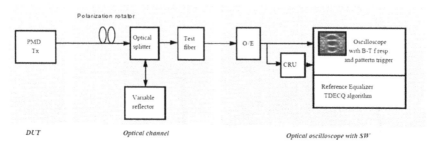

Figure 6.23   Example of TDECQ test setup.

Source: IEEE 802.3.

equalizer. The calculated TDECQ value is defined as follows:

$$TDECQ = 10\log_{10}\left(\frac{OMA_{outer}}{6} \times \frac{1}{Q_tR}\right) \tag{6.1}$$

Where:

OMAouter = difference between average optical launch power level 3 and power level 0 of PAM4 symbols

Qt = 3.414 (target BER = 2.4×10^{-4})

R = RMS noise term of the receiver

It's obvious from Equation (6.1) that TDECQ is defined as the ratio of the amount of noise a reference receiver could add to an ideal signal (with same OMAouter) to the noise it could add to the transmitter under test after transmission over a worst case fiber in order to achieve the same BER.

In other words, TDECQ value is actually a penalty vs ideal (simulated) transmitter in terms of dB. TDECQ = 0 dB is corresponding to a perfect transmitter. For a pass/fail criteria, note that there are now multiple limits for the TDECQ value depending on the applications. For example, for 400GBASE-DR4 transmit characteristics, the limit is 3.4 dB. Figure 6.24 shows the PAM4 TDECQ measurement methodology using defined histograms for floating time window at 0.45 UI and 0.55 UI from IEEE test specifications.

Ultimately, the key complicating factor here is the need to equalize the waveform at the receiver with a reference equalizer. Implementing reference equalization correctly is vital to obtaining accurate and repeatable results. Not getting it right adds many unknown variables to the measurement result, making it difficult to correctly evaluate a device. It's also critical to make sure that the reference equalizer algorithm compensates for the oscilloscope and any of the electrical-to-optical noise in the signal path. Getting oscilloscope effects out of the measurement and equalizing using a reference equalizer are

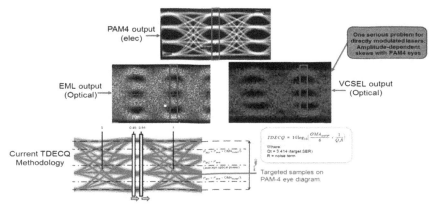

Figure 6.24 PAM4 TDECQ measurement methodology using defined histograms for floating time window at 0.45 UI and 0.55 UI from IEEE test specifications. An example of comparing EML and VCSEL eye diagrams is also shown.

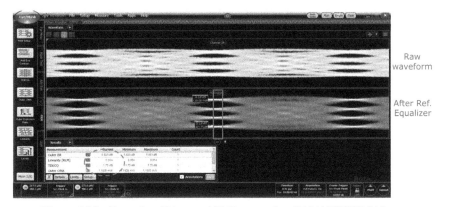

Figure 6.25 Example of Measured EML transmitter PAM4 TDECQ Values.

the biggest complicating factors in understanding accurate TDECQ performance. Figure 6.25 show the lab measurements on a 26 GBaud EML resulting in a TDECQ of 1.75 dB, which is considered as a solid, passing result. This follows the TDECQ measurement approach described above.

Figure 6.26 shows an example of 26 GBaud transmitter produced a passing TDECQ result of 2.98 dB for multimode transmitter. The left screen shows the measured PAM4 eye from the scope. The right screens show the equalized eye diagrams used to calculate the TDECQ value per the methodology illustrated in Figure 6.24.

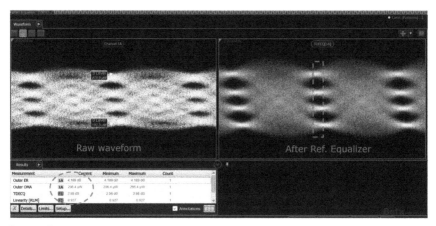

Figure 6.26   Example of Measured PAM4 TDECQ Values for directly modulated VCSEL transmitters.

6.7.2 Optical Stressed Receiver Sensitivity

The stressed receiver sensitivity (SRS) tests for PAM4 signaling are based on BER measurement for 400G. The basic principle is simple: the known transmitted bits are compared with the received bits over a transmission link including the device under test (DUT). The applied test data signal can be degraded with defined stress parameters, like transmission line loss, horizontal and vertical distortion to emulate worst-case operation scenarios at which the DUT has to successfully demonstrate error free data transmission. This test is obviously of fundamental importance for testing the robustness of the receivers, due to the manifold impairments occurring on optical transmission lines. Therefore, many optical transmission standards define such stressed receiver sensitivity on the basis of a BER measurement.

Setting-up a stressed eye compliant with the standard's specifications can be a very time consuming task because stressed eye parameters are interdependent and therefore several iterations of the optimization cycle are required to converge on the specific target. In addition, it is important that the setup is repeatable and remains stable from initialization of the stressed eye calibration to the end of the DUT measurements.

In conducting a SRS test, the purpose of the calibration is to generate a stable and repeatable stressed optical signal with specific characteristics, and send it to the receiver under test to collect the resultant bit error ratio.

The Stress Eye Closure Quaternary (SECQ) metric is adopted for calibrating the SRS test source. In principle, SECQ is identical to TDECQ but refers to the stress signal used for the receiver stress test, while TDECQ is a metric for the transmitter. For stress signal calibration, the SECQ measurement should be performed without test fiber using a SSPRQ test pattern captured by a sampling scope with a specific bandwidth and after digital equalization. The same digital reference equalizer is required to compute various signal metrics during stress signal calibration before receiver stress testing.

The calibration of specified stressed received conformance test signal with a given SECQ value is performed by creating a mixture of the following stress components:

- Inter-symbol interferences (ISI) by means of low-pass filter and frequency response of E/O converter
- Sinusoidal jitter (S.J)
- Sinusoidal amplitude interference noise (S.I)
- Gaussian noise (G.N) with a bandwidth of at least half the signal baud rate.

It is worthwhile to mention that various stress components has complex dependencies on the target metrics. So to carefully adjust the setting of the different stress components are required to generate the stress signal by iteration for the desired characteristics. The typical SRS setup and calibration procedures for 200GBASE-LR4/FR4 at 50 Gb/s per lane in the O-band is shown in Figure 6.27.

A calibrated reference transmitter example for receiver SRS testing is shown in Figure 6.28 (unstressed) and Figure 6.29 (fully stressed), respectively. The SECQ/TEDCQ values increased from 0.94 dB to 3.4 dB after calibrated stressed impairment signals are added.

The SRS tests for an actual 26GBaud ROSA and PAM4 receiver are depicted in Figure 6.30. The BER vs. OMAouter curves are plotted with the combination of different stress factors consisting of sinusoidal interference (S.I.), sinusoidal jitter (S.J.), Gaussian noise (G.N.) and low pass filter (LPF) for ISI. It is clear there exists complex dependencies and interplay of various stress component on the target metrics. Under the fully stressed condition, the optical stressed sensitivity is –10.4 dBm OMA with SECQ = 3.4 dB. This represents 5.2 dB margin beyond the –5.2 dB OMA as specified for 200 GBASE-LR4.

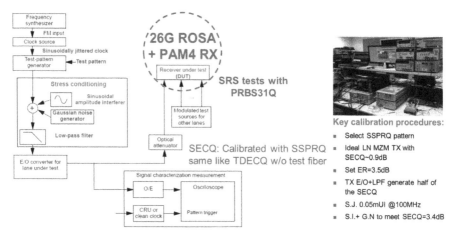

Figure 6.27   Optical SRS Testing Setup and Calibration Procedures. The inset shows a picture of the experimental setup.

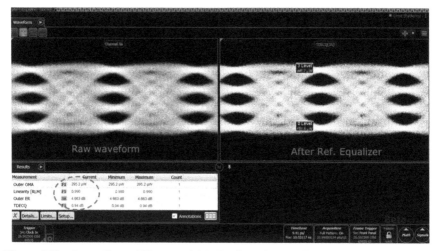

Figure 6.28   Scope Shots of Unstressed 26.5625GBaud PAM4 Signals with SSPRQ pattern.

6.8 Single Lambda PAM4

There has been extensive research on high baud rate PAM4 transmission around 100Gb/s per optical lane for recent several years. Single lambda 100Gb/s PAM4 has been experimentally demonstrated with off-line DSP processing using optical components spanning over MZM, EML, DML

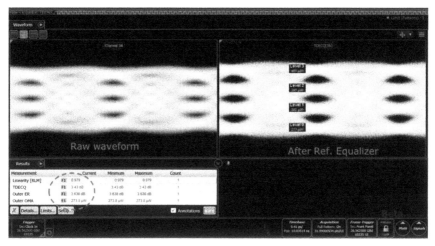

Figure 6.29   Scope shots of fully Stressed SRS Signals with SSPRQ pattern.

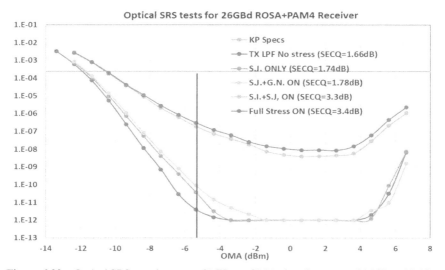

Figure 6.30   Optical SRS tests in terms of BER vs. OMA plots for an actual 26GBaud ROSA and PAM4 receiver using PRBS31Q pattern.

and VCSELs as shown in Table 6.5 [51–63]. For DML and VCSEL, 100G per channel was only demonstrated with offline processing so far using massive DSP and digital filtering due to their limited bandwidths.

Table 6.5 Single lambda 100G experimental survey

Paper title	Venue/Journal [Ref]	Year	Reach (km)	Mod. Format	Bitrate (Gb/s)	FEC threshold	Band (nm)	E/O technology	E/O bandwidth (GHz)	O/E Technology	O/E bandwidth (GHz)
Experimental study of 112 Gb/s short reach transmission employing PAM formats and SiP intensity modulator at 1.3 μm	Op Exp [51]	2014	10	PAM4	112	3.8×10^{-3}	1310	DFB + MZM	20	PIN + TIA	35
168 Gb/s Single Carrier PAM4 Transmission for Intra Data Center Optical Interconnects	PTL 29 [52]	2017	10	PAM4	168	2×10^{-4}	1310	DFB + MZM	28	BOA + PIN	50
Generation and Detection of a 56 Gb/s Signal Using a DML and Half-Cycle 16-QAM Nyquist-SCM	PTL 25 [53]	2013	4	16QAM Half-cycle Nyquist SCM	56	2×10^{-4}	1550	DML	35	PIN + TIA	N/A
100-Gbaud PAM-4 Intensity-Modulation Direct-Detection Transceiver for Datacenter Interconnect	ECOC [54]	2016	1 / 0.5	PAM4	168 / 200	3.8×10^{-3}	1550	DFB + MZM	35	PIN + TIA	40
Transmission of a 120-GBd PM-NRZ Signal Using a Monolithic Double-Side EML	PTL 28 [55]	2016	2	NRZ OOK	120	2×10^{-2}	1550	DFB + EAM	35	EDFA + PD	50
100 GHz EML for High Speed Optical Interconnect Applications	ECOC [56]	2016	B2B	OOK & PAM4 PAM8	116 / 105	3.8×10^{-3}	1550	DFB + EAM	100	EDFA + PD	90
Error-Free 100Gbps PAM-4 Transmission over 100m Wideband Fiber using 850nm VCSELs	ECOC [57]	2017	0.1	PAM4	110	2.4×10^{-4}	850	VCSEL+RC Shaping	20	OM5+PIN	N/A
55-GHz Bandwidth Short-Cavity Distributed Reflector Laser and its Application to 112-Gb/s PAM-4	OFC [58]	2016	2.2	PAM4	112	2×10^{-4}	1310	DML	55	PDFA + PD +TIA	38
4× 100Gbps VCSEL PAM-4 Transmission over 105m of Wide Band Multimode Fiber	OFC [59]	2017	0.1	PAM4	107	2.4×10^{-4}	850	VCSEL+RC Shaping	18	OM5+PIN	28
112 Gb/s PAM-4 Using a Directly Modulated Laser with Linear Pre-Compensation and Nonlinear Post-Compensation	ECOC [60]	2016	20	PAM4	112	4.6×10^{-3}	1310	DML	25	PIN + TIA	40
Experimental Investigation of Impulse Response Shortening for Low-Complexity MLSE of a 112-Gbit/s PAM-4 Transceiver	ECOC [61]	2016	7.5	PAM4	112	3.8×10^{-3}	1550	CW laser+MZM	35	PIN + TIA	30
Transmission of 214-Gbit/s 4-PAM signal using an ultrabroadband lumped-electrode EADFB laser module	OFC [62]	2016	10	PAM4	214	3.8×10^{-3}	1310	DFB + EAM	59	PDFA + PIN	50
140-Gb/s 20-km Transmission of PAM-4 Signal at 1.3 μm for Short Reach Communications	PTL 27 [63]	2015	20	PAM4	140	3.8×10^{-3}	1310	DFB + EAM	20	PIN + TIA	30

The industry is actively looking for true 100G PAM4 technologies utilizing commercialized ASIC silicon and will be moving to 400G in data centers during end 2019. Single-lambda 100G PAM4 offers the simplest architecture, higher reliability, and an easy upgrade path to 400Gb/s Ethernet, and it potentially enables lowest cost 100G transceivers.

There are several reasons behind this. Data-center customers ask for a steep downward trajectory in the cost of 100G pluggable transceivers, but existing 100G MSA module such as PSM4 and CWDM4 have limited capacity for cost reduction due to the cost of the fiber (PSM4) and the large number of components (both PSM4 and CWDM4). In another words, dual-lambda PAM4 and existing 100G Ethernet solutions such as PSM4 and CWDM4 will not be likely able to achieve the overall cost reductions demanded by data-center customers.

On the other hand, existing two-lambda PAM4 ($2 \times 50G$) trades off some optical components for more expensive DSP and may struggle to improve upon the cost of CWDM4. While the dual-lambda PAM4 architecture ($2 \times 50G$) currently uses components that appear to cost less, the lack of a "dual" ecosystem and the fact that the single-lambda ecosystem will mature quickly to support both 100 GbE and 400 GbE means that the volume for a $2 \times 50G$ solution may be never able to get truly materialized.

The first PAM4 ASIC chip fabricated by 16 nm FinFET CMOS was demonstrated operating at 53.125GBaud PAM4 links described [64] in Figure 6.31. The reference transmitter were based on EML at 1310 nm. RX sensitivity of ~ -8.5 dBm OMA were obtained for KP4 FEC threshold.

Intra-data-center links are scaling to 100 Gb/s per wavelength using PAM4 and direct detection, but a restrictive link budget makes it difficult to support the fiber link. We investigate receiver sensitivity improvements achievable using APDs. APD can be more cost-effective than optical pre-amplification followed by a PIN photodetector.

Figure 6.32 shows the same single lambda PAM4 ASIC operating at 53.125GBaud with reference to PIN and APD ROSAs. It's shown that APD provides more than 3.5 dB sensitivity improvement over PIN for the same reference system.

Table 6.6 shows the comparison of various single lambda 100G and 400G specifications. 100G and 400G Lambda MSA [65] actively explores to define 100G FR/LR and 400G FR4 options. APDs are promising alternatives to further improve receiver sensitivity. Important practical considerations such as cost, temperature sensitivity, and power consumption may nonetheless favor APDs in practical systems. Hence 400G LR4 could be enabled by APD approach.

Table 6.6   Comparison of single lambda 100G specifications

QSFP28 optical specs	IEEE 802.3ba		IEEE 802.3cd D3.0		IEEE 802.3bs D3.5		100G Lambda MSA D1.0		100G Lambda MSA D1.0	
	100GBASE-LR4		100GBASE-DR		400GBASE-DR4		100G-FR		100G-LR	
Fiber types	SMF duplex		SMF duplex		Parallel SMF 4 lane (8 fibers)		SMF duplex		SMF duplex	
Modulation format	NRZ		PAM4		PAM4		PAM4		PAM4	
Reach	2 m - 10 km		2 m - 500 m		2 - 500 m		2 - 2km		2 - 10 km	
Baud rate (GBd)	25.78125 ± 100 ppm		53.125 ± 100 ppm		53.125 ± 100 ppm		53.125 ± 100 ppm		53.125 ± 100 ppm	
	1294.53 to 1296.59									
	1299.02 to 1301.09									
	1303.54 to 1305.63		1304.5 to 1317.5 nm		1304.5 to 1317.5 nm		1304.5 to 1317.5 nm		1304.5 to 1317.5 nm	
Center or Lane wavelength (range)	1308.09 to 1310.19									
SMSR (min)	30dB		30dB		30dB		30dB		30dB	
BER*	1E-12		2.4E-4 (KP4)		2.4E-4 (KP4)		2.4E-4 (KP4)		2.4E-4 (KP4)	
Transmit (Tx)	OMA	AOP	OMAouter	AOP	OMAouter	AOP	OMAouter	AOP	OMAouter	AOP
Total lauching power (max)	+10.5 dBm									
Launching power, each lane (Max)	+4.5 dBm	+4.5 dBm	+4.2 dBm	+4 dBm	+4.2 dBm	+4dBm	+4.2 dBm	+4 dBm	+4.7 dBm	+4.5 dBm
Launching power, each lane (Min)	-1.3 dBm	-4.3dBm	-0.8dBm	-2.9 dBm	-0.8 dBm	-2.9dBm	0.2dBm	-1.9 dBm	1.2dBm	-0.9 dBm
OMA - TDECQ (min)	-2.3dBm		-2.2 dBm (ER≥5dB)		-2.2 dBm		-1.7 dBm (ER≥4.5dB)		-0.7 dBm (ER≥4.5dB)	
			-1.9 dBm (ER<5dB)				-1.6 dBm (ER<4.5dB)		-0.6 dBm (ER<4.5dB)	
TDECQ, each lane (max)	TDP, 2.2 dB		3.4 dB		3.4 dB		3.4 dB		3.4 dB	
ER (min)	4 dB		3.5 dB		3.5 dB		3.5 dB		3.5 dB	
Optical RL tolerance (max)	20 dB		15.5 dB		21.4 dB		16.5 dB		15.1 dB	
RINxOMA(max)	-130 dB/Hz		-136 dB/Hz		-136 dB/Hz		-136 dB/Hz		-136 dB/Hz	
AOP of OFF transmitter (max)			-15 dBm		-15 dBm		-15 dBm		-15 dBm	
Transmitter reflectancec (max)	-12 dB		-26 dB		-26 dB		-26 dB		-26 dB	
Receive (Rx)	OMAouter	AOP	OMAouter	AOP	OMAouter	AOP	OMAouter	AOP	OMAouter	AOP
Demage threshold (min)		+3.4 dBm		5 dBm		+5.0 dBm		+5.5 dBm		+5.5 dBm
Receive power, each lane (max)	+3 dBm	+2.4 dBm	+4.2 dBm	+4 dBm	4.2dBm	+4.0 dBm	4.7dBm	+4.5 dBm	4.7dBm	+4.5 dBm
Receive power, each lane (min)		-11 dBm		-5.9 dBm		-5.9 dBm		-5.9 dBm		-7.2 dBm
Receive reflectance (max)	-12 dB		-26 dB		-26 dB		-26 dB		-26 dB	
Rx sens (OMA), each laned (max)	-8.6 dBm		-4.4 dBm		-4.4 dBm		-5 dBm		-6.6 dBm	
SRS (OMA), each lane (max)	-6.8 dBm		-1.9 dBm		-1.9 dBm		-2.5 dBm		-4.1 dBm	
Condition for SECQ										
SECQ, each lane	VECP 1.8 dB		3.4 dB		3.4 dB		3.4 dB		3.4 dB	
OMAouter of each aggressor lane					4.2 dBm					
Optical path										
			6.5 dB (ER≥5dB)							
Power budget (for max TDECQ)	9.5 dB		6.8 dB (ER<5dB)		6.5 dB		7.6 dB		5.6 dB	
fiber distance	10 km*		500 m		500 m		2 km		10 km	
Channel IL (max)	6.3 dB		~3 dB (Table 140-12)		3 dB		4 dB		6.3 dB	
Discrete reflectance (max)			-35 dB		<-37 dB(Table 124-13)		<-37 dB(Table 2-5)		<-37 dB(Table 2-5)	
			6. 5- max cha IL (140-12)							
Allocated Penalities (for max TDECQ)			6. 8- max cha IL (140-12)		3.5 dB		2.6 dB		3.6 dB	
Additional insertion loss allowed			0 dB		0 dB		0 dB		0 dB	

*: Dispersion and DGD   * The use of KP4 from BER

100G Lambda MSA D1.0	IEEE 802.3bs D3.5	IEEE 802.3cd D3.0	IEEE 802.3cd D3.0	IEEE 802.3bs D3.5	IEEE 802.3bs D3.5
400G-FR4	200GBASE-DR4	50GBASE-FR	50GBASE-LR	200GBASE-FR4	200GBASE-LR4
SMF duplex	Parallel SMF 4 lane (8 fibers)	SMF duplex	SMF duplex	SMF duplex	SMF duplex
PAM4	PAM4	PAM4	PAM4	PAM4	PAM4
2 - 2 km	2 - 500 m	2 m - 2 km	2 m - 10 km	2 m - 2 km	2 m - 10 km
53.125 ± 100 ppm	26.5625 ± 100 ppm	26.5625 ± 100 ppm	26.5625 ± 100 ppm	26.5625 ± 100 ppm	26.5625 ± 100 ppm
1264.5 to 1277.5 nm				1264.5 to 1277.5	1294.53 to 1296.59
1284.5 to 1297.5 nm				1284.5 to 1297.5	1299.02 to 1301.09
1304.5 to 1317.5 nm	1304.5 to 1317.5 nm	1304.5 to 1317.5 nm	1304.5 to 1317.5 nm	1304.5 to 1317.5	1303.54 to 1305.63
1324.5 to 1337.5 nm				1324.5 to 1337.5	1308.09 to 1310.19
30dB	30dB	30dB	30dB	30dB	30dB
2.4E-4 (KP4)	2.4E-4 (KP4)	2.4E-4 (KP4)	2.4E-4 (KP4)	2.4E-4 (KP4)	2.4E-4 (KP4)
OMAouter AOP	OMAouter AOP	OMAouter AOP	OMAouter AOP	OMAouter AOP	OMAouter AOP
+9.3dBm				+10.7 dBm	+11.3 dBm
+3.7 dBm +3.5 dBm	+2.8 dBm +3 dBm	+2.8dBm +3 dBm	+4dBm +4.2dBm	+4.5dBm +4.7dBm	+5.1dBm +5.3dBm
-0.3dBm -3.3 dBm	-3 dBm -5.1 dBm	-2.5dBm -4.1 dBm	-1.5dBm -4.5 dBm	-1.2dBm -4.2 dBm	-0.4dBm -3.4 dBm
-1.7 dBm (ER≥4.5dB)	-4.4 dBm	-3.9 dBm	-2.9 dBm	-2.6 dBm (ER≥4.5dB)	-1.8 dBm (ER≥4.5dB)
-1.6 dBm (ER<4.5dB)				-2.5 dBm (ER<4.5dB)	-1.7 dBm (ER<4.5dB)
3.4 dB	3.4 dB	3.2 dB	3.4 dB	3.3 dB	3.4 dB
3.5 dB	3.5 dB	3.5 dB	3.5 dB	3.5 dB	3.5 dB
16.5 dB	21.4 dB	16.5 dB	15.1 dB	16.5 dB	15.1 dB
-136 dB/Hz	-132 dB/Hz	-132 dB/Hz	-132 dB/Hz	-132 dB/Hz	-132 dB/Hz
-15 dBm	-16 dBm	-16 dBm	-16 dBm	-30 dBm	-30 dBm
-26 dB	-26 dB	-26 dB	-26 dB	-26 dB	-26 dB
OMAouter AOP	OMAouter AOP	OMAouter AOP	OMAouter AOP	OMAouter AOP	OMAouter AOP
+4.5 dBm	+4.0 dBm	5.2dBm	5.2 dBm	+5.7 dBm	+6.3 dBm
3.7dBm +3.5 dBm	2.8dBm +3.0 dBm	+2.8 dBm +3 dBm	+4 dBm +4.2 dBm	+4.5 dBm +4 .7 dBm	+5.1 dBm +5.3 dBm
-7.9 dBm	-8.1 dBm	-8.1 dBm	-10.8 dBm	-8.2 dBm	-9.7 dBm
-26 dB	-26 dB	-26 dB	-26 dB	-26 dB	-26 dB
-5.1 dBm	-6.6 dBm	-7. dBm	-8.9 dBm	-6 dBm	-7.7 dBm
-2.6 dBm	-4.1 dBm	-5.1 dBm	-6.4 dBm	-3.6 dBm	-5.2 dBm
3.4 dB	3.4 dB	3.2 dB	3.4 dB	3.3 dB	3.4 dB
1.5 dBm	2.8 dBm			0.5 dBm	-1 dBm
				7.6 dB (ER≥4.5dB)	10.2 dB (ER≥4.5dB)
7.7 dB	6.5 dB	7.6 dB	10.3 dB	7.7 dB (ER<4.5dB)	10.3 dB (ER<4.5dB)
2.0 km	500 m	2 km	10 km	2 km	10 km
4 dB	3 dB	4 dB	<6.3	4 dB	6.3 dB
<-25 dB(Table 2-6)	<-37 dB(Table 121-11)	<-25 dB(Table 139-14)	<-22 dB(Table 139-14)	<-25 dB(Table 122-19)	<-22 dB(Table 122-19)
				3.6 (ER≥4.5dB)	3.9 (ER≥4.5dB)
3.7 dB	3.5 dB	3.6 dB	4 dB	3.7 (ER<4.5dB)	4.0 (ER<4.5dB)
0 dB	0 dB	0 dB	0 dB	0 dB	0 dB

*FR8 & LR8 assume the same 8 LAN-WDMs

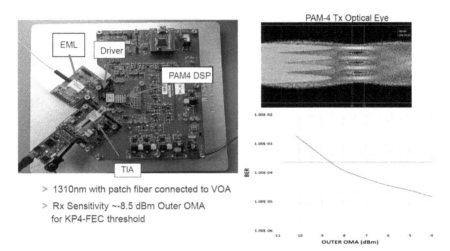

> 1310nm with patch fiber connected to VOA
> Rx Sensitivity ~-8.5 dBm Outer OMA
  for KP4-FEC threshold

Figure 6.31   First demonstration of single lambda PAM4 ASIC operating at 53.125GBaud.
Source: IEEE

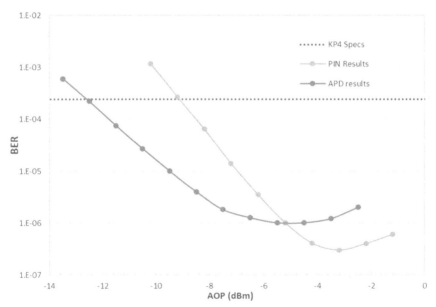

Figure 6.32   Single lambda PAM4 ASIC operating at 53.125GBaud with PIN and APD ROSAs.
Source: IEEE

6.9 Summary and Outlook

In this chapter, we demonstrated robust 50G/lane PAM4 operation over SMF linking the TOR, aggregation and core switches in data centers. PAM4 Ethernet SerDes enables 100Gb/s over reduced number of optical components. It provides the clear pathway to 50, 200 and 400 Gb/s as recognized by the IEEE standards.

In addition, PAM4 signaling were successfully applied for MMF space and OSNR limited systems. PAM4 target virtually "everywhere" for cloud data centers with clear roadmap forming within IEEE for 50 GbE (servers), 200 GbE and 400 GbE. PAM4 technology commercialized today, start shipping devices for all 50/100/200/400GbE deployments. It enables low power to support various MSA implementations such as SFP56, QSFP28, QSFP56, QSFP-DD, CFP8, or OSFP.

It's expected single wavelength 100 Gb/s ASICs will be next wave of Ethernet module developments before end of 2019. Measured performance to support 400G DR4/FR4/LR4 shows that single-lambda 100G PAM4 offers the simplest architecture, higher reliability, and an easy upgrade path to 400G Ethernet, and it enables the lowest-cost 100G transceiver.

Next-gen transceiver modules call for new enabling technology challenges to support 800G and 1.6T as shown below in Figure 6.33. It's predictive that 800+G data rates can be initiated after 2022.

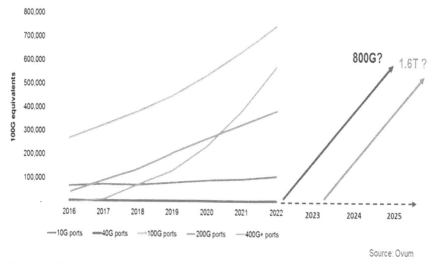

Figure 6.33 Prediction of next-gen transceiver module rates possibly at 800G and 1.6T.

To consider technology options from 400G to 800G/1.6T, it's expected the dramatic improvements may happen in the following areas:

1. On Board Optics (OBO)
   OBO can provide higher integration on E to O channels, lower power consumption per channel, better SI characteristics from chip to module. Possibly shorter trace length on host board.
2. Co-packaging
   Co-packaging is not a new concept. It require tighter corporations between key component vendors (Switch ASIC, TIA, LDD, modulator, DSP, LD/PD, etc) at more proprietary level.
3. ASIC integrated with Optics
   This may eliminate traditional transceiver module form factors. It requires much larger integration scale for ASIC + laser + PD (SiPho is one example). More advance IC development is required. High speed LD/PDs are traditionally not performing well in Si (CMOS, BiCMOS) base material system.

With increasing data rates, it is desirable to further improve receiver sensitivity while minimizing system power dissipation, cost and size. Stronger FEC codes could improve sensitivity, but would dramatically increase power consumption and latency.

With PAM4 approach already pushing the bandwidth limits of the supporting electro-optics, coherent detection could likewise improve sensitivity significantly, but it requires complex optical components and temperature-controlled lasers, while both these latter alternatives rely on costly, power-hungry digital signal processors. Nevertheless, OIF has initiated the next-generation "interoperable" 400Gb/s project to develop single wavelength coherent 16QAM solution near 60GBaud with hybrid SD-FEC for <120km application space as shown in Table 6.7 [66].

It's not the goal of this chapter to do a detailed comparison of direct detect vs. coherent technologies [67], but it's foreseeable that coherent technology may play a larger impact on sub-40 km applications beyond 400Gb/s.

Acknowledgements

The author would like to acknowledge many partners, not limited to Adva, Avago, Discovery, Keysight, Mitsubishi, Microsoft, NeoP, NEL, Oplink, OFS, SEDI, Source Photonics, UCSB for the collaborations.

Table 6.7 OIF developed 400ZR project for up to 120 km

	400ZR Loss-Limited Link	400ZR DWDM OSNR-Limited Link
PCS	400GE	400GE
Frequency Tolerance	+/- 20 ppm	+/- 20 ppm
Timing Transparency	Yes	Yes
Modulation	16QAM	16QAM
Baud Rate	~60Gbaud	~60Gbaud
Distance	40 to 80 km	80 to 120 km
Link Type	Unamplified	Amplified
Link Condition	12 dB loss	Mux/Dmux

- DSP is assumed to be based on 7nm CMOS technology   Source: OIF

This work has been the result of fruitful discussions with Bhoja S., Sheth S., Riani J., Hosagrahar I., Wu J., Herlekar S., Tiruvur A., Khandelwal P., Gopalakrishnan K, Nagarajan R., Furlong M., Nellis K., Paules G., and Mukherjee T.

References

[1] IEEE 802.3bs 200 Gb/s and 400 Gb/s Ethernet Task Force; and IEEE 802.3cd, 50 Gb/s, 100 Gb/s and 200 Gb/s Ethernet Task Force.

[2] R. Farjad-Rad et al, "A 0.3-um CMOS 8-Gb/s 4-PAM Serial Link Transceiver", IEEE Journal of Solid-State Circuits, Vol. 35, No. 5, May 2000 pp. 757–64.

[3] J. T. Stonick et al., "An Adaptive PAM-4 5 Gb/s Backplane Transceiver in 0.25 um CMOS"; IEEE Journal of Solid-State Circuits, Vol: 38, Issue: 3, March 2003, pp. 436–43.

[4] J. L. Zerbe et al., "Equalization and Clock Recovery for a 2.5 - 10 Gb/s 2-PAM/4-PAM Backplane Transceiver Cell", IEEE Journal of Solid-State Circuits, Vol: 38, Issue: 12, Dec. 2003, pp. 2121–2130.

[5] T. Toifl et al., "A 22-Gb/s PAM-4 Receiver in 90-nm CMOS SOI Technology"; IEEE Journal of Solid-State Circuits, Vol. 41, No. 4, April 2006, pp. 954–65.

[6] J. Lee et al., "Design and Comparison of Three 20-Gb/s Backplane Transceivers for Duobinary, PAM4, and NRZ Data", IEEE Journal of Solid-State Circuits, Vol. 43, No. 9, September 2008, pp. 2120–33.

[7] Broadcom unveils 40G/50G PAM-4 physical layer chip. http://www.lightwaveonline.com/articles/2014/12/broadcom-unveils-40g-50g-pam-4-physical-layer-chip.html

[8] P. Khandelwal et al., "100 Gbps Dual-channel PAM-4 transmission over Datacenter Interconnects"; DesignCon'16, January 2016.

[9] T. K. Chan et al., "112 Gb/s PAM4 Transmission Over 40 km SSMF Using 1.3 μm Gain-Clamped Semiconductor Optical Amplifier"; Proc. OFC'15, paper Th3A.4 (2015).

[10] K. Zhong et al., "Experimental Demonstration of 500 Gbit/s Short Reach Transmission Employing PAM4 Signal and Direct Detection with 25 Gbps Device"; Proc. OFC'15, paper Th3A.3 (2015).

[11] C. Chen et al., "Transmission of 56-Gb/s PAM-4 over 26-km Single Mode Fiber Using Maximum Likelihood Sequence Estimation"; Proc. OFC'15, paper Th4A.5 (2015).

[12] D. Sadot et al., "Single channel 112 Gb/s PAM4 at 56GBaud with digital signal processing for data center applications," Proc. OFC, Th2A.67, Los Angeles (2015).

[13] S. Bhoja et al., "Precoding proposal for PAM4 modulation," in *IEEE P802.3bj Task Force*, September 2011.

[14] F. Chang et al., "Link Performance Investigation of Industry First 100G PAM4 IC Chipset with Real-time DSP for Data Center Connectivity"; OFC'16, paper Th1G2, March 2016.

[15] K. Gopalakrishnan et al., "A 40/50/100Gb/s PAM-4 Ethernet Transceiver in 28 nm CMOS'; ISSCC 2016, San Francisco CA.

[16] K. H. Mueller and M. S. Muller, "Timing Recovery in Digital Synchronous Data Receivers," *IEEE Trans. on Communications*, Vol. COM-24, pp. 516–531, May 1976.

[17] M. Nada et al., "High-linearity Avalanche Photodiode for 40-km Transmission with 28-GBaud PAM4"; Proc. OFC'15, paper M3C.2, Los Angeles, CA (2015).

[18] J. C. Campbell, "Recent Advances in Avalanche Photodiodes"; Proc. OFC'15, paper M3C.1, Los Angeles, CA (2015).

[19] T. Saeki et al., "Compact Optical Transmitter Module with Integrated Optical Multiplexer for 100 Gbit/s"; Proc IPC'13 paper TuG3.2 Bellevue, WA (2013).

[20] M. Mazzini, et al., "25 GBaud PAM-4 error free transmission over both single mode fiber and multimode fiber in a QSFP form factor based on Silicon Photonics," Proc. OFC, PDP Th5B.3, Los Angeles, CA (2015).

[21] R. Nagarajan et al., "Silicon Photonics-Based 100 Gbit/s, PAM4, DWDM Data Center Interconnects"; J. OPT. COMMUN. NETW. Vol. 10, No. 7 pp. B25–36 (2018).

[22] SWDM alliance online and references therein for duplex MMF at 40 and 100Gb/s. http://www.swdm.org

[23] R. Shubochkin et al., "Next-gen wideband multimode for data centers," Proc. SPIE 977504, San Francisco, CA (2016).

[24] K. Szezerba et al., "60 Gbits error-free 4-PAM operation with 850 nm VCSEL," *Elec Lett*, Vol. 49 (15), p. 953 (2013).

[25] S. K. Pavan et al., "Experimental demonstration of 51.56 Gbit/s PAM-4 at 905 nm and impact of level dependent RIN," ECOC 2014, W1F.5.

[26] J. Castro et al., "48.7-Gb/s 4-pam transmission over 200 m of high bandwidth mmf using an 850-nm vcsel," *IEEE Photon. Technol. Lett.*, Vol. 27, pp. 1799–1801 (2015).

[27] Y. Sun et al., "51.56 Gb/s SWDM PAM4 transmission over next generation wide band multimode optical fiber," OFC 2016, paper Tu2G.3.

[28] C. Chen et al., "Transmission of 56-Gb/s PAM-4 over 26-km single mode fiber using maximum likelihood sequence estimation," OFC 2015, paper Th4A.5.

[29] S. M. R. Motaghiannerzam et al., "104 Gbps PAM4 transmission over OM3 and OM4 fibers using 850 and 880 nm VCSELs," CLEO 2016, paper SW4F.8, San Jose, CA.

[30] S. M. R. Motaghiannezam, et al., "180 Gbps PAM4 VCSEL transmission over 300m wideband OM4 fiber," OFC 2016, paper Th3G.2, Anaheim, CA.

[31] M. Huang et al. "Breakthrough of 25 Gb/s Ge on Silicon Avalanche PD," OFC 2016, paper Tu2D.2, Anaheim, CA.

[32] Y. Kang et al,"Epitaxially-grown Ge/Si avalanche photodiodes for 1.3μm light detection"; Optics Express, Vol. 16, No. 13, pp. 9365–71 (2008).

[33] F. Chang et al., "First Demonstration of PAM4 Transmissions for Record Reach and High-capacity SWDM Links Over MMF Using 40G/100G PAM4 IC Chipset with Real-time DSP"; OFC'2017, Paper Tu2B.2, March 2017.

[34] Y. Sun et al., "SWDM PAM4 Transmission from 850 to 1066 nm over NG-WBMMF using 100G PAM IC Chipset with Real-time DSP"; J. Lightwave Technology, Vol. 35, Iss. 15. pp. 3149–58, August 2017.

[35] IEEE802.3, 400 Gb/s over Multimode Fiber Task Force, http://www.ieee 802.org/3/cm/index.html, 2018.

[36] Y. Sun et al., "SWDM PAM4 Transmission over Next Generation Wide Band Multimode Optical Fiber"; J. Lightwave Technology, Vol. 35, Iss. 4, Pages: 690–697 (2017).

[37] J. Man, L. Zeng, and W. Zhou, "Investigation of 56Gbps PAM4 based bi-directional architecture for 400 GbE", in *IEEE P802.3bs Task Force,* May 2014.

[38] F. Chang and S. Bhoja, "New Paradigm Shift to PAM4 Signaling at 100/400G for Cloud Data Centers: A Performance Review"; ECOC 2017, Invited Paper W.1.A.5, September 2017, Gothenburg, Sweden.

[39] A. Dochhan, H. Griesser, N. Eiselt, M. H. Eiselt, and J. P. Elbers; "Solutions for 80 km DWDM Systems"; J. Lightwave Technology" Vol. 34, Issue 2, pp. 491–499 (2016).

[40] A. Dochhan, "Flexible bandwidth 448 Gb/s DMT Transmission for Next Generation Data Center Inter-Connects", Proc. ECOC, P. 4.10, Cannes (2014).

[41] N Kaneda, J. Lee, Y. Chen; "Nonlinear equalizer for 112-Gb/s SSB-PAM4 in 80-km dispersion uncompensated link'; OFC2017, paper Tu2D.5, Los Angeles, CA.

[42] A. Li, W. Peng, Y. Cui, and Y. Bai, "Single-λ 112 Gbit/s 80-km Transmission of PAM4 Signal with Optical Signal-to-Signal Beat Noise Cancellation", OFC2018, paper Tu2C.5, San Diego, CA.

[43] S. Yin, et al., "100 km DWDM Transmission of 56 Gb/s PAM4 per λ via a Tunable Laser and 10 Gb/s InP MZM", IEEE Photonics Technology Letters, Vol: 27, Iss: 24, Dec. 15, 2015, pp. 2531–4 (2015).

[44] G. Brodnick et al., "Extended Reach 40 km Transmission of C-Band Real-Time 53.125 Gbps PAM-4 Enabled with a Photonic Integrated Tunable Lattice Filter Dispersion Compensator"; OFC'2018, Paper W2A.30, March 2018, San Diego, CA.

[45] E. Nicklas et al.; "First Real-Time 400G PAM-4 Demonstration for Inter-Data Center Transmission over 100 km of SSMF at 1550 nm"; Paper W1K.5 OFC 2016, Anaheim, CA.

[46] M. Filer et al., "Demonstration and Performance Analysis of 4 Tb/s DWDM Metro-DCI System with 100G PAM4 QSFP28 Modules'; OFC 2017, Paper W4D.4, March 2017.

[47] N. Eiselt, J. Wei, H. Griesser, A. Dochhan, M. H. Eiselt, J. P. Elbers, J. J. V. Olmos, and I. T. Monroy, "Evaluation of Real-Time 8 $\times$ 56.25 Gb/s (400G) PAM-4 for Inter-Data Center Application Over 80 km of

SSMF at 1550 nm"; Journal of Lightwave Technology, Vol. 35, No. 4, pp. 955–62 (2017).

[48] J. Petrilla, P. Dawe, G. LeCheminant , "New metric offers more accurate estimate of optical transmitters impact on multimode fiber-optic links" Design Con 2015.

[49] K. Jonathan et al, "TDECQ (Transmitter Dispersion Eye Closure Quaternary) Replaces Historic Eye-mask and TDP Test for 400 Gb/s PAM4 Optical Transmitters"; OFC2017, paper W4D.1, Los Angeles, CA.

[50] K. Tan and P. Thota. "TDECQ measurements replace mask testing in PAM4 optical signals"; EDN, February 23, 2018.

[51] M. Chagnon et al.; "Experimental study of 112 Gb/s short reach transmission employing PAM formats and SiP intensity modulator at 1.3 µm"; Opt Express. 2014 Aug 25; 22(17): 21018–36.

[52] E. El-Fiky et al; "168-Gb/s Single Carrier PAM4 Transmission for Intra-Data Center Optical Interconnects"; IEEE Photonics Technology Letters, Vol 29, Iss: 3, Feb.1, 1 2017, Page(s): 314–317.

[53] A. S. Karar and J. C. Cartledge; "Generation and Detection of a 56 Gb/s Signal Using a DML and Half-Cycle 16-QAM Nyquist-SCM'; IEEE Photonics Technology Letters, Vol: 25, Iss: 8, April 15, 2013, Page(s): 757–760.

[54] M. A. Mestre et al; "100-GBaud PAM-4 intensity-modulation direct-detection transceiver for datacenter interconnect"; ECOC 2016, Düsseldorf, Germany.

[55] K. Zhong et al.; "Transmission of a 120-GBd PM-NRZ Signal Using a Monolithic Double-Side EML"; IEEE Photonics Technology Letters, Vol: 28, Iss: 20, 30 June 2016, Page(s): 2176–9.

[56] O. Ozolins et al.; "100 GHz EML for High Speed Optical Interconnect Applications"; ECOC 2016, Düsseldorf, Germany.

[57] J. Lavrencik et al.; "Error-Free 100Gbps PAM-4 Transmission over 100m Wideband Fiber using 850nm VCSELs"; ECOC 2017, Gothenburg, Sweden.

[58] Y. Matsui; "55-GHz bandwidth short-cavity distributed reflector laser and its application to 112-Gb/s PAM-4"; OFC, 2016.

[59] J. Lavrencik, "4λ × 100Gbps VCSEL PAM-4 Transmission over 105m of Wide Band Multimode Fiber"; OFC, 2017 Los Angeles, CA.

[60] Y. Gao et al., "112 Gb/s PAM-4 Using a Directly Modulated Laser with Linear Pre-Compensation and Nonlinear Post-Compensation"; ECOC 2016; Dusseldorf, Germany.

[61] S. V. D. Heide et al., "Experimental Investigation of Impulse Response Shortening for Low-Complexity MLSE of a 112-Gbit/s PAM-4 Transceiver"; ECOC 2016; Dusseldorf, Germany.

[62] S. Kanazawa et al; "Transmission of 214-Gbit/s 4-PAM signal using an ultra-broadband lumped-electrode EADFB laser module"; OFC 2016, Anaheim, CA.

[63] K. Zhong et al.; "140-Gb/s 20-km Transmission of PAM-4 Signal at 1.3 µm for Short Reach Communications"; IEEE Photonics Technology Letters, Vol: 27, Iss: 16, Aug.15, 2015 Page(s): 1757–1760.

[64] B. Zeydel, F. Caggioni, and T. Palkert "SERDES for 100 Gbps"; http://www.ieee802.org/3/ad_hoc/ngrates/public/17_05/palkert_nea_02_0517.pdf; IEEE 802.3 NEA Ad Hoc, May 2017.

[65] 100G Lambda MSA, http://100glambda.com/

[66] OIF2016.400.04, 400G ZR interop (online), https://www.oiforum.com/

[67] F. Chang, "Current Status of 400G and Its Path towards 1T System in Optical Networking"; Invited Talk to FOE 2018, Presented in April, 2018, Tokyo, Japan.

7

Discrete Multitone for Metro Datacenter Interconnect

Gordon Ning Liu, Tianjian Zuo and Liang Zhang

Huawei Technologies Co. Ltd., Shenzhen, China

Abstract

For the Metro Datacenter Interconnects, the intensity modulation and direct detection (IM/DD) offers the advantages of low cost and low complexity. In this chapter, we will review an important IM/DD scheme, discrete multitone (DMT). After the introduction of DMT's application scenarios and history, the conventional operational principle of DMT will be illustrated in detail. Then, we will review some advanced DMT architectures for Metro DCI applications which can overcome the dispersion and interference challenges by employing some advanced modulators. Single sideband (SSB) DMT is very simple, while chromatic dispersion (CD) calculation and pre-compensation are not necessary. Electrical dispersion compensation (EDC) double sideband (DSB) DMT owns 3-dB better performance than the SSB-DMT. Twin-SSB-DMT has the highest spectral efficiency (SE). Furthermore, a spectrally efficient signal to signal beating interference (SSBI) cancellation system based on the guard-band twin-SSB-DMT will be demonstrated. These schemes make DMT capable of tens of kilometers of transmission and suitable for Metro DCI applications.

7.1 Introduction

The approaching services such as mobile broadband, high-definition 4K/8K video, augmented reality/virtual reality, cloud computing, and Internet of things would imply an enormous amount of datacenters at the metro networks for connecting content to end users directly. Moreover, the convergence ratio from the access networks to the metro networks will be greatly reduced compared with the traditional one. The architecture can lower down the latency and improve the quality of experience. However, it also leads to a severe transmission capacity crunch in Metro datacenter interconnect (DCI), even higher than that from the backbone network. Therefore, 100 Gbps or beyond per wavelength is the trend for those Metro DCI applications with tens of kilometers of transmission reach.

Although beyond 100 Gbps per wavelength coherent techniques have been commercialized in long-haul systems, they are expensive, large footprint, and power hungry. There will be tremendous amount of equipment in Metro DCI compared with the long-haul transmission systems and will induce large cost and huge energy consumption. Thus, for Metro DCI, the power and cost efficiency of optical transceivers should be key considerations besides the transmission performance. Fortunately, with the development of high-speed analog-to-digital converter/digital-to-analog converter (DAC), and powerful digital signal processing (DSP) capability, the intensity modulation and direct detection (IM/DD) technologies can be introduced to boost the line rate with the limited bandwidth of optoelectronic devices. As a successful IM/DD scheme in the copper access area, discrete multitone (DMT) also plays an important role in Metro DCI.

In this chapter, we will introduce the DMT format and its application in Metro DCI. In Section II, the history of DMT will be introduced briefly. Then, Section III describes the operational principle of a conventional DMT. In Section IV, we will summarize some new DMT architectures for Metro DCI applications which can overcome the dispersion and interference challenges. Finally, Section V summarizes and concludes this chapter.

7.2 A Brief History of DMT

The multicarrier modulation technique has been originally applied in Collins Kineplex system for more than 60 years [1]. The idea is dividing a high-speed data stream into several parallel low-speed data streams and then modulating each low speed one to a subcarrier. The multicarrier modulation technique has the low equalization complexity due to its flat frequency response within the narrow subcarrier. Moreover, it is immune to narrow-band noise because it can introduce the water pouring method.

Orthogonal frequency division multiplexing (OFDM) is a widely deployed multicarrier modulation technique in wireless communications. As a baseband version of OFDM, DMT has been demonstrated as a cost-efficient way in wireline communications. It was first proposed in digital subscriber line (DSL) in 1991 [2]. It was chosen by the American National Standards Institute in 1993 and published in 1995 [3] as the US standard for DSL. DMT was also applied in some other copper access scenarios such as cable modem systems [4] and power line systems [5].

In recent years, DMT has been proposed for high-speed optical fiber transmission. The investigations start from those highly dispersive media such as multimode fibers [6] and plastic optical fibers [7]. Then it was proposed in single-mode fiber (SMF) transmission for improving the line rate [8].

7.3 How DMT Works

In this section, the basic function of a typical OFDM system for optical communications is presented. Figures 7.1 and 7.2 show the block diagram of the transmitter and the receiver of a typical OFDM system [9–11]:

At the transmitter side, the bit sequence is first distributed and mapped to different subcarriers. After serial to parallel (S/P) conversion, the inverse fast Fourier transform (IFFT) [12, 13] process transfers the signal from the

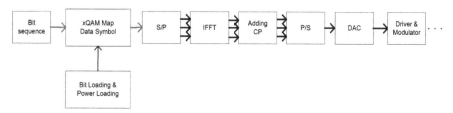

Figure 7.1 Block diagram of the transmitter of the OFDM system.

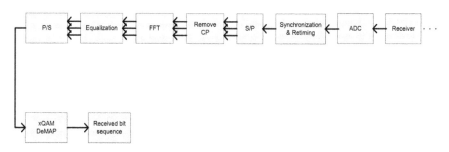

Figure 7.2 Block diagram of the receiver of the OFDM system.

frequency domain to the time domain. Cyclic prefix (CP) [14, 15] is added to remove the intersymbol interference and intercarrier interference. After parallel to serial (P/S) conversion, the signal is sent to DAC [16].

At the receiver side, the start of a symbol of OFDM symbols is first indicated by the synchronization process [17, 18]. Retiming is used to mitigate the frequency and phase mismatch between clocks at the transmitter and the receiver. After the S/P and CP removing, the signal is converted to the frequency domain using fast Fourier transform (FFT). A one-tap equalization [19, 20] compensates linear distortions of the system. After the P/S and the demapper, the transmitted bit sequence will be recovered.

7.3.1 FFT/IFFT

Before the depiction of other blocks, functions of the IFFT and FFT are introduced first. FFT and IFFT are the main components of an OFDM system. These are the most distinctive difference between an OFDM and single carrier systems.

For an OFDM system, the input of IFFT $X(k)$ is complex vectors carrying the data mapped and distributed to subcarriers, where $k = (0, 1, 2, \ldots, N - 1)$. N represents the size of the IFFT. Each of the elements of X is a particular quadrature amplitude modulation (QAM) constellation. The output of the IFFT x_n is also the complex vector. The expression of the inverse discrete Fourier transform of X is:

$$x(n) = \sum_{k=0}^{N-1} X(k) \exp\left(j\frac{2\pi k}{N}n\right) \quad (0 \leq n \leq N - 1) \qquad (7.1)$$

where n is the index of the time-domain samples.

At the receiver side, frequency-domain data can be recovered using the FFT, and the expression is given by:

$$X(n) = \sum_{k=0}^{N-1} x(k) \exp\left(j\frac{2\pi k}{N}m\right) \quad (0 \leq m \leq N - 1) \qquad (7.2)$$

where m is the index of the subcarriers.

DMT is the real-valued OFDM [21, 22], i.e., the output of IFFT is real-valued signal, which can be transmitted and received using the low-cost IM/DD optics. However, it achieves only half of the spectrum efficiency in contrast to OFDM. The real-valued signaling is realized by utilizing an

N-point IFFT, where the first half and the second half of the subcarrier satisfy the conjugate symmetry, namely, the Hermitian symmetry property:

$$X_{2N-k} = X_k^*$$ (7.3)

where the index of subcarrier $k = 1, 2, \ldots, N - 1$, and the imaginary part of X_0 and X_N is equal to zero.

7.3.2 Cyclic Prefix

As a DD scheme, in contrast to the single-carrier modulation format, DMT has two main advantages including better dispersion tolerance [23, 24] and better spectrum efficiency. The ability of dispersion tolerance is achieved by the CP. An example of the DMT symbol with CP is shown in Figure 7.3.

After the IFFT, in most DMT systems, a CP is added to the start of each time-domain DMT symbol before transmission. Namely, a number of samples from the end of the symbol are added to the start of the symbol. Therefore, if the duration length of the impulse response of the end-to-end system is less than the length of the CP, an FFT window can be found, whose samples are originated from the same DMT symbol except that a phase shift is applied to subcarriers. This phase shift can be eliminated by the receiver side equalization.

Based on the IFFT, the DMT symbol can be expressed by:

$$x(n) = \sum_{k=0}^{N-1} X(k) \exp\left(j\frac{2\pi k}{N}n\right)$$ (7.4)

One subcarrier of the DMT symbol g_k is given by:

$$g_k(n) = \exp\left(j\frac{2\pi k}{N}n\right) = \exp\left(j\frac{2\pi k}{N}n + j2\pi ki\right)$$
$$= \exp\left(j\frac{2\pi k}{N}(n + Ni)\right)$$ (7.5)

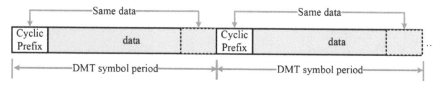

Figure 7.3   DMT symbol with cyclic prefix.

Therefore, each subcarrier is periodic functions. Besides, the range of time index is from 0 to $N-1$, and hence

$$
\begin{aligned}
g_k &= [g_k\,(0)\,,g_k\,(1)\,,g_k\,(2)\,,\ldots g_k\,(N-1)] \\
&= \left[\exp\left(j\frac{2\pi k}{N}0\right),\exp\left(j\frac{2\pi k}{N}1\right),\exp\left(j\frac{2\pi k}{N}2\right),\ldots,\right. \\
&\quad\left.\exp\left(j\frac{2\pi k}{N}\,(N-1)\right)\right]
\end{aligned}
\tag{7.6}
$$

Adding a CP with a length of m, the transmitted signal g'_k becomes:

$$
\begin{aligned}
g'_k &= [g_k\,(N-m)\,,g_k\,(N-m+1)\,,g_k\,(N-m+1)\ldots g_k\,(0)\,, \\
&\quad g_k\,(1)\,,g_k\,(2)\,,\ldots g_k\,(N-1)]
\end{aligned}
\tag{7.7}
$$

Assuming g''_k is the first N samples of the received signal:

$$
\begin{aligned}
g''_k &= [g_k\,(N-m)\,,g_k\,(N-m+1)\,,g_k\,(N-m+1)\ldots g_k\,(0)\,, \\
&\quad g_k\,(1)\,,\ldots g_k\,(N-m-1)] \\
&= \left[\exp\left(j\frac{2\pi k}{N}\,(N-m)\right),\exp\left(j\frac{2\pi k}{N}\,(N-m+1)\right),\ldots\right. \\
&\quad\left.\exp\left(j\frac{2\pi k}{N}0\right),\exp\left(j\frac{2\pi k}{N}1\right),\ldots,\exp\left(j\frac{2\pi k}{N}\,(N-m-1)\right)\right] \\
&= \left[\exp\left(j\frac{2\pi k}{N}\,(-m)\right),\exp\left(j\frac{2\pi k}{N}\,(-m+1)\right),\ldots\right. \\
&\quad\left.\exp\left(j\frac{2\pi k}{N}0\right),\exp\left(j\frac{2\pi k}{N}1\right),\ldots,\exp\left(j\frac{2\pi k}{N}\,(N-m-1)\right)\right] \\
&= \exp\left(j\frac{2\pi k}{N}m\right)\left[\exp\left(j\frac{2\pi k}{N}0\right),\exp\left(j\frac{2\pi k}{N}1\right),\ldots\exp\left(j\frac{2\pi k}{N}m\right),\right. \\
&\quad\left.\exp\left(j\frac{2\pi k}{N}\,(m+1)\right),\ldots,\exp\left(j\frac{2\pi k}{N}\,(N-1)\right)\right] \\
&= \exp\left(j\frac{2\pi k}{N}m\right)g'_k
\end{aligned}
\tag{7.8}
$$

Therefore, adding CP can be considered as an additional phase shift. This phase shift can be compensated by the receiver side equalizer.

7.3.3 Loading Algorithm

In optical systems, an adaptive constellation mapping can be employed in DMT modulation according to the signal-to-noise ratio (SNR) of the transmission channel [25], e.g., subcarriers with a high SNR load the higher order QAM, but the low-order QAM is mapped for subcarriers with a low SNR instead. The detailed loading procedures [22] are listed below:

1. The training DMT signal of qudarture phase shift keying (QPSK) modulation for all subcarriers with an equal power is first transmitted for SNR testing.
2. SNR is computed using the received signal constellations and transmitted signal.
3. Calculate the number of bits for each subcarrier $b(i)$ and difference between the computed number of bits and closest integer number of bits $diff(i)$:

$$b(i) = \log_2(1 + SNR(i)/\Gamma \cdot \gamma_{m\,\mathrm{arg}\,in}) \tag{7.9}$$

$$diff(i) = b(i) - round[b(i)] \tag{7.10}$$

where Γ is the SNR gap [*].
4. The total number of bits per DMT symbol is given by:

$$B_{total} = \sum_{i=1}^{N} \hat{b}(i). \tag{7.11}$$

5. According to the total number of bits, the system margin can be calculated using the expression:

$$\gamma_{margin} = \gamma_{margin} + 10\ln \cdot B_{total} - B_{target} \tag{7.12}$$

where B_{target} is the target number of bits per DMT symbol.
6. Itinerate steps 3–5 to make B_{total} close to B_{target}.
7. According to the iteration result, if $B_{total} > B_{target}$, the number of bits shoud be reduced for the subcarriers with low $diff(i)$. If $B_{total} < B_{target}$, the number of bits shoud be increased for the subcarriers with high $diff(i)$.

Figure 7.4 shows the probed SNR and bit-loading results of back to back (BTB) and 40-km SMF transmissions. Constellations varying from binary phase shift keying to 64-QAM are mapped. Significant power fading points are observed for double sideband transmission over 40 km in Figure 7.4b leading to a heavy SNR decrement at the fading points [26]. Therefore,for the subcarriers at the fading points, the lower order QAM is mapped.

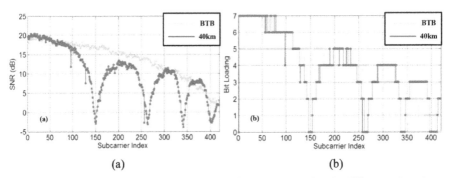

(a)   (b)

Figure 7.4   (a) Probed SNR for different subcarriers; (b) bit- loading for different subcarriers.

7.3.4 PAPR Suppression

The high peak-to-average power ratio (PAPR) is a major drawback of OFDM systems, but the PAPR has a slightly different effect on the optical system compared with the radio frequency (RF) system. In the optical system, because the optical amplifier has a slow response time, regardless of its input signal power, the optical amplifier can maintain the linear output. The PAPR induced penalty mainly comes from the nonlinearity in the optical system, e.g., the nonlinearity of the modulator. The control of the PAPR is designed to improve the system nonlinear margin as much as possible.

The PAPR suppression algorithm is mainly divided into the signal distorted PAPR suppression and the signal undistorted PAPR suppression. The former is achieved by the limiting method. It has an advantage of simple structure except some extra signal distortion. The latter is the conversion of the original waveform into a new waveform with relative low PAPR including the selected mapping [27, 28], optimization algorithm [29], modified signal constellation [30], and active constellation extension [31]. However, these methods have higher complexity.

The limiting method is achieved by clipping the signal envelope (Figure 7.5). A limiter automatically detects the peak value of the signal and limits it according to the signal power, that is, it specifies a threshold value A that limits the signal beyond the threshold so that the maximum output amplitude is equal to the threshold value. The expression of clipping process is given by

$$y_n = \begin{cases} -A, |x_n| < -A \\ x_n, -A \leq |x_n| \leq A \\ A, |x_n| > A \end{cases} \qquad (7.13)$$

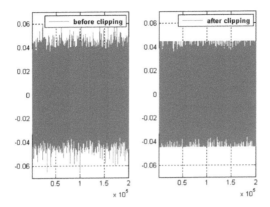

Figure 7.5 DMT signal with hard clipping.

where y_n is output of the limiter, x_n is the input of the limiter, and A is the threshold.

The acquisition of the clipping threshold A is obtained by both the signal power and the clipping ratio. The energy of each subcarrier is assumed to be equal. When subcarriers satisfy the Hermitian symmetry property, the average power P_{avg} of the transmitted signal and its corresponding average amplitude A_{ref} is

$$P_{avg} = 2n_{sc}/N \qquad (7.14)$$
$$A_{ref} = \sqrt{P_{avg}} \qquad (7.15)$$

where n_{sc} is the number of subcarriers with bit loading. A_{ref} is used as the reference value of the clipping ratio:

$$CR = 20 \log_{10} \left(\frac{A}{A_{ref}} \right) \qquad (7.16)$$

The method has the advantages of simplicity and convenience, but it will produce the large in-band distortion and out-of-band noise. In-band distortion will reduce the system error rate, while the out-of-band noise will affect the adjacent channel signal and reduce spectral efficiency (SE). We need a filter and filter out the out-of-band components of the spectrum. After filtering, the PAPR of the OFDM signal will be picked up. In order to avoid this problem, the proper clipping ratio and FFT/IFFT size should be chosen.

7.3.5 Synchronization

The symbol synchronization algorithm uses a training sequence for synchronous estimation. Several synchronization methods are described below. The biggest difference between these synchronization methods is the training sequence design. Synchronization detection principle is basically the same. Block diagram for synchronization detection is shown in Figure 7.6.

Based on HAO's method, the training sequence Ts with length M is:

$$Ts = [x(1), x(2), \ldots x(M/2), x(M/2), x(M/2 - 1), \ldots x(2), x(1)]$$

$$(7.17)$$

That is, the i-th data are the same as the $(N + 1 - i)$ th data. So the synchronization location can be identified by the autocorrelation using a sliding with a size of N. If the beginning of sliding window is located at d, the output value M_fft_d is:

$$M\_fft_d = \frac{\sum_{m=1}^{N/2} x_{d+m-1} \cdot x_{d+N-m}^*}{\sum_{m=1}^{N/2} x_{d+m-1} \cdot x_{d+m-1}^*}$$

$$(7.18)$$

By scanning no less than one DMT symbol, the maximum value of M_fft identifying the synchronization position can be found.

Another design of training sequence is based on the Golay pseudonoise sequence [32]. The sequence is characterized by the fact that the cross correlation of the sequence is constant, and the algorithm uses the received signal to correlate with the training sequence, i.e., xcorr$\{r\_sig, Ts\}$, where xcorr$\{\}$ operator denotes the cross correlation, r_sig is the receiver signal, and Ts is the training sequence. The maximum value of the cross correlation represents the synchronization position.

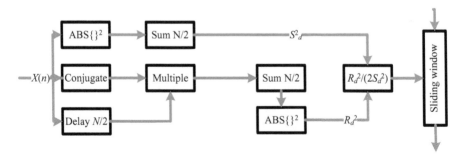

Figure 7.6 Block diagram for synchronization detection.

Table 7.1 An example of training pattern design

L	Pattern
4	− + − −
	+ + + −
8	+ + − − + − − −
	− + + − − − + −
16	+ − − + + + − − + − + + − + − −
	− − + + − + + − − − − + + + + −

Moreover, the training sequence can be designed with repeated pattern based on MINN algorithm [33] to achieve synchronization. The synchronization can be estimated using Λ:

$$\Lambda_\varepsilon(d) = \left(\frac{L}{L-1} \frac{|P(d)|}{E(d)} \right)^2 \tag{7.19}$$

$$P(d) = \sum_{k=0}^{L-2} b(k) \bullet \sum_{m=0}^{M-1} r^*(d + kM + m) \cdot r(d + (k+1)M + m) \tag{7.20}$$

$$E(d) = \sum_{i=0}^{M-1} \sum_{k=0}^{L-1} |r(d + i + kM)|^2 \tag{7.21}$$

$$b(k) = p(k)p(k+1), k = 0, 1, 2, \ldots, L-2 \tag{7.22}$$

where $\{p(k): k = 0, 1, \ldots L - 1\}$ represents the symbol of each repeating pattern and L is the length of the pattern. An example of pattern design is given in Table 7.1.

7.3.6 Channel Equalization

The channel equalization is based on the training sequence. The designed training sequence is $[t_{1x}, t_{2x}, \ldots, t_{Mx}]$, where $t_{1x} = t_{2x} = \cdots = t_{Mx}$ and M is the length of the training sequence. Each t_x is a random sequence composed of QPSK signals. According to the different link situation, the length of the training sequence can be designed with different lengths.

The channel transfer function for each subcarrier of each DMT symbol is obtained based on the least square algorithm:

$$H_{m,n} = R_{m,n}/t_{mx} (m = 1, \ldots M, n = 1, 2, \ldots, N) \tag{7.23}$$

where $R_{m,n}$ is the received training sequence.

After obtaining the N transfer matrices, it is possible to perform the sliding window for $H_{m,n}$ in both time and frequency domains to improve the accuracy of the channel estimation or to perform the sliding-window-filtering process in the time domain alone. Frequency-domain sliding window is given by:

$$H'_m(k') = \frac{1}{\min(k_{\max}, k' + l) - \max(k_{\min}, k' - l) + 1} \sum_{k=k'-l}^{k=k'+l} H_m(k)$$

(7.24)

The domain sliding window can be expressed as:

$$H'(k') = \frac{1}{\min(m_{\max}, m' + l) - \max(m_{\min}, m' - l) + 1} \sum_{m=m'-l}^{m=m'+l} H_m(k')$$

(7.25)

When filling the training sequence, it is also possible to fill the data only on the odd subcarrier to improve the accuracy of the odd subcarrier channel estimation, and the even subcarriers are obtained by interpolation.

7.4 Advanced DMT Techniques for Metro DCI

For short reach interconnections of 2, 10, and even 40 km, the directed modulated laser (DML) or electro-absorption modulated laser (EML) at 1310 nm (O-band) will dominate, due to the low transmitter cost and low fiber chromatic dispersion (CD) [34–37]. However, for the O-band transmission, the fiber attenuation is higher than the one in the C-band, and low-cost optical amplifiers are unavailable. Therefore, the transmission distance is limited [38, 39]. On the other hand, the coherent modules can cover the transmission distance from short reach (several meters) to long haul (thousands of miles). But the cost and power consumption are big issues. Currently, for the inter-DCI, the transmission distance can be higher than 80 km and the data rate is about 50–100 Gbps. This application requires low cost and low power consumption. Because of its simple architecture and low component requirement, the DD system is much cheaper than the coherent detection system. However, the main challenge for DD system in the C-band is the frequency-related power fading induced by the CD after the square law photo-detector (PD). Generally, CD can be (partially) compensated in either optical domain using dispersion-compensated modules (DCMs: fiber or waveguide) or electrical domain or both. Optical compensation is especially useful for high-speed single-carrier IM/DD schemes such as PAM-4. For example, 56- [40] and 112-Gbps [41] PAM-4 are transmitted over 80 km (100 km) SMF transmission using DCFs. The drawback of DCFs is that it increases system cost and link loss, reduces system flexibility, and complicates link configuration. An alternative approach is single sideband (SSB) via simple optical filtering or Hilbert signal modulation to increase the CD tolerance, which has been implemented in many schemes together with DSP [42–46]. DSPs are very efficient for combating CD through either precompensation or postcompensation or their combination [47–52]. In-phase/quadrature Mach–Zehnder modulators (IQ-MZM) or dual-driver MZM had been used for precompensation [48–51]. Although PAM4 [42], CAPs [43], and DMT can be used in the DD system, DMT is preferred since bit loading and power loading can be used to optimize bandwidth utilization of DMT subcarriers. Recently, several DMT-based methods are proposed to cope with the fiber dispersion in the DD system: (1) SSB-DMT [54, 55]; (2) electrical dispersion compensation (EDC)-DSB-DMT [26]; (3) twin-SSB-DMT [56–58]; and (4) SSBI-free DMT.

7.4.1 The Principle of CD-induced Power Fading

Generally, an optical signal can be expressed by:

$$E(t) = e^{iw_0 t} \left(a e^{jwt} + b e^{-jwt} + C \right) \tag{7.26}$$

where ω_0 is the frequency of optical carrier, a and b are the data signals in positive and negative frequency, and C is the direct current (DC). When the optical signal is input into a PD, the output can be illustrated by:

$$
\begin{aligned}
I(t) &\propto E^*(t) E(t) \\
&= \left(a e^{jwt} + b e^{-jwt} + C \right) \left(a^* e^{-jwt} + b^* e^{jwt} + C^* \right) \\
&= aa^* + ab^* e^{j2wt} + C^* a e^{jwt} + a^* b e^{-j2wt} + bb^* + C^* b e^{-jwt} \\
&\quad + Ca^* e^{-jwt} + Cb^* e^{jwt} + CC^* \\
&= \left(aa^* + bb^* + CC^* \right) + \left(Cb^* + C^* a \right) e^{jwt} \\
&\quad + \left(Ca^* + C^* b \right) e^{-jwt} + ab^* e^{j2wt} + a^* b e^{-j2wt} \tag{7.27}
\end{aligned}
$$

The first term contains signal to signal beating interference (SSBI) and DC, which can be processed by DSP-based algorithm and DC block. The second and third terms are the required signal. The other two terms are high-order components, which can be suppressed by a bandwidth-limited filter.

It is known that the transfer function of fiber dispersion can be expressed by:

$$H\left(\omega\right) = \exp\left(-j\frac{\lambda^2}{4\pi c}DL\omega^2\right) = \exp\left(-j\frac{1}{2}\beta_2 L\omega^2\right) \tag{7.28}$$

where λ is the carrier wavelength, D is the dispersion coefficient in the unit of ps/(nm·km), L is the fiber length, c is the speed of light, and β_2 is known as the group velocity dispersion parameter.

When the optical signal is transmitted over fiber, Equation (7.26) is illustrated by:

$$E\left(t\right) = e^{iw_0 t}\left(ae^{jwt}e^{\left(-j\frac{\lambda^2}{4\pi c}DL\omega^2\right)} + be^{-jwt}e^{\left(-j\frac{\lambda^2}{4\pi c}DL\omega^2\right)} + C\right) \tag{7.29}$$

Similar to Equation (7.27), after PD detection, one can get many terms. For simplicity, we consider only the first-order term with the frequency of ω:

$$\begin{aligned} D\left(\omega\right) &= Cae^{\left(j\frac{\lambda^2}{4\pi c}DL\omega^2\right)} + C^*ae^{\left(-j\frac{\lambda^2}{4\pi c}DL\omega^2\right)} \\ &= 2a\Re\left(Ce^{\left(j\frac{\lambda^2}{4\pi c}DL\omega^2\right)}\right) \\ &= 2aC\cos\left(\frac{\pi c}{\omega_0^2}DL\omega^2\right) \end{aligned} \tag{7.30}$$

For the wavelength of 1550 nm, the dispersion coefficient is $D = 16.8$ ps/(nm·km). Based on simulation, the power fading against frequency is shown in Figure 7.7.

7.4.2 Generations of SSB-DMT

Generally, there are two ways to generate the SSB-DMT signal. The first one is optical filter-based SSB-DMT and the other is E/O modulator-based SSB-DMT.

7.4.2.1 Optical filter-based SSB-DMT

As shown in Figure 7.8a, an optical DSB-DMT is generated when an EML or DML is driven by an electrical DSB-DMT. Then an optical filter (MUX or DEMUX) is used to select one sideband and the other sideband is filter out.

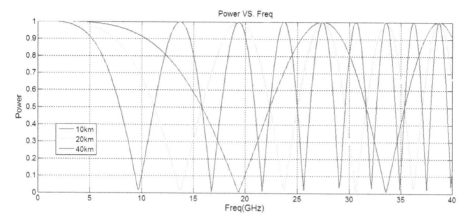

Figure 7.7   The power fading results of different fiber lengths.

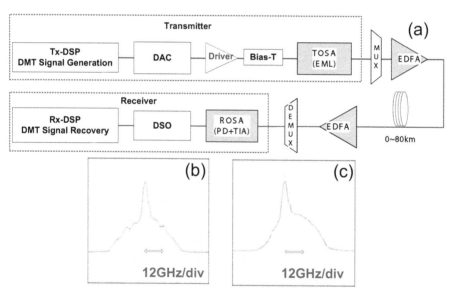

Figure 7.8   (a) Optical filter-based SSB-DMT system. (b) Spectrum of DSB-DMT. (c) Spectrum of SSB-DMT.

Thus, an optical SSB-DMT is realized [53]. The optical spectra of DSB-DMT and SSB-DMT are shown in Figures 7.8b and c. It is noted that wavelength control should be used to adjust the frequency detuning between the center wavelengths of optical signal and optical filter.

7.4.2.2 E/O modulator-based SSB-DMT

7.4.2.2.1 *DD-MZM-based SSB-DMT*

Figure 7.9 shows the principle of generation of optical SSB-DMT modulation based on a dual-driver Mach–Zender modulator (DD-MZM), which is composed of two parallel phase modulator (PM) with independent RF and bias ports [54, 55]. For a complex electrical signal $Txwave(t) = I(t) + j * Q(t)$, the real ($I(t)$) and imaginary ($Q(t)$) components are used to drive two PMs with a bias difference of $V_\pi/2$. The output of the DD-MZM can be expressed as:

$$E_{out} = \frac{\sqrt{2}}{2} E_{in} * \left\{ e^{j*[\frac{\pi}{V_\pi} I(t) - \frac{\pi}{2}]} + e^{j*[\frac{\pi}{V_\pi} Q(t)]} \right\}$$

$$= \frac{\sqrt{2}}{2} E_{in} * \left\{ -j * e^{j*[\frac{\pi}{V_\pi} I(t)]} + e^{j*[\frac{\pi}{V_\pi} Q(t)]} \right\} \qquad (7.31)$$

In Equation (7.31), E_{in} is the field of the input optical signal and V_π is the half-wave voltage of the PM. When the $I(t)$ and $Q(t)$ are signals with low amplitude, Equation (7.31) can be appropriated to be:

$$E_{out} \approx \frac{\sqrt{2}}{2} E_{in} * \left\{ -j*[1 + j * \frac{\pi}{V_\pi} I(t)] + [1 + j * \frac{\pi}{V_\pi} Q(t)] \right\}$$

$$\approx \frac{\sqrt{2}}{2} E_{in} * \left\{ \frac{\pi}{V_\pi} * [\underline{I(t) + j * Q(t)}] + 1 - j*V_\pi/2 \right\}. \qquad (7.32)$$

From Equation (7.32), it is observed that the electrical complex signal $Txwave(t) = I(t) + j * Q(t)$ is linearly converted to optical domain. Thus, an optical SSB-DMT signal can be obtained if the real and imaginary components of an electrical SSB-DMT are used to drive a DD-MZM, which is biased at the quadrature point. It is known that an electrical SSB signal is easily to be generated by DSP-based Hilbert transfer.

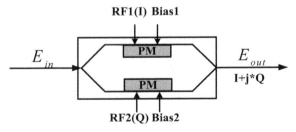

Figure 7.9   Principle of DD-MZM-based SSB-DMT.

Figure 7.10 shows the probed SNR of BTB and 80-km SMF transmission. Power fading is observed for the DSB-DMT signal after 80-km transmission, while no power fading is observed for the SSB-DMT signal even after transmission. From the results, it is found that SSB-DMT is an effective method to cope with the power fading issue, enabling a higher capacity system.

7.4.2.2.2 *IQ-MZM-based SSB-DMT*

It is known that IQ-MZM can also be used to generate SSB-DMT with a similar method to DD-MZM. As shown in Figure 7.11, the IQ-MZM consists of two independent MZMs, which are controlled by bias-1 and bias-2, respectively. Bias-3 is used to adjust the relative phase difference between two MZMs. The output of IQ-MZM can be expressed by:

$$
E_{out} = \frac{\sqrt{2}}{2} E_{in} * \left\{ \cos \left(\frac{\pi}{V_\pi} I(t) + \frac{\pi}{V_\pi} \cdot V_{bias1} \right) + e^{j * \frac{\pi}{V_\pi} V_{bias3}} \right.
$$
$$
\left. \cdot \cos \left(\frac{\pi}{V_\pi} Q(t) + \frac{\pi}{V_\pi} \cdot V_{bias2} \right) \right\} \tag{7.33}
$$

When the drive signals $I(t)$ and $Q(t)$ are small signals and bias-3 is $V_\pi/2$, Equation (7.33) can be approximately illustrated by:

$$
E_{out} = \frac{\sqrt{2}}{2} \cdot E_{in} \cdot \frac{\pi}{V_\pi} [V_{bias1} + j \cdot V_{bias2} + \underline{I(t) j \cdot Q(t)}] \tag{7.34}
$$

As shown in Equation (7.34), the electrical signal is linearly converted to optical domain, and thus an optical SSB-DMT can be realized if the IQ-MZM is driven by an electrical SSB-DMT signal.

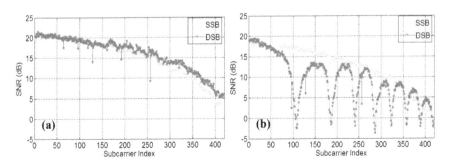

Figure 7.10   Probed SNRs for: (a) back to back and (b) 80-km SMF.

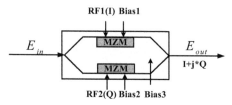

Figure 7.11   Principle of IQ-MZM-based SSB-DMT.

7.4.3 Generation of EDC-DSB-DMT

As shown in Equation (7.28), the CD is a linear lossless channel response, whose phase curve is illustrated in the blue curve of Figure 7.12. When an electrical DSB-DMT is predispersed in DSP, the EDC-DSB-DMT can be expressed by

$$X(\omega) = S(\omega) \cdot H^{-1}(\omega) = S(\omega) \cdot \exp\left(-j\frac{\lambda^2}{4\pi c}DL\omega^2\right) \qquad (7.35)$$

where $S(w)$ is the DSB-DMT signal in the frequency domain, $H^{-1}(w)$ is achieved by inverting the sign of phase delay introduced by dispersion, which is shown in the red curve of Figure 7.12.

When a DD-MZM or IQ-MZM is used, the EDC-DSB-DMT is linear converted from the electric field X(w) to the optical field E(w). Then the optical signal E(w) is transmitted over fiber with the dispersion of H(w). At the receiver side, the optical signal can be expressed by:

$$G(w) = E(w)H^{-1}(w)H(w) = E(w) \qquad (7.36)$$

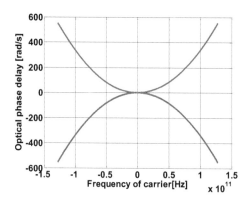

Figure 7.12   The optical phase delay with respect to the frequency.

From Equation (7.36), it is found that EDC-DSB-DMT can be used to mitigate the fiber dispersion successfully.

In reference [26], the results of EDC-DSB-DMT and SSB-DMT are compared. No power fading is observed after 80-km SMF transmission for both systems. As shown in Figure 7.13, the EDC-DSB-DMT has better SNR and required optical SNR performance, since both of the sidebands have useful information. More details can be found in reference [26].

7.4.4 Generation of Twin-SSB-DMT

Figures 7.14a and b show the schematic diagrams of conventional SSB and twin-SSB schemes, respectively [56–58]. In Figure 7.14a, the electrical data A are a real signal and its Hilbert pair is $\hat{A}$. According to the Hilbert theorem, the electrical data of $A+j\cdot\hat{A}$ and $A-j\cdot\hat{A}$ are right-sideband and left-sideband

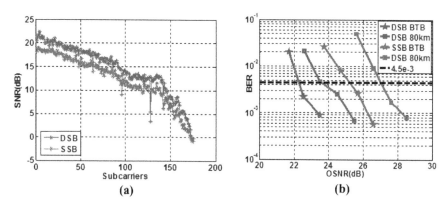

Figure 7.13   (a) SNR curves of DSB-DMT and SSB-DMT. (b) Required optical SNR results of DSB-DMT and SSB-DMT.

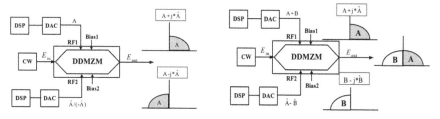

Figure 7.14   Architectures of (a) conventional SSB-DMT transmitter and (b) twin-SSB-DMT transmitter.

SSB signals, respectively. When electrical data A and Â are used to drive the upper and lower arms of the DD-MZM biased at quadrature point, the output of the DD-MZM can be expressed by [54]:

$$E_{out} = E_{in} * (A + j * \hat{A}) \tag{7.37}$$

From Equation (7.37), it is observed that the electrical right-sideband signal is linearly converted to optical domain. Thus, an optical right-sideband signal is obtained. Similarly, an optical left-sideband signal can be achieved when $-\hat{A}$ is employed to drive the lower arm of the DD-MZM. In reference [56], a twin-SSB scheme was proposed to increase the SE with the schematic diagram shown in Figure 7.14b. Two independent real-value electrical data A and B are combined to drive the upper arm of DD-MZM and their Hilbert pairs Â and B̂ are subtracted to drive the lower arm of DD-MZM biased at the quadrature point. The output of the DD-MZM can be expressed by:

$$E_{out} = E_{in} \cdot [(A + B) + j \cdot (\hat{A} - \hat{B})]$$
$$= E_{in} \cdot [\underline{(A + j \cdot \hat{A})} + \underline{(B - j \cdot \hat{B})}] \tag{7.38}$$

In the DSP, two independent electrical DMT signals (DMT-A and DMT-B) are generated. From Equation (7.38), one can find that both of the right-sideband-DMT signal and left-sideband-DMT signal are linearly converted to optical domain. An optical twin-SSB-DMT signal is realized with independent data DMT-A and DMT-B at the right and left sidebands. Thus, the SE of the twin-SSB-DMT modulation is twice as that of the conventional SSB-DMT modulation.

For the twin-SSB-DMT modulation, due to the limited sideband suppression ratio, the leakage between the two sidebands will cause background interference, so-called class-I crosstalk. The twin-SSB-DMT can be received using a low-cost integrated receiver, where a dual-channel optical filter is used to separate the two-sideband signal and two Receiver Optical Sub-Assemblys (ROSAs) with the same performance are utilized to detect the left and right sidebands simultaneously. Since the filters are not ideal brick-wall shape with flat top and steep slope, class-II crosstalk will incur at the edge of the two sidebands, as shown in insets (i) and (ii) of Figure 7.15a. A frequency-domain multiinput multioutput (MIMO)-array DSP can be used to mitigate both types of crosstalks, and the SNR is improved by 10 and 2 dB for the low-frequency and middle-frequency domains, respectively. Based on the twin-SSB-DMT, 224-Gbps data were generated and transmitted over 80 km using DDs. Similarly, 300-Gbps data were realized using the time-domain crosstalk reduction method [57].

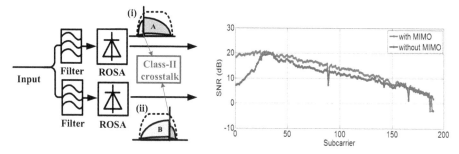

Figure 7.15 Architectures of (a) a twin-SSB-DMT receiver and (b) SNR improvement with a MIMO array.

7.4.5 Generation of SSBI-free Twin-SSB-DMT

The same as the traditional DSB-DMT, the SSB-DMT also suffers from SSBI, which is introduced into the received signal after square-law detection [59]. Several solutions have been proposed to eliminate SSBI, such as guard-band SSB-DMT [60] and guard-band-shared SSB-DMT [61]. However, only half of the bandwidth was used in reference [60] and several expensive transmitters were required in reference [61]. Moreover, DSP-based SSBI cancellations were also proposed to improve the system performance [59], where several iterations were required.

Recently, a spectrally efficient SSBI cancellation system based on the guard-band twin-SSB-DMT was experimentally demonstrated. Figure 7.16a shows the schematic diagram. At the transmitter side, an electrical DMT signal A and its Hilbert pair Â are used to drive a DD-MZM to generate an optical SSB-DMT signal. When the bandwidth of the optical SSB-DMT is B_S, a guard band with a bandwidth of B_S is required to accommodate the SSBI [60]. Thus, only half of the bandwidth of the electrical devices, modulators, and PD are used in this scheme.

To increase the SE, an SSBI cancellation technique based on twin-SSB DMT was proposed in reference [62], with the schematic diagram shown in Figure 7.16b. Two independent electrical DMT data A and B are combined to drive the upper arm of DD-MZM and their Hilbert pairs are subtracted to drive the lower arm of the DD-MZM, which is biased at the quadrature point. An optical twin-SSB signal is realized with independent data A and B at the right and left sidebands [56]. At the receiver side, two filters and PDs are used to detect the DMT-A and DMT-B. From insets (iv) and (v) of Figure 7.16(b), it is observed that the SSBI is located at the low-frequency band from 0 to

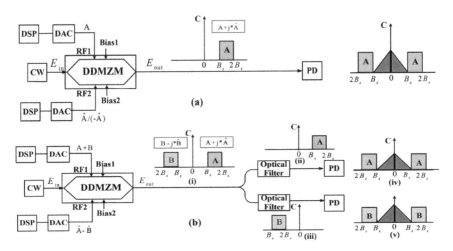

Figure 7.16   Architectures of (a) a twin-SSB-DMT receiver and (b) SNR improvement with a MIMO array.

B_S, which will not induce interference to the data occupying the frequency from B_S to $2B_S$. It is noted that the proposed SSBI-free twin-SSB-DMT doubles the SE of the conventional SSBI-free SSB-DMT, without changing the transmitter structure.

An experiment was implemented and the optical spectrum of SSBI-free twin-SSB-DMT is shown in Figure 7.17a, with the DMT-A at the right sideband and DMT-B at the left sideband. At the receiver side, a 3-dB optical coupler and two tunable optical filters are employed to select the left and right sidebands of the SSBI-free twin-SSB-DMT, whose optical spectra are shown in Figures 7.17b and c. Since 15.7-GHz guard band is allocated, the filters

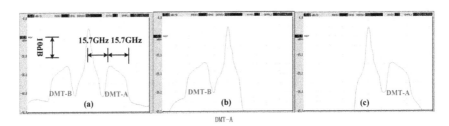

Figure 7.17   Optical spectra of (a) SSBI-free twin-SSB-DMT; (b) filtered left-sideband of twin-SSB-DMT; and (c) filtered right-sideband of twin-SSB-DMT.

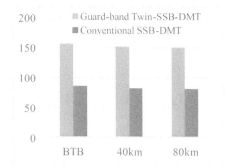

Figure 7.18 System capacity versus SMF length.

with sharp slope are not necessary and the interference between the left and right sidebands is negligible.

The capacity versus transmission distance at an FEC threshold of 4.5×10^{-3} is shown in Figure 7.18. For the conventional guard-band SSB-DMT, the capacities are 85 and 80 Gbps for BTB and 80-km SMF, which are significantly improved to 156 and 150 Gbps using the proposed SSBI-free twin-SSB-DMT.

7.5 Summary

DMT was introduced to boost the line rate with the limited bandwidth of optoelectronic devices. It plays an important role in Metro DCI. In this chapter, the Metro DCI application scenarios are described at first. Then, followed by a brief history of DMT, the conventional operational principle of DMT has been illustrated in detail, including FFT/IFFT, CP, loading algorithm, PAPR suppression synchronization, and channel equalization. Finally, we have reviewed some new DMT architectures for Metro DCI applications which can overcome the dispersion and interference challenges by employing some advanced modulators. Within these schemes, SSB-DMT is very simple, while CD calculation and precompensation are not necessary. EDC-DSB-DMT owns 3-dB better performance than SSB-DMT. CD can be calculated according to the power fading and the result is sent back to the transmitter for precompensation. Twin-SSB-DMT has the highest SE. More than 200-Gbps data have been successfully generated and transmitted over 80-km SMF. Furthermore, a spectrally efficient SSBI cancellation system based on the guard-band twin-SSB-DMT was demonstrated. These new schemes make DMT capable of tens of kilometers of transmission and suitable for Metro DCI applications.

References

[1] Doelz, M. L., Heald, E. T., and Martin, D. L. (1957). "Binary data transmission techniques for linear systems," *in Proceedings of the IRE* (New York, NY: IEEE), 45, 656–661.

[2] Cioffi, J. M. (1991). "A Multicarrier Primer," Clearfield, FL: ANSI Contribution T1E1.4/91-157.

[3] ANSI T1.413-1995 (1995). *Telecommunications – Network and Customer Installation Interfaces – Asymmetric Digital Subscriber Line (ADSL) Metallic Interface.* Issue 1, ANSI T1.413-1995.

[4] Jacobsen, K. S., Bingham, J. A. C., and Cioffi, J. M. (1995). "Synchronized DMT for multipoint-to-point communications on HFC networks," *in Proceedings of the IEEE Global Telecommunications Conference, 1995* (Singapore: IEEE), 2, 963–966.

[5] Esmailian, T., Gulak, P. G., Kschischang, F. R. (2000). "A discrete multitone power line communications system," *in Proceedings of ICASSP'2000* (Istanbul: IEEE), 2953–2956.

[6] Lee, S. C. J., Breyer, F., Randel, S., Schuster, M., Zeng, J., Huijskens, F., et al. (2007). "24-Gb/s transmission over 730 m of multimode fiber by direct modulation of an 850-nm VCSEL using discrete multi-tone modulation," *in Proceedings of National Fiber Optic Engineers Conference* (Anaheim, CA: Optical Society of America), paper PDP6.

[7] Randel, S., Lee, S. C. J., Spinnler, B., Breyer, F., Rohde, H., Walewski, J., et al. (2006). "1 Gbit/s transmission with 6.3 bit/s/Hz spectral efficiency in a 100m standard 1 mm step-index plastic optical fibre link using adaptive multiple sub-carrier modulation," *in Proceedings of the 32nd European Conference on Optical Communication (ECOC 2006)* (Cannes: SEE), PDP Th4.4.1.

[8] Tanaka, T., Nishihara, M., Takahara, T., Li, L., Tao, Z., and Rasmussen, J. C. (2012). "50 Gbps class transmission in single mode fiber using discrete multi-tone modulation with 10G directly modulated laser," *in Proceedings of the Optical Fiber Communication Conference 2012* (Los Angeles, CA: Optical Society of America), paper OTh4G.3.

[9] Chang, R. (1966). Synthesis of band-limited orthogonal signals for multichannel data transmission. *Bell Sys. Tech. J.* 45, 1775–1796.

[10] Armstrong, J. (2008). "OFDM: from copper and wireless to optical," *in Proceedings of the OFC/NFOEC 2008* (San Diego, CA: IEEE).

[11] Zou, W. Y., and Wu, Y. (1995). "COFDM: An overview," *in Proceedings of the IEEE Transactions on Broadcasting (*New York, NY: IEEE), 41, 1–8.

[12] Weinsten, S., Ebert, P. (1971). "Data transmission by frequency-division multiplexing using the discrete Fourier transform," *in Proceedings of the IEEE Transactions on Communication Technology*, 19, 628–634.

[13] Salz, J., and Weinstein, S. B. (1969). "Fourier transform communication system," *in Proceedings of the First ACM Symposium on Problems in the Optimization of Data Communications Systems* (Pine Mountain, GA: ACM).

[14] Peled, A., and Ruiz, A. (1980). "Frequency domain data transmission using reduced computational complexity algorithms," *in Proceedings of the IEEE International Conference on ICASSP '80* (Denver, CO: IEEE), III, 964–967.

[15] Duhamel, P., and Hollmann, H. (1984). Split-radix FFT algorithm. *IET Elect. Lett.* 20, 14–16.

[16] Dardari, D. (2006). "Joint clip and quantization effects characterization in OFDM receivers," *in Proceedings of the IEEE Transactions on Circuits and Systems I: Regular Papers I: Fundamental Theory Applications* (New York, NY: IEEE), 53, 1741–1748.

[17] van de Beek, J. J., Sandell, M., and Börjesson, P. O. (1997). "ML estimation of time and frequency offset in OFDM systems," *in Proceedings of the IEEE Transactions on Signal Processing* (Piscataway, NJ: Institute of Electrical and Electronics Engineers), 45, 1800–1805.

[18] Schmidl, T. M., and Cox, D. C. (1997). "Robust frequency and timing synchronization for OFDM," *in Proceedings of the IEEE Transactions on Communications*, 45, 1613–1621.

[19] Armstrong, J. (2009). OFDM for optical communications. *J. Lightwave Technol.* 27, 189–204.

[20] Shieh, W., Djordjevic, I. (2010). *OFDM for Optical Communications*. Boston: Elsevier.

[21] Bingham, J. A. C. (1990). "Multicarrier modulation for data transmission: an idea whose time has come," *in Proceedings of the IEEE Communications Magazine,* 28, 5–14.

[22] Chow, J. S., Tu, J. C., and Cioffi, J. M. (1991). "A discrete multitone transceiver system for HDSL applications," *in Proceedings of the IEEE*

Journal on Selected Areas in Communications (Piscataway, NJ: Institute of Electrical and Electronics Engineers), 9, 895–908.

[23] Schmidt, B. J. C., Lowery, A. J., and Armstrong, J. (2007). "Experimental demonstrations of 20 Gbit/s direct-detection optical OFDM and 12 Gbit/s with a colorless transmitter," *in Proceedings of the Optical Fiber Communication Conference 2007* (Anaheim, CA: Optical Society of America), PDP1.

[24] Jansen, S. L., Morita, I., and Tanaka, H. (2007). "16×52.5-Gb/s, 50-GHz spaced, POLMUX-CO-OFDM transmission over 4,160 km of SSMF enabled by MIMO processing," *in Proceedings of the 33rd European Conference and Exhibition of Optical Communication-Post-Deadline Papers* (Berlin: VDE), PD 1.3.

[25] Chow, P. S., Cioffi, J. M., and Bingham, J. A. C. (1995). "A practical discrete multitone transceiver loading algorithm for data transmission over spectrally shaped channels," *in Proceedings of the IEEE Transactions on Communications*, 43, 773–775.

[26] J. Zhou, et al. (2016). "Transmission of 100-Gb/s DSB-DMT over 80-km SMF using 10-G class TTA and direct-detection," *in Proceedings of the 42nd European Conference on Optical Communication (ECOC 2016)* (Dsseldorf: VDE), Tu3.F.

[27] Bauml, R., Fischer, R. F. H., and Huber, J. B. (1996). Reducing the peak-to-average power ratio of multicarrier modulation by selected mapping. *Electron. Lett.* 32, 2056–2057.

[28] Wang, C.-L., Ku, S.-J., and Yang, C.-J. (2010). "A low-complexity PAPR estimation scheme for OFDM signals and its application to SLM-based PAPR reduction," *in Proceedings of the IEEE Journal of Selected Topics in Signal Processing*, 4, 637–645.

[29] Koussa, B., Bachir, S., Perrine, C., Duvanaud, C., and Vauzelle, R. (2012). "A comparison of several gradient based optimization algorithms for PAPR reduction in OFDM systems," *in Proceedings of the 2nd International Conference on Communications, Computing and Control Applications 2012 (CCCA12),* (Marseilles: IEEE).

[30] Yoshizawa, R., and Ochiai, H. (2017). "Trellis-Assisted Constellation Subset Selection for PAPR Reduction of OFDM Signals," *in Proceedings of the IEEE Transactions on Vehicular Technology*, 66, 2183–2198.

[31] Zhong, J., Yang, X., and Hu, W. (2017). "Performance-improved secure OFDM transmission using chaotic active constellation extension," *in Proceedings of the IEEE Photonics Technology Letters*, 29, 991–994.

[32] Zhi-nan, L., Jin-jin, Z., Jiang, Z., and Er-yang, Z. (2012). "Golay complementary pair aided time synchronization method for OFDM systems," *in Proceedings of the IEEE 14th International Conference on Communication Technology* (Chengdu: IEEE), 166–170.

[33] Wei, J., Hu, J., and Chen, J. (2014). "An improved algorithm based on training symbol for OFDM symbol synchronization," *in Proceedings of the 10th International Conference on Wireless Communications, Networking and Mobile Computing* (Beijing: IET), 105–108.

[34] Tao, L., Ji, Y., Liu, J., Lau, A. P. T., Chi, N., and Lu, C. (2013). "Advanced modulation formats for short reach optical communication systems," *in Proceedings of the IEEE Network*, 27, 6–13.

[35] Wei, J., Cheng, Q., Penty, R. V., White, I. H., Cunningham, D. G. (2015). "400 gigabit Ethernet using advanced modulation formats: performance, complexity, and power dissipation," *in Proceedings of the IEEE Communications Magazine*, 53, 182–189, 2015.

[36] Chan, T., and Way, W. I. (2015). "112 Gb/s PAM4 transmission over 40 km SSMF using 1.3 μm gain-clamped semiconductor optical amplifier," *in Proceedings of the Optical Fiber Communications Conference and Exhibition (OFC)* (Los Angeles, CA: IEEE), Th3A.4.

[37] Xu, X., Zhou, E., Liu, G. N., Zuo, T., Zhong, Q., Zhang, L., et al. (2015). Advanced modulation formats for 400-Gbps short-reach optical interconnection. *Optics Express, 23*, 492–500.

[38] Chan, T., and Way, W. I. (2015). "112 Gb/s PAM4 transmission over 40 km SSMF using 1.3 μm gain-clamped semiconductor optical amplifier," *in Proceedings of the Optical Fiber Communications Conference and Exhibition (OFC)* (Los Angeles, CA: IEEE), Th3A.4.

[39] Xu, X., Zhou, E., Liu, G. N., Zuo, T., Zhong, Q., Zhang, L., et al. (2015). Advanced modulation formats for 400-Gbps short-reach optical interconnection. *Optics Express 23*, 492–500.

[40] Eiselt, N., Wei, J., Griesser, H., Dochhan, A., Eiselt, M. H., Elbers, J.-P., et al. (2017). Evaluation of real-time 8 × 56.25 Gb/s (400G) PAM-4 for inter-data center application over 80 km of SSMF at 1550 nm. *J. Lightwave Technol.* 35, 955–962.

[41] Eiselt, N., et al. (2016). "Experimental demonstration of 112 Gbit/s PAM-4 over up to 80 km SSMF at 1550 nm for inter-DCI applications," *in Proceedings of the 42nd European Conference on Optical Communication (ECOC 2016)* (Düsseldorf: VDE), P130.

[42] Lee, J., Kaneda, N., and Chen, Y.-K. (2016). "112-Gbit/s intensity-modulated direct-detect vestigial-sideband PAM4 transmission over an

80-km SSMF link," *in Proceedings of the 42nd European Conference on Optical Communication (ECOC 2016)* (Düsseldorf: VDE), P136.

[43] Wei, J., Eiselt, N., Sanchez, C., Du, R., and Griesser, H. (2016). 56 Gb/s multi-band CAP for data center interconnects up to an 80 km SMF. *Opt. Lett.* 41, 4122–4125.

[44] Li, Z., Erkılınç, M. S., Shi, K., Sillekens, E., Galdino, L., Thomsen, B. C., et al. (2017). SSBI mitigation and the Kramers–Kronig Scheme in single-sideband direct-detection transmission with receiver-based electronic dispersion compensation. *J. Lightwave Technol.* 35, 1887–1893.

[45] Dochhan, A., Griesser, H., Eiselt, N., Eiselt, M., and Elbers, J.-P. (2016). "Optimizing discrete multi-tone transmission for 400G data center interconnects," *in Proceedings of Photonic Networks; 17. ITG-Symposium* (Leipzig: VDE).

[46] Peng, W. R., Wu, X., Feng, K.-M., Arbab, V. R., Shamee, B., Yang, J.-Y., et al. (2009). Spectrally efficient direct-detected OFDM transmission employing an iterative estimation and cancellation technique. *Opt. Express* 17, 9099–9111.

[47] Randel, S., Pilori, D., Chandrasekhar, S., Raybon, G., and Winzer, P. (2015). "100-Gb/s discrete-multitone transmission over 80-km SSMF using single-sideband modulation with novel interference-cancellation scheme," *in Proceedings of the 2015 European Conference on Optical Communication (ECOC)*, Valencia.

[48] Li, Z., Erkılınç, M. S., Maher, R., Galdino, L., Shi, K., Thomsenet, B. C., et al. (2016). "Reach enhancement for WDM direct-detection subcarrier modulation using low-complexity two-stage signal-signal beat interference cancellation," *in Proceedings of the 42nd European Conference on Optical Communication* (Düsseldorf: VDE), P103.

[49] Li, Z., Erkılınç, M. S., Bouziane, R., Thomsen, B. C., Bayvel, P., and Killeyet, R. I. (2016). Simplified DSP-based signal-signal beat interference mitigation techniques for direct detection OFDM. *J. Lightwave Technol.* 34, 866–872.

[50] Okabe, R., Liu, B., Nishihara, M., Tanaka, T., Takahara, T., Li, L., et al. (2015). "Unrepeated 100 km SMF transmission of 110.3 Gbps/lambda DMT signal," *in Proceedings of the 2015 European Conference on Optical Communication (ECOC)* (Valencia: IEEE), P. 5.18.

[51] Chen, X., Antonelli, C., Chandrasekhar, S., Raybon, G., Sinsky, J., Mecozzi, A., et al. (2017). "218-Gb/s single-wavelength, single-polarization, single-photodiode transmission over 125-km of standard

singlemode fiber using Kramers–Kroning detection," *in Proceedings of the Optical Fiber Communications Conference and Exhibition (OFC)* (Los Angeles, CA: IEEE), Th5B.6.

[52] Liu, Z. (2017). "300-km Transmission of Dispersion Pre-compensated PAM4 Using Direct Modulation and Direct Detection" *in Proceedings of the Optical Fiber Communications Conference and Exhibition (OFC)* (Los Angeles, CA: IEEE), Th3D.6.

[53] Zhang, Q., Fang, Y., Xu, X., Zhou, E. (2014). "C-band 56Gbps transmission over 80-km SMF without CD compensation by IM-DD," *in Proceedings of the ECOC 2014*, paper P.5.19.

[54] Zhang, L., Zhou, E., Zhang, Q., Xu, X., Liu, G. N., and Zuo, T. (2015). "C-band single wavelength 100-Gb/s IM-DD transmission over 80-km SMF without CD compensation using SSB-DMT," *in Proceedings of the Optical Fiber Communication Conference* (Los Angeles, CA: Optical Society of America), Th4A2.

[55] Zhang, L., Zuo, T., Mao, Y., Zhang, Q., Zhou, E., Liuet, G. N., et al. (2016). Beyond 100-Gbps transmission over 80-km SMF using direct-detection SSB-DMT at C-band. *J. Lightwave Technol.* 34, 723–729.

[56] Zhang, L., Zuo, T., Zhang, Q., Zhou, J., Zhou, E., and Liu, G. N. (2016). "Single wavelength 248-Gbps transmission over 80-km SMF based on Twin-SSB-DMT and direct detection," *in Proceedings of the 42nd European Conference on Optical Communication (ECOC)* (Dsseldorf: VDE), M2D2.

[57] Wang, Y., Yu, J., Chien, H.-C., Li, X., Chi, N. (2016). "Transmission and direct detection of 300-Gbps DFT-S OFDM signals based on O-ISB modulation with joint image-cancellation and nonlinearity-mitigation," *in Proceedings of the 42nd European Conference on Optical Communication (ECOC)* (Dsseldorf: VDE), P142.

[58] Zhu, Y., Ruan, X., Chen, Z., Jiang, M., Zou, K., Li, C., et al. (2017). "4×200Gbps Twin-SSB Nyquist subcarrier modulation WDM transmission over 160km SSMF with direct detection," *in Proceedings of the Optical Fiber Communications Conference and Exhibition (OFC)*, 1–3.

[59] Randel, S., Pilori, D., Chandrasekhar, S., Raybon, G., and Winzer, P. (2015). "100-Gb/s discrete-multitone transmission over 80-km SSMF using single-sideband modulation with novel interference-cancellation scheme," *in Proceedings of the Optical Communication (ECOC), 2015 European Conference,* 1–3.

[60] Schmidt, B. J., Lowery, A. J., and Armstrong, J. (2008). Experimental demonstrations of electronic dispersion compensation for long-haul transmission using direct-detection optical OFDM. *J. Lightwave Technol.* 26, 196–203.

[61] Zhang, X., Li, Z., Li, C., Luo, M., Li, H., Li, C., et al. (2014). Transmission of 100G DDO-OFDM over 320-km SMF with a single photodiode. *Opt. Express* 22, 12079–12086.

[62] Zhang, L., Zuo, T., Zhang, Q., Zhou, J., Zhou, E., and Liu, G. N. (2016). 150-Gb/s DMT over 80-km SMF transmission based on spectrally efficient SSBI cancellation using guard-band Twin-SSB technique," *in Proceedings of the ECOC 2016; 42nd European Conference on Optical Communication* (Dsseldorf: VDE), M2D2.

8

A Duobinary Approach Toward High-speed Short-reach Optical Interconnects

Xin (Scott) Yin, Guy Torfs and Johan Bauwelinck

imec - Ghent University, IDLab, Gent, Belgium

8.1 Introduction

The continuous growth of the Internet traffic is boosting the requirement of ultra-high-speed optical interconnects in datacenters. This growth is largely driven by cloud computing, mobility, and the Internet of Things. Currently, the evolution from 100-Gbps Ethernet to 400 Gbps is under discussion within the IEEE P802.3bs 400-Gigabit Ethernet Task Force [1]. Among different approaches, the four-lane 100-Gbps scheme is particularly attractive for 500-m and 2-km single-mode (SM) fiber applications as it allows lower lane counts and thus offers higher spatial efficiency. For the intra-datacenter communication, avoiding complex transceivers is crucial in terms of cost and power consumption. Consequently, intensity modulation and direct detection (IMDD) links are preferred rather than coherent transmission technologies. Non-return to zero (NRZ) on–off keying (OOK) [2, 3] keeps the optical hardware simple but presents a big challenge to the bandwidth of transceivers in applications using 100 Gbps and beyond. Advanced modulation formats, such as four-level pulse-amplitude modulation (PAM-4) [4, 5] discrete multitone (DMT) [6, 7], and electrical duobinary (EDB) [8–10], overcome this limitation by improving the spectral efficiency while maintaining the benefits of direct detection. However, most of the realized 100-Gbps class DMT, PAM-4, and EDB demonstrations [3–8] are based on offline digital signal processing (DSP). The realization of 100-Gbps class real-time DMT transmission [11] is hindered by the huge amount of power-consuming calculations. In [12], a real-time 112-Gbps PAM-4 optical link over 2-km standard single-mode fiber (SSMF) was demonstrated with a SiGe BiCMOS transceiver [including clock and data recovery (CDR)] but with a high power consumption of ∼8.6 W.

Recently, real-time implementation of the alternative modulation format, three-level EDB, has gained a lot of attention [9, 10]. Compared to NRZ, the three-level signaling in duobinary narrows the optical spectral bandwidth which improves the chromatic dispersion (CD) tolerance and the equalization requirements of the high-speed serial link [9]. Meanwhile, thanks to the simple conversion between duobinary and NRZ (also see the next section), an EDB transmission system can provide backward compatibility with existing NRZ systems. In this chapter, we present an overview of the EDB modulation format and our recent progress on real-time EDB/NRZ optical links using high-speed EDB/NRZ transmitter (TX) and receiver (RX) ICs.

8.2 Three-Level Electrical Duobinary Modulation

Duobinary (DB) was first introduced in 1963 [13] as a modulation format that uses known inter-symbol interference to reduce the required channel bandwidth. This makes it a specific form of partial response signaling. In short, to convert an NRZ signal to a duobinary signal (see Figure 8.1), the current NRZ symbol ($+1$ or -1) is added to the previous symbol by a delay-and-add filter $H(z) = (1+z^{-1})/2$, which results in three levels (-1, 0, and $+1$). As a consequence, this duobinary signal does not contain any direct transitions from $+1$ to -1 or vice versa; the signal will always pass through the intermediate 0-symbol. As such, the fastest rising edges of the signal are effectively removed. In addition, it is also impossible that a duobinary signal has alternate symbols, e.g., (0, +1, 0, +1, ...) in adjacent levels. The fastest alternate symbol stream has a periodicity of three symbols and is of the form

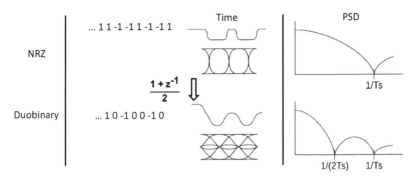

Figure 8.1 Three-level duobinary modulation using a delay-and-add filter.

(0, 0, +1, 0, 0, +1, ...). This reduces the required signal bandwidth compared to the original NRZ signals.

For high-speed optical links, there are two different schemes to map these three levels (−1, 0, and +1) into modulated signals, namely, optical duobinary (ODB) and EDB. The ODB signal has two levels ("on" and "off"), and the phase of the "on" signal takes two values of 0 and π [14]. The ODB receiver can be a simple binary IM-DD receiver, and modulation is typically achieved by 100% overdriving a Mach–Zehnder modulator (MZM) with duobinary encoded electrical signals. In EDB systems, the intensity of the optical signals uses three levels corresponding to a duobinary signal, and therefore a three-level decision circuit is needed at the receiver. Compared to ODB, several benefits can be found in the usage of EDB. The EDB scheme enables the usage of components and packaging technologies with lower bandwidth, reducing link cost and power consumption. Moreover, the ODB transmitter requires an increased complexity (e.g., a dual-drive MZM) compared to more conventional and simple intensity modulation in EDB. However, for the ODB receiver, a single-threshold receiver can be used, yet requiring an increased bandwidth, compared to a two-threshold receiver for EDB. In terms of CD tolerance, ODB shows a better performance than EDB. However, for the considered short ranges (or in case of using the zero-dispersion wavelength window), the reduced complexity of EDB modulation offers a larger benefit than the limited increase in CD tolerance of the ODB.

In the rest of the chapter, we will focus on three-level duobinary or EDB schemes for high-speed optical interconnects. First, we briefly review and compare the characteristics of NRZ, EDB, and PAM-4 in this section.

8.2.1 Nyquist Frequency

A common way to compare different modulation formats is based on Nyquist frequency (f_N). This is the highest sinusoid frequency component transmitted using a certain modulation format. For NRZ and PAM-4, these are well known and equal to 1/2 and 1/4 of the bit rate, respectively. As shown in Figure 8.2, for EDB, the highest sinusoidal signal frequency is dependent on the signal swings. The f_N reaches one-third of the bit rate when the signal swing of the Nyquist component is only two-third of the full signal swing. However, the highest frequency component of an EDB signal that reaches the full signal swing is, similar to PAM-4, equal to one-fourth of the data rate.

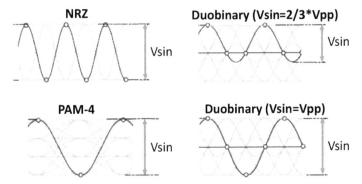

Figure 8.2 Highest sinusoid frequency component using non-return to zero (NRZ), four-level pulse-amplitude modulation (PAM-4), and electrical duobinary (EDB).

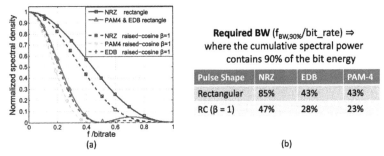

Figure 8.3 (a) Power spectral density and (b) required bandwidth of NRZ, EDB, and PAM-4 for both rectangular and raised-cosine (RC) $\beta = 1$ pulse shaping.

8.2.2 Power Spectral Density

Another common measure to express the required bandwidth is the power spectral density. Figure 8.3a shows the calculated power spectral density of NRZ, EDB, and PAM-4 signals, using different pulse shapes. It is clear from Figure 8.3 that the EDB and PAM-4 need considerably less bandwidth compared to NRZ independent of the pulse shape. When comparing EDB to PAM-4, the bandwidth is identical for a rectangular pulse shape and slightly higher for EDB when using a raised-cosine pulse. In order to quantify the required bandwidth, the frequency where the cumulative spectral power contains 90% of the bit energy is defined as the required bandwidth and this is expressed relative to the bit rate frequency. This occurs at 43% of the bit rate for both PAM-4 and EDB and 85% of the bit rate for NRZ assuming a rectangular pulse shape (Figure 8.3b). With a raised-cosine $\beta=1$ pulse shape, the 90% bit energy is located at 47% of the bit rate for NRZ and at 28% and 23% for EDB and PAM-4, respectively.

8.2.3 Vertical and Horizontal Eye Openings

The vertical and horizontal eye openings are important measures that show the tolerance of the modulated signal against crosstalk, jitter, and noise [15, 16]. In the horizontal direction, NRZ modulation is the most symmetrical scheme, and the ideal eye width is 100% of the bit width (= 1/data_rate). The maximal eye width of EDB is equal to NRZ; however, at the optimal sampling point, where the vertical opening of the eye is maximized, it is reduced to ~66% of the bit width. Because PAM-4 symbols are transmitted at half of the symbol rate of NRZ, one would expect that the eye width would be doubled. However, due to multiple crossings of different levels, only ~108% of the NRZ eye width is obtained at the optimal sampling point.

For the vertical eye opening comparison, there are quite a few methods that have been well established for electrical backplane applications [16–18]. For instance, one can compare different schemes by taking the ratio of the smallest eye height to the maximum signal swing and relating this to the attenuation of a backplane channel at the Nyquist frequency [17]. This approach has been extended in [16]; analytical formulas of transmission penalty can be attained by assuming a perfect linear (on log scale) loss profile of a backplane channel. However, for high-speed short-reach optical links, usually the link is not limited by fiber channels, rather by the bandwidth constraints of optical or electronic devices. Those device bandwidth constraints can be better modeled as a filter transfer function, e.g., Bessel–Thomson or Butterworth are two types of reference transfer functions commonly used in optical communication compliance standards and research. With such models, we can look at the maximum amplitude of the sinuous wave at the Nyquist frequency that occurs in the data stream, and calculate the penalty as the ratio between this maximum amplitude and the normalized eye height.

First, we look at the penalty at the Nyquist frequency without taking into account the difference of the fundamental sinusoid components (see Figure 8.2). The penalty of NRZ is 0 dB because the vertical eye height is equal to the maximum voltage swing of the sine wave at the Nyquist frequency. When looking at PAM-4, the eye height has a ratio of one-third with respect to the maximum signal swing which results in a 9.54 dB (= $-20 \cdot \log_{10}[1/3]$) penalty compared to NRZ. For EDB, the Nyquist frequency component has a smaller amplitude than the total voltage swing (i.e., 2/3), as can be seen in Figure 8.2. The loss present in the link is in fact used to partly generate the EDB waveform, which means that a certain amount of loss does not need to be compensated. As a result, the penalty of EDB compared to NRZ amounts to only 2.5 dB. The 2.5 dB represents the ratio between the vertical eye height (1/2) and the amplitude of the Nyquist component (2/3) in the EDB signal, i.e., 2.5 dB $= -20 \cdot \log_{10}[3/4]$. On the other hand, at half the NRZ Nyquist frequency (equal to f_{N_PAM4}), the EDB signal has a full swing frequency component, also shown in Figure 8.2. Therefore, at low filter losses, the penalty at the f_{N_PAM4} will dominate for the EDB eye height reduction. In this case, up to f_{N_PAM4}, EDB will undergo an equal relative reduction in eye height as PAM-4. Of course, it should be remarked that the initial eye height of EDB is -6.02 dB (1/2) while the PAM-4 eyes are -9.54 dB (1/3). Hence, up to 19-dB attenuation at half the bit rate, EDB is expected to have a larger eye height than PAM-4.

There is, however, a second effect that restricts the eye height: the maximum peak-to-peak voltage (Vpp) of the transmitter, which limits the signal swing after a pre-emphasis filter. When looking at a perfect rectangular pulse train at the Nyquist frequency, the fundamental tone for an alternating bit sequence $(+1, -1, +1, -1, \ldots)$ is $4/\pi$ times higher (2.1 dB) than Vpp. As long as the loss at the NRZ Nyquist frequency is less than 2.1 dB, the eyes of an NRZ clock signal remain maximal. Ideally, PAM4 modulation also transmits rectangular pulses. As a result, a similar gain as for NRZ, between 0 and 2.1 dB, is expected. For low channel attenuation, the maximum eye height of EDB will be dominated by the attenuation around f_{N_PAM4}. In this case, the previous reasoning also holds for EDB and a gain between 0 and 2.1 dB is expected. However, with rising attenuation, the maximum eye height of EDB is determined by the attenuation around $\frac{2}{3}f_{N\_NRZ}$. In contrast to NRZ and PAM4, here the fundamental tone is generated with a $(+1, -1, -1, +1, -1, -1, \ldots)$ sequence of which the fundamental tone is $2\sqrt{3}/\pi$ times higher (0.85 dB) than Vpp.

Taking both effects into account, we are ready to calculate the eye heights of the different modulation formations. For a given filter frequency response $H(j\omega)$, we can find the relative maximum eye-height *MEH* as

Relative maximum eye height MEH

$$
= \begin{cases}
\min\left(0,\ G_{NRZ} + 2.1\right)\ \text{dB}, & for\ NRZ \\
\min\left(-9.5,\ G_{PAM4} + 2.1 - 9.5\right)\ \text{dB}, & for\ PAM4 \\
\min\left(-6,\ G_{PAM4} + 2.1 - 6,\ G_{EDB} + 0.85 - 2.5\right)\ \text{dB}, & for\ EDB
\end{cases}
$$

in which, G_{NRZ}, G_{PAM4}, and G_{EDB} are the decibel gains of the normalized frequency response $|H(j\omega f|)$ at defined frequencies

$$G_{NRZ} = 20 \cdot \log_{10}|H(j\omega_{NRZ})|, \quad \omega_{NRZ} = \frac{1}{2}2\pi f_b$$

$$G_{PAM4} = 20 \cdot \log_{10}|H(j\omega_{PAM4})|, \quad \omega_{PAM4} = \frac{1}{4}2\pi f_b$$

$$G_{EDB} = 20 \cdot \log_{10}|H(j\omega_{EDB})|, \quad \omega_{EDB} = \frac{1}{3}2\pi f_b$$

where f_b is the bit rate of transmitted modulation formats.

The presented theory allows us to evaluate, which is the optimal modulation format for a given frequency response and required data rate. For instance, we have plotted the estimated maximum eye-heights for second-and fourth-order Bessel–Thomson responses in Figure 8.4. For the fourth-order Bessel–Thomson response (Figure 8.4b), it is clear that for high bandwidth links, NRZ remains the most optimal approach and PAM-4 will be beneficial for very high loss channels. However, for a large range of applications in terms of $\frac{f_b}{f_{3dB}}$ ratio, EDB out-performs both NRZ and PAM-4 modulations. This is even more profound in case of the second-order Bessel–Thomson response (Figure 8.4a). There, EDB outperforms PAM-4 in all the practical bandwidth-limited situations, i.e., $\frac{f_b}{f_{3dB}} \leq 8$.

8.3 100-Gbps EDB/NRZ Transmitter and Receiver Chipset

High-speed optical links up to serial 100 Gbps demand careful design of both the electrical and optical devices. The design of the separate components

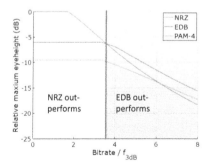 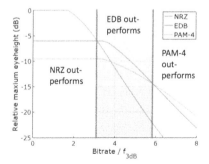

Figure 8.4 Comparison of relative maximum eye-heights of NRZ, EDB, and PAM-4: assuming the link channel response is (a) second-order Bessel–Thomson transfer function or (b) fourth-order Bessel–Thomson transfer function.

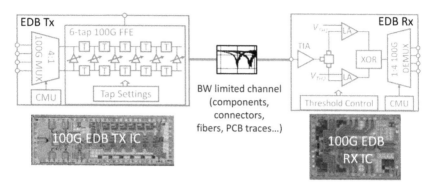

Figure 8.5   Block diagram and die photograph of SiGe BiCMOS EDB/NRZ TX and RX chipset.

will influence power and signal quality to great extents. In this section, the EDB/NRZ transceiver chipset is presented, which provides the capability to equalize the transmitted signals and decode EDB/NRZ signals in real-time.

Figure 8.5 shows the block diagram and photograph of our 100-Gbps EDB TX and RX chipset. Both EDB TX and RX chips were designed in-house and fabricated in a 0.13-um SiGe BiCMOS technology. The TX IC consists of a 4-to-1 MUX and a six-tap feed forward equalizer (FFE). The 4-to-1 MUX is capable of multiplexing four signal streams up to 25 Gbps into a serial 100-Gbps NRZ signal. Inside the TX, the high-speed master–slave flip-flops are custom designed and clocked at half the serial data rate, i.e., 50 GHz for a 100-Gbps serial rate. The six-tap FFE is designed as a tapped delay line (~10-ps delay spacing) where each gain cell can be independently tuned with a resolution of 8 bits. Ideally, a baseband cosine filter with 1/4-rate 3-dB bandwidth is needed to create duobinary signaling from an NRZ signal [19]. As shown in Figure 8.5, this cosine filter can be approximated by the bandwidth-limited channel, along with an appropriately adapted TX FFE response. As such, the real-time FFE in the TX should compensate only a part of the channel to create an equivalent overall channel response that transforms the NRZ from the MUX output into three-level duobinary at the RX input. The TX IC occupies 1555 μm × 4567 μm including IOs. At a serial rate of 100 Gbps, the chip consumes about 1 W of which 0.64 W is used to power the MUX and 0.36 W is used by the FFE. The required equalization was chosen to be implemented in the TX. This choice is driven by the ease of the input design of the FFE, which can be nonlinear due to the NRZ input and the ease of the RX TIA design where some degree of non-linearity is tolerable. The exclusion of the noise shaping at the receiver by an FFE is an additional benefit. This choice challenges the automatic optimization in a practical implemented system. For this purpose, a (low-speed) back channel will be necessary. Updates itself can be guided by an LMS engine located at the receiver [20].

To decode the EDB signal, a custom receiver with two separate threshold levels and DEMUX is used (Figure 8.5). Controlling the levels of both thresholds V_{TH1} and V_{TH2} shifts the eyes up or down resulting in an extraction of the upper and lower eyes of the EDB signal. Sending both singals through an XOR gate restores the transmitted NRZ stream under the condition of a precoded transmitted NRZ signal. The EDB RX IC consists of a TIA input stage with a bandwidth of 41 GHz, two parallel level-shifting limiting amplifiers, an XOR stage, and a 1-to-4 DEMUX. The decoding and DEMUX operations are combined to interleave the decoding step, significantly reducing power consumption. The EDB RX can be reconfigured into an NRZ RX with four demuxed outputs. This NRZ decoding is also used in our experiments for comparison with NRZ signaling. The RX IC occupies 1926 μm × 2585 μm including IOs and consumes 1.2 W at 100 Gbps. The combined electrical transmit and receive power will be 2.2 W resulting in 22 pJ/bit in the electrical domain (without CDR). At the time of writing this chapter, a new design of the transceiver chips with lower power consumption and added functionality (i.e., CDR) is under test. Compared to the 77 pJ/bit in [12] (with CDR), a clear increase in power efficiency could be achieved.

8.4 EDB/NRZ Transmission with DFB-TWEAM

In [10], we presented, for the first time, a real-time 100-Gbps EDB transmission over 2-km SSMF, employing a traveling-wave (TW) electro-absorption modulator (EAM) with an integrated distributed feedback (DFB) laser. The DFB-TWEAM was grown by metal organic vapor phase epitaxy (MOVPE) on n-doped InP substrate [21]. The gain section of the laser consists of seven quantum wells (QWs), each 7-nm thick, whereas the modulator has 12 QWs with a thickness of 9 nm. The two components are integrated using a butt-joint technique. The DFB-laser exhibits a threshold of 25 mA and an output slope of 0.4 W/A, which allows an output power of 3.2 mW with 120-mA driving current. The integrated DFB laser emitted at 1,548.7 nm in the experiment.

The experimental setup is shown in Figure 8.6. A Xilinx FPGA board is adopted to generate four electrical 25-Gbps NRZ shifted 2^7-1 PRBS signal streams. The EDB TX IC multiplexes 4×25 Gbps streams into a

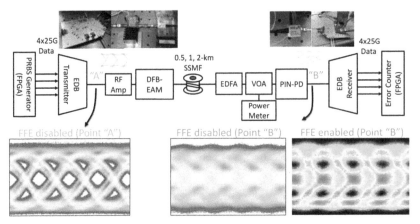

Figure 8.6 Experimental setup of real-time 100-Gbps EDB optical link and eye-diagrams with FFE disabled and enabled.

serial 100-Gbps stream. To obtain a serial stream which is again the same periodic $2^7 - 1$ PRBS signal, the four streams are shifted with a delay of, respectively, 0, 32, 64, and 96 bits. The TX IC pre-equalizes the serialized NRZ signal with the analog equalizer discussed in Section 1.3. The pre-emphasized signal was amplified by a 50-GHz amplifier and used to drive the TWEAM. The optical output power of the EML was approximately 0 dBm. After transmission over an SSMF, the received optical signal was detected by a PIN-photodiode (PIN-PD) and the custom EDB RX IC. An erbium doped fiber amplifier (EDFA), a variable optical attenuator (VOA), and a power meter were used before the PIN-PD to adjust/record the received optical power for measurement purposes. The EDFA can be removed from the setup when a higher power EML becomes available. The PIN-PD is a high-speed InP-based O/E converter packaged prototype photodetector from U^2T with a responsivity of 0.5 A/W and a bandwidth >67 GHz (limited by measurement bandwidth of the available network analyzer). The single-ended output of the diode is directly connected to the RX. The resulting received maximum peak-to-peak voltage was around 90 mVpp which is high enough to be detected by the RX IC. The RX has a maximum input range of ±250 mVpp such that a 9-dB extra-linear amplification can be useful. However, due to bandwidth limitations of the available amplifier, inferior performance is observed and this option is not considered. The RX IC demodulates three-level signals with two separate threshold levels as discussed in the previous section. The levels are independently tuned to obtain the lowest bit error rate (BER). Due to the

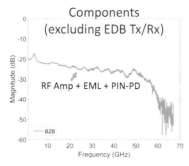
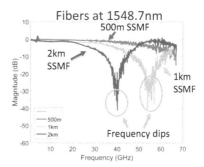

Figure 8.7 Measured frequency responses of (a) E/O/E components excluding the EDB ICs and PCBs, and (b) the optical channel response of fibers with different lengths.

absence of CDR circuitry in the RX IC, a delayed version of the TX half-rate clock is used to demodulate the signal. The delay is tuned manually for each BER measurement to obtain the lowest BER.

The demodulated signal is de-serialized on-chip into 4×25 Gbps NRZ outputs which can be used for real-time error detection. The error detection is implemented on the same FPGA board used for the PRBS generation by comparing the incoming data to a locked PRBS reference stream. Only one of the four outputs of the RX IC is used for error detection in this experiment. Due to the properties of (de)multiplexed PRBS streams, the measured BER will be representative for the full rate BER.

The measured frequency responses of the E/O/E components excluding the EDB ICs and PCBs are shown in Figure 8.7a. The frequency responses of fibers with a length of 500 m, 1 km, and 2 km are measured and plotted in Figure 8.7b. As can be seen in Figure 8.7a, in the optical back-to-back (B2B) case, the bandwidth is mainly limited by the electrical amplifiers, interfaces, and ICs. However, SSMF operating in the C-band severely degenerates the flatness of the frequency response, especially at 2 km showing the increased challenges compared to [8].

The tap pulse responses were measured and used for tap coefficient optimization by fitting the FFE response to the ideal duobinary impulse response, taking the channel response into account. Figure 8.6 shows the electrical 100-Gbps NRZ eye-diagram after the TX IC without FFE (interface point "A"), which is clearly open. However, the received optical eye at the PIN-PD (interface point "B") was completely closed due to the bandwidth limitation as indicated in Figure 8.7. By enabling the real-time six-tap FFE in the EDB TX IC, the received optical eye has been shaped into three-level duobinary and two separate eyes were clearly formed after the PIN-PD. For fair comparison, the TX FFE is always enabled in the experiment and separately optimized for each rate.

The measured BER curves for 50-and 70-Gbps NRZ signals are shown in Figure 8.8b. The NRZ signaling works relatively well in B2B up to 70 Gbps. However, after 2 km of SSMF, we noticed a decreased performance at 70 Gbps [4.4 dB penalty compared to the B2B at 7% hard-decision forward error correction (HD-FEC) limit]. The different NRZ eye-diagrams for 70-Gbps communication are shown in Figure 8.8a. The longer fiber lengths with the resulting CD introduce significant inter-symbol interference (ISI) and close the NRZ eye gradually. In contrast, we can utilize this ISI effect and create the three-level EDB signal by optimizing the TX FFE. The measured BER for the 70-Gbps EDB link over 2 km can reach a BER of 1E-10, which is well below the HD-FEC with 7% overhead (BER=3.8E-3 [22]) or RS (544, 514, 10) FEC (BER=3.09E-4), an FEC code defined for 802.3bs 400-Gbit Ethernet Task Force [1, 23]. In addition, using EDB modulation, we obtained a sensitivity improvement of 3.3 dB at 7% HD-FEC limit [or 4.5 dB at the RS (544, 514, 10) limit] with respect to the 70-Gbps NRZ transmission over 2-km distance. The optimized three-level duobinary eye in Figure 8.8b for the 2-km fiber shows the two open eyes.

Final, 100-Gbps EDB measurements were performed with different fiber lengths. When the rate is increased to 100 Gbps, NRZ communication becomes nearly impossible due to the bandwidth limitations of the cascade as shown in Figure 8.7. In case of EDB, the resulting eye-diagrams for B2B, 500-m, 1-km, and 2-km SSMFs are depicted in Figure 8.9a. The real-time

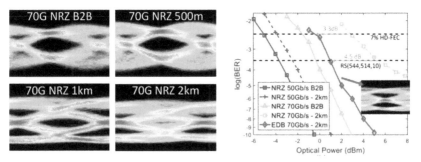

Figure 8.8   Measured (a) eye-diagrams of 70-Gbps NRZ transmission, and (b) NRZ bit-error-rate (BER) curves up to 70-Gbps and 2-km standard single-mode fiber (SSMF), compared to 70-Gbps EDB modulation.

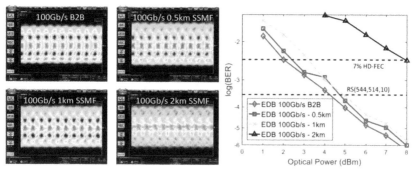

Figure 8.9 Measured (a) eye-diagrams of 100-Gbps EDB transmission, and (b) real-time BER curves for back-to-back (B2B), 500-m, 1-km, and 2-km SSMF at 100 Gbps.

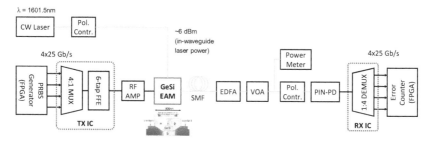

Figure 8.10 Experimental setup of real-time 100-Gbps EDB optical link with an integrated GeSi EAM implemented on a silicon photonics platform.

100-Gbps EDB BER measurements are shown in Figure 8.9b. For 500-m applications, negligible penalty was observed as could be expected from the measured frequency response. Up to 1-km SSMF, for powers higher than 5 dBm, a BER below the RS (544, 514, 10) FEC threshold is achieved. Over 2-km SSMF, the CD starts to introduce extra losses at frequencies below 50 GHz which influence the EDB performance significantly. Nonetheless, a 100-Gbps EDB modulated signal can be received with a BER of 3.7E-3, below the 7% HD-FEC threshold.

8.5 NRZ-OOK Transmission with GeSi EAM

In a follow-up experiment, the TX and RX ICs were used with a lumped GeSi EAM in a 200-mm silicon photonics platform for a 100-Gbps NRZ-OOK link [2]. Compared to the TW EAM, the lumped EAM removes the required 50-Ω termination and can further improve the power efficiency of the link.

The experimental setup is shown in Figure 8.10. The GeSi EAM consists of a 600-nm-wide and 80-μm-long germanium waveguide with embedded lateral p-i-n junction, connected via tapers to silicon waveguides, and was fabricated on imec's 200-mm silicon photonics platform. Modulation is obtained through the Franz–Keldysh effect, which shifts the bandgap edge of GeSi by applying an electrical field [24]. Light is coupled in and out of the waveguide device by fiber-to-chip grating couplers which have an insertion loss of ~6 dB/coupler. An electrical RF probe (without 50-Ω termination) is used to apply the bias voltage and high-speed signal. The same EDB/NRZ TX-RX chipset has been used to generate and receive the 100-Gbps NRZ-OOK signal.

The measured eye-diagrams and BER curves are shown in Figure 8.11. For B2B transmission, the hard-decision forward error coding limit (BER = 3.8E-3 for a 7% overhead) is reached for an average power of −0.6 dBm (BER < 6E-9 is achievable at higher input power). Since the GeSi EAM is operating in the L-band (1,601.5 nm), the resulting frequency responses for longer fiber spans are degraded severely due to high levels of CD. Nevertheless, we still manage to obtain transmission over more than 500 m of SSMF with a BER down to 2E-5. Sub-FEC operation is achieved for powers >1.5 dBm, resulting in a penalty of 2 dB with respect to B2B. Finally, transmission over 2 km of dispersion shifted fiber (DSF, with ~8 ps/nm.km dispersion) assuming 7% HD-FEC is also achieved, saturating in an error-floor of 1E-3.

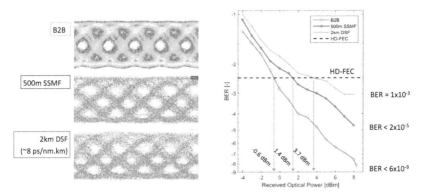

Figure 8.11   Measured (a) eye-diagrams after the PIN-PD, and (b) BER curves at B2B, 500-m SSMF, and 2-km dispersion shifted fiber (DSF).

8.6 SM LW-VCSEL EDB Links

Another interesting optical transmitter for future HPC/datacenters is based on long-wavelength (LW) SM VCSELs. It offers potentially an energy-efficient and future-proof solution for >500-m applications, and operating at LW facilitates using wavelength division multiplexing to reduce fiber counts. So far, the highest reported data rate for real-time SM VCSEL links is ~56 Gbps NRZ [25, 26], which is still lower than the multimode (MM) counterpart (e.g., 71-Gbps NRZ [27]). This is also confirmed by the measured small-signal bandwidth of such LW SM VCSELs, i.e., max. 22 GHz [25] versus 30 GHz for MM VCSELs [28]. To study the effectiveness of the proposed TX FFE and EDB modulation, we have evaluated our EDB transceiver chipset with an 18-GHz SM VCSEL (1,544 nm). The InP-based VCSEL uses an ultra-short semiconductor cavity (~1.5 μm) and two distributed Bragg reflectors (DBRs) to enhance its bandwidth. The test setup is shown in Figure 8.12. The measured frequency response showed that the link bandwidth is limited by the VCSEL and the real-time BER with PRBS7 at 80 Gbps was below 7% HD-FEC threshold. This is the highest data rate demonstrated in real-time for an SM LW-VCSEL without complex offline DSP [29].

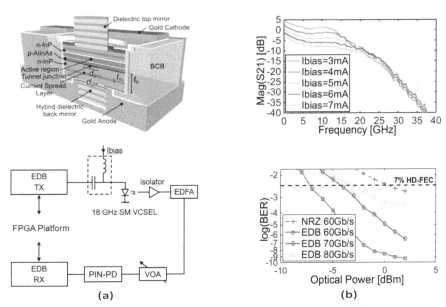

Figure 8.12 (a) Experimental setup of the VCSEL link and (b) measured frequency response and BER.

8.7 Conclusion

Recent advances in integrated opto-electronic devices and frontend circuits have made it possible to efficiently transmit very high data rates over optical links for HPC/datacenter applications. This chapter reviews our current progress toward ultrahigh-speed optical interconnects, with an emphasis on three-level duobinary (i.e., EDB) modulation. At a data-rate of 100 Gbps, transmission of EDB over various fiber lengths is demonstrated. Over 2-km SSMF, a BER below the 7% HD-FEC threshold is achieved in real-time. Using the same EDB/NRZ TX-RX chipset, we have also demonstrated 100-Gbps NRZ compatibility with a silicon photonic EAM, and a 80-Gbps EDB link with a low-cost SM LW-VCSEL-based transmitter. All signal processing is done on-chip in real-time and without any offline DSP, which proves the possibility for power-efficient ultra-high serial rate optical links.

Acknowledgment

The authors would like to thank the support from UGent IOF/BOF, the FWO, and the European Commission for Optical Interconnects related projects such as FP7 ICT MIRAGE, PhoxTrot, H2020 ICT STREAMS, WIPE, PICTURE, and Teraboard. The authors also thank various project partners such as CMST and Optical IO group of imec, BiFAST, KTH, RISE, TUM and Nokia Bell-Labs for their collaboration and valuable inputs.

References

[1] "IEEE P802.3bs 400 Gigabit Ethernet Task Force." [Online]. Available: http://www.ieee802.org/3/bs/

[2] Verbist, J., Verplaetse, M., Srivinasan, S. A., De Heyn, P., De Keulenaer, T., Pierco, R., et al. (2015). "First real-time 100-Gbps NRZ-OOK transmission over 2 km with a silicon photonic electro-absorption modulator," *in Proceedings of the Optical Fiber Communication Conference (OFC)*, Valencia.

[3] Ozolins, O., Olmedo, M. I., Pang, X., Gaiarin, S., Kakkar, A., Udalcovs, A., et al. (2016). "100 GHz EML for high speed optical interconnect applications," *in Proceedings of the European Conference on Optical Communication (ECOC)*, Dusseldorf.

[4] Sadot, D., Dorman, G., Gorshtein, A., Sonkin, E., and Vidal, O. (2015). "Single channel 112 Gbit/sec PAM4 at 56 Gb aud with digital signal

processing for data centers applications," *in Proceedings of the Optical Networking and Communication Conference*, Los Angeles, CA.

[5] Yang, C., Hu, R., Luo, M., Yang, Q., Li, C., Li, H., et al. (2016). IM/DD-based 112-Gbps/lambda PAM-4 transmission using 18-Gbps DML. *IEEE Photon. J.* 8, 1–7.

[6] Kai, Y., Nishihara, M., Tanaka, T., Okabe, R., Takahara, T., Rasmussen, J. C., et al. (2015). "130-Gbps DMT transmission using silicon Mach-Zehnder Modulator with chirp control at 1.55-μm," *in Proceedings of the Optical Fiber Communications Conference and Exhibition (OFC)*, Los Angeles, CA.

[7] Zhong, K., Zhou, X., Gui, T., Tao, L., Gao, Y., Chen, W., et al. (2015). Experimental study of PAM-4, CAP-16, and DMT for 100 Gbps short reach optical transmission systems. *Opt. Exp.* 23, 1176–1189.

[8] Lee, J., Kaneda, N., Pfau, T., Konczykowska, A., Jorge, F., Dupuy, J. Y., et al. (2014). "Serial 103.125-Gbps transmission over 1 km SSMF for low-cost, short-reach optical interconnects," *in Proceedings of the Optical Fiber Communications Conference and Exhibition (OFC)*, San Francisco, CA.

[9] Yin, X., Blache, F., Moeneclaey, B., Van Kerrebrouck, J., Brenot, R., Coudyzer, G., et al. (2016). "40-Gbps TDM-PON downstream with low-cost EML transmitter and 3-level detection APD receiver," *in Proceedings of the Optical Fiber Communications Conference and Exhibition (OFC)*, Anaheim, CA.

[10] Yin, X., Verplaetse, M., Lin, R., Van Kerrebrouck, J., Ozolins, O., De Keulenaer, T., et al. (2016). "First demonstration of real-time 100 Gbit/s 3-level duobinary transmission for optical interconnects," *in Proceedings of the European Conference on Optical Communication (ECOC)*, Dusseldorf, 1–3.

[11] Xiao, X., Li, F., Yu, J., Xia, Y., Chen, Y., et al. (2014). "Real-time demonstration of 100 Gbps class dual-carrier ddo-16qam-dmt transmission with directly modulated laser," *in Proceedings of the Optical Networking and Communication Conference*, San Francisco, CA.

[12] Lee, J., Shahramian, S., Kaneda, N., Baeyens, Y., Sinsky, J., Buhl, L., et al. (2015). "Demonstration of 112-Gbit/s optical transmission using 56 GB aud PAM-4 driver and clock-and-data recovery ICs," *in Proceedings of the European Conference on Optical Communication*, Valencia.

[13] Lender, A. (1963). The duobinary technique for high-speed data transmission. *Trans. Am. Inst. Electr. Eng., Part I: Comm. Electron.* 82, 214–218.

[14] Yonenaga, K., Kuwano, S., and Norimatsu, S. (1995). Optical duobinary transmission system with no receiver sensitivity degradation. *Electron. Lett.* 31, 302–304.

[15] Yin, X., Van Kerrebrouck, J., Coudyzer, G., and Bauwelinck, J. (2016). "Multi-level high speed burst-mode receivers," *in Proceedings of the 2016 21st OptoElectronics and Communications Conference (OECC) held jointly with 2016 International Conference on Photonics in Switching (PS),* Niigata, 1–3.

[16] Van Kerrebrouck, J., et al. "High-speed electrical interconnects: NRZ, duobinary or PAM4?," accepted in *IEEE Microw Mag.*

[17] Yamaguchi, K., Sunaga, K., Kaeriyama, S., Nedachi, T., Takamiya, M., Nose, K., et al. (2005). "12Gbps duobinary signaling with x2 oversampled edge equalization," *in Proceedings of the ISSCC 2005 IEEE International Digest of Technical Papers. Solid-State Circuits Conference*, 70–71.

[18] Zhang, G., Zhang, H., Asuncion, S., and Jiao, B. (2016). *A Tutorial on PAM4 Signaling for 56G Serial Link Applications*. Santa Clara, CA: DesignCon, 19–21.

[19] Moeneclaey, B. et al. (2017). 40-Gbps TDM-PON downstream link with low-cost EML transmitter and APD-based electrical duobinary receiver. *J. Lightwave Technol.* 35, 1083–1089.

[20] Verplaetse, M. De Keulenaer, T., Vyncke, A., Pierco, R., Vaernewyck, R., Van Kerrebrouck, J., et al. (2017). "Adaptive transmit-side equalization for serial electrical interconnects at 100 Gbps using duobinary," *in Proceedings of the IEEE Transactions on Circuits and Systems I: Regular Papers*, 64, 1865–1876.

[21] Chacinski, M., Westergren, U., Stoltz, B., Thylén, L., Schatz, R., and Hammerfeldt, S. (2009). Monolithically integrated 100 GHz DFB-TWEAM. *J. Lightwave Technol.* 27, 3410–3415.

[22] ITU-T Recommendation (2014). "*ITU-T Recommendation G.975.1,*" Appendix I.9. Available at: http://handle.itu.int/11.1002/1000/7069

[23] "Investigation on technical feasibility of stronger RS FEC for 400GbE." [Online]. Available: http://www.ieee802.org/3/bs/public/15_01/wang_x_3bs_01a_0115.pdf

[24] Srinivasan, S. A., Verheyen, P., Loo, R., De Wolf, I., Pantouvaki, M., Lepage, G., et al. (2016). "50Gbps C-band GeSi waveguide electro-absorption modulator," *in Proceedings of the Optical Fiber Communication Conference, OSA Technical Digest (online) (Optical Society of America, 2016)*, Tu3D.7.

[25] Spiga, S., Soenen, W., Andrejew, A., Schoke, D. M., Yin, X., Bauwelinck, J., et al. (2017). Single-mode high-speed 1.5-μm VCSELs. *J. Lightwave Technol.* 35, 727–733.

[26] Kuchta, D. M., Huynh, T. N., Doany, F. E., Schares, L., Baks, C. W., Neumeyr, C., et al. (2016). Error-free 56 Gbps NRZ modulation of a 1530-nm VCSEL link. *J. Lightwave Technol.* 34, 3275–3282.

[27] Kuchta, D. M., Rylyakov, A. V., Doany, F. E., Schow, C. L., Proesel, J. E., Baks, C. W., et al. (2016). "70+Gbps VCSEL-based multimode fiber links," *in Proceedings of the 2016 IEEE Compound Semiconductor Integrated Circuit Symposium (CSICS)*, Austin, TX, 1–4.

[28] Haglund, E., Westbergh, P., Gustavsson, J. S., Haglund, E. P., and Larsson, A. (2016). "High-speed VCSELs with strong confinement of optical fields and carriers," *in Proceedings of the IEEE Journal of Lightwave Technology*, 34, 269–277.

[29] Yin, X., Verplaetse, M., Breyne, L., Van Kerrebrouck, J., De Keulenaer, T., Vyncke, A., et al. (2017). "Towards efficient 100 Gbps serial rate optical interconnects: A duobinary way," *in Proceedings of the 2017 IEEE Optical Interconnects Conference (OI)*, Santa Fe, NM, 33–34.

9

LiNbO$_3$ Mach-Zehnder Modulator

Hirochika Nakajima[1] and Yuya Yamaguchi[2]

[1]Waseda University, Tokyo, Japan
[2]National Institute of Information and Communications Technology (NICT), Tokyo, Japan

9.1 Introduction

This chapter describes the basic principles and present industrial features of LiNbO$_3$ (LN) Mach–Zehnder Modulator (MZM) which has been success-fully operated in long-distance and high-capacity optical fiber transmission systems for over 30 years.

9.2 Physical Properties of LN (LiNbO$_3$) Crystal

The LN ferroelectric crystal with pseudo-ilmenite structure – point group 3m and trigonal crystal structure – was first synthesized in 1949 [1]. Figure 9.1 shows a schematic layout of the Li and Nb atoms surrounded by the oxygen octahedron along the c-axis which demonstrates strong electric polarization. LT (LiTaO$_3$) can be obtained when Ta replaces Nb and demonstrates similar ferroelectric and electro-optic properties to LN. Table 9.1 shows the basic physical properties of LN/LT.

In 1965, LN/LT single-crystal growth was first demonstrated using the Czochralski (Cz) method [2] similar to Si and III–V compound semicon-ductor crystal growth. However, as-grown crystal bowls have a random ferroelectric domain; therefore, the so-called poling process should be fol-lowed by the application of a DC electric field along the c-axis of the crystal near its Currie temperature (T_c). LT has a relatively low T_c of approximately 600°C compared to more than 1,100°C for LN as shown in Table 9.1. This difference is exceedingly important because it is the reason why we have not yet obtained LT-MZM instead of LN-MZM, which can maintain its single domain during Ti diffusion even at high temperatures of approximately 1,000°C as mentioned in the subsequent section.

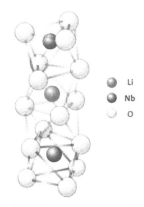

Li
Nb
O

Figure 9.1   Crystal structure of LiNbO$_3$.

Table 9.1   Physical properties of LN/LT

		LN	LT
Lattice Constant [nm]	a-axis	1.3864	1.3781
	c-axis	0.51499	0.51543
Melt temperature [°C]		1257	1650
Currie temperature [°C]		113	600
Density [g/cm^3]		4.65	7.46

LiNbO$_3$ (LT) has a strong electric–mechanic interactive coefficient similar to quartz; therefore, it is easy to generate surface acoustic wave (SAW). The SAW device consumer electronics markets such as clock generator, analog TV tuner, and mobile phone filter have been successfully created and expanded since the 1970s. Today, we have numerous SAW filters in our smart phones, tablets, and PCs. Figure 9.2 shows LN single-crystal bowls manufactured by a Japanese LN/LT crystal supply company in the year of around 2000.

LiNbO$_3$ has an exceedingly unusual phase diagram [3] for the solid/liquid interface as shown in Figure 9.3. Generally, the melt source composition is selected as the so-called "congruent" composition with Li:Nb = 48.5:51.5 because this ratio is maintained through the Cz growth process. This imbalance from the stoichiometric composition results in the development of intrinsic crystal defects [4], which may assist the metallic Ti diffusion into the LN bulk as described in the subsequent section.

LiNbO$_3$ and LiTaO$_3$ possess satisfactory transparency in the visible and near infrared wavelength regions and are significantly useful in several optical applications including non-linear optics. Table 9.2 shows their anisotropic

Figure 9.2 Bowls of LN (6" and 3" diameter, courtesy of YAMAJU CERAMIKS CO., LTD.).

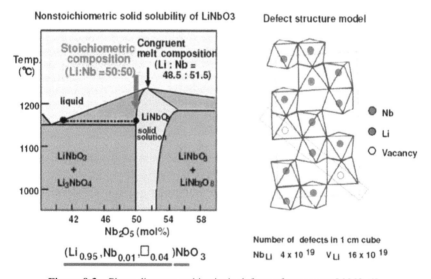

Figure 9.3   Phase diagram and intrinsic defects of congruent LN [3, 4].

optical properties. They have essentially polarization-dependent characteristics even for bulk applications. They demonstrate a relatively large Pockels effect as a linear electro-optic effect [5]. The refractive index change under an electric field **E** is formulated as shown in Equation (9.1)

$$\Delta n = \frac{1}{2} n^3 r_{ij} E \qquad (9.1)$$

Table 9.2   Optical properties of LN/LT

	LN	LT
n_e	2.2	2.18
n_o	2.286	2.176
r_{33}[pm/V]	30.8	35.8
r_{13}[pm/V]	8.6	7.9
r_{51}[pm/V]	28	20
r_{22}[pm/V]	3.4	1

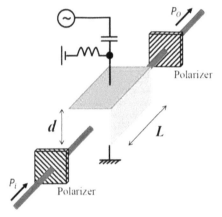

Figure 9.4   Optical intensity modulators using LN single crystals.

Here, the Pockels coefficient r_{ij} is a tensor, and the maximum one is the r_{33} ($\approx$30 pm/V), which is often used in both bulk and waveguide.

In the 1960s, LN/LT bulk crystals with electrodes were used as optical phase/intensity modulators for CW gas lasers such as He–Ne (Figure 9.4). In such a configuration, the half-wave voltage V_π is given as Equation (9.2). Here, λ_0 is an incident light wavelength in vacuum.

$$V\pi = \frac{d\lambda_0}{Lrn^3} \tag{9.2}$$

We can easily calculate and observe that V_π exceeds multi-100 V; therefore, the driving electric circuit and modulation frequency bandwidth are limited.

Problem (a)

Further, during laser light beam irradiation, photorefractive phenomenon occurs in the crystal that disturbs the laser beam collimation through an LN crystal.

Problem (b)

Moreover, long-term application of a DC electric field somewhat drifts the modulation characteristics exceedingly slowly.

Problem (c)

These three critical problems prevent the practical use of an LN bulk optical modulator for visible lasers.

9.3 Low-loss Ti-diffused Waveguides on LN Since 1974

Significant reduction of $V\pi$ in electro-optic modulators was achieved via the introduction of the waveguide structure on LN [6] in the mid-1970s to solve *problem (a)* described in the previous section and for realization of the concept of "integrated optics" [7]. Moreover, a relatively simple waveguide fabrication method was invented at the same time in the Bell Labs. Specifically, attempts were made to diffuse certain metals from the LN crystal surface to the bulk at high temperatures $(< T_c)$ to increase the refractive index similar to a core of a single-mode optical fiber. Among several efforts, metallic Ti was observed to be the most suitable source for thermal in-diffusion for LN crystals [8] and even now this technique has been continuously used in developing low-loss channel waveguides in the LN-MZM wafer process.

Refractive indices n_e and n_o of LN increase with Ti doping [9]. The increasing ratio of the refractive index owing to Ti diffusion is less than 0.1%. Therefore, narrow fabricated waveguides with less than 10 μm width readily match to the single-mode condition for 1.5 μm wavelengths. The thermal diffusion condition to obtain the single-mode waveguides depends on both the crystal cut of the wafer and the polarization of the input lightwave.

However, strong anisotropy and essential intrinsic defects of the congruent LN crystal structure shown in Figure 9.1 suggest that diffusion constants also have strong anisotropy which makes remarkable differences in the index profile and diffusion conditions between Z-cut and X-cut of the LN wafers. Specifically in the X-cut, a diffusion coefficient to the horizontal direction is larger than one for the depth direction. Therefore, the refractive index profile of the X-cut is relatively flat compared with the one for the Z-cut.

Here, LiO_2 is out diffused from the crystal surface under high temperatures of approximately 1,000°C during the process of Ti diffusion. To eliminate this out-diffusion phenomenon [10], bubbling wet carrier gas is

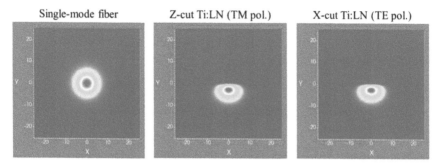

Single-mode fiber   Z-cut Ti:LN (TM pol.)   X-cut Ti:LN (TE pol.)

Figure 9.5   Typical mode profile of SMF and Ti:LN waveguides.

often used throughout the Ti diffusion process [11]. Selecting an appropriate Ti stripe width and thickness along with other diffusion parameters, Ti:LiNbO$_3$ waveguides can readily demonstrate satisfactory single-mode propagation characteristics at 1.5-μm wavelength regions with a low propagation loss of 0.1 dB/cm and good matching of the optical field profile with both conventional single-mode fibers (SMFs) and polarization-maintaining fibers (PMFs), as shown in Figure 9.5. Therefore, a satisfactory coupling loss of less than 1 dB/facet is readily obtained in both direct and lens coupling of fiber/waveguide, respectively [12]. Moreover, the optical damage free at C-/L-band wavelength region solves *problem (b)*.

9.4 Mach–Zehnder (MZ) Guided-wave Circuit with Y-branches on LN

Since 1974, several waveguide elements and their combinations have been studied by numerous authors/organizations to realize functional devices including optical switch/modulators [13–20]. Table 9.3 shows a schematic at a glance of the typical waveguide elements and circuits.

The guided lightwave propagating in all elements and circuits shown in Table 9.3 can be analyzed using the beam propagation method (BPM) with paraxial approximation because of low Δn and small angle of propagation direction change. If we intend to connect two waveguides on different axes, a bending waveguide is required. A bending radius of over 10 mm is generally used in simple Ti-diffused waveguides on LN because of small increase in the refractive index. It is one of the reasons for the large chip length of LN devices. Each circuit element demonstrates the following characteristics:

Table 9.3 Elements and circuit components of Ti:LN

	Elements	Circuits
	1. Straight line	
	2. Y-branch	
Directional Coupler	3. Reversal $\Delta\beta$	
	4. Complete coupling	
	5. Mach-Zehnder with Y-branch	
	6. Balanced-bridge	
	7. Symmetric X	
	8. Asymmetric X	

(1) A straight waveguide is the most basic element and constructs a phase modulator along with an electrode, but could not operate as an intensity modulator or a spatial/time switch by itself. Therefore, the use of functional circuits with multi-waveguides has been attempted.

(2) The most simple method to divide an input guided light power is using Y-branch. Directional coupler also possesses a 3-dB coupler function selecting a critical condition. Directional couplers with electrodes were used as SW/modulators in the early R&D era. At first, (3) $\Delta\beta$-reversal type switch was attempted to realize bar and cross states without complete coupling length information. However, crosstalk characteristics were essentially poor. Therefore, (4) complete coupling length type was attempted under relatively critical design and fabrication conditions compared with $\Delta\beta$-reversal type. They were also used in early demonstration of external modulation experiments.

(5) With regard to only intensity modulator, a one-port I/O interferometer circuit with two Y-branches and electrodes was introduced. This is an original MZ circuit.

(6) Balanced bridge type with two-port I/O interferometers was also attempted as an MZ modulator possessing two 3-dB directional couplers instead of Y-branch.

(7) Symmetric X and (8) asymmetric X were studied as elements for space switch matrices.

Figure 9.6 shows the analogically explained principle and schematics of (a) spatial Mach–Zehnder interferometers and (b) waveguide MZ mentioned previously as (6) followed by (c) the guided-wave interference mechanism for brief and essential understanding.

In 1988, F. Koyama and K. Iga demonstrated that Y-branch MZI has the least chirping parameter α compared with others as shown in Table 9.4 [22]. This means the significant advantage potential against large fiber dispersions over several 1,000 km with multi-Gbps at that time in the format of NRZ IM-DD, now OOK. Subsequently, almost all LN-MZ external modulators have been based on the Y-branch MZ structure for long-distance and high-bit-rate transmission applications [23].

We can describe the static modulation curve for applied voltage as typically shown in Figure 9.7. This is formulated by the following:

Here, we consider the input lightwave as $E_{in} \cos(w_0 t)$. After splitting using a front Y-branch of the Mach–Zehnder interferometer, the lightwaves E_1 and E_2 in the two optical paths are described as

$$\begin{cases} E_1 = \frac{1}{\sqrt{2}} E_{in} \cos(\omega_0 t + \phi) \\ E_2 = \frac{1}{\sqrt{2}} E_{in} \cos(\omega_0 t - \phi) \end{cases} \tag{9.3}$$

where we assume the push–pull type modulator, which means that the same amount but inverse sign of the induced phase changes is provided to the two paths or arms, and ϕ is the induced phase change owing to the application of voltage based on the Pockels effect. The output from the modulator calculated as a summation of Equation (9.3) by the rear Y-branch as a 3-dB coupler is described as

$$E_{out} = \cos\phi \cdot E_{in} \cos(\omega_0 t) \tag{9.4}$$

Thus, we observe that the MZMs can control the amplitude of the lightwaves. In addition, MZM can also provide binary phase changes as a sign inversion of the lightwaves. In the intensity domain, the output is expressed by

$$I_{out} = \left| E_{out}^2 \right| = \cos^2\phi \cdot \left| E_{in}^2 \right| \tag{9.5}$$

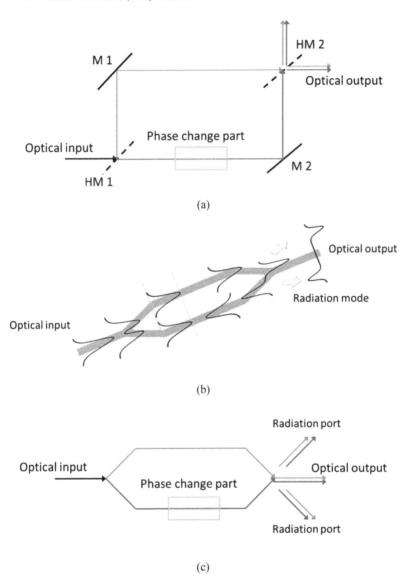

Figure 9.6   (a) Spatial Mach–Zehnder (MZ) interferometer and (b) waveguide MZ interferometer with Y-branches followed by (c) the illustrative interference mechanism of (b) as an analogy for (a).

Table 9.4   Superior chirp parameters of MZ compared to others [21]

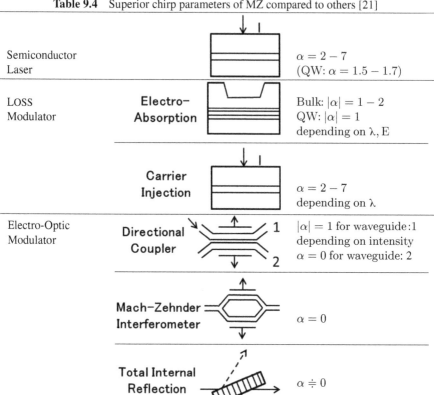

Semiconductor Laser		$\alpha = 2 - 7$ (QW: $\alpha = 1.5 - 1.7$)
LOSS Modulator	Electro– Absorption	Bulk: $|\alpha| = 1 - 2$ QW: $|\alpha| = 1$ depending on λ, E
Electro-Optic Modulator	Carrier Injection	$\alpha = 2 - 7$ depending on λ
	Directional Coupler	$|\alpha| = 1$ for waveguide:1 depending on intensity $\alpha = 0$ for waveguide: 2
	Mach–Zehnder Interferometer	$\alpha = 0$
	Total Internal Reflection	$\alpha \doteq 0$

Here, we assume that the phase change ϕ is given by the electrical signal V as

$$\phi = \frac{a}{2} \cdot V \sin(\omega_m t) - \pi/4 \qquad (9.6)$$

where a is a constant; under the condition of quadrature bias point, the modulated lightwave is described as

$$I_{out} = \frac{1}{2} + \frac{1}{2} \sin(a \cdot V \sin(\omega_m t)) \qquad (9.7)$$

In small-signal modulation, we can apply the approximation of $\sin \theta \approx \theta$, and then the output is

$$I_{out} = \frac{1}{2} + \frac{1}{2}(a \cdot V \sin(\omega_m t)) \qquad (9.8)$$

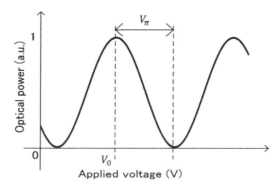

Figure 9.7　Modulation characteristics of LN-MZM (static).

Thus, the electrical signal waveform is transferred to the intensity of the lightwave. As shown in Figure 9.7, the voltage required to change the output power from minimum to maximum is called half-wave voltage ($V\pi$).

9.5 Velocity Matching Between Lightwave and Electric Signal

Almost all digital optical fiber communications require bandwidths from DC up to several tenths of gigahertz, also valid for traveling-wave electrodes in LN-MZM. A conventional LN-MZM has an electrode of several centimeters length, and subsequently, the velocity matching between the lightwave and the electrical signal is a key technique to achieve high-speed or broadband modulation. When the refractive indices of the lightwave and the electric signal are different, they propagate at different velocities in the device chip, and the effective applied voltage to the lightwave depends on the average of the applied voltages when propagating. To maximize the effective applied voltage, the traveling-wave modulator shown in Figure 9.8 was first proposed [13]. In the modulator, the lightwave and the electrical signal respectively propagate in optical waveguide and electrode with about the same velocity [24].

In the traveling-wave modulator, the applied voltage at the position $x(0 \leq x \leq L)$ in the electrode can be described as

$$V(x) = V_{in} \exp\left\{i(\beta_o - \beta_e)(L - x)\right\} \qquad (9.9)$$

where β_o and β_e are propagation constants for lightwave and electrical signal, respectively, and L is the length of the electrode. Therefore, the effective applied voltage is

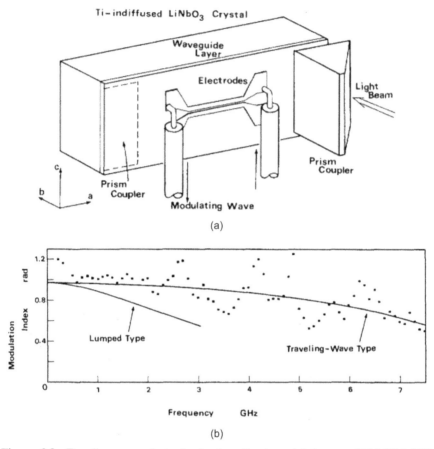

(a)

(b)

Figure 9.8 Traveling-wave electrode for broadband modulation on LN-MZM [13]. (a) Sketch of the traveling-wave modulator using a LiNbO3 optical waveguide. (b) Measured and calculated frequency response of the modulator at 250-mW drive power.

$$V_{eff} = \frac{1}{L} \int_0^L V_{in} \exp \left\{ i(\beta_o - \beta_e)(L - x) \right\} dx \qquad (9.10)$$

However, the microwave signals and guided lightwaves through the MZ circuit have different refractive indices with respect to each other. Therefore, a velocity mismatch exists between the microwaves and guided lightwaves. Bandwidth limitation of modulation Δf is

$$\Delta f = \frac{2c}{\pi L |n_m - n_o|} \qquad (9.11)$$

Here, n_m is the index for microwaves and n_o is the refractive index for optical waves, respectively. And c is the optical wave velocity in vacuum.

To avoid optical absorption of the propagating lightwave by the metal of the electrode, a buffer layer of SiO_2 is used. The coplanar waveguide (CPW) structures [25–27] are common as the electric field intensity near the electrode edge becomes the strongest (Figure 9.9).

Regarding the Pockels effect r_{33} in Z-cut wafers (a) and X-cut (b), in the case of waveguide propagation, we have to consider the TM (a) and TE (b) wave polarization. To match the speed, it is necessary to adjust the buffer layer thickness and the electrode plating thickness. Here, usually, a characteristic impedance of 50 Ω in the drivetrain and the terminator has been practically selected. Also, the microwave electrode loss reduction of even broadband traveling-wave modulator is one of the most key problems. Figure 9.10 shows an example of totally broadband modulated optical response in recent X-cut LN-MZM [28].

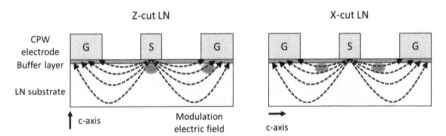

Figure 9.9 Electrode and waveguide; cross section.

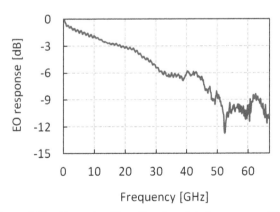

Figure 9.10 Example of broadband modulation response [28].

9.6 Stabilization of LN-MZM Operation

There are several unstable factors in LN-MZM; however, optical damage is never a problem even for Ti-diffused waveguides in the C/L bands, but the remaining key problems are still serious. LN/LT has essentially strong pyroelectric properties along the c-axis owing to its ferroelectric crystal structures. Therefore, charge-up occurs at the high-resistive surface of the substrate crystal with temperature changes. If charge distributions become un-uniform, an undesirable electric field occurs and shifts the modulation characteristics. We call this phenomenon "thermal drift" and Figure 9.11 schematically explains the surface charge distribution changes on a waveguide arm of a LN-MZM caused by temperature change. To eliminate thermal drift problems, a relatively low resistive layer such as non-doped amorphous Si between electrodes [29] was introduced as shown in Figure 9.12, and prototype LN-MZM was successfully operated in laboratory subsystem experiments [30, 31].

The remaining drift of the modulation characteristics is caused by space charge in the crystal over a long-term period. We can overcome this DC-drift

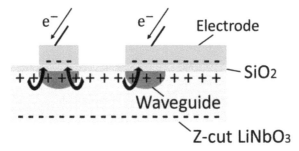

Figure 9.11   Mechanism of thermal drift caused by the pyro-electric effect.

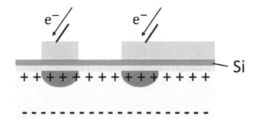

Figure 9.12   Elimination of thermal drift by Si-coating.

phenomenon by using an automatic bias control method [32] as shown in Figure 9.14, since the early 1990s. These efforts have successfully guaranteed more than 15 years of operation of 10-Gbps intensity modulators reported in early 1990s [33] and successfully operated in commercial field services.

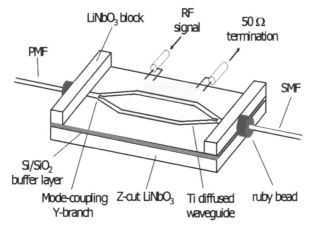

Figure 9.13 Early stabilized LN-MZM against for high dispersion [30].

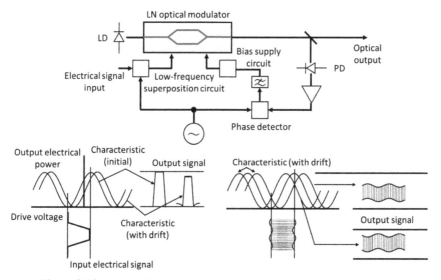

Figure 9.14 Automatic bias control methods for DC-drift suppression [32].

9.7 External Modulation by LN-MZM Accompanied with EDFA Repeating

In 1987, the first 1.5-μm MZ external modulator consisting of two Y-branches and a traveling-wave electrode on Ti:LN was demonstrated for a few multi-Gbps transmission experiments (Figure 9.14) [34]. Around the same period, the potential of Er-doped fiber amplifiers (EDFA) appeared. Soon, this couple was used by R&D organizations globally and 10-Gbps repeating test results were reported at IOOC'89 in Kobe, Japan.

In contrast to 3R repeating, 1R repeating with EDFA requires an exceedingly chirp parameter α for the external modulator owing to large fiber dispersions accumulated over a long distance. At that moment, the already stabilized LN-MZM was the only available external modulator for this purpose. LN-MZMs were introduced to terrestrial 10G (NTT) in the early 1990s and 5G (TPC-5N) in the mid-1990s.

EDFA amplification wavelength bandwidth can cover the C- and/or L-band completely; therefore, dense wavelength division multiplexing/demultiplexing (DWDM) was readily introduced via coupling with several types of optical filters, especially arrayed waveguide gratings (AWG), which occurred in the early 1990s. LN-MZM is essentially wavelength independent in EDFA bandwidth. The reasons of wavelength insensitivity in C-(L-)band are as follows:

- Y-branch characteristics with essentially wavelength insensitive
- Wavelength insensitive Pockels effect (r_{33}) of LN
- Wavelength insensitive refractive index n_e or n_o of LN
- Wide transparent wavelength regions including C-/L-band

Therefore, LN-MZM could satisfy the requirements even for DWDM and more than 1 Tbps was recorded with C/L DWDM using LN-MZM in 1996 (Figure 9.15). Practical DWDM systems were soon commercially available and the optical IT bubble occurred during the millennium years and soon the bubble collapsed. However, FTTx and digital wireless communications for subscribers expanded in this period coupled with new services, which encouraged research once more for large capacity traffic by not only the so-called heavy users but also by a huge amount of people worldwide.

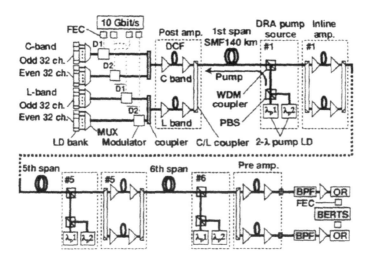

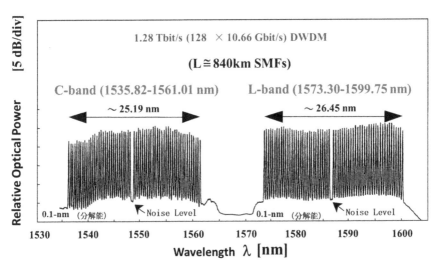

Figure 9.15 Example of DWDM using LN-MZM [35].

9.8 Vector Modulation with LN-MZM for Digital Coherent Optical Communications

In the 20th century, the research and development for post-10-Gbps communication systems with on–off keying gained significant attention. Figure 9.16 shows the representative schemes for high-capacity communications.

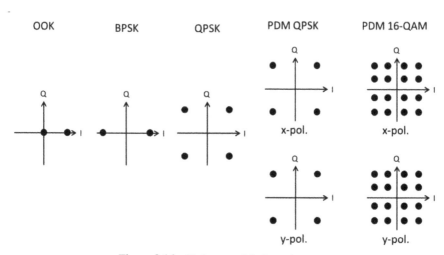

Figure 9.16   Various modulation schemes.

At the same time, the LSI technology, based on Moore's law which had primarily supported optical fiber communications, enabled high-speed signal processing at the transmitter and receiver ends.

The coherent detection was once considered to be applied to the practical systems in the latter half of the 1980s, but it was postponed as a future technology owing to the development of EDFA. Subsequently, however, the digital coherent technology based on LSI signal processors was proposed and attempted for the post-10-Gbps communication systems. In the early years of digital coherent technology, the digital signal processing was used to compensate for frequency differences and fluctuations between the laser source in the transmitter and the local laser oscillator in the receiver. Nowadays, however, the digital signal processors will be applied to both the transmitter and the receiver for equalizing the characteristics of the numerous elements constituting the communication links.

In 2002, optical DQPSK modulation is demonstrated as vector modulation [36]. Subsequently, many experiments on vector modulation with digital coherent detection have been reported. IQ modulator [37], which has dual parallel MZ modulators multiplexed in quadrature-phase carrier, can generate a vector signal expressed in the phaser domain as shown in Figure 9.17. The signal trajectory in the IQ plane or constellation map is shown in the figure, but in this case, a frequency difference exists between the master laser and the local laser for coherent detection, the constellation rotates, and it is

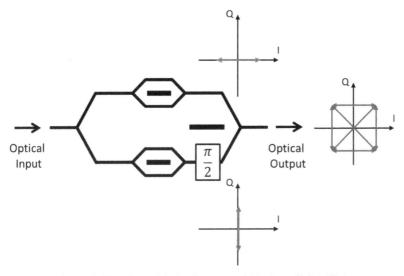

Figure 9.17 IQ modulation by a nested (dual parallel) MZM.

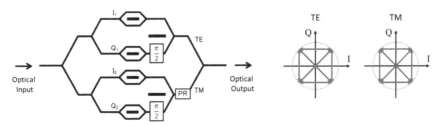

Figure 9.18 Nested LN-MZM (DP-QSK) for digital coherent communications.

difficult to determine the phase of the optical signals. However, the phase of the optical carrier can be estimated using the digital coherent detection, and it also supports a stable operation.

In the recent commercial 100-Gbps/channel communication, DP-QPSK modulator consisting of two IQ modulators multiplexed in polarization is used. The configuration is shown in Figure 9.18. In several cases, modulators have the polarization-multiplexing part with a polarization rotator (PR) and a polarization multiplexer outside the chip.

The advanced technology of the optical modulation based on the history of optical fiber communications is considered to be applied to another application such as radio-over-fiber (RoF) technology. An example of the method to generate a radio signal on the optical carrier is shown in Figure 9.19.

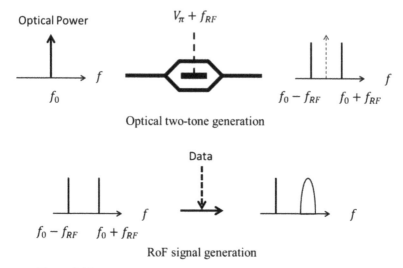

Figure 9.19   Nested and equivalent LN-MZM for analog applications.

This method consists of two steps of RF-carrier generation and data modulation. Using double-sideband suppressed carrier (DSB-SC) modulation that uses LN-MZM driven by a single-tone RF signal, an optical two-tone signal consisting of a first-order modulation sideband is obtained. Subsequently, one of the frequency components is data-modulated. After the transmission on fiber, the signal is received with a photodiode in square-law detection, and then, the modulated radio signal can be generated.

9.9 Current Status of LN-MZM and Future Potential

Certain discussions for standardizations of optical transceivers for long-haul optical fiber communications are progressing primarily on the Optical Internetworking Forum (OIF). The standardization of transmitters for today's commercial 100-Gbps communication systems was completed in 2015. In the standardization, the modulation format is DP-QPSK with 32 Gbaud. The DP-QPSK modulator consists of a pair of dual MZ modulators, also called IQ modulators, multiplexed in polarization, as shown in Figure 9.20. Several specifications such as bandwidth, half-wave voltage, and insertion loss of the modulator for commercial use were standardized by the OIF [38].

Table 9.5 shows the standardized specifications. In general, modulators require having a wider bandwidth than the frequency of the data signals by

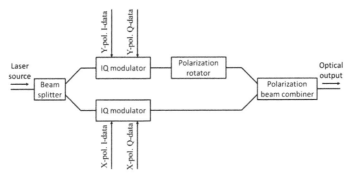

Figure 9.20 DP-QPSK LN-MZM in 100G digital coherent link.

Table 9.5 100G standard spec @ OIF (2015)

Parameter	Minimum	Typical	Maximum
3dB E/O Bandwidth [GHz]	23	—	—
S_{11} Electrical return loss [dB]	10 (<25GHz) 8 (<32GHz)	—	—
V_π @ 32Gbaud [V]	—	—	3.5
RF impedance [Ohm]	—	50	—
Optical insertion loss [dB]	—	—	14
MZI extinction ratio [dB]	20	—	—

approximately 0.7 × baud-rate. The baud-rate in the commercial 100-Gbps communication system is 32 Gbaud; thus, a 23 GHz bandwidth is required as the modulator's 3 dB E/O bandwidth. LN modulators are applied for the commercial 100-Gbps communication system because high-bandwidth LN modulators were already developed owing to past results of commercial 10-Gbps systems. Moreover, LN modulators are selected as an option to be used for digital coherent communications in not only long-haul but also metro networks. The discussion on the standardization for future 400-Gbps systems has already started, and thereby, 64-Gbaud DP-16 QAM is considered as the modulation format.

To generate the 16 QAM signal, it is necessary to input 4-ary electric signals as the I-component data signal and the Q-component data signal of the IQ modulator, respectively, and the modulator driver (electric amplifier) is required to possess high linearity. In addition, as the symbol rate increases, the bandwidth required for the electric driver also increases, and the demand for the amplifier becomes even more significant. Furthermore, higher performance is required for the modulator as well. Broadband modulation of the modulator is indispensable in order to cope with the increased symbol rate,

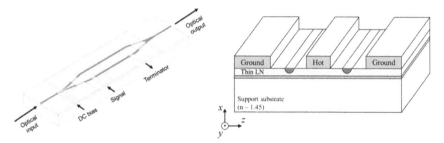

Figure 9.21   Low VπL thin-film LN-MZM.

and based on OIF, 35 GHz is regarded as one criterion for the 3 dB E/O bandwidth as the next generation of high-speed modulators. Moreover, since the output power of electric amplifiers (drivers) with broadband and high linearity is generally small, lowering the driving voltage of the modulator (reduction of Vπ) is also an important problem.

To respond to these demands, research and development is still being actively conducted regarding LN-MZM. Although it was already known since the 1990s that thinning of the LN substrate is effective for widening the bandwidth of LN-MZM [39], the LN substrate is difficult to process precisely and its performance is limited. In recent years, however, the development of ridge waveguides and thinning of substrates has become possible owing to the development of processing technologies (RIE, ICP, ion slicing, polishing, etc.). For example, a modulator, where a ridge waveguide is formed on a Z-cut LN substrate and the back surface of the thinned substrate is air, has been confirmed to have a wideband modulation up to 300 GHz [40]. Research and development of ultra-wideband modulators is progressing, and research on reducing the driving voltage has also attracted attention. For example, in a modulator where the X-cut LN substrate is thinned, the overlap between the modulating electric field and the lightwave becomes large, and thus, the drive voltage can be reduced. In the case where the modulator's waveguide is formed by proton exchange on the substrate thinned owing to the ion slicing method, a VπL ∼6 V·cm has been reported, and VπL ∼2 V·cm is predictable with further structural optimization.

As a trend of progress with respect to optical modulation, introduction of higher order modulation such as QAM is considered to maximize the transmission capacity at the given SNR. To generate optical multi-level signals, MZM modulators are driven by multi-level electrical signals. Subsequently, it is well known that the modulator's extinction ratio and chirp parameter are key parameters that affect the distortion of optical QAM. For example, the calculated relation between the modulator's extinction ratio and distortion of constellation diagram of 256-QAM is shown in Figure 9.22. Higher order modulation requires higher modulator's extinction ratio not to degrade the signal quality. As an advanced modulator for the higher order modulation, we proposed a high-extinction-ratio IQ modulator as shown in Figure 9.22. The modulator has tunable splitters at Y-branch sections, so-called active Y-branch, and the three active Y-branches

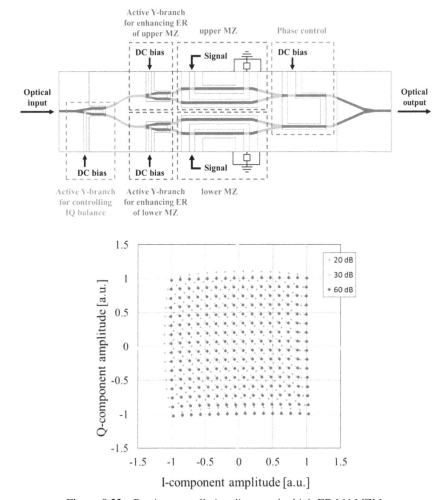

Figure 9.22 Precise constellation diagrams by high-ER LN-MZM.

can enhance the extinction ratio of each Mach–Zehnder interferometer by controlling the optical power balance. Using the modulator, high extinction ratio over 60 dB was experimentally demonstrated [41].

9.10 Summary

In summary, LN-MZM first established in the late 1980s for 10-Gbps IM-DD (OOK) was successfully introduced in the practical field of long-distance high-capacity trunk line in the 1990s, and is suitable for C-/L-band DWDM wavelength regions without any design options. Furthermore, it can operate as a vector IQ modulator with a nested MZ and satisfy digital coherent 100 Gbps initially. In the 20th century, one of the most critical problems of LN devices was polarization-dependent characteristics caused by birefringent and Pockels tensor. However, currently, DP-QPSK modulators utilize this disadvantage for polarization division multiplexing.

The reasons for the past strength of LN-MZM are reviewed as follows:

1. EO modulation based on Pockels effect only and no optical absorption (or imaginary part of refractive index) at C-/L-band in LN. These two simplify the design rule of the MZ circuit.
2. Furthermore, the real part of the refractive index has the same value in the C-/L-band. Therefore, essentially, the MZ circuit in LN possesses flat wavelength characteristics.
3. Ti-diffused single-mode waveguide with index profile similar to conventional SMF can be easily coupled at low loss without any mode conversion.
4. There are few wafer/chip processes of Ti diffusion and electrode fabrication. Moreover, lithography accuracy is not critical.
5. Thermal drift and DC drift are suppressed without electric coolers. There are significant reliability data in the practical fields stocked.
6. High-quality and large-size substrates of LN have been continuously supplied. Huge consumer market of SAW devices supports optical LN wafer supply chain.

However, compact size with higher frequency LN-MZM is strongly demanded especially in the field of application related to the data center. There are no solutions to satisfy such a high-level demand only using the legacy Ti:LN technologies. New technologies [39–44] are expected to be introduced in the production of next-generation LN-MZM.

References

[1] Matthias, B. T., Remeika, J. P., Ferroelectricity in the ilmenite structure. *Phys. Rev.* 76, 1886, 1949.

[2] Ballman, A. A., Growth of piezoelectric and ferroelectric materials by the Czochralski technique. *J. Am. Ceram. Soc.* 48, 112–113, 1965.

[3] Lerner, P., Legras, C., Duman, J. P., Stoechiométrie des monocristaux de métaniobate de lithium. *J. Crys. Growth* 3/4, 231–235, 1968.

[4] Kitamura, K., Furukawa, Y., Private communications.

[5] Pockels, F., Lehrbuch der Kristalloptik, *Teubner,* 1906.

[6] Kaminow, I. P., Ramaswamy, V., Schmidt, R. V., Turner, E. H., Lithium Niobate ridge waveguide modulator. *Appl. Phys. Lett.* 24, 622–624, 1974.

[7] Miller, S. E., Integrated optics: an introduction. *Bell Sys. Tech. J.* 48, 2059–2069, 1969.

[8] Schmidt, R. V., Kaminow, I. P., Meatal-diffused optical waveguides in $LiNbO_3$. *Appl. Phys. Lett.* 25, 458, 1974.

[9] Minakata, M., Saito, S., Shibata, M., Miyazawa, S., Precise determination of refractive-index changes in Ti-diffused $LiNbO_3$ optical waveguides. *J. Appl. Phys.* 49, 4677–4682, 1978.

[10] Ranganath, T. R., Wang, S., Suppression of Li_2O out-diffusion from Ti-diffused optical waveguide. *Appl. Phys. Lett.* 30, 376–379, 1977.

[11] Nozawa, T., Noguchi, K., Miyazawa, H., Kawano, K., Water vapor effects on optical characteristics in Ti:$LiNbO_3$ channel waveguides. *Appl. Opt.* 30, 1085–1089, 1991.

[12] Ramaswamy, V., Alferness, R. C., Divino, M., High efficiency single-mode fibre to Ti:$LiNbO_3$waveguide coupling. *Electron. Lett.* 18, 3031, 1982.

[13] Izutsu, M., Yamane, Y., Sueta, T., Broad-band travelling-wave modulator using a $LiNbO_3$ optical waveguide. *IEEE J. Quantum. Electron.* QE-13, 287–290, 1977.

[14] Mikami, O., Noda, J., Fukuma, M., Directional coupler type light modulator using $LiNbO_3$waveguides. *Trans. IEICE Japan* E-61, 144–147, 1978.

[15] Alferness, R. C., Schmidt, R. V., Turner, E. H., Characteristics of Ti-diffused $LiNbO_3$ optical directional couplers. *Appl. Opt.* 18, 401–418, 1979.

[16] Leonberger, F. J., High-speed operation of $LiNbO_3$ electro-optic interferometric waveguide modulators. *Opt. Lett.* 5, 312–314, 1980.

[17] Alferness, R. C., Waveguide electro-optic modulators. *IEEE Trans. Microwave Theory Tech.* 23, 57–70, 1982.

[18] Korotky, S. K., Eisenstein, G., Gnauk, A. H., Kasper, B. L., Veselka, J. J., Alferness, R. C., et al., 4Gb/s transmission experiment over 117 km of optical fiber using a Ti: LiNbO$_3$ external modulator. *J. Lightwave Technol.* 3, 1027–1031, 1985.

[19] Kondo, M., Tanizawa, Y., Ohta, Y., Aoyama, T., Ishikawa, R., Low-drive-voltage and low-loss polarization-independent LiNbO$_3$ optical waveguide switches. *Electron. Lett.* 23, 1167–1169, 1987.

[20] Sawaki, I., Shimoe, T., Nakamoto, H., Iwama, T., Yamane, T., Nakajima, H., Rectangularly configured 4x4 Ti:LiNbO$_3$ matrix switch with low drive voltage. *IEEE J. Selected Comm.* 6, 1267–1272, 1988.

[21] Thylen, L., Integrated optics in LiNbO$_3$: Recent developments in devices for telecommunications. *J. Lightwave Technol.* 6, 847–861, 1988.

[22] Koyama, F., Iga, K., Frequency chirping in external modulators. *J. Lightwave Technol.*, 6, 87–93, 1988.

[23] Noguchi, K., Lithium Niobate modulators, *Broadband Optical Modulators*, chapter 6, eds. E. Murphy, A. Chen, CRC Press, 151–172, 2012.

[24] Kubota, K., Noda, J., Mikami, O., Traveling wave optical modulator using a directional coupler LiNbO$_3$ waveguide. *IEEE J. Quantum. Electron.* 16, 754–760, 1980.

[25] Kawano, K., Kitoh, T., Jumonji, H., Nozawa, T., Yanagibashi, M., New traveling wave electrode Mach-Zehnder optical modulator with 20 GHz bandwidth and 4.7 V driving voltage at 1.52 µm wavelength. *Electron. Lett.* 25, 1382–1383, 1989.

[26] Dolfi, D., Ranganath, T. R., 50 GHz velocity-matched, broad wavelength LiNbO$_3$ modulator with multimode active section. *Electron. Lett.* 28, 1197–1198, 1992.

[27] Noguchi, K., Miyazawa, H., Mitomi, O., Frequency-dependent propagation characteristics of a coplanar waveguide electrode on a 100-GHz-Ti:LiNbO$_3$ optical modulator. *Electron. Lett.* 34, 661–662, 1998.

[28] Yamaguchi, Y., Kanno, A., Yamamoto, N., Kawanishi, T., Nakajima, H., "High extinction ratio LN modulator with low half-wave voltage and small chirp by using thin substrate," in *Proceedings of the Microoptics Conference (MOC2017)*, A-2, 2017.

[29] Sawaki, I., Nakajima, H., Seino, M., Asama, K., "Thermally stabilized z-cut Ti:LiNbO$_3$ waveguide switch," in *Technical digest of Conference on Laser and Electro-Optics*, *CLEO'86*, MF2, 46–47, 1986.

[30] Nakajima, H., "Temperature-stable, low-loss Ti:LiNbO$_3$ devices," in *Proceedings of the IOOC'89*, 19D3-3 (invited), 164–165, 1989.

[31] Nakajima, H., "Integrated optics devices for high bit rate applications," in *Proceedings of the OFC'90*, TUH6 (invited), 1990.

[32] Suzuki, K., Onishi, M., Nishimoto, H., Amemiya, I., Kuwata, N., Okushima, H., Miyauchi, A. S., "A compact 2.5-Gbit/sec optical transmitter module with a LiNbO$_3$ mach-Zehnder modulator," in *Proceedings of the OFC'92*, WM3, 1992.

[33] Seino, M., Nakazawa, T., Kubota, Y., Doi, M., Yamane, T., Hakogi, H., "A low DC-drift Ti:LiNbO$_3$ modulator assured over 15 years," in *Proceedings of the OFC'92* PD-3, 325–328, 1992.

[34] Okiyama, T., Nishimoto, H., Touge, Seino, M., T., Nakajima, H., "Optical fiber transmission over 132 km at 4 Gb/s using Ti:LiNbO$_3$ Mach-Zehnder modulator," in *Proceedings of the ECOC'87*, 55–58, 1987.

[35] Terahara, T., Hoshida, T., Kumasako, J., Onaka, H., "128×10.66 Gbit/s Transmission over 840-km standard SMF with 140-km optical repeater spacing (30.4-dB loss) employing dual-band distributed Raman amplification," in *Proceedings of the OFC*, PD28, 2000.

[36] Griffin, R. A., Carter, A. C., "Optical differential quadrature phase-shift key (oDQPSK) for high capacity optical transmission," in *Proceedings of the OFC2002*, WX6, 2002.

[37] Kawanishi, T., Sakamoto, T., Izutsu, M., High-speed control of lightwave amplitude, phase and frequency by use of electrooptic effect. *IEEE J. Selected Topics in Quantum Electronics*, 13, 79–91, 2007.

[38] OIF Document, Implementation Agreement for Integrated Polarization Multiplexed Quadrature Modulated Transmitter. *OIF-PMQ-TX-01.2*, May 15, 2015.

[39] Minataka, M., "Recent Progress of 40 GHz high-speed LiNbO$_3$ optical modulator," in *Proceedings of the SPIE*, vol. 4532, 16–27, 2001.

[40] Macario, J., Yao, P., Shi, S., Zablocki, A., Harrity, C., Martin, R. D., et al., Full spectrum millimetre-wave modulation. *Opt. Express* 20, 23623–23629, 2012.

[41] Yamaguchi, Y., Nakajima, S., Kanno, A., Kawanishi, T., Izutsu, M., Nakajima, H., High extinction ratio characteristics of over 60 dB Mach-Zehnder modulator with asymmetric power-splitting Y-branches on X-cut Ti:LiNbO₃. *Japan. J. Appl. Phys.* 53, 08MB03, 2014.

[42] Doi, M., Sugiyama, S., Tanaka, K., Kawai, M., Advanced optical modulators for broadband optical communications. *IEEE J. Selected Topics in Quantum Electronics*, 12, 745–750, 2006.

[43] Wang, C., Zhang, M., Stern, B., Lipson, M., Loncar, M., Nanophotonic lithium niobate electro-optic modulators. *Opt. Express* 26, 1547–1555, 2018.

[44] Zhang, M., Wang, C., Chen, Xi., Bertrand, M., S.-Ansari, A., Chandrasekhar, S., Winzer, P., Lončar, M., "Ultra-high bandwidth integrated Lithium niobate modulators with record-low Vπ," in *Proceedings of the OFC 2018*, Th4A5,

10

Silicon Photonics Based PAM4, DWDM Datacenter Interconnects

Radhakrishnan Nagarajan[1] and Mark Filer[2]

[1]Inphi Corporation, Santa Clara, CA, USA
[2]Microsoft Corporation, Redmond, WA, USA

10.1 Introduction

Although datacenters themselves are physical locations, limited in land area, the optical interconnects between them could span anywhere from tens of kilometers of terrestrial distances to thousands of kilometers of subsea routes.

Figure 10.1 shows the worldwide footprint of the Microsoft datacenter network as of mid-2017 [1]. This supports 42 Azure regions (which themselves consist of hundreds of datacenters), 4,500 points of presence (or peering points), and over 130 edge locations [2]. All these facilities are interconnected over vast distances.

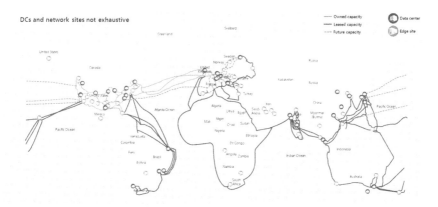

Figure 10.1   Worldwide footprint of the Microsoft datacenter network.

Optical transmission distances of more than 2 km are generally inter-datacenter interconnects, external to the physical datacenter, and are abbreviated as Datacenter Interconnect (DCI) in this chapter. In general, DCI have the following characteristics.

1. Point-to-point interconnects with little optical add/drop or switching.
2. Ethernet is the framing protocol of choice.
3. Dense wavelength-division multiplexed (DWDM) is the optical transmission format.
4. Cover a wide range of geographic distances including submarine links.

Direct detection and coherent are the choices of optical technology for the DCI. Both are implemented using the DWDM transmission format in the C band, 192–196 THz window, of the optical fiber. Direct-detection modulation formats are amplitude modulated, have simpler detection schemes, consume lower power, cost less, and in most cases, need external dispersion compensation. Coherent modulation formats are amplitude and phase modulated, need a local oscillator for detection, need sophisticated digital signal processing (DSP), consume more power, have a longer reach and are more expensive. It is not the goal of this chapter to do a detailed comparison of these two formats – they both have applications in the DCI. We will only limit ourselves to the discussion of the direct detection-based optical transmission.

Practical direct-detection schemes could either employ a binary level, Non-Return-to-Zero (NRZ) modulation or a 4-level Pulse Amplitude Modulation (PAM4) [3, 4]. PAM4 modulation format has twice the capacity of the NRZ. In this chapter, we will discuss the application of the PAM4-based direct-detection modulation format for a 100 Gbit/s DCI optical link.

Figure 10.2 shows the currently preferred optical technology, as a function of link distance. In the fiber-rich environment inside the datacenter, where minimizing impairments due to chromatic dispersion (CD) and meeting tight link budgets without amplification are key design criteria, 1300 nm links are commonly deployed. The pluggable modules for the 100 Gbit/s links inside the datacenter, like the PSM4, LR4 and CWDM4, employ the direct-detect NRZ format. As the data rate progresses to 400 Gbit/s for DR4 and FR4 modules, PAM4 direct-detect formats have become the standard [5].

Outside and in-between datacenters, as the transmission distances progressively increase and where maximizing spectral density is important, 1550 nm DWDM links are commonly deployed. The DWDM links are either direct-detect PAM4 or coherent mQAM. For transmission distances of less

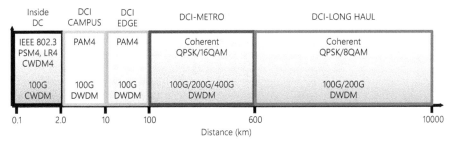

Figure 10.2 Optical technology as a function of distance.

Figure 10.3 Microsoft datacenter in Quincy, WA, United States [1].

than 120 km, up to the limit of the DCI-EDGE links PAM4 is a suitable modulation format.

Datacenters are large deployments in physical space, with the infrastructure, like power and cooling, needed to support them. Figure 10.3 shows the buildings at a Microsoft datacenter facility in Quincy, WA, United States [1]. The longest fiber links inside these building can be up to a kilometer or so. Datacenter geographic disaggregation is becoming more common as it is increasingly difficult to build single, large, contiguous mega-datacenters. Disaggregation is also critical in metropolitan areas where land area maybe at a premium, and for disaster recovery where geographically diverse locations

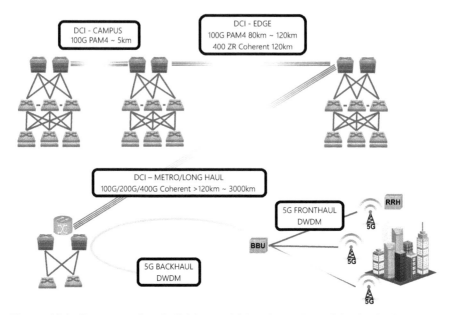

Figure 10.4   Dense wavelength-division multiplexed (DWDM) links in the Datacenter Interconnects including possible DWDM deployments in wireless 5G.

are needed. Large-bandwidth interconnects are essential to connect these datacenters.

Figure 10.4 shows how the commonly used nomenclature in Figure 10.2 for the various DCI reaches are related.

1. DCI-Campus: These connect datacenters which are close together, as in a campus environment. The distances are typically limited to between 2 and 5 km, which may be easily covered without the need for external dispersion compensation at 28 Gbaud symbol rates. There is also an overlap of CWDM and DWDM links over these distances, depending on fiber availability in the environment.
2. DCI-Edge: The reaches for this category range from 2 to 120 km. These are generally latency-limited and used to connect regional, distributed datacenters.
3. DCI-Metro/Long Haul: The DCI-Metro and DCI-Long Haul, as a group, lumps fiber distances beyond the DCI-Edge up to 3,000 km, for terrestrial links, and longer for subsea ones. Coherent modulation format is used for these, and the modulation type may be different for the different distances. A typical deployment scenario may be as follows.

(a) 16QAM: <1,000 km
(b) 8QAM: >1,000–3,000 km
(c) QPSK: >3,000 km

4. Although not part of the DCI infrastructure, the wireless networks are also becoming integrated into the datacenter network.

10.2 Datacenter Interconnect–Edge

The DCI space has become an area of increased focus for traditional DWDM system suppliers over the last few years. The growing bandwidth demands of cloud service providers (CSPs) offering SaaS, PaaS, and IaaS capabilities have in turn driven demand for optical solutions to connect switches and routers at the different tiers of the CSP's datacenter network. Today, this requires solutions at 100 Gbps, which inside the datacenter, can be met with direct attach copper cabling (DAC), active optical cables (AOC), or 100G "gray" optics (e.g., CWDM4, PSM4). For links connecting datacenter facilities (campus or edge/metro applications), the only choice that was available until recently was full-featured coherent transponder solutions, which are sub-optimal, as discussed below.

Along with the transition to the 100G ecosystem, there has been a shift in datacenter network architectures from more traditional datacenter models where all the datacenter facilities reside in a single, large "mega datacenter" campus. Most CSPs have converged on distributed regional architectures to achieve the required scale and provide cloud services with high availability as shown in Figure 10.5.

Datacenter regions are typically located in close proximity to large metropolitan areas with high population densities in order to provide the best possible service (with regard to latency and availability) to end customers nearest those regions. Regional architectures vary slightly among CSPs but fundamentally consist of redundant regional "gateways" or "hubs" which connect to the CSP's WAN backbone (and possibly to edge sites for peering, local content delivery, or subsea transport). Each regional gateway connects to each of the region's datacenters, where the compute/storage servers and supporting switching fabrics reside. As the region needs to scale, it becomes a simple matter of procuring additional center facilities and connecting them to the regional gateways. This enables rapid scaling and growth of a region, compared to the relatively high expense and long construction times of

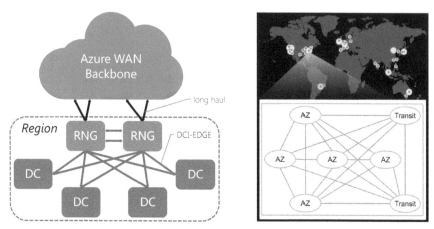

Figure 10.5 Basic datacenter architecture of Microsoft Azure (left) and Amazon Web Services (right).

building new mega datacenters, and has the side benefit of introducing the concept of diverse "availability zones (AZ's)" [1, 6, 7] within a given region.

The transition to regional from mega datacenter architectures introduce some additional constraints which must be considered when choosing the locations of gateways and datacenter facilities. One is that, to ensure the same customer experience (from a latency perspective), the maximum distance between any two datacenters (via the common gateway) must be bounded. For Microsoft Azure, this is driven by a maximum round-trip server-to-server latency in the few millisecond range. This is also true for the Amazon Web Services datacenter architecture, where the AZ's are <2 ms apart [6]. This latency is primarily dominated by the physical fiber propagation time in the DCI portion of the network. Telecom-grade optical fibers have latencies of approximately 5 μs/km; to meet the latency SLAs, the maximum distance of a gateway to any datacenter is around 80 km. Requirements for other CSPs may differ slightly in this regard, but the consensus is that DCI applications typically do not practically exceed 120 km. Supporting this statement, a distribution of fiber distances for DCI applications in Microsoft's network can be seen in Figure 10.6.

Another consideration is that the spectral efficiency of gray optics is too low for interconnecting physically disparate datacenter buildings within the same geographic region. In the 100G era, this problem was typically solved using DWDM coherent QPSK transponders which offer soft-decision

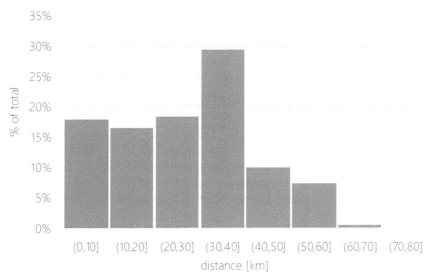

Figure 10.6 Distribution of Datacenter Interconnect (DCI) fiber distances in Microsoft's network.

FEC-enabled performance down to 11 dB optical signal to noise ratio (OSNR), subsea-capable CD compensation of 250,000 ps/nm, power efficiencies on the order of 100W per 100G, and capacities of 8–9.6 Tbps per fiber pair. For DCI links that are latency-constrained to 120 km or less (<2,400 ps/nm), line systems can easily deliver OSNRs in the low 30dB range. Rack space and power are typically limited and costly in these facilities, making power and space efficiency critical design goals. While fiber is not as abundant in these metro environments as within the datacenter itself, typically tens of fiber pairs are available at reasonable costs, relaxing ultimate spectral efficiency as a primary design criterion. With these considerations, today's coherent solutions are not an ideal fit for DCI applications.

In response, low-power, low-footprint direct-detect solutions have been conceived employing PAM4 modulation format [8]. By utilizing silicon photonics technology, a dual-carrier transceiver featuring a PAM4 ASIC, with integrated DSP and forward error correction (FEC), was developed and packaged into a QSFP28 form-factor. The resulting switch-pluggable module enables DWDM transmission over typical DCI links at 4 Tbps per fiber pair, and electrical power consumption of 4.5 W per 100G. The following sections cover this development in detail.

10.3 Switch Pluggable 100Gbit/s DWDM Module

The switch pluggable QSFP28 module (SFF-8665 MSA compatible) is based on a highly integrated silicon photonics optical chip and a PAM4 ASIC as shown in Figure 10.7 [9, 10]. The four KR4-encoded 25.78125-Gbit/s inputs are first stripped of their host FEC. Then the two 25-Gbit/s streams are combined into a single PAM4 stream and a more powerful line FEC called IFEC is added, making the line rate 28.125 GBaud per wavelength. The four 25-Gbit/s Ethernet streams are thus converted to two 28.125-GBaud PAM4 streams. In the 40Gbit/s use case, there are four 10-Gbit/s inputs, a single 22.5-GBaud PAM4 stream is generated.

The line system configuration for the DCI link is shown Figure 10.8. This is essentially a single-span DWDM link with external (chromatic) dispersion compensation module (DCM). It has a booster EDFA (Erbium-Doped Fiber Amplifier) at the transmit side and pre-amplifier EDFA on the receive side. These are labeled as OA (optical amplifier) in Figure 10.8.

The pluggable optical modules are meant to operate with commercially available external multiplexers and demultiplexers on the 100 GHz ITU grid. The two wavelengths from the module are spaced 50 GHz apart, but they

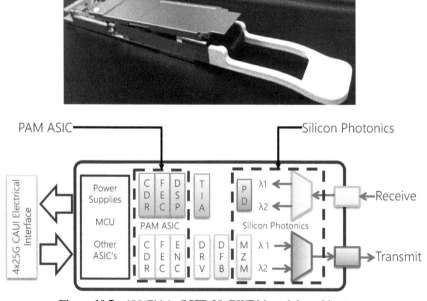

Figure 10.7   100 Gbit/s, QSFP-28, DWDM module architecture.

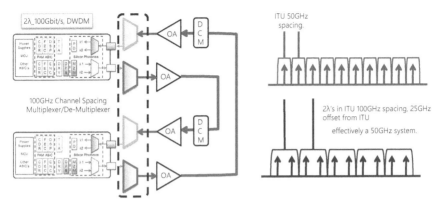

Figure 10.8 Typical configuration of point-to-point DCI-Edge link.

are not on the 50 GHz ITU grid. To transmit both wavelengths within the 100 GHz passband, the wavelengths are offset 25 GHz to either side of the standard ITU frequency on the 100 GHz grid. This enables the transmission of up to 40, 100-Gbit/s channels or equivalently 80, 50-Gbit/s carriers at 50 GHz spacing, within the C band of the optical fiber transmission spectrum. The channel schema is depicted in the right half of Figure 10.8.

The dispersion compensation could either be fixed or tunable [11], and is broadband to compensate all wavelengths simultaneously. The tunable DCM's are commonly based on either the tunable fiber Bragg grating (FBG) or tunable etalon technologies, both of which are channelized to compensate 50 GHz-spaced optical carriers. These modules could be easily incorporated into the mid-stage of the transmit or receive EDFA with minimal impact on the link OSNR.

10.4 PAM4 DSP ASIC

The PAM4 ASIC [12–14] is the electrical heart of the module. The ASIC architecture and its electrical performance are shown in Figure 10.9. The ASIC supports a dual wavelength 100 Gbit/s mode (with 4 × 25 Gbit/s inputs), and a single wavelength 40 Gbit/s mode (with 4 × 10 Gbit/s inputs). The ASIC has MDIO and I2C management interfaces for diagnostics and device configuration. There is also an on-chip microcontroller to initialize the device, sequence DSP, and monitor link health.

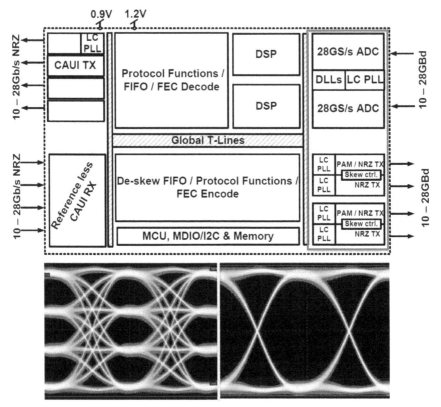

Figure 10.9 Architecture and electrical performance of the 4-level Pulse Amplitude Modulation (PAM4) digital signal processing (DSP) ASIC.

The ASIC has a CML (current-mode logic) driver with CMOS backend as shown in the top of Figure 10.9. The transmit block is clocked at half rate and the receive block is clocked at 1/4 sampling rate.

The ASIC also provides the CAUI 4 host interface to the switch or another host. On the transmit side 4 × 25.78125 Gbit/s input data are converted into 2 × 28.125 Gbaud PAM4 streams with the appropriate FEC overhead. These are then used to drive the silicon photonics modulator via an external driver. The ASIC is capable of a combined PAM4 output or MSB/LSB constituent outputs of the PAM4 signal.

On the receive side, the output of the transimpedance amplifier (TIA) is sampled in the PAM ASIC using 28-GS/s 7-bit ADCs with a SAR (successive approximation register) core. The receive DSP has multi-tap FFE (feed forward equalizer) and DFE (decision feedback equalizer), with calibration, to recover the PAM4 signal. The equalizer taps are automatically adapted using a least-mean-squared (LMS) algorithm. The FEC decoder then corrects the errors and generates the original Ethernet data.

The PAM4 ASIC generates the clock from the input CAUI data, and does not need an internal reference.

The bottom of Figure 10.9 shows the electrical eye diagrams from the PAM4 ASIC's transmitter block. The NRZ and PAM4 outputs have the same swing and rate (25 Gbaud). The measured electrical SNDR (signal to noise distortion ratio) to be better than 33 dB [14].

10.5 Silicon Photonics

InP and Si are the commonly used platforms for large-scale integration of optical components [15], but the Si CMOS platform allows a foundry-level access to the optical component technology at larger, 200- and 300-mm wafer sizes [16, 17]. Although the Si absorption is in the <1 μm wavelength region, photodetectors (PDs) in the 1,300- and 1,550-nm wavelength range can be built by adding Ge epitaxy to the standard Si CMOS platform. Further, silica-based components may be integrated to fabricate low-index contrast and temperature-insensitive optical components [15, 18].

The Si photonics technology is now sufficiently mature and the design methodology sufficiently well-established that various tools are commercially available in this field [19, 20].

The highly integrated Si photonics chip that is at the optical heart of the pluggable module contains a pair of traveling wave Mach-Zehnder modulators (MZM) in the output optical path, one for each wavelength. These are standard depletion mode CMOS structures [21], with a small-signal bandwidth of about 25 GHz as shown on the left of Figure 10.10. The two wavelength outputs are then combined on-chip using an integrated 2:1 interleaver which functions as the DWDM multiplexer. The Si photonic circuit schematic is shown in Figure 10.7.

The same Si MZM may be used for both NRZ and PAM4 modulation formats, with different drive signals, as shown for the 25.78-Gbit/s eye diagrams on the right in Figure 10.10. The PAM4 eye diagram was measured after an EDFA. The signal-spontaneous beat noise causes the 11 level to broaden more than the 00 level.

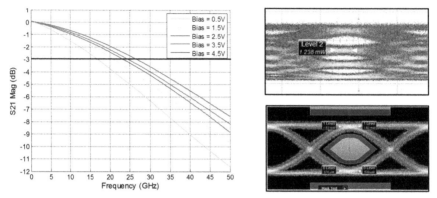

Figure 10.10   Performance of the silicon photonics Mach-Zehnder modulator.

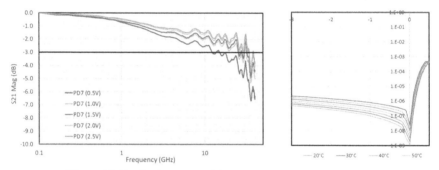

Figure 10.11   Performance of the high-speed Ge photo detector.

There are a pair of integrated high-speed Ge PDs in the receive path. The PD has a small signal bandwidth of more than 25 GHz as shown in Figure 10.11. The dual wavelength receive signal is separated using a de-interleaver structure which is similar to the one used in the transmitter. The reverse dark current at $-2V$ bias is typically less than 10 μA at 85°C. This is within the performance requirements for a PAM4 link.

Both the MZM driver amplifier and the PD TIA are wire bonded to the Si photonics chip. We use a direct coupled MZM differential driver configuration to minimize electrical power consumption [22].

The DFB lasers are external, and edge coupled to the Si photonics chip. Unlike the vertical grating couplers, edge couplers have a wide optical bandwidth with low insertion loss [23]. The optical signal output and input are likewise fiber coupled to the Si photonics chip.

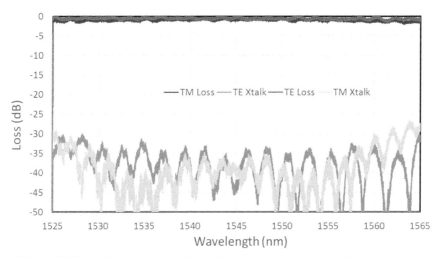

Figure 10.12 Extinction ratio and loss of the polarization diverse path in the receiver.

Since the polarization state of the incoming optical signal is not deterministic, the receive path is designed to be polarization diverse to eliminate polarization-induced signal fading [24]. This is accomplished by using a low-loss polarization beam splitter (PBS). After the PBS, TE and TM paths are processed independently, and then combined again at the PD's. The insertion loss of the PBS is <0.5 dB and the crosstalk isolation >25 dB across all of C band, as shown in Figure 10.12.

10.6 Module and Transmission Performance

The PAM4 ASIC supports two IEEE standard, 802.3bj 100GBASE-KR4 and 100GBASE-KP4 FEC schemes. In addition, it also supports a proprietary IFEC mode. The details of the data rate overheads and the coding schemes for the three FEC modes are in Figure 10.13. The ASIC is also capable of operating in the bypass mode, where the host FEC and framing are preserved.

IFEC is a low-power, multi-level, iterative code [25]. Its 10.5 dB coding gain is high compared to the 5 to 7 dB coding gains for the various optical and copper-standard FECs. For the IFEC, the theoretical correction limit BER is 1E-2, compared to 1E-5 for the KR4 FEC. The relative OSNR performances of the three FEC modes are given in Figure 10.13. IFEC has more than 2 dB operating margin at a BER of 1E-3.

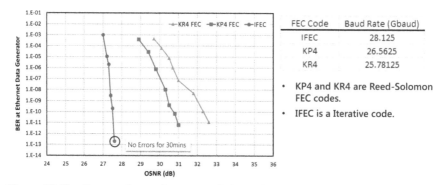

Figure 10.13 Comparative performance of the various forward error correction (FEC) modes in the PAM4 ASIC.

The PAM4 format, like all optical modulation formats, is susceptible to CD-induced impairments during fiber propagation. In coherent transmission systems, the recovery of the phase and amplitude information at the receiver enables the digital signal processor to compensate for the effects of CD, which is largely a linear impairment. The phase information is lost in the much simpler direct-detect PAM4 systems due to the nature of square-law detection. The intrinsic dispersion tolerance of the 28.125 Gbaud PAM4 format is shown in Figure 10.14. It has a ±100 ps/nm window. This is in line with the previously reported data [26]. The typical OSNR penalty in that window is about 0.8 dB.

In general, except for links that are <5 km (within the intrinsic dispersion tolerance of the 28.125 Gbaud PAM4 format) of standard single mode fiber (SSMF), external CD compensation is needed for PAM4 systems as discussed in Section 10.3. Figure 10.15 shows the results when using a tunable DCM to compensate for CD impairments up to 80 km of SSMF.

Two implementation cases are discussed in Figure 10.15. For 50 and 80-km links, the case where all the dispersion is compensated with the tunable dispersion compensation module (TDCM) is compared to the case where 40 km of fixed DCM is used together with a TDCM. In this test, the complete link configuration with the EDFA's external multiplexers, demultiplexers, and TDCM are used. Fiber Bragg grating-based TDCM is used in this experiment.

The baseline back-to-back BER is 1.1E-3. The back-to-back BER with the TDCM set to zero is 1.8E-3. There is some residual impairment due to the limited bandwidth and phase ripple of the FBG-based TDCM. Within the tolerance of the BER measurement, there is negligible penalty using either

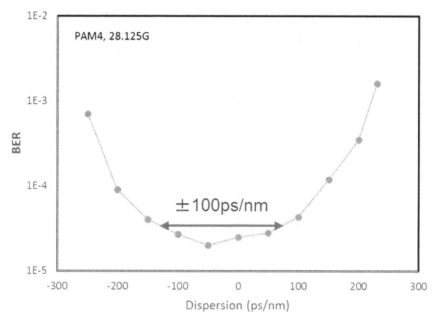

Figure 10.14 Dispersion tolerance of the 28 GBaud PAM4 signal.

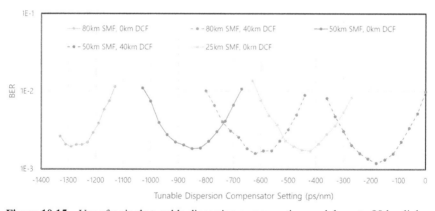

Figure 10.15 Use of a single tunable dispersion compensation module up to 80 km links.

dispersion compensation strategies. A single TDCM unit may be used to compensate for all the dispersion impairment due to 80 km of SSMF.

Fiber Bragg gratings and etalons both have free-spectral ranges (FSR) equal to the spacing of the optical carriers but differ in slightly their characteristics. Both offer linear dispersion compensation across all channels.

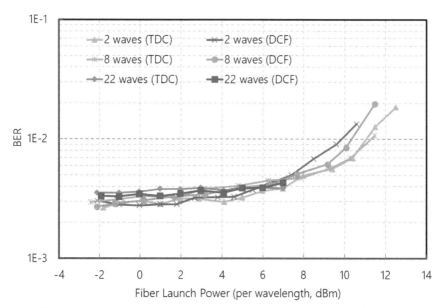

Figure 10.16   Non-linear tolerance of the PAM4 modulation format in standard single mode fiber (SSMF).

In addition to the linear compensation, FBGs offer dispersion slope compensation across the channels. However, this comes at the expense of a filter passband width which becomes narrower with increasing dispersion compensation. This can have the undesired effect of clipping the edges of the modulated carriers, causing an ISI penalty at the receiver. Additionally, FBGs generally have higher uncontrolled group delay ripples compared to etalons, which can cause additional penalties. In contrast, Etalons apply minimal passband filtering on the modulated signals – their FSR is primarily a phase effect and only has minimal impact on the passband, but they lack the dispersion slope compensating abilities of the FGBs. For the single-span DCI-Edge applications, this is typically sufficient. Additionally, etalons' dispersion compensation range is more limited than that of FBGs, and symmetric around 0 ps/nm, and so, for longer spans, must be combined with a fixed-DCM technology.

The PAM4 format was measured to have <0.8dB OSNR penalty for PMD (polarization mode dispersion) up to 10ps.

The non-linear tolerance of the PAM4 format was first investigated in SSMF (Figure 10.16). DWDM optical signals experience several non-linear impairments, as a function of launch power, during fiber propagation [27].

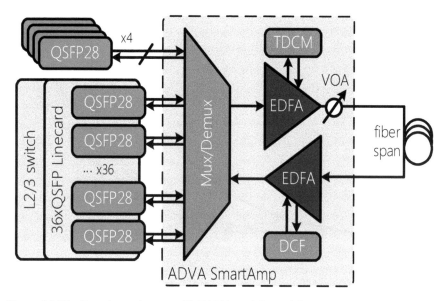

Figure 10.17 Experimental setup with PAM4 modules and line system and different fiber types.

Several different types of fibers have been developed over the years to mitigate the impact of non-linear fiber propagation.

In Figure 10.16, the channels were running at 28.125 GBaud over 80 km of SSMF. There are two sets of dispersion compensation conditions considered in the experiment. Each set of propagating wavelengths, after 80 km, was compensated either using a fixed DCF (dispersion compensation fiber) or a tunable DCM. The dispersion compensation was incorporated in the mid-stage of the receiver EDFA. The non-linear impact on performance starts increasing after about 6–8 dBm launch power per wavelength, irrespective of the number of wavelengths. This shows that the non-linear interactions between the wavelengths are negligible. The effect is solely governed by the individual launch power which shows that the direct-detect PAM4 format seems to be primarily impacted by self-phase modulation. This is largely in line with the data that has been reported to date on the non-linear tolerance of the PAM4 modulation format [26, 28].

Then a series of tests was performed over five widely-deployed fiber types to quantify the impact of non-linearity [29]. The experimental link, shown in Figure 10.17, consisted of a single fiber span with ADVA line system, comprised of a 100-GHz arrayed-waveguide grating mux/demux pair of

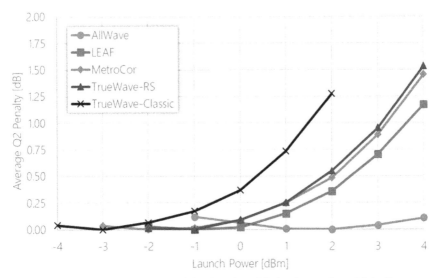

Figure 10.18   Non-linear tolerance of the PAM4 modulation format in multiple fiber types.

EDFAs, etalon-based TDCM, and optional DCF. The system was fully loaded with 40 100-Gbit/s QSFP28 transceivers (for a total of 80 wavelengths).

Tolerance to fiber non-linearity was tested over five different fiber types: OFS AllWave, TrueWave-RS, and TrueWave-Classic, and Corning LEAF and MetroCor. The spans ranged from 50 to 60 km with 11–13 dB loss (including connectors), such that the average received OSNR was in the same range (∼34 and 35 dB, depending on launch power). For AllWave, a 40 km DCF was used, while for the other types no DCF was used and the TDCM performed all compensation.

Results are shown in Figure 10.18 where the average Q^2 penalty across all monitored channels is plotted versus the average launch power per wavelength into the fiber span. There is a small increase of ∼1 dB in received OSNR when moving from lowest to highest launch powers. For AllWave, the impact from fiber non-linearity is negligible up to the maximum launch power of +4 dBm/λ. This non-linear tolerance is higher than previously reported for similar systems, despite the significantly larger number of co-propagating channels [26]. The other NZ-DSF types all exhibit some non-linear penalty as launch powers increase. LEAF, MetroCor, and TrueWave-RS all show similar performances, with non-linear impact becoming significant starting at launch power +2 dBm/λ. TrueWave-Classic shows the strongest non-linearity due to its small effective mode area and very low fiber dispersion. Significant impact

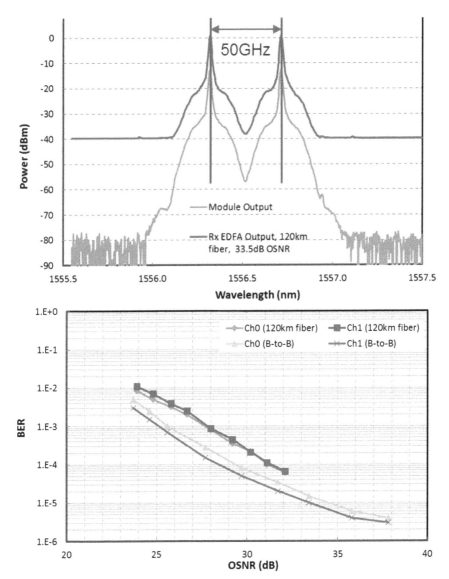

Figure 10.19 Optical signal-to-noise ratio (OSNR) performance of the longest link closed to date.

starts at launch power 0 dBm/λ, yet even at this condition, all channels had >1 dB Q² margin.

For a simple point-to-point transmission system with two EDFAs, the ultimate transmission distance is limited by the launch power at the transmitter. Figure 10.19 shows the results for a pair of wavelengths (100 Gbit/s combined), a 120-km link of SSMF. The bottom of Figure 10.19 shows the BER vs. OSNR curve for the 120 km transmission for a total fiber loss of 26.2 dB. The launch power per wavelength was pushed close to 10 dBm at the transmit EDFA to achieve this non-linearity-limited transmission distance. The FEC correction limit is about 24 dB OSNR. Since the saturation output power of EDFAs typically used in commercial applications is limited to about 24 dBm, the number of wavelengths that can be launched at such high powers will also be limited. There is a tradeoff between reach and density in single span links limited by the transmit EDFAs used.

With dispersion compensation, there is ~1 dB OSNR penalty after 120 km transmission. The overlay on the optical spectrum plot (Figure 10.19 top) shows the shot noise limited output spectrum of the module, and the optical spectrum at the output of the receive EDFA. The OSNR at the output of the receive EDFA was 33.5 dB, and the signal was then noise loaded to generate the BER vs. OSNR curve. Since the link had excess OSNR, one could have, in principle, increased the overall transmission distance.

The DCI-Campus and Edge links are shown in Figure 10.20. A DCI-Edge configuration has two EDFAs and a DCM (which may be part of the mid-stage of the transmit or receiver EDFA). A DCI-Campus configuration can operate off a single receive EDFA without any external dispersion compensation, since the intrinsic dispersion tolerance of ±100 ps/nm is sufficient for this distance. The inset to the bottom of Figure 10.20 shows the optical spectrum for a fully loaded 40-channel system for a maximum of 4 Tbit/s capacity in a fiber pair. The rack to the left in Figure 10.20 shows a single blade of the Arista 7504 chassis switch with 36 QSFP28 ports, fully loaded. A single blade of the switch can support 3.6 Tbit/s of traffic in any one direction. The ADVA SmartAmp line system is used for this configuration with 80 km of SSMF in each direction.

10.7 Live Datacenter Deployments

There have been several live DCI deployments of systems like the one in Figure 10.20 using different switch and line system configurations. Figure 10.21 shows a performance snapshot of one such system. In this deployment, there are four fiber pairs between datacenters, each carrying 3.2 Tbit/s of data. Including both sides of the link, there is a total of 256 100-Gbit/s pluggable modules in this deployment. The total bisectional data rate between the datacenters is 25.6 Tbit/s.

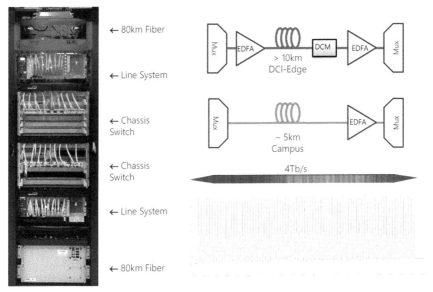

Figure 10.20   Typical switch and line system configuration for DCI-Campus/Edge.

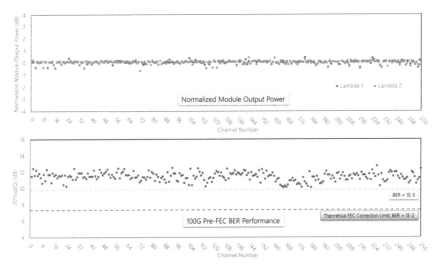

Figure 10.21   Live deployment of a DCI-Edge system.

The top chart in Figure 10.21 shows the normalized output power for both wavelengths, for all 256 modules. The distribution is tight and is less than ±1 dB.

The bottom chart in Figure 10.21 shows the pre-FEC BER (equivalent Q^2 values) at the receive end. Pre-FEC BER is one of the diagnostics that can be read directly from the module. Although the FEC correction limit is 1E-2, the deployed BERs are all better than 1E-3.

10.8 Evolution to Switch Pluggable 400-Gbit/s DWDM Module

As bandwidth demands in datacenter networks continue to grow, Moore's law dictates that advances in switching silicon will enable switch and router platforms to maintain switch chip radix parity while increasing capacities per port. At the time of this manuscript's writing, the next generation of switch chips are all targeting per-port capacities of 400G. Accordingly, work has begun to ensure the optical ecosystem timelines coincide with the availability of next-gen switches and routers.

Toward this end, a project has been initiated in the Optical Internetworking Forum (OIF), currently termed 400ZR [30], to standardize next-gen optical DCI solutions and create a vendor-diverse optical ecosystem. The concept is in principle like that of the DWDM PAM4 technology described throughout, but scaled up to support 400 Gbit/s requirements. The trade offs in spectral efficiency and complexity of this type of approach vs a single laser, single modulator coherent approach make using a coherent solution at 400 Gbit/s and beyond the obvious choice, especially when such a coherent solution can be developed to still fit in a faceplate pluggable form-factor at < 4W/100 Gbit/s.

With 100-Gbit/s serial PAM4 already pushing the bandwidth limits of the undergirding electro-optics, it was quickly determined in the OIF that next-gen solutions at 400 Gbit/s would need to leverage coherent technologies. However, the drawbacks of coherent mentioned in Section 10.2 dictate that a different approach needs be taken with the sub-120 km application space in mind. A dedicated, DCI-targeted coherent chipset with very limited features (and therefore power consumption) would need to be developed independently from the "swiss-army knife" coherent solutions currently available – namely, a coherent DSP with limited dispersion and PMD compensation, baud rate options, flex-modulation support, and FEC modes. Additionally, to make it viable as a switch-pluggable solution like today's PAM4 modules, it was decided that the optics and electronics (including DSP chip) need to fit in a power envelope of 15 W and package sizes to achieve parity with today's solutions ($\sim\geq$32 ports per 1RU). Finally, to ensure a robust supplier ecosystem, this technology must all be interoperable (including modulation, framing, pilot tones, and most importantly FEC), which is why the project is being carried out in the OIF.

Work on the OIF Implementation Agreement (IA) is underway at the time of writing, with several of the larger hurdles toward interoperability already passed, and good support from the supplier community in validating the assumptions and models around product feasibility. At a high level, the solution will be based on coherent 16QAM near 60 Gbaud with hybrid SD-FEC achieving moderate NCG (Net Coding Gain), but sufficient to meet the requirements of the application space. Form factor possibilities include OSFP, QSFP-DD, and COBO options, but there is much work to be done there to ensure thermal and signal-integrity concerns around the desired form-factor are fully vetted out in the required time frames.

10.9 Conclusion

In this chapter, we discussed the100 Gbit/s, DWDM PAM4 datacenter inter-connections that span regional distances of up to 120 km. The combination of Si photonics for the highly integrated optical components, and high-speed Si CMOS for signal processing is critical for the implementation of low-cost, low-power, switch pluggable optical modules that enable massive interconnections between vast regional datacenter deployments today.

There is work underway currently to extend this concept to low-power 400 Gbit/s switch pluggable 16QAM coherent modules using the next-generation Si photonics and Si CMOS.

Acknowledgments

Radhakrishnan Nagarajan would like to thank the extended engineering and operations team, and the management at Inphi without whom the work on the module would not have been possible. There were major contributions to Si photonics from Masaki Kato, to PAM4 ASIC/DSP from Sudeep Bhoja, to module engineering from James Stewart, to firmware engineering from Todd Rope, and to optical testbed measurements from Yang Fu.

Mark Filer would also like to thank the optical network and architecture teams at Microsoft, and in particular, Jeff Cox, for his uncanny ability to see into the future and execute in transforming the DCI landscape. Additionally, thanks to Steven Searcy, Sorin Tibuleac, and the rest of the team at ADVA Optical Networking for the majority of PAM4 transmission results presented within.

References

[1] Russinovich, M. (2017). *Inside Microsoft Azure Datacenter Hardware and Software Architecture*. Orlando, FL: Microsoft Ignite. Available at: https://myignite.microsoft.com/videos/54962

[2] Zander, J. (2017). *Cloud Infrastructure: Enabling New Possibilities Together*. Orlando, FL: Microsoft Ignite. Available at: https://channel9.msdn.com/Events/Ignite/Microsoft-Ignite-Orlando-2017/GS05

[3] Zhang, H., Jiao, B., Liao, Y., and Zhang, G. (2016). *PAM4 Signaling for 56G Serial Link Applications – A Tutorial*. Santa Clara, CA: DesignCon.

[4] Bhoja, S. (2017). "PAM4 signaling for intra-data center and Data center to data center connectivity (DCI)," *in Proceedings of the Optical Fiber Communications Conference and Exhibition (OFC)*, W4D.5.

[5] IEEE (2017). *IEEE 802.3bs, 400GbE standard.*

[6] Vogels, W. (2015). "The Pace of Innovation at AWS," *Amazon Web Services Summit*. Available at: https://d0.awsstatic.com/events/aws-hosted-events/2015/israel/datacenter-innovation.pdf

[7] Microsoft Azure (2017). *Overview of Availability Zones in Azure*. Available at: https://docs.microsoft.com/en-us/azure/availability-zones/az-overview

[8] ACG Research (2017). *Connecting Metro-Distributed Data Centers: The Economics and Applicability of Inphi's ColorZ^{TM} Technology*. Available at: http://www.acgresearch.net/wp-content/uploads/2017/03/INPHI_COLORZ_for_DCI_ACG.pdf

[9] Nagarajan, R., Bhoja, S., and Issenhuth, T. (2016). "100Gbit/s, 120km, PAM 4 based switch to switch, layer 2 silicon photonics based optical interconnects for datacenters," *in Proceedings of the Hot Chips 28 Symposium (HCS)* (Cupertino, CA: IEEE), HC28.23.521.

[10] Nagarajan, R., et al. (2017). "100Gbit/s, switch pluggable, silicon photonics based PAM4 DWDM modules for 4Tbit/s inter-datacenter links," *in Proceedings of the CLEO-PR/OECC/PGC/Photonics SG*, 3-4E-1 (Invited).

[11] TeraXion/Inphi Webinar (2017). *Data Center Interconnect: Taking the Complexity out of Dispersion Management*. Available at: http://www.teraxion.com/en/events

[12] Khandelwal, P., Riani, J., Farhoodfar, A., Tiruvur, A., Hosagrahar, I., Chang, F., et al. (2016). *100Gbps Dual-Channel PAM-4 Transmission over Datacenter Interconnects*. Santa Clara, CA: DesignCon.

[13] Chang, F., Bhoja, S., Riani, J., Hosagrahar, I., Wu, J., Herlekar, S., et al. (2016). "Link performance investigation of industry first 100G PAM4 IC chipset with real-time DSP for data center connectivity," *in Proceedings of the OFC/NFOEC*, Th1G.2.

[14] Gopalakrishnan, K., Farhood, A., Ren, A., Tan, A., Tiruvur, A., Helal, B., et al. (2016). "A 40/50/100Gbps PAM-4 Ethernet Transceiver in 28 nm CMOS," in *Proceedings of the ISSCC*, 3.4.

[15] Nagarajan, R., Doerr, C., and Kish, F. (2013). "Semiconductor photonic integrated circuit transmitters and receivers," in *Optical Fiber Telecommunications VIA: Components and Subsystems*, Chapter 2, eds I. Kaminow, T. Li, and A. Willner (Cambridge, MA: Academic Press), 25–88.

[16] Izhaky, N., Morse, M., Koehl, S., Cohen, O., Rubin, D., Barkai, A., et al. (2006). Development of CMOS-compatible integrated silicon photonics devices. *IEEE J. Sel. Topics Quant. Elec.* 12, 1688–1698.

[17] Liow, T., Ang, K., Fang, Q., Song, J., Xiong, Y., Yu, M., et al. (2010). Silicon modulators and germanium photodetectors on SOI: monolithic integration, compatibility, and performance optimization. *IEEE J. Sel. Topics Quant. Elec.* 16, 307–315.

[18] Tsuchizawa, T., Yamada, K., Watanabe, T., Park, S., Nishi, H., Kou, R. et al. (2011). Monolithic integration of silicon-, germanium-, and silica-based optical devices for telecommunications applications. *IEEE J. Sel. Topics Quant. Elec.* 17, 516–525.

[19] Vivien, L., and Pavesi, L. (eds). (2013). *Handbook of Silicon Photonics*. Boca Raton, FL: CRC Press.

[20] Chrostowski, L., and Hochberg, M. (2015). *Silicon Photonics Design*. Cambridge: Cambridge University Press.

[21] Reed, G., Mashanovich, G., Gardes, F., and Thomson, D. (2010). Silicon optical modulators. *Nature Photon.* 4, 518–526.

[22] Pobanz, C. (2015). *Direct-Coupled Driver for Mach-Zehnder Optical Modulators*. US Patents no. 8948608.

[23] Shastri, K. (2009). "CMOS photonics," in *Proceedings of the Asia Communications and Photonics Conference*, TuR3.

[24] Zhang, J., Zhang, H., Chen, S., Yu, M., Lo, G., and Kwong, D. (2011). "A polarization diversity circuit for silicon photonics," in *Proceedings of the OFC/NFOEC*, JThA19.

[25] Farhood, A. (2012). *Optimal Unipolar PAM Solutions for 100G SMF link from Channel Capacity Perspective*. Available at: http://www.ieee80 2.org/3/bm/public/sep12/farhood_01_0912_optx.pdf

[26] Eiselt, N., Wei, J., Griesser, H., Dochhan, A., Eiselt, M., Elbers, J., et al. (2016). "First real-time 400G PAM-4 demonstration for inter-data center transmission over 100 km of SSMF at 1550nm," in *Proceedings of the OFC/NFOEC*, W1K.5.

[27] Agrawal, G. (2012). *Nonlinear Fiber Optics*, 5th Edn. Cambridge, MA: Academic Press.

[28] Yin, S., Chan, T., and Way, W. (2015). 100-km DWDM transmission of 56-Gb/s PAM4 per λ via tunable laser and 10-Gb/s InP MZM. *IEEE Photonics Tech. Lett.* 27:2531.

[29] Filer, M., Searcy, S., Fu, Y., Nagarajan, R., and Tibuleac, S. (2017). "Demonstration and performance analysis of 4Tbps DWDM Metro-DCI system with 100G PAM4 QSFP28 Modules," in *Proceedings of the OFC/NFOEC*, W4D.4.

[30] OIF (2016). *OIF2016.463, 400ZR Interoperability*. Available at: https://www.oiforum.com/get/48077

11

Low-Loss Photonic Integration: Applications in Datacenters

**Demis D. John[1], Grant Brodnik[1], Sarat Gundavarapu[1],
Renan L. Moreira[1], Michael Belt[2], Taran Huffman[3]
and Daniel J. Blumenthal[1]**

[1]University of California Santa Barbara, California, USA
[2]Honeywell Inc., California, USA
[3]GenXComm Inc., Austin, TX, USA

11.1 Datacenters and Photonic Integrated Circuits

Interconnects in communication systems play a major role in contributing to the global carbon footprint, especially in data centers and cloud-computing applications where switch fabric capacity is expected to grow exponentially [1]. The power consumption of these systems is very high, with a typical large single-rack router consuming over 10 kW of power [2, 3]. Architecting electronic core routers with higher capacities continues to burden all aspects of system design and underlying technologies including switch fabric capacity and packet processing, such as forwarding, queuing, and buffering. To reduce the power density, and push aggregate capacities to 100s of Tb/s, multi-rack core designs are employed to spread the system power over several racks. However, these systems require as many as six optoelectronic conversions per input/output, in addition to multirack configurations dominated by interface cards. For example, a 25 Tb/s router with 128 40 Gb/s I/O ports can require 768 40 Gb/s actual or equivalent optoelectronic/electro-optic (OE/EO) conversions, whose power dissipation and footprint increase with number of ports and bit rate per port. One reason for the tradeoff in system bandwidth with physical size/density (where system bandwidth is defined as router capacity or throughput with a given packet loss rate and offered load at the input), even with faster and denser electronics, is related to the power-spreading problem.

431

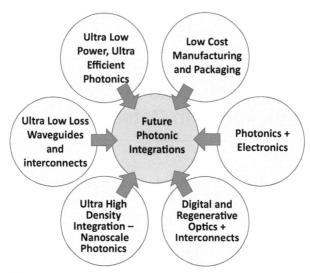

Figure 11.1   Illustration of advances required in new generation PICs to realize future high speed, low cost, power efficient photonic SOCs.

Owing to the exponentially increasing demand of capacity/IP traffic, and the limitations of power density/dissipation in high capacity multi-rack systems, commercial and research interest has surged in realizing high speed, low power, power/temperature efficient optical/photonic interconnects. Key to maximizing the benefits of photonics technology is the realization of low cost, foundry compatible, highly functional, system-on-chip (SOC) solutions, thereby relaxing the power and thermal constraints. Figure 11.1 illustrates the various capabilities that a PIC must be able to perform to meet the aforementioned needs.

Indium phosphide (III–V) [4], silicon [5], hybrid silicon/III–V [6], and ultra-low loss (ULL) silicon nitride (Si_3N_4) [7–9] waveguide platforms are among the commercially mature photonic technologies that enable various system-on-chip functionalities in these high capacity applications.

11.2 InP, Si, and Si_3N_4 Waveguide Platforms

The choice of the integrated waveguide platform is driven by the functional requirements of the device and the underlying (active/passive) components available on each platform. Figure 11.2 lists the realizable optical components available on each platform, and Figure 11.3 illustrates the cross-sectional microchip structure of each.

Building Block	InP	Si	Si$_3$N$_4$
Passive Components	●	● ●	● ● ●
Lasers	● ● ●	○	○
Modulators	● ● ●	● ●	●
Switches	● ● ●	● ● ●	●
Optical Amplifiers	● ● ●	○	○
Detectors	● ● ●	● ● ●	○

Performance	
● ● ●	Very Good
● ●	Good
●	Modest
○	Challenging

	InP	Si	Si$_3$N$_4$
Footprint	● ●	● ● ●	●
Chip Cost	●	● ●	● ●
CMOS Compatibility	○ ○	● ●	●
Low Cost Packaging	○	○^{1}/● ● ^{2}	● ●

^{1} Endfire Coupling (low refl.)
^{2} Vertical Coupling (med. refl.)

Figure 11.2   Key features of three major photonic integration technologies: InP, silicon, and Si$_3$N$_4$ waveguide platforms. From [95].

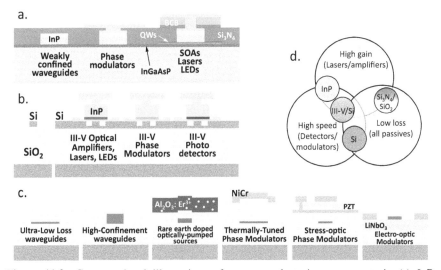

Figure 11.3   Cross-sectional illustrations of common photonic components in (a) InP, (b) III-V/Si and (c) Si$_3$N$_4$ waveguide platforms; (d) Functional capabilities/limitations of each platform.

In general, three of the primary attributes vital to the performance of optical communications devices are: high gain, high speed, and low loss. Figure 11.3(d) illustrates how each platform performs in each of these areas.

As the only direct bandgap material, indium phosphide (InP) provides the best available gain performance by far, but also has higher loss due to electrical doping for current injection, necessitating the use of integrated optical amplifiers [10]. In addition, the InP platform is the most mature in terms of optical system-on-chip complexity, with chips demonstrating upwards of a hundred components.

Silicon photonics leverages the maturity of silicon processing technology to enable high speed and relatively low loss components, but as an indirect bandgap material, native gain on silicon has been difficult to realize. The platform sits in between the InP and ULL Si_3N_4 platforms by providing some active detection but no source, while providing optical losses in between that of InP and Si_3N_4.

Research at U.C. Santa Barbara and LiONiX B.V. resulted in ULL Si_3N_4/glass planar lightwave circuits (PLCs) with propagation lower than 0.1 dB/m [7], orders of magnitude lower than other waveguide platforms. This enables high performance passive components such as long delay lines, AWGRs and optical hybrid splitters, but cannot be electrically pumped for gain due to the insulating nature of the Si_3N_4/silica glass waveguide. Recent work, however, has demonstrated optical gain using erbium-doped glass waveguides [11, 12], showing promise for active component integration on the platform.

Of all the platforms, silicon and silica utilize the most mature materials, as they can leverage existing technology from the silicon electronics & MEMS industries to enable both commercial products and robust foundry services, while indium phosphide has a maturity in between the two, having many commercial devices available and a nascent foundry model. Proper architectural choices can combine the advantages of these different integrated photonic technologies with electronics to achieve optimized performance, power, footprint, and cost.

11.3 The Ultra-Low Loss Si_3N_4/SiO_2 Platform

This chapter discusses the advantages and areas of application of the ultra-low loss Si_3N_4 waveguide platform [7, 8] that can enable a suite of high power handling [13], CMOS compatible, power efficient photonic components including true time delay lines [14], tunable dispersion compensators [15], and comb generators [16] that can be used in highly thermal/power efficient optical interconnects for high capacity communication systems. The Ultra-Low Loss (ULL) SiO_2/Si_3N_4 waveguide platform provides an ideal combination of optical properties desired in passive photonic components such as low loss, high power handling with negligible thermal dissipation, as shown in Figure 11.4.

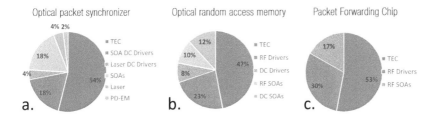

Figure 11.4 Estimated power consumption breakdown for components in a typical high capacity system illustrating the high-power requirements for thermal cooling of a system based on InP PICs: (a) Optical packet synchronizer (b) Optical random-access memory (c) Packet forwarding chip. From [3]. TEC: Thermo-Electric Cooler; SOA: Semiconductor Optical Amplifier; PD-EAM: Photodetector Electro-Absorption Modulator.

The ability to provide active functionality on an ultra-low loss platform can lead to significant power savings. Figure 11.4 illustrates the massive power draw required by the TEC Themro-Electric Coolers that are necessary for optical amplification and laser gain regions to operate in the InP platform. The ULL platform removes the need for amplification, potentially reducing power consumption by significant amounts.

11.4 Integration Building Blocks on the ULL Silicon Nitride Platform

A wide variety of building blocks have been realized in the Si_3N_4 platform. Basic elements including bends, crossings, gain blocks and directional couplers, and are described in further detail in [17, 18]. These building blocks, summarized in Table 11.1, can be used to realize higher level photonic functions and circuits, described in the later sections, and make it possible to:

- convert optical mode sizes using spot-size converters,
- modify the phase of the light using thermal/stress-optic/electro-optic actuators,
- spectrally filter the optical signal via grating filters and optical resonators,
- add optical gain to the waveguides,
- realize 3D photonic integrated circuits using multiple waveguide layers, and
- modulate and switch the optical signal via Mach-Zehnder interferometers

Table 11.1   Various photonic components demonstrated in ULL Si_3N_4 platform

Building Block	Key Terms	References
Mode converters and transition elements, polarizers	Spotsize converters High extinction polarizers	[18–21]
Waveguide Actuation: Thermal, electro-optic, and piezo-electric (PZT)	Thermal tuning, stress-optic tuning, electro-optic tuning	[22–26]
Optical filters, resonators	Single bus and double bus Microring resonators. Asymmetric MZI, Bragg grating filters	[27–30]
Optical Gain	Rare earth doped layers, colloidal quantum dots	[11, 23, 31–34]
3D Multiple Layer circuits	Vertically integrated silicon nitride PLCs	[35–39]

Table 11.1 includes some of the work to realize active devices such as lasers and modulators on the silicon nitride platform. Given that the modulation techniques proposed on this platform are predominantly thermal/stress-optic in nature, they do not meet the modulation speed requirements of high speed communication systems. Realization of lasers on silicon nitride platform can have huge benefits in realizing narrow linewidth, power/thermal efficient optical sources that can be integrated with a suite of high performance passive devices already reported in this platform. Erbium-doped silicon nitride waveguides and colloidal quantum dots on nitride waveguides offer a promising solution in this direction. Besides these, nonlinear optical comb generators [16] using high confinement nitride resonators are poised to have significant improvement in the field of coherent terabit communications.

Finally, ULL Si_3N_4 waveguides can be used as highly effective, low coupling loss interconnects between components in a multi-chip photonic integrated circuit, thereby greatly improving the power efficiency of the device. Coupling losses as low as (0.4 ± 0.2) dB per transition between silicon and ULL waveguides have been demonstrated in [39]. Such integrated coupling between active/passive components would provide a solution to alleviate the high fiber to waveguide coupling loss through seamless chip scale integration of all photonic components (with hybrid III–V/Si actives with ULL Si_3N_4 delay lines) and realize a fully integrated energy efficient communication system-on-chip. In the following sections, the device applications of ULL waveguides in high capacity communication systems, such as dispersion compensators and delay lines will be discussed in further detail.

11.4.1 Available PIC Platforms

The backbone of a Photonic Integrated Circuits (PICs) is the waveguide – the medium by which light is routed around the photonic circuit. Analogous to the electrical traces that route currents and voltages around an electrical microchip, the optical waveguide is the critical enabling component for the photonic integrated circuits that perform the transmission and reception of optical signals in the modern telecom link or datacenter. Many capabilities of the PIC are determined by the choice of the proper waveguide platform, and different waveguide platforms may be better suited to certain aspects of the optical system.

In particular, the optical loss of the waveguide determines how long a device can be made before optical amplification is required, how completely two optical signals can be interfered (ie. the quality factor of resonators and extinction ratio of modulators/switches), and due to power constraints, ultimately limits the total number of optical components that can be integrated onto a single chip.

In addition, the materials and fabrication techniques used to build the waveguide platform determine the types of optical devices that can be integrated. For example, only a few material systems allow for the generation of coherent laser light on a PIC via direct electrical injection, while that same electrical conductivity produces high waveguide losses.

Three primary commercially available PIC waveguide platforms are in current commercial use, each with very different capabilities and applications.

1. The Indium Phosphide (InP) PIC platform offers the benefits of many electro-optical effects, such as electrically-pumped gain & absorption, electro-optic refractive index control, and high refractive index contrast for very short devices and tight bends. However, this high refractive index contrast & electrical conductivity also exacerbate the waveguide's optical loss, limiting lengths to tens of millimeters before optical amplification is required, thus drawing more power and generating more heat. The InP platform is widely used in optical communications due to the vast array of complex PICs able to be produced, with direct electrical injection/detection. Lasers and optical amplifiers on the PIC require thermal cooling to maintain high optical gain. This is often performed

by thermoelectric coolers (TECs), which draw significant power up to one or two Watts.

2. The Silicon Photonics (SiP) waveguide platform leverages the existing infrastructure of the CMOS electronics industry, making high-quality and high-volume wafer fabrication available for photonics devices. The platform requires highly specialized procedures to introduce electrically-pumped optical gain, although this major hurdle has been overcome by bonding of InP to Silicon in various ways, or by high-accuracy packaging for edge-coupling of laser chips to the SiP chip. This platform also allows for small devices/tight bends due to a high refractive index contrast between Silicon and Air, and typically exhibits lower optical waveguide loss than InP due to the removal of electrical conductivity from the waveguide layers. Device lengths are able to reach up to a few centimeters before optical amplification is required.

3. Recent advances in Silicon-Nitride "Glass Waveguide" platforms, aka. Planar Lightwave Circuits (PLCs), have enabled record low losses for an integrated chip-scale waveguide [7]. Glass waveguides have been available for decades, typically based on low-refractive index SiO_xN_y materials. Commercial applications have primarily involved optical routing for telecom applications, such as Arrayed Waveguide Grating Routers for de/multiplexing various optical wavelengths from/onto single fibers. With the advent of ultra-low losses, within a few orders of magnitude of long distance optical fibers, optical delay lines and extremely high-Q resonators which were previously considered entirely unfeasible are now easily produced. For communications applications, the high-Q of low-loss resonators enables low linewidth lasers, high accuracy wavelength discrimination/filters and high-extinction-ratio switching. Ultra-low propagation loss reduces power consumption by eliminating the need for amplifiers and the associated thermoelectric coolers. Device lengths of up to tens of meters have been demonstrated with no amplification [14, 24, 40, 41]. Although amorphous glasses are not conducive to producing on-chip electrically-pumped lasers (due to the lack of electrical conductivity), optically-pumped on-chip lasers can be generated using off-chip pump lasers, the former of which can thus take advantage of ultra-high-Q resonators.

4. Lithium Niobate ($LiNbO_3$) photonic devices have been in use for optical modulation, polarization control and non-linear effects for decades. The material platform has a strong electro-optic and nonlinear effect owing to the ferroelectric nature of the crystal. However, their application in

photonic integrated circuits has been limited by the lack of suitable bending/turning capabilities of the waveguides while maintaining the electro-optic control. Thus, this platform will not be discussed further in this chapter, as it is not yet commercially available as a fully-integrated PIC platform, although some impressive research been shown on the subject [42–44].

The glass waveguide platforms, based on some of the most mature materials in the semiconductor industry (SiO_2, Si_3N_4, Si), are highly manufacturable with high yield in a relatively large number of commercial foundries. Fabrication complexity is generally low, and yet optical performance in regards to loss, quality factor, fiber coupling and transparency bandwidth is the highest of all three platforms. The following sections will discuss these ultra-low-loss (ULL) PICs as applied to datacenter interconnects.

11.5 Ultra-Low Loss PIC Components for Datacom

11.5.1 Low-Loss PICs and Optical Delays

As the reach of datacenter links increases, dispersion compensation is an increasingly important capability. Section 11.5.2 will discuss the specifics of dispersion compensation on an integrated chip-scale platform. Of note is the fact that the compensator requires optical delays of 10–30 cm, often implemented in fiber coils which must be temperature and vibration stabilized while adding significant bulk to the system.

Many photonic applications benefit greatly from the ability to achieve fiber-like losses on an integrated platform, in particular, such systems that require optical delay lines. Delaying an optical signal is a valuable function for many applications including optical buffering [45], packet synchronization [46], optical beamforming [47], microwave signal processing [48], optical coherence tomography [49], and optical gyroscopes [50]. One of the simplest ways to realize a time delay on a non-integrated system is to propagate light through a physical distance, typically done via a fiber coil or free-space propagation region. To achieve such optical delays on an integrated device becomes challenging, because propagation losses and bend losses on most platforms limit the total length and footprint of integrated "fiber" coils. Low propagation loss is required to achieve long delay lengths without significant signal attenuation (without integrating amplifiers), while a small bend radius allows for small footprints.

The ULL waveguide platform plays a major role in the integration of such devices as loss is typically the limiting factor in the total delay achievable.

Low attenuation alone is not sufficient when considering an integration platform for optical delay line circuits. If a planar lightwave circuit (PLC) is to be used as part of an overall optical system then the PLC interfacing with fiber and/or other devices must be taken into account if the processed signal is to be extracted. Optical mode-field mismatch between the fundamental mode of the integrated waveguide and that of the optical fiber can cause a significant amount of loss. Device coupling loss becomes of great importance as the refractive-index contrast (Δ) of the integration platform increases. Index contrast can be calculated based on the refractive index of the core (n_{core}) and the refractive index of the cladding (n_{clad}) as following:

$$\triangle n = \frac{n_{core}^2 - n_{clad}^2}{2n_{core}^2}$$

Waveguide optical confinement is also directly related to the index contrast. The lower the index contrast, the easier it is to match the optical mode between the waveguide and the optical fiber due to the weak confinement. However, decreasing Δn also increases the minimum bend radius and thus the overall device footprint. As a result, three parameters must be investigated before selecting the proper platform: coupling loss, bend radius, and attenuation.

Table 11.2 summarizes all 6 major integration platforms with respect to these three parameters. Index contrast is also included as a comparison metric. The table is not meant to compare state-of-the-art devices but instead the list attempts to provide overall average values for all the platforms. It

Table 11.2 Index contrast, coupling loss, bend loss, and attenuation for all 6 major integration platforms. (SSC: Spot-size converter)

Platform	Si_3N_4	Doped Silica	Polymer	$LiNbO_3$	InP	Silicon
Index contrast (%)	23	0.7	0.7	1	10	40
Coupling Loss (dB)	<1	<1	<1	<1	2 (complex SSC)	2 (complex SSC)
Bend loss (mm)	1	7	7	0.2	0.1	0.02
Attenuation (dB/cm)	0.001	0.05	0.05	0.2	2.5	2

is crucial to mention that some of the very low spot-size converter (SSC) losses come at the cost of complex fabrication and such approaches must be evaluated when selecting the proper platform. Overall, the silicon nitride planar lightwave circuit platform provides an ultra-low propagation loss with coupling losses < 1 dB with bend losses around 1 mm; and therefore, outweighs all the other platforms in regards to the integration of optical delay line circuits.

Without proper optimization of the integration platform for waveguide loss and bend radius, the integration of optical delay line circuits becomes impractical. A sacrifice of propagation loss or footprint can degrade the system performance in such a way that most benefits from photonics integration (stability, power consumption, and cost) become outweighed.

11.5.2 Integrated Dispersion Compensation

Dispersion compensation becomes a critical function in an optical communication system as the transmission length and bitrate (or bitrate-distance product) increase. Optical dispersion causes the spreading in time & space of modulated pulses, impairing the error-rate of the link. Advanced modulation formats are especially sensitive to such optical degradation. PAM-4, a four-level modulation format with higher capacity than standard two-level on-off keying (OOK), tends to be limited to $\sim$5 km due to dispersion-related signal distortion, and there is currently much interest in increasing the reach of this technique via dispersion compensating schemes that can accommodate multiple wavelength channels.

Reach extension beyond the dispersion limited tolerance of 5 km SMF-28 for direct detection was previously focused on using the low dispersion window at 1310 nm [51–53]. However, due to fiber scarcity in these applications, there has been increased interest to move PAM-4 to the 1550 nm C-band in order to increase the link capacity. A tradeoff in moving to C-band has been a need for discrete dispersion compensation technologies like dispersion compensating fibers (DCFs) or Fiber Bragg gratings (FBGs), with which demonstration of real-time 28 GBd WDM C-Band to 80 km and 100 km has been demonstrated [54–56].

Compact photonic integrated technologies address the cost, size, weight, and form factor of these fiber-based dispersion compensation methods. An integrated solution would need to mitigate PAM-4 dispersion for multiple WDM channels, while at the same time satisfying the strict OSNR requirements.

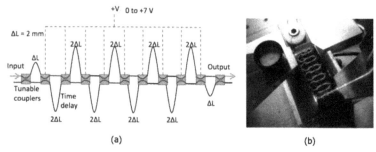

(a)   **(b)**

Figure 11.5   (a) Integrated programmable lattice filter architecture with $\Delta L = 2$ mm for 100 GHz channel spacing. (b) Photograph of the fabricated 10-stage tunable dispersion compensating lattice filter (red light used for fiber-coupling). From [59].

11.5.2.1 Design of an Integrated Dispersion Compensator

The tunable lattice filter for dispersion equalization is based on a generalized optical finite impulse response (FIR) filter, which is the same filter architecture also demonstrated by Doerr et al. and Geheler et al. [57, 58].

A schematic of the device architecture for a programmable 10-stage filter is shown in Figure 11.5. The filter is composed of cascaded alternating symmetric and asymmetric Mach-Zehnder interferometers (MZI). The symmetric MZIs are designed to function as tunable couplers for choosing the path the signal will take, while the asymmetric MZIs function as the dispersive elements and set the filter order. When the MZI couplers are set to 100/0 (bar state) the entire device will function as a large symmetric interferometer with flat transmission and zero dispersion. As the coupling ratio starts to deviate from 100/0, the outer delays act as a mux/demux and all the middle delays act as a long wavelength dependent delay due to the alternating pattern. The maximum delay is achieved when the MZI couplers are at a 50/50 splitting ratio.

As mentioned above in Section 11.5.1, the integration platform restricts the total optical delay achievable. The time delays are chosen to be integer multiples of the unit delay, ΔL, thus making the filter discrete in the time domain. This assures the filters periodicity in the frequency domain, which can then be used to simultaneously compensate multiple WDM channels.

The filter frequency response can be simulated though the use of a T-matrix formalism, as described in [15]. Figure 11.6 shows the simulated dispersion for the different filter designs, allowing us to determine the required delay lengths ΔL. The Figure shows that dispersion is proportional

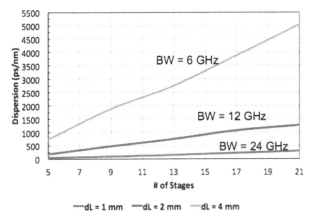

Figure 11.6 Simulated filter dispersion for the generalized lattice filter as a function of the number of stages for 3 different unit delay length. The associated bandwidth (BW) is shown above each curve. From [15].

to ΔL^2 for a set number of stages, and it is therefore desirable to make ΔL as large as possible. However, the filter passband and FSR scale with $1/\Delta L$ and a tradeoff must be considered. Then, for a given delay length, the only way to increase dispersion without compromising bandwidth and FSR is by increasing the number of filter stages, which necessitates the use of a low-loss waveguide platform. By using the design curves in Figure 11.6, a unit delay length of $\Delta L = 2$ mm was chosen, corresponding to an FSR of 100 GHz. As a result, in order to achieve 500 ps/nm of compensation a 10-stage filter was chosen.

The fabricated waveguide platform, shown in Figure 11.7, was designed to accommodate a smaller $\sim$500 μm bend radius while maintaining C-Band losses below 0.001 dB/cm. The ultra-low loss enables the required propagation length >10cm without amplification. Core dimensions of 2.8 μm x 100 nm provide single mode operation at 1550 nm while minimizing the optical mode overlap with the sidewall roughness and maintaining the low-loss characteristics, as described in [60]. Figure 11.7 (Right) shows the measured TE and TM propagation losses as a function of wavelength over the range from 1530 nm to 1600 nm, as determined by optical frequency domain reflectometry (OFDR) measurements on a separate 1 meter spiraled waveguide test structure fabricated on the same wafer. The lowest measured waveguide losses were 0.058 dB/cm and 0.018 dB/cm for TE and TM, respectively. Although the lower confinement of the fundamental TM mode

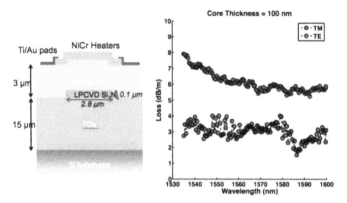

Figure 11.7   (Left) Schematic of the fabricated waveguide cross-section. (Right) OFDR Optical loss spectrum of a spiral test-structure. From [15].

generates lower losses than that of the fundamental TE mode, the TM mode experiences high loss at 500 µm bend radius, so this device will operate in the TE mode only. With the measured propagation losses and the simulated bend losses, the platform described here provides a vast advantage over its doped SiO_2 core PLC counterpart, which provides similar propagation losses but at a much higher minimum bend radius of > 5 mm [61].

The MZI switches utilize NiCr resistive heaters as thermally-actuated optical phase shifters. The NiCr heaters were 10 μm wide and 1000 μm in length – the complete switch characterization is shown in [62].

Figure 11.8 shows the photo mask layout as well as the actual fabricated device. With the minimum bend radius of 0.5 mm, the device dimensions were 9.89 mm $\times$ 22.5 mm, which is equivalent to a footprint of only 2.23 cm^2.

To analyze the device's ability to compensate dispersion, we then analyze the filter's full response at a single channel. The device is measured at a total of seven different bias points from 0 to 7 Volts with a maximum power dissipation of 723 mW. The filter's group delay response is controlled by adjusting the coupling ratio of the tunable coupler via the thermo-optic effect by thermally-tuning a single arm of the MZI, so the MZI switch response was varied from 0/100 to 100/0 to characterize the tunable dispersion properties.

Figure 11.9 displays the transmission and group delay response for a single channel for seven different bias voltages. The transmission plot (left) is normalized to the lowest loss transmission in order to highlight the difference

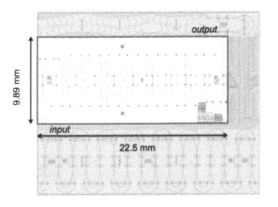

Figure 11.8 (Left) Mask layout showing the dimensions for the filter (9.89 mm × 22.5 mm). (Right) Optical microscope picture of the final fabricated device. From [15].

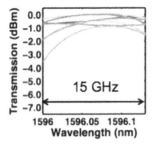
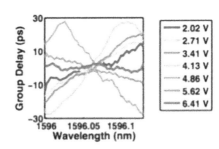

Figure 11.9 Complete filter characterization for a single passband, showing transmission and group delay results for seven different heater bias settings. From [15].

in transmission between each bias setting. As can be seen from the plot, the transmission remains above 3 dB for all of the bias settings with a measured "useful" bandwidth of 15 GHz. Figure 11.9 (right), shows a linear group delay across the transmission bandwidth for all the different biases. Dispersion is defined as the rate of change of group delay over wavelength, which is calculated from the slope of a linear fit of the group delay vs wavelengh. The plot also shows different slope signs for various settings, which correspond to the filter's ability to compensate positive and negative dispersion.

The measured dispersion as a function of heater voltage is shown in Figure 11.10 (right). From the measured results, the filter displays the ability to compensate ± 550 ps/nm with biases between 0 and 7 V.

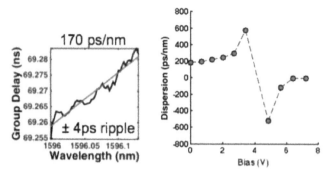

Figure 11.10 (Left) Example of a linear fit of group delay data showing a measured dispersion of 170 ps/nm and a group delay ripple of +/- 4 ps. (Right) Measured dispersion as a function of voltage bias. From [15].

11.5.2.2 Demonstration of 40 Gbps NRZ-OOK Dispersion Compensation

Applying the device to real-world modulated data, transmission measurements were performed on a 40 Gbits/sec NRZ-OOK signal according to the setup in Figure 11.11.

A 10-kilometer single-mode fiber spool was used to introduce dispersion. For the given fiber length, the calculated dispersion expected is approximately 170 ps/nm. The received data for the uncompensated link shows a distorted and closed eye diagram as seen in Figure 11.12 (left). The dispersion equalizer is introduced and biased accordingly to fully compensate the dispersion. Figure 11.12 (right) shows the dispersion-compensated eye

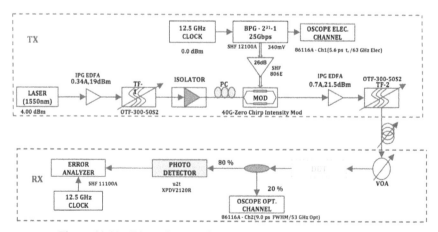

Figure 11.11 Dispersion equalizer transmission testbed. From [15].

uncompensated   compensated

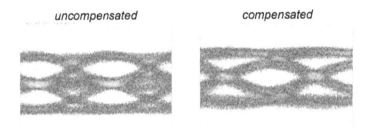

Figure 11.12 Eye diagrams for the uncompensated transmission link (left) and compensated link (right). From [15].

diagram demonstrating the signal improvement over the uncompensated case. The result then assures the filter's proper functionally by compensating a 40 Gb/sec signal. It is important to recall that dispersion tolerance is inversely proportional to bit-rate square, and for a 40 Gb/sec signal, that is equivalent to 4 km of uncompensated propagation on single-mode fiber. Therefore, the filter presented here provides sufficient dispersion compensation for high-speed communications. However, PAM-4 modulation is a more commercially-applicable modulation format, and the following section applies the integrated dispersion compensator to this more modern usage.

11.5.2.3 Demonstration of 40 Gbps PAM-4 Dispersion Compensation

NRZ modulation has become increasingly inadequate for meeting the capacity needs of the modern datacenter interconnect (DCI). In addition, the inter-campus and metro region length-scales are becoming increasingly important with as much as 50% in the 10–40 km distance range for certain deployments. The penetration of PAM-4 as a solution for today's datacenter interconnects has increased dramatically, and the push to apply this interface beyond the original dispersion limited reach of 5 km is of great interest. The same integrated dispersion compensator demonstrated in the previous section can be used to extend the reach of PAM-4 modulated data up to 40 km, as described in the following section.

The transmission experimental setup is shown in Figure 11.13. Inphi PAM-4 Phy IC transmitter and receiver boards were used to generate and detect real-time PAM-4 PRBS-31 patterns, generate histograms and count BER. The real time PAM-4 boards were interfaced to a C-band 40 Gbps Mach-Zehnder modulator-based transmitter and a 32 GHz linear receiver.

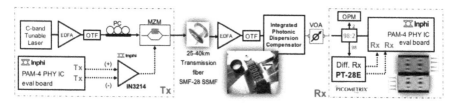

Figure 11.13   Experimental setup for transmission of real time C-band 53.125 Gbps PAM-4 transmission using 10-stage photonic integrated programmable lattice-filter dispersion compensator. From [59].

With the integrated dispersion compensation chip included in the system, the OSNR (optical signal-to-noise ratio) was increased to >40 dB via transmission booster and receiver optical amplifiers and optical filters. The transmitter electrical block consists of an Inphi PAM-4 PHY IC board generating at KR4 (25.78125 GBd) and KP4 (26.5625 GBd) standard IEEE baud rates [52]. The PAM-4 IC board outputs differential signals that drive a single ended Sumitomo MZM intensity modulator using a differential to single-ended linear amplifier (IN3214). A C-band tunable laser and booster EDFA (Erbium-doped fiber amplifier) were input to the modulator at a power level of +7 dBm to transmit 53.125 Gbps data channels onto a 100 GHz C-band ITU grid. A single-ended optical to differential electrical output linear receiver (Picometrix PT-28E) suitable for 28 GBd direct detection enabled measurements with received power as low as –16 dBm. Receiver-side DSP and signal recovery was performed on-board the PHY-IC PAM-4 unit, with built-in BER and SNR analytics, before Forward Error-Correction (FEC).

For each real-time pre-FEC BER measurement only the laser frequency and EDFA filter center frequencies were changed, in increments of 100 GHz. For each BER measurement all other transmission components were kept constant including the booster EDFA, the MZM and PAM-4 drive electronics, the receiver EDFA, the linear receiver gain and bandwidth, and the PAM-4 receiver board equalization settings. All BER measurements are pre-FEC and shown relative to FEC threshold. Back to back optical pre-FEC BER measurements using the experimental setup in Figure 11.13, were made with no transmission fiber in place at 1560 nm, and baseline error detection better than the FEC threshold limit is shown in Figure 11.14(a). Transmission measurements were performed using SMF-28 standard single mode fiber with approximate dispersion of 18 ps/nm · km at 1560 nm and loss of ~0.2 dB/km.

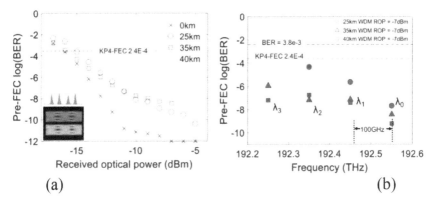

Figure 11.14 Measured Pre-FEC BER for (a) λ_0 back-to-back and over 25 km, 35 km and 40 km of SMF-28 SSMF with the KP4-FEC 2.4E-4 threshold indicated and (b) four 100 GHz spaced channels λ_0 - λ_3 with received optical power (ROP) = −7 dBm also showing performance well below the KP4-FEC 2.4E-4 threshold (channel λ_3 was not recovered for 40 km due to total dispersion exceeding compensated + residual). From [59].

Fiber lengths of 40 km, 35 km and 25 km were used for transmission measurements with resulting pre-FEC BER for channel 1560 nm shown in Figure 11.14(a) below the KP4-FEC error threshold of 2.4×10^{-4} BER. Also shown is a sample eye diagram at the receiver differential output and a post-DSP 4-level histogram. The 40 km length corresponds to dispersion of ~720 ps/nm. An EDFA was incorporated before the photonic chip to overcome device loss and enable transmission measurements up to −5 dBm received optical power. Optical signal-to-noise ratio (OSNR) was maintained >40 dB for all measurements to ensure link performance was not limited by the OSNR margin for PAM-4 optical transmission. Measurement of the real-time pre-FEC BER multiple channels at 100 GHz spacing for the three fiber lengths at four wavelengths is shown in Figure 11.14(b). One of the four channels, λ_3, for the 40 km case, was not plotted due to the accumulated dispersion of 720 ps/nm and the combination of 170 ps/nm residual dispersion tolerance at the PAM-4 receiver board with the 500 ps/nm integrated photonic circuit compensation falling just short of the required compensation at that wavelength. The photonic circuit was operated at 80 mA and 5.9 V = 470 mW.

Test results show 100 GHz spaced C-band channels to be transmitted error-free.

Thus the integrated compensator tunable dispersion range of +/-500 ps/nm was able to support the transmission of three out of four channels over 40 km and all four channels over 35 km and 25 km, below the KP4-FEC 2.4E-4 BER threshold. The dispersion for the fourth channel at 40 km was just under the receiver dispersion tolerance combined with the compensator maximum and therefore was not able to be recovered while keeping all transmission components constant while only the transmission wavelength was changed. It should be noted that this fourth channel was recoverable by adjusting its transmission frequency slightly, however we left the data point off to show only true 100 GHz grid spaced channels. All measurements were made with dispersion limited transmission and OSNR >40 dB. The integrated photonic lattice filter had an insertion loss of 20 dB, due to uncompensated bend-offsets and heater losses [15], which did not limit the link OSNR, but in next generation filters can be lowered to <4 dB using fiber taper couplers on-chip [63] and thicker upper cladding moving the thermal tuning metal further away from the optical mode. These are relatively straight-forward and simple device improvements, and the above demonstration shows that the ULL photonic integrated platform is well-suited to low-power datacenter interconnect applications, especially those requiring amplifier-free optical delays.

11.5.3 Grating Filters

Bragg grating structures are an integral part of the diverse array of components that comprise modern optical systems. These include distributed tunable Bragg reflector (DBR) lasers, distributed feedback (DFB) lasers, dispersion compensators, and wavelength-division multiplexed (WDM) channel filters. With the current push towards coherent transmission and detection schemes, stable and extremely narrow linewidth lasers have become increasingly important. A low loss grating capable of providing high reflectivity, while at the same time maintaining an extremely selective passband, is necessary to realize a new generation of low-linewidth coherent optical devices.

There have been prior results reported on sidewall grating technologies that include patterning through reactive-ion etching [64] and direct-write spatial-phase-locked electron beam lithography [65]. Prominent results in these approaches have passbands on the order of 0.12 nm with coupling constant values of roughly 4.5 cm^{-1} [66]. The limitations to these approaches include their reliance on strict etch tolerances, as well as their requirements for multiple lithographic patterning and etch steps [65, 66]. The limitations of prior approaches are primarily due to a combination of both design constraints and fabrication technology.

This section covers the design of a collection of sidewall gratings in ultra-low-loss Si_3N_4 planar waveguides. Through a precise design approach that leverages lithographically defined waveguide geometry, the presented sidewall gratings can achieve a wide range of coupling constants, including the very low values required for an exceptionally narrow passband. Since the gratings are all defined in a single lithographic step, multiple types of gratings can follow one another along the same waveguide section. This creates an opportunity for integrated circuits with complex filter functions. The requirement of only a single lithography step means that such structures are ideal candidates for integration within the Ultra-Low-Loss waveguide platform.

Figure 11.15 gives a schematic representation of a standard linear Bragg grating implemented within the waveguide sidewall framework.

Here the Bragg period of the structure (Λ), or the total length of a single repeating cell of the structure, controls the Bragg frequency ω_o (or equivalently the Bragg wavelength λ_o). A fabricated and simulated stop-band can be seen in Figure 11.16. The width difference between the wide and narrow sections controls the grating modulation strength (κ). As a qualitative matter of intuition, a greater width difference between grating half-periods will induce a greater magnitude of reflection per unit length. Quantitative measurements describing the extent of the width difference's effect on measured κ values are shown in Figure 11.17 [67].

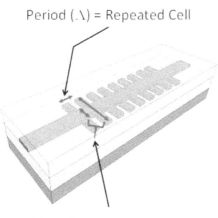

Period (Λ) = Repeated Cell

Width Difference = $w_{wide} - w_{narrow}$

Figure 11.15 Schematic representation of a linear Bragg grating structure with periodic width perturbations along the waveguide propagation length. The period of the structure is denoted as Λ.

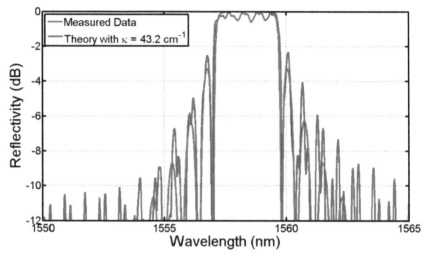

Figure 11.16   Measured and simulated (FIMMWAVE and T-matrix formalism [68] grating bandwidth for 1000 μm long gratings under TE or TM excitation. The plot also gives fitted and simulated coupling constant values for the same set of gratings. The nominal waveguide width is 2.8 μm and the waveguide core thickness is 80 nm. For an increased core thickness is it expected that these measured κ values would decrease. This is because as the core thickness increases, a changing waveguide width has less impact on an overall effective index change for the structure.

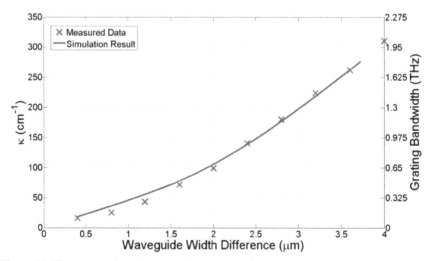

Figure 11.17   Measured grating coupling constants for 1 cm long gratings under either TE or TM excitation. The nominal waveguide width is 2.8 μm and the waveguide core thickness is 80 nm. The plot also gives fitted coupling constant values (κ) for the same set of gratings. Figure taken from [67].

Additionally, for equivalent length gratings, a larger coupling constant will yield a large grating bandwidth. Such design is readily amenable to high volume nanophotonics manufacturing, as the entirety of the structure's properties of interest are most heavily influenced at the highly repeatable lithography stage, rather than the etching stage, which traditionally imparts a higher degree fabrication process variability.

11.5.4 Ring Resonator Filters

A resonant device, at its most basic, is a structure that couples the mode of propagating light back into itself. A ring resonator does this by physically looping a waveguide on itself to form a ring. As light travels around the ring and interferes with itself it will either constructively or destructively interfere determined by the wavelength of light and the effective length of the cavity, expressed as $n_{eff} * L = N * \lambda$, where n_{eff} is the effective index of the mode, L is the length of the ring, N is the mode number, and λ is the wavelength. Light that constructively interferes resonates within the cavity and builds up to a higher intensity. This wavelength is said to be "on resonance." Light that that destructively interferes is rejected from the ring, and is described as "off resonance." If the ring has two buses, resonant light will be transmitted between the buses, shown in Figure 11.18.

We use the transfer function to define the energy that is passed between two ports. The main properties that impact the transfer function is the effective length within the resonator, propagation loss within the resonator, and the coupling between the buses and the resonator, which we refer to as L, α, and κ respectively. Various transfer functions between the input port and the opposite bus, called the drop port, are shown in Figure 11.19 as well as the transfer function of a single bus ring. These plots were derived using Mason's rule following [69].

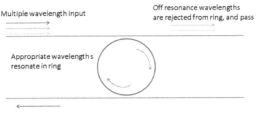

Figure 11.18   Illustration of wavelength selectivity of an add/drop ring resonator.

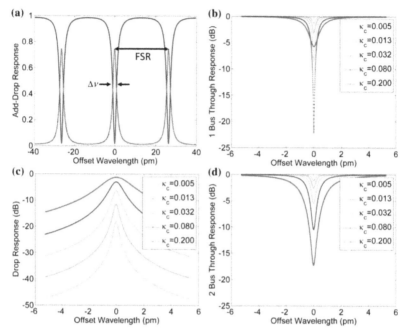

Figure 11.19   Ring resonator spectra at 1550 nm for R = 9.8 mm, α = 5 dB/m, n = 1.48. (a) Add-drop spectra highlighting the FSR and FWHM, and varying k_c for (b) single bus ring, (c) drop port of 2 bus ring ($k_{c1} = k_{c2}$), and (c) through port of 2 bus ring ($k_{c1} = k_{c2}$). Image taken from [70].

There are multiple ways to evaluate a resonator, depending on the application. The Q factor of the resonator is one of the most common metrics. The Q factor is the amount of energy stored within the resonator divided by the power lost per optical cycle (round trip time of resonating light).

$$Q = \omega \frac{T_{rt}}{L}$$

Here ω is the angular frequency of the resonating light, T_{rt} is the round trip time of the resonator, and L is the fractional power loss of the resonator per round trip. We can further define Q into intrinsic Q, Q_i, or loaded Q, Q_1. The loaded Q, Q_1, is the Q of the resonator including the loss of the couplers. It is also called the external Q. Q_i is the Q of the resonator isolated resonator, if it had no couplers. The same ring resonator with varying couplers would have varying Q_1 but the same Q_i.

The Q of the resonator can be expressed in many ways [71]. In the frequency domain Q_1 is represented by the operation: wavelength divided by the full width half maximum (FWHM) of the resonator. In the time domain it relates to the photon lifetime, as it relates to the decay of optical intensity in the cavity. The Q is a measure of sharpness of the resonance, and is relevant for reference cavities, rapid switching, and non-linear ring resonators.

The extinction ratio (ER) of a resonator is the ratio of the highest and lowest detected power at the port. For a single bus resonator, there is, notably, a critical coupling where the extinction ratio is, theoretically, infinite. This is when the propagation loss and coupling loss of the resonator ring destructively interfere perfectly with light in the bus waveguide.

The insertion loss (IL) of the resonator is the ratio of the input power over the output power, at the wavelength of highest transmission. This is independent of any other losses of the measurement, such as coupling loss or system loss. The insertion loss can become significant for extremely low coupling values relative to the loss of the resonator.

The free spectral range (FSR) of the resonator is the span between two resonances, measured in frequency or wavelength calculated as:

$$FSR = \frac{\lambda^2}{n_g L}$$

Finesse of a cavity, F, a measure of the sharpness of the resonance compared to the density of resonances, is defined by the ratio of the FSR over the FWHM. The Q and finesse of a resonator can be related to each other by:

$$Q = \frac{n_{eff} L}{\lambda} F$$

11.5.5 High-Extinction Filters

By serially coupling multiple rings together we can create an extremely high extinction ratio for a variety of applications including separating pump and Stokes signals for Brillouin scattering [72], filtering of idler signals in four-wave mixing (FWM) processes for nonlinear micro-resonators and non-magnetic optical isolation [73], and quantum communications and computing that employ frequency conversion [74]. For many of these filtering applications a larger FSR is required. For this reason, the filter here was fabricated using a 175 nm thick core to allow bend radii as low as 500 μm, enabling FSRs on the order of 50 GHz.

The filter consists of three coupled-ring waveguide resonators interfaced via directional couplers to input and output bus waveguides as illustrated in Figure 11.20(a). S-bends to the directional couplers are used to isolate the

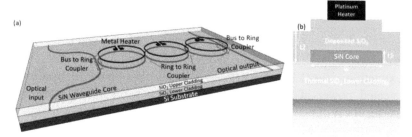

Figure 11.20   (a) Third-order ring filter design; (b) Schematic cross section of Si_3N_4 low loss waveguide. In this device, we use nitride core thickness $t1 = 175$ nm, core width $w = 2.2$ μm and upper cladding thickness $t2 = 6.8$ μm. Thermal oxide lower cladding thickness is 15 microns. From [75].

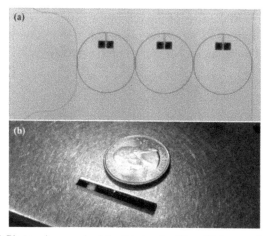

Figure 11.21   (a) Photo micrograph of the third-order filter. (b) Image of a 3.5 mm wide bar of 5 third-order filters relative to a quarter. From [75].

fiber-coupled input and output buses from the ring resonators. The waveguide schematic presented in Figure 11.20(b) incorporates an additional metal layer for filter tuning. A completed third order filter is shown in Figure 11.21(a) and a wafer diced into 3.5 mm columns, each holding 5 third-order filters, is shown in Figure 11.21(b).

A wavelength swept laser source was used to measure the passband and align and tune the rings. However, this approach limits the measurement of the filter stop band to the ER of the laser being used to test. In order to measure a stopband ER greater than 70 dB, an Agilent 86140B optical

Figure 11.22   Schematic representation of measurement setup. From [75].

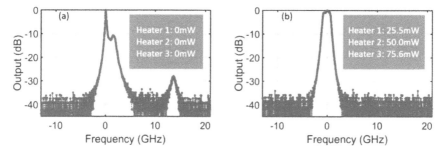

Figure 11.23   Wavelength sweeps of third-order filters. The measurement is limited by photodetector dynamic range. (a) shows a filter initially out of resonance, (b) shows the same filter tuned to resonance. From [75].

spectrum analyzer with sensitivity of −90 dBm used in combination with the tunable laser and EDFA as shown schematically in Figure 11.22.

Each ring within the filter is fabricated with an independently controllable platinum heater. Due to small variations in individual rings within the filter, tuning is required to properly align the resonances as shown in Figure 11.23(a) and enables optimization of both the stopband and the passband as shown in Figure 11.23(b). Filter tuning is achieved through small heater changes as the filter transmission is measured, a technique that has been automated for up to fifth-order filters as reported in [76]. The best shape factor and ripple are ultimately determined by the designed coupling between the resonators. Using this aligning technique, we found device yields greater than 90% across a single wafer.

Relative optical power transmission of the third-order ring filter is shown in Figure 11.24 by plotting the ratio of power at the input vs. output facets (the I/O facets are illustrated in Figure 11.20). The filter extinction ratio is measured to be 80 dB as shown in Figure 11.24(a). Analytically fitting these transfer functions gives coupling values of $\kappa_1 = 0.125$ and $\kappa_2 = 0.005$, very close to the targeted values of 0.13 and 0.006 respectively. The filter 3 dB bandwidth and 20 dB bandwidth were measured to be 1.60 GHz and 3.12 GHz respectively. The filter input loss was measured using a laser set to the passband of the filter and received by a photodetector. The power measured at the

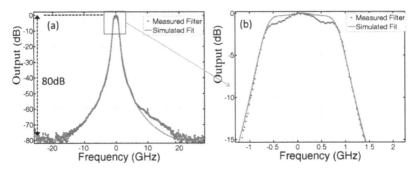

Figure 11.24   (a) Third-order filter function, with an extinction ratio of 80 dB and FSR 48.2 GHz. The analytical fit yields κ_1 fit = 0.125, κ_2 fit = 0.005. (b) Third-order filter passband with a shape factor of 0.437 and no ripple. From [75].

facet was 5.6 dBm, and the power at the detector was 11.7 dBm. The average coupling loss measured on straight waveguide test structures was 2.4 dB. Removing the coupling loss from the power loss in the filter gives an insertion loss of 1.3 dB.

Previously, the heaters were used to independently align the rings to realize a third-order filter. If the power dissipated in the heaters is increased uniformly, such that the differences in power between each heater from the alignment are maintained, the filter can be tuned over its full FSR while maintain the filter shape. Tuning the rings in this manner results in an efficiency of 0.461 GHz per mW of power per ring, equivalent to 0.105 W/FSR, shown in Figure 11.25.

To evaluate metal layer induced losses, we compare the losses of two identical first-order rings, one with a metal tuning layer deposited and the other without a metal tuning layer, shown in Figure 11.26. Fitting the two filter functions to the theoretical model, we find the additional loss of the metal layer to be 1.7 dB/m at 1550 nm.

11.5.6 C-band Lasers on the Si_3N_4 Platform

Optically pumped monolithic rare-earth-ion-doped waveguides leveraging an Al_2O_3 host material on oxidized silicon substrates demonstrate significant market potential due to their wide gain bandwidths [31], high degree of thermal stability [77], low-cost highly repeatable back-end deposition method [12], and capability of handling high bit rate optical communication data streams [78]. The issue of such a gain material requiring an

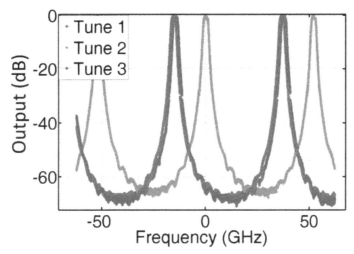

Figure 11.25 A third-order ring filter is tuned over its full FSR. Tune 1 represents no thermal tuning, tune 2 represents 50 mW of thermal tuning, and tune 3 represents 110 mW of tuning. From [75].

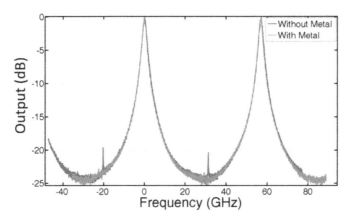

Figure 11.26 First order rings with and without a metal layer are compared. Fitting the two curves to the theoretical model yields and additional loss, due to the metal layer of, 1.7 dB/m.

external optical pump laser has been reduced to merely a pump source/on-chip coupling issue as a consequence of recent cost reductions in the mass production of laser diode components. One such high power GaAs-based laser source could potentially pump many active devices in a PIC through a single fiber-to-chip connection.

Exceptional quality optical amplifiers capable of handling high optical signal powers are key components of any telecommunications and integrated optical system. While on the surface it may seem a straightforward endeavor to translate the concepts of rare-earth-ion-doped fiber amplifiers into a waveguide configuration, scaling down the amplifier's dimension from a few meters of fiber to a few centimeters or less of waveguide requires a proportional magnitude concentration increase of dopant ion to achieve the same optical gain. When doing so, many of the physical processes that were unimportant in fiber amplifiers play a significant role in determining the net optical gain in such waveguide amplifiers.

Rare-Earth-Ion-Doped Waveguide Amplifiers and Lasers

The net modal gain (g_{mod}) of a rare-earth-doped amplifier is dependent upon the fractional overlap (Γ) of the signal with the gain structure (material) through the following relationship [79]:

$$g_{mod} = \Gamma g_{mat} - \alpha_i = \Gamma(\sigma_{em}(\lambda)N_2 - \sigma_{abs}(\lambda)N_1) - \alpha_i$$

where $\sigma_{abs}(\lambda)$ and $\sigma_{em}(\lambda)$ are the stimulated absorption and emission cross-sections [80, 81], $N_2 - N_1$ is the difference between the population densities of the excited state and the ground state, and α_i is the optical propagation loss to extraneous scattering and absorption events. Using this equation as a guide, it is clear that for a high gain on-chip rare-earth-ion-doped waveguide amplifier the following four properties are preferred: 1) a high degree of overlap of the signal with the excited gain structure (high signal/pump overlap factor) to maximize Γ, 2) low optical propagation loss to minimize α_i, 3) a high stimulated emission cross section, and finally 4) a large density of excited ions. The optimization of both properties 3) and 4) is mainly a material engineering challenge. The optimization of both properties 1) and 2) requires an appropriately engineered waveguide structure, as is considered below.

Figures 11.27(a) and Figure 11.27(b) give a schematic representation of the cross section of a shallowly etched waveguide amplifier, while Figure 11.27(c) shows the simulated (via FIMMWAVE by Photon Design [82]) transverse electric (TE) polarized optical mode profile for the 1550 nm wavelength. Here the high aspect ratio Si_3N_4-core design featured in the ULL waveguide acts as a lateral guide for the optical mode.

Such a waveguide design as presented obtains >85% confinement factor for all wavelengths longer than 800 nm and >90% intensity overlap factor with either an 808 nm or 980 nm optical pump over the entire near-infrared

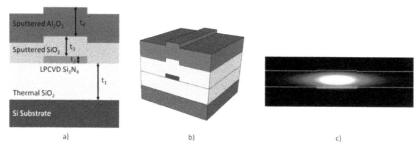

a)   b)   c)

Figure 11.27   (a) Cross-sectional geometry of an on-chip Si_3N_4-core/SiO_2/Al_2O_3-clad rare-earth-ion-doped waveguide amplifier. The Al_2O_3 layer will act as a host for any dopant ions incorporated during the deposition process. The respective layer thicknesses are as follows: t_1 = 15.0 μm, t_2 = 0.08 μm, t_3 = 0.1 μm, t_4 = 1.5 μm. The width of the Si_3N_4-core is 2.8 μm. It is this Si_3N_4-core that provides the majority of the lateral guiding of the optical mode. (b) Three-dimensional schematic image of the geometry. (c) Simulated (via FIMMWAVE) optical mode profile of the fundamental TE waveguide mode at the 1.55 μm wavelength. Here the effective index of the mode is 1.60.

wavelength range, demonstrating insensitivity of mode profiles at various wavelengths. This design also ensures single transverse-mode operation at the pump (including 800 nm, 980 nm, and 1480 nm) and signal wavelengths (1330 nm and 1550 nm) and a strong interaction between the pump and signal light fields and the rare-earth ions [83]. The drawback of this design is that the minimum bend radius is greater than 10 mm.

Figure 11.28 gives a graphical representation of the entirety of the waveguide fabrication process.

Waveguide fabrication begins with either a 500 μm or a 1 mm thick 100 mm diameter silicon substrate upon which 15 microns of silica is thermally grown by wet oxidation for the lower cladding. Next, the Si_3N_4 waveguide core is deposited by low pressure chemical vapor deposition (LPCVD).

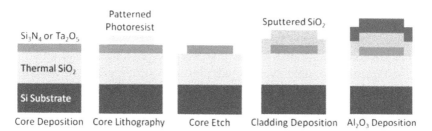

Core Deposition   Core Lithography   Core Etch   Cladding Deposition Al_2O_3 Deposition

Figure 11.28   Schematic overview of the fabrication process for the rare-earth-ion-doped waveguide amplifier designs.

This core layer is then patterned using a photoresist mask by 248 nm stepper lithography and an optimized $CH_3/CF_4/O_2$ inductively coupled plasma etch. The etching chamber had $CH_3/CF_4/O_2$ gas flows of 35/5/10 cm^3/min, a pressure of 0.5 Pa, an RF source power of 500 W, and an RF bias of 50 W. The patterning and etch steps are followed by a blanket sputter deposition of the upper SiO_2 cladding. At this stage in the fabrication, the wafer will undergo an anneal protocol of 7 hours long at 1050°C in 3.0 SLPM N_2 atmosphere. Following annealing, the rare-earth-ion dopant layer within the Al_2O_3 host material will next be deposited by a reactive co-sputtering process [12]. Finally, the wafer is diced into separate die and a mechanical polishing process conditions the device facets. This process is essential to ensure maximum fiber-to-chip coupling efficiency and waveguide-to-waveguide uniformity.

Low cost, high performance laser integration technologies that establish power efficient, temperature stable, and large scale multi-wavelength on-chip arrays are critical for a variety of important applications including coherent optical communications, integrated analog photonics, microwave signal generation, and high spectral resolution light detection and ranging (LIDAR). Integrated waveguide lasers are of particular importance for the realization of compact, rigid, and robust optical devices, since the entire laser cavity along with the optical feedback elements can be fabricated on the same substrate.

When compared to rare-earth-ion-doped dielectric materials, semiconductor gain media exhibit relatively wide lasing linewidths, high amplifier noise figures, and low temperature stability. Since rare-earth-ion-doped dielectric materials do not exhibit an amplitude-phase coupling mechanism as large as that observed in semiconductor lasers, these materials can be used to realize linewidth values that would otherwise be unobtainable with standard semiconductor designs. Furthermore, the high gain in semiconductor lasers makes it difficult to maintain single-longitudinal-mode operation for Bragg-grating-based cavities with relatively strong gratings, since the achievable gain in the cavity also supports the operation of higher order longitudinal modes. In rare-earth-ion-doped lasers, however, single longitudinal-mode operation is typically possible even for cavities with strong gratings, which allows high-quality cavities to be demonstrated due to the high grating reflectivity [84]. The waveguide sidewall Bragg gratings previously detailed in the passive section are integrated together with the rare-earth-ion-doped waveguide amplifiers just discussed to create a novel set of high performance distributed feedback and distributed Bragg reflector lasers.

Laser Characterization

Figure 11.29 depicts the experimental setup used to characterize the lasers. Pump light from a 974 nm laser diode is passed through the 980 nm port of a 980/1550 nm wavelength division multiplexer (WDM) and subsequently coupled onto the device die using a 5 μm spot size (at the $1/e^2$ level) lensed fiber. The lasing signal is collected from the device facet and passed through the 1550 nm port of the WDM, after which the output power is quantified using a power meter while the spectrum is recorded by an optical spectrum analyzer (OSA). The coupling loss for the TE-polarized 1550 nm lasing signal and the 974 nm pump laser diode are approximately 6.3 and 5.4 dB, respectively. The device chip was left uncooled throughout the measurements.

Figure 11.30(a) shows the single-sided lasing output power as a function of pump laser input power for a DBR operating at 1560 nm (grating period of 486 nm). The lasing threshold is observed at 11 mW of launched pump power, and a maximum on-chip pump power of 55 mW generates an on-chip laser power of 2.1 mW. This corresponds to a pump-to-signal conversion efficiency (η) of 5.2%. Such a low operating threshold and high slope efficiency is a consequence of our strongly reflecting cavity design, as well as the low

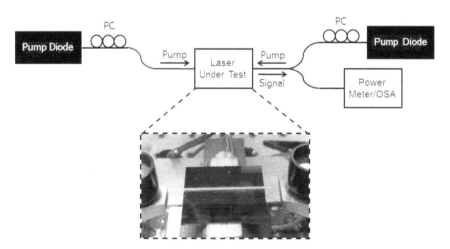

Figure 11.29   Measurement setup of the experiment. The inset photo shows the device under 974 nm excitation. For the DBR devices signal light was collected from the side with the low reflectivity mirror. The green emission seen in the waveguide is due to the cooperative upconversion process the erbium atoms experience when under pump excitation [85]. Figure taken from [86].

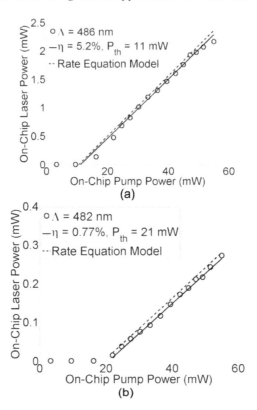

Figure 11.30 (a) DBR laser power as a function of launched pump power for the device operating at 1546 nm. and (b) DFB laser power as a function of launched pump power for the device operating at 1560 nm. Figure taken from [83].

propagation loss of the LPCVD Si_3N_4. Figure 11.30(b) shows the single-sided lasing output power as a function of pump laser input power for a DFB operating at 1546 nm (grating period of 482 nm). Here, the lasing threshold is observed at 21 mW of launched pump power, and for a maximum on-chip pump power of 55 mW we obtain an on-chip laser power of 0.27 mW. This corresponds to a pump-to-signal conversion efficiency of 0.77%, which is a factor of more than 130 times improvement over our design reported in [87]. The main contribution to this improvement in efficiency came from the extension of the total cavity length from 7.5 mm to 21.5 mm, allowing for sufficient pump light absorption to provide useful lasing gain.

Figure 11.31(a) gives the spectra of five different DBR lasers as recorded by the OSA. A simple modification of the grating period within the Si_3N_4

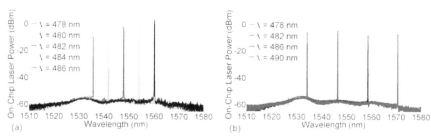

Figure 11.31 (a) Superimposed DBR output laser spectra. (b) Superimposed DFB output laser spectra. Figure taken from [83].

core layer from 478 to 486 nm causes the lasers to output light at 1535, 1541, 1547, 1554, and 1560 nm wavelengths. As is shown, the SMSR for all devices exceeds 50 dB. Figure 11.31(b) gives the spectra of four different DFB lasers as recorded by the OSA. Here, devices with grating periods between 478 and 490 nm operate at 1534, 1546, 1558, and 1570 nm wavelengths. Again the side-mode suppression ratio (SMSR) for all structures is in excess of 50 dB. The differences in output power seen between the devices can be attributed to differences in the gain threshold and the maximum small signal gain spectrum of the erbium-doped active layer.

11.6 Silicon-Nitride Waveguide Design

In order to achieve the optical delays required for dispersion compensation and enable datacenter-scale transmission distances (without amplification), the optical loss of the passive waveguide must be lowered to the below the dB/cm level.

For a feasible and practical low loss waveguide technology platform, there are three essential requirements that must be met [88]. First, in the interest of device packaging, application performance and cost, the on-chip footprint of a photonic integrated circuit should be as flexible as possible. With the current high cost estimates for multi-project wafer (MPW) fabrication runs, available waveguide designs should be able to conserve real estate (through minimization of bend radius) when required, while at the same time achieving as low of a propagation loss as possible. Secondly, the fabrication of a planar lightwave circuit should be both uniform and reproducible over this same range of on-chip area. Specific target applications, such as integrated waveguide optical gyroscopes [41], receive a performance benefit from an increased device area. In such a device, optical waveguides must have uniform loss and scattering characteristics over the entirety of the centimeter scale diameter

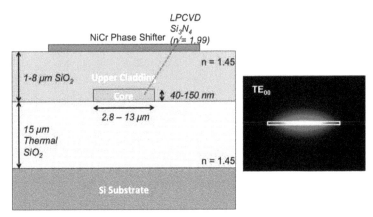

Figure 11.32   (Left) Schematic of ULL "Ultra-Low-Loss" Waveguide structure. (Right) Electrical field intensity for the fundamental mode for a 2.8 μm × 90 nm waveguide. The waveguide structure is outlined in white showing an optical confinement < 10%. Image taken from [89].

of the circular gyroscope coil. Finally, to meet demands of complex higher functionality, multiple passive components should exist and be integrable with one another.

A schematic of the cross-sectional structure of an ultra-low loss waveguide is shown in Figure 11.32. The waveguide consists of a buried silicon nitride core (n = 1.93–1.99) and a silicon dioxide cladding (n = 1.45). The index contrast between the waveguide core and the waveguide cladding is one of the fundamental parameters of an integrated optical waveguide. To achieve a high-density of photonic integration, a high-index contrast is mandatory for the tight confinement required for the small bend radii. On the other hand, high-index contrast usually means high propagation loss due to roughness at the waveguide boundaries between the core and the cladding. The main sources of losses in an optical waveguide are fabrication-induced roughness at the waveguide sidewalls followed by material absorption. Therefore, in order to achieve a low propagation loss at a tight bend radius, a high index contrast core is required without compromising the loss.

When the electric field propagates down the waveguide, photons can be absorbed, scattered, or radiated forming the three fundamental types of losses on an optical waveguide. The major contributor of loss on the ULL platform comes from scattering loss due to surface/sidewall roughness of the waveguide core arising from deposition and etching. Scattering loss occurs when electromagnetic waves interact with roughness "centers" of a size smaller

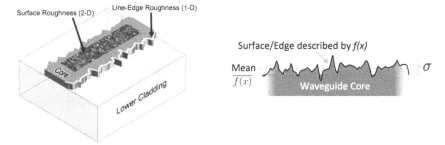

Figure 11.33 (Left) Illustration of waveguide sidewall roughness. (Right) Line edge roughness is the standard deviation, σ, from the ideal waveguide or "Mean" flat edge. From [92].

than the wavelength [88]. Waveguide root-mean square (RMS) roughness, also referred to as line edge roughness (LER), is defined as the deviation from the ideal edge as seen from the top-down, as shown in Figure 11.33. LER originates from the sidewall striations of photoresist and the subsequently etched feature.

Payne and Lacey developed a theoretical upper bound expression on the scattering loss for planar waveguides based on the values of the line edge roughness [90].

$$\alpha \leq \frac{\sigma^2}{k_0 d^4 n_1} \kappa$$

where k_0 is the free-space wave number ($2\pi/\lambda$), n_1 is the core effective index, d is the core half-width, and κ is a parameter that depends on the waveguide geometry and roughness statistical distribution (Gaussian, exponential, etc.). For most practical waveguide geometries κ is typically on the order of unity [91]. The equation above then predicts that the best method of reducing waveguide scattering loss is by reducing the waveguide roughness (σ).

Figure 11.34 shows the simulated scattering loss per unit length as a function of RMS roughness for the waveguide geometry of 2.8 μm $\times$ 90 nm. The simulation shows the quadratic dependence of the loss on roughness and predicts that in order to achieve losses below 1 dB/m, an RMS roughness of less than 3 nm is required.

After scattering loss is considered, the second limiting factor of waveguide loss is the material absorption. This type of loss is due primarily to the molecular vibration caused from impurities incorporated into the waveguide material. Of particular interest are the overtone absorptions caused by hydrogen impurities, which have strong absorption lines around 1550 nm. Si-H and N-H for example have absorption lines centered at 1480 nm and 1510 nm,

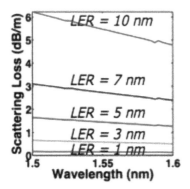

Figure 11.34 Simulated scattering loss for different waveguide RMS roughness for a 2.8 um × 90 nm waveguide geometry. Image taken from [89].

respectively. Although the peaks are not centered at the operating wavelength of 1550 nm, the broad tails of the absorption lead to high losses in the region of interest.

Hydrogen bonds can be eliminated and the hydrogen diffused out of the material by annealing the films at a high temperature typically around 1100°C [93, 94]. Figure 11.35 shows the measured waveguide loss before and after anneal (1050°C for 7 hours) for both types of waveguide cladding, sputter and PECVD. For both cases, the high temperature anneal reduced the propagation loss. The PECVD upper cladding had greater loss reduction with a high temperature anneal due to its higher hydrogen content before anneal.

The final contribution to propagation loss is bend loss. Figure 11.36 (Left) shows the schematic of a fundamental mode of a waveguide propagating in a circular bend of radius R. The dashed lines on the plot indicate the constant phase plane, and if the phase front is to remain linear, the tangential speed must increase as the radial distance is increased (indicated by the arrow sizes in the plot). Therefore, as R increases the tangential speed approaches the speed of light, and as a result the tail of the mode must "leak" away from the guided mode resulting in bend loss.

The bend loss is determined by how confined the mode is to the core. A thicker core means the light will be more confined, and can bend much tighter before experiencing significant bend losses. We define 0.1 dB/m as a critical threshold for bend loss contribution. We then define a critical bend radius for each waveguide geometry.

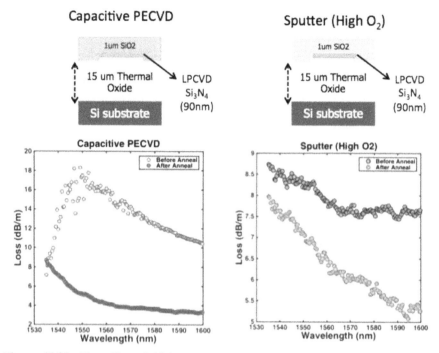

Figure 11.35 The effect of high temperature anneal (1050 C, 7 hrs) on waveguide propagation loss. From [89].

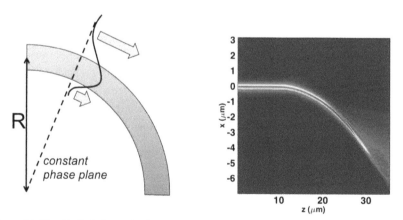

Figure 11.36 (Left) Schematic view of the fundamental mode propagating in a circular bend. Arrows indicate the tangential speed of the mode. (Right) Simulation of the fundamental mode using Beam Propagation Method showing mode leakage as it propagates through the bend. From [89].

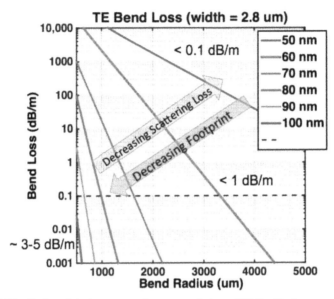

Figure 11.37 Series of design curves for proper design of PLCs. Each curve corresponds to waveguide geometry and the bend loss is plotted for each geometry as a function of bend radius. The minimum bend radius is defined as the point where the bend losses reach 0.1 dB/m and it is shown as the black dashed line. Image taken from [89].

Figure 11.37 shows the simulated bend losses as a function of bend radius for multiple waveguide core thicknesses. The critical bend limit is designated by the dotted black line. As can be seen from the figure, as the core thickness decreases the minimum bend radius becomes larger, which is a direct result of "squeezing" out the mode so that it becomes weakly confined. Scattering loss on the contrary, as explained in the previous section, decreases as the core thickness becomes thinner. A trade-off arises between scattering loss and footprint, and the proper geometry must be selected according to the application requirements for footprint and insertion loss. Figure 11.37 is thus an essential tool in determining the proper design geometry.

11.7 Summary

A few primary platforms for Photonic Integrated Circuits are available for optical data transmission, each offering some combination of advantages and drawbacks. The InP PIC platform offers the advantage of maturity owing to widespread use in long-haul Datacom links. Complex circuits including

combinations of active and passive devices enable highly sophisticated devices, utilizing cheap monolithic fabrication processes. The platform does require significant power draw, especially in the cooling of active components. The Silicon Photonics platform is very promising owing to the low cost and ability to be produced by existing CMOS foundries, and many critical optical components have been demonstrated, although integrated light sources are not yet in the fully mature stage.

As power consumption is a critical factor in the datacenter, the ultra-low-loss Si_3N_4 or Ta_2O_5 waveguide platform has the potential to enable low power consumption, amplifier and cooling-free PICs due to optical losses being 2–5 orders of magnitude lower than the two aforementioned platforms. Although the primary drawback is the requirement of off-chip optical sources, the low-loss properties potentially reduce overall power consumption, while enabling powerful datacenter-applicable technologies previously unfeasible on PICs. Optical delay lines, enabling integrated dispersion compensation and longer-distance transmission, are only available on a low-loss platform such as this. In addition, high-Q resonators and optical filters, enabling low-linewidth lasers and add/drop switches, have much higher performance on the ultra-low loss platform.

As datacenter requirements advance in terms of transmission length, bit rate, and importantly, reduced power consumption, these photonic integrated circuit technologies are expected to play a critical role in the system capacity as a whole.

Acknowledgements

We would like to acknowledge our funding sources: DARPA MTO "IPhOD" (grant no. HR0011-09-C-0123) and DARPA MTO "E-PHI" (grant no. HR0011-12-C-0006), Keysight Technologies.

References

[1] M. K. Patterson, D. G. Costello, P. F. Grimm, and M. Loeffler, "Data center TCO; a comparison of high-density and low-density spaces," Intel Corp., Presented at THERMES 2007, Santa Fe, NM, USA, 2007.

[2] D. J. Blumenthal, J. Barton, N. Beheshti, J. E. Bowers, E. Burmeister, L. A. Coldren, et al., "Integrated photonics for low-power packet networking," *IEEE Journal on Selected Topics in Quantum Electronics,* vol. 17, pp. 458–471, 2011.

[3] J. P. Mack, "Asynchronous Optical Packet Routers," Dept of Electrical and Computer Engineering, Univ. of California Santa Barbara, Santa Barbara, CA, USA, 2009.

[4] M. Smit, X. Leijtens, H. Ambrosius, E. Bente, J. Van Der Tol, B. Smalbrugge, et al., "An introduction to InP-based generic integration technology," *Semiconductor Science and Technology,* vol. 29, 2014.

[5] B. Jalali and S. Fathpour, "Silicon Photonics," *Journal of Lightwave Technology,* vol. 24, pp. 4600–4615, 2006.

[6] A. W. Fang, H. Park, O. Cohen, R. Jones, M. J. Paniccia, and J. E. Bowers, "Electrically pumped hybrid AlGaInAs-silicon evanescent laser," *Optics Express,* vol. 14, p. 9203, 2006.

[7] J. F. Bauters, M. Heck, D. D. John, J. S. Barton, C. M. Bruinink, A. Leinse, et al., "Planar waveguides with less than 0.1 dB/m propagation loss fabricated with wafer bonding," in *OpEx* vol. 19, ed, 2011, pp. 24090–24101.

[8] P. Munoz, J. D. Domenech, C. Dominguez, A. Sanchez, G. Mico, L. A. Bru, et al., "State of the art of Silicon Nitride photonics integration platforms," *2017 19th International Conference on Transparent Optical Networks (ICTON),* pp. 1–4, 2017.

[9] K. Wörhoff, R. G. Heideman, A. Leinse, and M. Hoekman, "TriPleX: A versatile dielectric photonic platform," *Advanced Optical Technologies,* vol. 4, pp. 189–207, 2015.

[10] L. A. Coldren, S. C. Nicholes, L. Johansson, S. Ristic, R. S. Guzzon, E. J. Norberg, et al., "High performance InP-based photonic ICs – A tutorial," *Journal of Lightwave Technology,* vol. 29, pp. 554–570, 2011.

[11] M. Belt, & Blumenthal, D. J. (2014). Erbium-doped waveguide DBR and DFB laser arrays integrated within an ultra-low-loss Si_3N_4 platform. Optics Express (2014), "Erbium-doped waveguide DBR and DFB laser arrays integrated within an ultra-low-loss Si_3N_4 platform," *Optics Express,* vol. 22, p. 10655, 2014.

[12] K. Wörhoff, J. D. B. Bradley, F. Ay, D. Geskus, T. P. Blauwendraat, and M. Pollnau, "Reliable Low-Cost Fabrication of Low-Loss $Al_2O_3:Er^{(3+)}$ Waveguides With 5.4-dB Optical Gain," in *Quantum Electronics, IEEE Journal of* vol. 45, ed, 2009, pp. 454–461.

[13] M.-C. Tien, J. F. Bauters, M. J. R. Heck, D. J. Blumenthal, and J. E. Bowers, "Ultra-low loss Si_3N_4 waveguides with low nonlinearity and high power handling capability.," *Optics express,* vol. 18, pp. 23562–23568, 2010.

[14] R. L. Moreira, J. Garcia, W. Li, J. Bauters, J. S. Barton, M. J. R. Heck, et al., "Integrated ultra-low-loss 4-Bit tunable delay for broadband phased array antenna applications," *IEEE Photonics Technology Letters,* vol. 25, pp. 1165–1168, 2013.

[15] R. Moreira, S. Gundavarapu, and D. J. Blumenthal, "Programmable eye-opener lattice filter for multi-channel dispersion compensation using an integrated compact low-loss silicon nitride platform," *Optics Express,* vol. 24, p. 16732, 2016.

[16] J. Pfeifle, V. Brasch, M. Lauermann, Y. Yu, D. Wegner, T. Herr, et al., "Coherent terabit communications with microresonator Kerr frequency combs," pp. 1–11, 2013.

[17] D. J. Blumenthal, "Ultra-Low Loss SiN Waveguide Platform for Integrated Passive and Active Components for Next Generation Photonic Integrated Circuits," *Summer Topical Meeting Series,* 2016.

[18] C. G. H. Roeloffzen, M. Hoekman, E. J. Klein, L. S. Wevers, R. B. Timens, D. Marchenko, et al., "Low-loss Si_3N_4 triplex optical waveguides: Technology and applications overview," *IEEE Journal of Selected Topics in Quantum Electronics,* vol. 24, 2018.

[19] Y. Lin, C. Browning, R. B. Timens, D. H. Geuzebroek, C. G. H. Roeloffzen, D. Geskus, et al., "Narrow linewidth hybrid InP-TriPleX photonic integrated tunable laser based on silicon nitride micro-ring resonators," in *Optical Fiber Communication Conference* vol. 1, ed. Washington, D.C.: OSA, 2018, p. Th2A.14.

[20] J. F. Bauters, M. Heck, D. Dai, D. D. John, J. Barton, D. Blumenthal, et al., "High Extinction, Broadband, and Low Loss Planar Waveguide Polarizers," *Advanced Photonics Congress,* p. ITu2B.2, 2012.

[21] J. F. Bauters, M. J. R. Heck, D. Dai, J. S. Barton, D. J. Blumenthal, and J. E. Bowers, "Ultralow-loss planar Si_3N_4 waveguide polarizers," *IEEE Photonics Journal,* vol. 5, pp. 0–7, 2013.

[22] J. P. Epping, D. Marchenko, A. Leinse, R. Mateman, M. Hoekman, L. Wevers, et al., "Ultra-low-power stress-optics modulator for microwave photonics," *Proc. SPIE, Integrated Optics: Devices, Materials, and Technologies XXI,* vol. 10106, 2017.

[23] E. S. Hosseini, J. D. B. Bradley, J. Sun, G. Leake, T. N. Adam, D. D. Coolbaugh, et al., "Distributed Feedback Laser," vol. 39, pp. 3106–3109, 2014.

[24] A. Leinse, R. G. Heideman, M. Hoekman, F. Schreuder, F. Falke, C. G. H. Roeloffzen, et al., "TriPleX waveguide platform: low-loss technology

over a wide wavelength range," *Proc. of SPIE Photonics West,* vol. 8767, p. 87670E, 2013.

[25] S. Miller, Y.-h. D. Lee, J. Cardenas, A. L. Gaeta, and M. Lipson, "Electro-optic effect in silicon nitride," *OSA Technical Digest (online),* pp. 3–4, 2015.

[26] S. Jin, L. Xu, H. Zhang, and Y. Li, "LiNbO3 Thin-Film Modulators Using Silicon Nitride Surface Ridge Waveguides," in *Phot. Tech. Lett.* vol. 28, ed, pp. 736–739.

[27] T. Huffman, G. M. Brodnik, C. Pinho, S. Gundavarapu, D. Baney, and D. J. Blumenthal, "Integrated Resonators in an Ultra-Low Loss Si_3N_4/SiO_2 Platform for Multifunction Applications," *IEEE Photonics Society (IPS) Journal of Selected Topics in Quantum Electronics (JSTQE),* 2018.

[28] S. Gundavarapu, M. Puckett, T. Huffman, R. Behunin, T. Qiu, G. M. Brodnik, et al., "Integrated Waveguide Brillouin Laser," pp. 1–15, 2017.

[29] P. Del'Haye, A. Schliesser, O. Arcizet, T. Wilken, R. Holzwarth, and T. J. Kippenberg, "Optical frequency comb generation from a monolithic microresonator," *Nature,* vol. 450, pp. 1214–1217, 2007.

[30] M. Belt, J. Bovington, R. Moreira, J. F. Bauters, M. J. R. Heck, J. S. Barton, et al., "Sidewall gratings in ultra-low-loss Si_3N_4 planar waveguides," *Optics express,* vol. 21, pp. 1181–8, 2013.

[31] J. D. B. Bradley, L. Agazzi, D. Geskus, F. Ay, K. Worhoff, and M. Pollnau, "Gain bandwidth of 80 nm and 2 dB/cm peak gain in $Al_2O_3{:}Er^{3+}$ optical amplifiers on silicon," in *J. Opt. Soc. Am. B* vol. 27, ed, 2010, p. 187.

[32] E. H. Bernhardi, H. A. G. M. V. Wolferen, L. Agazzi, M. R. H. Khan, and C. G. H. Roeloffzen, "Distributed feedback waveguide laser in $Al_2O_3{:}Er^{3+}$ on silicon," vol. 35, pp. 2394–2396, 2010.

[33] Purnawirman, E. S. Hosseini, A. Baldycheva, J. Sun, J. D. B. Bradley, T. N. Adam, et al., "Erbium-doped laser with multi-segmented silicon nitride structure," *Conference on Optical Fiber Communication, Technical Digest Series,* vol. 1, pp. 5–7, 2014.

[34] Y. Zhu, W. Xie, P. Geiregat, S. Bisschop, and T. Aubert, "Hybrid Colloidal Quantum Dot Silicon Nitride Waveguide Gain Measurement Based on Variable Stripe Length Method," pp. 1–2, 2016.

[35] D. D. John, M. J. R. Heck, J. F. Bauters, R. Moreira, J. S. Barton, J. E. Bowers, et al., "Multilayer platform for ultra-low-loss waveguide applications," *IEEE Photonics Technology Letters,* vol. 24, pp. 876–878, 2012.

[36] R. Moreira, J. Barton, M. Belt, T. Huffman, and D. Blumenthal, "Optical Interconnect for 3D Integration of Ultra-Low Loss Planar Lightwave Circuits," *Advanced Photonics 2013,* p. IT2A.4, 2013.

[37] K. Shang, S. Pathak, B. Guan, G. Liu, and S. J. B. Yoo, "Low-loss compact multilayer silicon nitride platform for 3D photonic integrated circuits," *Optics Express,* vol. 23, p. 21334, 2015.

[38] W. D. Sacher, J. C. Mikkelsen, P. Dumais, J. Jiang, D. Goodwill, X. Luo, et al., "Tri-layer silicon nitride-on-silicon photonic platform for ultra-low-loss crossings and interlayer transitions," *Optics Express,* vol. 25, p. 30862, 2017.

[39] J. F. Bauters, M. L. Davenport, M. J. R. Heck, J. K. Doylend, A. Chen, A. W. Fang, et al., "Silicon on ultra-low-loss waveguide photonic integration platform," *Optics Express,* vol. 21, pp. 544–555, 2013.

[40] J. F. Bauters, M. J. R. Heck, D. D. John, J. S. Barton, C. M. Bruinink, A. Leinse, et al., "Planar waveguides with less than 0.1dB/m propagation loss fabricated with wafer bonding," *Optics Express,* vol. 19, pp. 24090–24101, 2011.

[41] S. Gundavarapu, T. Huffman, M. Belt, R. Moreira, J. Bowers, and D. Blumenthal, "Integrated Ultra-Low-Loss Silicon Nitride Waveguide Coil for Optical Gyroscopes," in *Optical Fiber Communication Conference*, ed: OSA, 2016, p. W4E.5.

[42] G. Poberaj, M. Koechlin, F. Sulser, and P. Günter, "High-density integrated optics in ion-sliced lithium niobate thin films," in *Integrated Optics: Devices, Materials, and Technologies XIV* vol. 7604, ed: International Society for Optics and Photonics, 2010, p. 76040U.

[43] J. Thomas, M. Heinrich, P. Zeil, V. Hilbert, K. Rademaker, R. Riedel, et al., "Laser direct writing: Enabling monolithic and hybrid integrated solutions on the lithium niobate platform," in *physica status solidi (a)* vol. 208, ed: Wiley-Blackwell, 2010, pp. 276–283.

[44] P. O. Weigel, M. Savanier, C. T. DeRose, A. T. Pomerene, A. L. Starbuck, A. L. Lentine, et al., "Lightwave Circuits in Lithium Niobate through Hybrid Waveguides with Silicon Photonics," in *Scientific Reports* vol. 6, ed: Nature Publishing Group, 2016, p. 22301.

[45] E. F. Burmeister, D. J. Blumenthal, and J. E. Bowers, "A comparison of optical buffering technologies," in *Optical Switching and Networking* vol. 5, ed, 2008, pp. 10–18.

[46] J. P. Mack, K. N. Nguyen, J. G. N. F. Optic, and 2010, "Asynchronous 2 × 2 optical packet synchronization, buffering, and forwarding," in *ieeexplore.ieee.org*, ed.

[47] J. L. Corral, J. Marti, and J. Fuster, "True time-delay scheme for feeding optically controlled phased-array antennas using chirped-fiber gratings," *IEEE Photonics Journal*, 1997.

[48] J. Capmany, B. Ortega, and D. Pastor, "A tutorial on microwave photonic filters," *Journal of Lightwave Tech.*, vol. 24, p. 201, 2006.

[49] E. S. Choi, J. Na, S. Y. Ryu, G. Mudhana, and B. Lee, "All-fiber variable optical delay line for applications in optical coherence tomography: feasibility study for a novel delay line," *Optics Express*, vol. 13, pp. 1334–1345, 2005.

[50] C. Ciminelli, F. DellOlio, C. E. Campanella, and M. N. Armenise, "Photonic technologies for angular velocity sensing," in *Advances in Optics and Photonics* vol. 2, ed, 2010, pp. 370–404.

[51] M. Birk, L. E. Nelson, G. Zhang, C. Cole, C. Yu, M. Akashi, et al., "First 400GBASE-LR8 interoperability using CFP8 modules," in *Optical Fiber Communication Conference*, ed: OSA, 2017, p. Th5B.7.

[52] Y. F. Chang, "Link Performance Investigation of Industry First 100G PAM4 IC Chipset with Real-time DSP for Data Center Connectivity," in *Optical Fiber Communication Conference*, ed: OSA, 2016, p. Th1G.2.

[53] T. Chan and W. I. Way, "112 Gb/s PAM4 Transmission Over 40 km SSMF Using 1.3 μm Gain-Clamped Semiconductor Optical Amplifier," in *Optical Fiber Communication Conference*, ed: OSA, 2015, p. Th3A.4.

[54] M. Filer, S. Searcy, Y. Fu, R. Nagarajan, and S. Tibuleac, "Demonstration and Performance Analysis of 4 Tb/s DWDM Metro-DCI System with 100G PAM4 QSFP28 Modules," in *Optical Fiber Communication Conference*, ed: OSA, 2017, p. W4D.4.

[55] N. Eiselt, J. Wei, H. Griesser, A. Dochhan, M. H. Eiselt, J.-P. Elbers, et al., "Evaluation of Real-Time 8 $\times$ 56.25 Gb/s (400G) PAM-4 for Inter-Data Center Application Over 80 km of SSMF at 1550 nm," in *Jnl. Lightwave Tech.* vol. 35, ed, 2017, pp. 955–962.

[56] N. Eiselt, J. Wei, H. Griesser, A. Dochhan, M. H. Eiselt, J.-P. Elbers, et al., "First Real-Time 400G PAM-4 Demonstration for Inter-Data Center Transmission over 100 km of SSMF at 1550 nm," in *Optical Fiber Communication Conference*, ed: OSA, 2016, p. W1K.5.

[57] C. R. Doerr, M. Cappuzzo, A. Wong-Foy, L. Gomez, E. Laskowski, and E. Chen, "Potentially Inexpensive 10-Gb/s Tunable Dispersion Compensator With Low Polarization Sensitivity," in *Phot. Tech. Lett.* vol. 16, ed, 2004, pp. 1340–1342.

[58] J. Gehler, R. Wessel, F. Buchali, G. Thielecke, A. Heid, and H. Blow, "Dynamic adaptation of a PLC residual chromatic dispersion compensator at 40 Gb/s," in *OFC 2003 - Optical Fiber Communication Conference and Exhibition*, ed: IEEE, 2003, pp. 750–751 vol.2.

[59] G. M. Brodnik, C. Pinho, F. Chang, and D. J. Blumenthal, "Extended Reach 40 km Transmission of C-Band Real-Time 53.125 Gbps PAM-4 Enabled with a Photonic Integrated Tunable Lattice Filter Dispersion Compensator," in *Optical Fiber Communication Conference*, ed: OSA, 2018, p. W2A.30.

[60] J. F. Bauters, M. Heck, D. D. John, D. Dai, M. C. Tien, J. S. Barton, et al., "Ultra-low-loss high-aspect-ratio Si_3N_4 wavequides," in *OpEx* vol. 19, ed, 2011, pp. 3163–3174.

[61] R. Adar, M. Serbin, and V. Mizrahi, "Less than 1 dB per meter propagation loss of silica waveguides measured using a ring resonator," in *Jnl. Lightwave Tech.* vol. 12, ed, 1994, pp. 1369–1372.

[62] R. L. Moreira, J. Garcia, W. Li, J. F. Bauters, J. S. Barton, M. J. R. Heck, et al., "Integrated Ultra-Low-Loss 4-Bit Tunable Delay for Broadband Phased Array Antenna Applications," *Phot. Tech. Lett.,* vol. 25, pp. 1165–1168, Jun 15 2013.

[63] T. Zhu, S. Veilleux, J. Bland-Hawthorn, and M. Dagenais, "Ultra-broadband High Coupling Efficiency Using a Si<inf>3</inf>N<inf>4</inf>/SiO<inf>2</inf>wave guide on silicon," in *2016 IEEE Photonics Society Summer Topical Meeting Series (SUM)*, ed: IEEE, 2016, pp. 92–93.

[64] V. V. Wong, "Ridge-waveguide sidewall-grating distributed feedback structures fabricated by x-ray lithography," in *J. Vac. Sci. Technol. B* vol. 11, ed, 1993, p. 2621.

[65] J. T. Hastings, M. H. Lim, J. G. Goodberlet, and H. I. Smith, "Optical waveguides with apodized sidewall gratings via spatial-phase-locked electron-beam lithography," in *J. Vac. Sci. Technol. B* vol. 20, ed, 2002, p. 2753.

[66] T. E. Murphy, J. T. Hastings, and H. I. Smith, "Fabrication and characterization of narrow-band Bragg-reflection filters in silicon-on-insulator ridge waveguides," in *Jnl. Lightwave Tech.* vol. 19, ed, 2001, pp. 1938–1942.

[67] M. Belt, J. Bovington, R. Moreira, J. F. Bauters, M. J. R. Heck, J. S. Barton, et al., "Sidewall gratings in ultra-low-loss Si_3N_4 planar waveguides," in *OpEx* vol. 21, ed: Optical Society of America, 2012, pp. 1181–1188.

[68] L. A. Coldren, S. W. Corzine, and M. L. Mashanovitch. (2012, Diode lasers and photonic integrated circuits. *Wiley Series in Microwave and Optical Engineering*.

[69] C. Chaichuay, P. P. Yupapin, and Saeung, *The serially coupled multiple ring resonator filters and Vernier effect.* vol. 39, 2009.

[70] D. T. Spencer, "Ultra-Narrow Bandwidth Optical Resonators for Integrated Low Frequency Noise Lasers," Department of Electrical and Computer Engineering, University of California, Santa Barbara, 2016.

[71] D. G. Rabus, "Integrated ring resonators," ed: Springer-Verlag Berlin Heidelberg, 2007.

[72] E. A. Kittlaus, H. Shin, and P. T. Rakich, "Large Brillouin amplification in silicon," in *Nature Photon* vol. 10, ed, 2016, pp. 463–467.

[73] S. Hua, J. Wen, X. Jiang, Q. Hua, L. Jiang, and M. Xiao, "Demonstration of a chip-based optical isolator with parametric amplification," in *Nature Communications* vol. 7, ed, 2016, p. 13657.

[74] J. W. Silverstone, D. Bonneau, K. Ohira, N. Suzuki, H. Yoshida, N. Iizuka, et al., "On-chip quantum interference between silicon photon-pair sources," in *Nature Photon* vol. 8, ed, 2013, pp. 104–108.

[75] T. Huffman, D. Baney, and D. J. Blumenthal, "High Extinction Ratio Widely Tunable Low-Loss Integrated Si_3N_4 Third-Order Filter," in *OpEx*, ed, 2017.

[76] J. C. C. Mak, W. D. Sacher, T. Xue, J. C. Mikkelsen, Z. Yong, and J. K. S. Poon, "Automatic Resonance Alignment of High-Order Microring Filters," in *Quantum Electronics, IEEE Journal of* vol. 51, ed, pp. 1–11.

[77] M. Belt and D. J. Blumenthal, "High Temperature Operation of an Integrated Erbium Doped DBR Laser on an Ultra-Low-Loss Si_3N_4 Platform," in *Optical Fiber Communication Conference*, ed, 2015.

[78] J. D. B. Bradley, M. Costa e Silva, M. Gay, L. Bramerie, A. Driessen, K. Wörhoff, et al., "170 Gbit/s transmission in an erbium-doped waveguide amplifier on silicon," in *OpEx* vol. 17, ed, 2009, p. 22201.

[79] D. Geskus, S. Aravazhi, S. M. García-Blanco, and M. Pollnau, "Giant Optical Gain in a Rare-Earth-Ion-Doped Microstructure," in *Adv. Mater.* vol. 24, ed, 2011, pp. OP19-OP22.

[80] J. D. B. Bradley, "AI_2O_3:ER^{3+} as a gain platform for integrated optics," Department of Electrical Engineering, Mathematics,and Computer Science, University of Twente, 2010.

[81] J. Yang, "Neodymium-doped waveguide amplifiers and lasers for integrated optical applications," Department of Electrical Engineering, Mathematics,and Computer Science, University of Twente, 2010.

[82] P. D. Inc., "FIMMWAVE," 4.06 ed, 2011.

[83] D. J. Blumenthal and M. Belt, "Erbium-doped waveguide DBR and DFB laser arrays integrated within an ultra-low-loss Si_3N_4 platform," in *OpEx* vol. 22, ed: Optical Society of America, 2014, pp. 10655–10660.

[84] M. Belt, "Optically Pumped Lasers on an Ultra-Low Loss Planar Waveguide Platform," Department of Electrical and Computer Engineering, University of California Santa Barbara, Santa Barbara, California, U.S.A., 2017.

[85] G. N. van den Hoven, E. Snoeks, A. Polman, C. van Dam, J. W. M. van Uffelen, and M. K. Smit, "Upconversion in Er-implanted Al_2O_3 waveguides," in *Journal of Applied Physics* vol. 79, ed, 1996, pp. 1258–1266.

[86] M. Belt and D. J. Blumenthal, "Erbium-doped waveguide DBR and DFB laser arrays integrated within an ultra-low-loss Si_3N_4 platform," in *OpEx* vol. 22, ed, 2014, p. 10655.

[87] M. Belt, T. Huffman, M. L. Davenport, W. Li, J. S. Barton, and D. J. Blumenthal, "Arrayed narrow linewidth erbium-doped waveguide-distributed feedback lasers on an ultra-low-loss silicon-nitride platform," in *Opt. Lett.* vol. 38, ed, 2013, p. 4825.

[88] J. F. Bauters, "Ultra-Low Loss Waveguides with Application to Photonic Integrated Circuits," Dept. of Electrical and Computer Engineering, University of California Santa Barbara, 2013.

[89] R. L. Moreira, "Integrated Optical Delay Line Circuits on a Ultra-low Loss Planar Waveguide Platform," Department of Electrical and Computer Engineering, University of California, Santa Barbara, Santa Barbara, CA, U.S.A., 2016.

[90] F. Payne and J. Lacey, "A theoretical analysis of scattering loss from planar optical waveguides," in *Opt Quant Electron* vol. 26, ed, 1994, pp. 977–986.

[91] S. Janz, "Silicon-Based Waveguide Technology for Wavelength Division Multiplexing," in *Silicon Photonics* vol. 94, ed: Springer Berlin Heidelberg, 2004, pp. 323–360.

[92] D. D. John, "Etchless Core-Definition Process for the Realization of Low Loss Glass Waveguides," Doctoral, Dept. of Electrical and Computer Engineering, University of California Santa Barbara, Santa Barbara, CA, USA, 2012.

[93] G. L. Bona, R. Germann, and B. J. Offrein, "SiON high-refractive-index waveguide and planar lightwave circuits," in *IBM J. Res. & Dev.* vol. 47, ed, 2003, pp. 239–249.

[94] R. Germann, H. W. M. Salemink, R. Beyeler, G. L. Bona, F. Horst, I. Massarek, et al., "Silicon Oxynitride Layers for Optical Waveguide Applications," in *Journal of The Electrochemical Society* vol. 147, ed, 2000, p. 2237.

[95] L. Augustin, M. Smit, N. Grote, M. Wale, and R. Visser, "Standardized Process Could Revolutionize Photonic Integration," *EuroPhotonics*, vol. 18, no. 3, 2013, pp. 30–35.

12

Advanced Optical Measurements for Data Centers

Steve Yao[1], Wajih Daab[1], Gang He[2] and Daniel Gariépy[2]

[1]General Photonics Corporation, California, USA
[2]EXFO Inc., Quebec, Canada

12.1 Introduction

The rapid advancement of data center technologies has blurred the boundaries among short reach, access, metro, long haul, and trans-oceanic fiber optic transmission networks. The data center communication links are no longer limited to short reach distance of less than 2 km for intra-data center communications between different servers, but have expanded to access, metro, long haul, and even trans-oceanic distances for intra-data center transmissions between data centers located in different buildings, different cities, different states, different countries, or even different continents. For such a large diversity of transmission distances, there have been adopted different data transmission technologies optimized for different distances, which require different test and measurement technologies to ensure adequate performances of the corresponding fiber optic communication systems. For example, for the short reach transmissions, direct modulated fiber optic links with PAM-4 modulation format are often adopted, which require much larger Optical Signal to Noise Ratio (OSNR) than what is required for the coherent detection systems of long-haul transmissions.

Optical power and insertion loss are the most basic tests for any fiber optic systems and components, which will not be discussed here. In addition to transmission loss, polarization dependent loss (PDL) and polarization mode dispersion (PMD) often impair the signal quality and cause transmission uncertainties, in both coherent and non-coherent communication systems, due to the non-deterministic nature of polarization in an optical fiber. Moreover, optical coherence detection systems generally involve two orthogonal polarizations with multi-level phase and amplitude modulations, and therefore tests relating to polarization, phase, amplitude, and their relative relationship are important. Such signals with multi-level phase and amplitude modulations and polarization division multiplexing (PDM) are often referred to as vector-modulated signals. Therefore, advanced optical measurements in this book include 1) polarization related parameters, such as PDL, PMD, the state of polarization (SOP), the polarization dependent responsivity (PDR) and the degree of polarization (DOP); 2) WDM channel bandwidth, crosstalk and ripple of the components; 3) the in-band OSNR of the optical signals propagating in the systems, and 4) parameters relating to the vector-modulated signals, such as error vector magnitude (EVM), bit-error-rate (BER), and OSNR. In addition, the tests of system tolerance to different polarization related impairments in the systems, such as PDL, PMD, SOP variations, are also included.

12.2 Polarization Related Tests

Polarization is one of the five fundamental properties of light and the other four are power, wavelength (frequency), phase, and coherence. In classical physics, light is modeled as a sinusoidal electromagnetic wave in which an oscillating electric field and an oscillating magnetic field propagate through space. Polarization is defined in terms of the pattern traced out in the transverse plane by the electric field vector as a function of time as shown in Figure 12.1. For unpolarized light, the plane of polarization fluctuates randomly around the direction of light beam propagation. Therefore, on average, no direction is favored. The rate of the fluctuation is so fast that an "observer" or a detector cannot tell the state of polarization (SOP) at any instant in

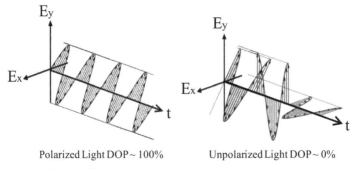

Polarized Light DOP ~ 100%   Unpolarized Light DOP ~ 0%

Figure 12.1   Different polarization traces and their degree of polarization.

time. For example, natural light (sunlight, firelight, etc.) is unpolarized. In any other case, the light beam can be considered to consist of partially polarized or fully polarized light. Degree of Polarization (DOP) is used to describe how much of the total light intensity is polarized:

$$DOP = \left(\frac{I_{\max} - I_{\min}}{I_{\max} + I_{\min}} \right) \times 100\% \qquad (12.1)$$

where $I_{\min}$ and $I_{\max}$ are the minimum and maximum intensities of the light passing through a polarizer, respectively. For totally polarized light, the DOP is one. For completely unpolarized light, the DOP is zero. For partially polarized light, this can also be written as:

$$DOP = \left(\frac{I_p}{I_p + I_{un}} \right) \times 100\% \qquad (12.2)$$

where I_p is the intensity of the polarized portion and I_{un} is the intensity of the unpolarized portion.

Most high performance lasers used in long-haul communication systems are polarized light sources. The polarization of light beams is an important factor in high speed optical communication network system design. As the bit rate increases, fiber-optic communication systems become increasingly sensitive to polarization related impairments [1]. Such impairments include polarization mode dispersion (PMD) in optical fibers, polarization dependent loss (PDL) in passive optical components, polarization dependent modulation (PDM) in electro-optic modulators, polarization dependent gain (PDG) [2, 3] in optical amplifiers, polarization dependent center wavelength (PDW) in WDM filters, polarization dependent response (PDR) in receivers, and polarization dependent sensitivity (PDS) in detectors and coherent communication systems' receivers.

Polarization related impairments have become the major obstacles to the increase of transmission rates in WDM systems. The cause for these polarization impairments is the imperfection of the optical fibers and the manufacturing process of optical components. If the fibers were perfect, the state of polarization (SOP) of the light signal transmitting in the fiber would remain constant and the effects of PMD, PDL, PDM, and PDG could easily be eliminated. Unfortunately, the SOP of light propagating in a length of standard communication fiber varies along the fiber due to the random birefringence changes induced by the thermal stress, mechanical stress, pressure, irregularities of the fiber core, and other environmental variations, making polarization related impairments time dependent.

Polarization Mode Dispersion (PMD)

Before the emerging of optical coherence detection, PMD is often cited as a critical hurdle for the high bit rate transmission systems (10 Gb/s and higher) after chromatic dispersion and fiber nonlinearity impairments are successfully managed [4]. As illustrated in Figure 12.2, a fiber link can be considered as a concatenation of many, randomly oriented retardation plates with different birefringences placed in serial. In the absence of PDL or PDG in the fiber link, these retardation plates are optically equivalent to a single retardation plate with an effective DGD and a pair of effective orthogonal principal axes (can either be linear or circular) for a given optical frequency[5]. Upon entering the retardation plate, an optical pulse is decomposed into two polarization components along the two axes. Because the two components travel with different speeds in the retardation plate, they exit with a relative time delay called Differential Group Delay (DGD). When DGD is comparable with the bit separation of a data stream, bit error rate may significantly increase.

The rms value of the DGD is often referred to as first order PMD. PMD is proportional to the square-root of the number of cascaded retardation plates, or equivalently, to the square-root of the fiber length. Contrary to the case of a true retardation plate, the DGD and the principal axes of the fiber link depend on the wavelength and fluctuate in time as a result of temperature variations and external constraints. Consequently, the corresponding pulse broadening is random, both as a function of wavelength at a given time and as a function of time at a given wavelength. Moreover, because the optical pulse is composed of many wavelengths, the dependency of Principal State of Polarization (PSP)

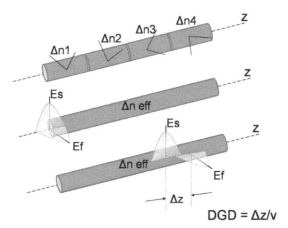

$$DGD = \Delta z / v$$

Figure 12.2   An optical pulse broadens because fibre has slow and fast propagating axes.

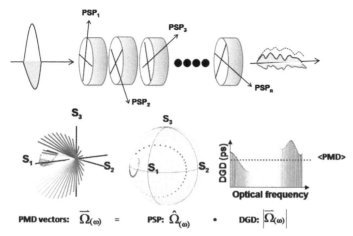

Figure 12.3   2nd order PMD representation.

and DGD on wavelength could cause further pulse spreading, and is referred to as second-order PMD (SOPMD), as shown in Figure 12.3 [6].

It should be noted that the first term of the PMD vector (the direction of the vector dependence on wavelength) has the dominant effect. The polarization states of the spectrum of an input signal will disperse around the PSP, and thus causing spectral depolarization, effectively decreasing the signal's DOP – and hence the OSNR. This will affect the PMD compensation circuitry at the coherent receiver side as well, because the polarization demultiplexing function is no longer done perfectly. Unlike the effects of chromatic dispersion and fiber nonlinearity, which are deterministic and stable in time, the PMD-induced penalty can be totally absent at any given moment and adversely large enough several days later to cause an unacceptable bit-error-rate for no apparent reason. To ensure an acceptable outage probability of the fiber optic system, PMD compensation must be dynamic in nature and adaptive to the random time variations. With the advancement of optical coherence detection, PMD can be compensated digitally in a polarization multiplexed system, however, the degree and speed of compensation are still need to be characterized.

Polarization Dependent Loss (PDL)

The PDL of an optical component is defined as the difference between the maximum and the minimum insertion losses for all possible input SOPs. Optical components with PDL act as partial polarizers with two orthogonal axes (either linear or circular). A light signal experiences a maximum loss

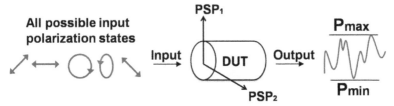

Figure 12.4 Output power fluctuates as a result of PDL.

if its SOP is aligned with one axis and a minimum loss if aligned with the other axis as shown in Figure 12.4. Almost all fiber optic components have PDL and the causes may be different for different components, as shown in Table 12.1. First of all, when light passes from an optical medium with an index of n_1 to another medium with an index n_2, reflection occurs. The reflection coefficients for the polarization states perpendicular and parallel to the plane of incidence are different if the angle of incidence is not normal (many fiber components have angled input and output surfaces for increased return loss.)

Such a difference in reflection results in a difference in transmission loss or PDL. For example, an 8° angle polished connector (FC/APC or SC/APC) has a PDL of 0.022 dB. Fiber grating based devices may also exhibit PDL if the grating is not normal to the fiber longitudinal axis. Second, for many optical components, such as isolators and circulators, birefringent crystals are often used. Because a birefringent crystal has two principal axes with different indexes of refraction, n_o and n_e, the Fresnel reflection coefficients of two polarization states perpendicular and parallel to the principal axes are

Table 12.1 Typical PDL values of common optical components

Component	Typical PDL Value
1 meter single mode fiber	<0.02 dB
10 km single mode fiber	<0.05 dB
PC type connector	0.005 ∼ 0.02 dB
APC type connector	0.02 ∼ 0.06 dB
50% fused coupler, single window	0.1 ∼ 0.2 dB
50% fused coupler, dual window	0.15 ∼ 0.3 dB
90/10 fused coupler, through path	0.02 dB
90/10 fused coupler, −10 dB path	0.1 dB
Isolator	0.05 ∼ 0.3 dB
3-port circulator	0.1 ∼ 0.2 dB
DWDM	0.05 ∼ 0.15 dB
Polarizer	30 ∼ 50 dB

different even at normal incidence, resulting in different transmission loss. The corresponding PDL is thus [5]:

$$PDL = 20 \times \log \left| \frac{(n_e - 1)(n_o + 1)}{(n_o - 1)(n_e + 1)} \right| \tag{12.3}$$

Anti-reflection (AR) coatings can greatly reduce the reflection, however, they may not totally eliminate PDL because the optimal coating layer thickness is determined by the refractive index of the coated material: it is either optimized for n_o or n_e. Diffraction grating based optical components or instruments generally have high PDL because the diffraction efficiencies for the two polarization states perpendicular and parallel to the plane of incidence are different. Finally, any fiber component containing a dichroic material also has PDL. A dichroic material has two principal axes with different absorption or attenuation coefficients. The principal axes can either be linear (linear dichroism) or circular (circular dichroism). For example, the LiNbO3 waveguide made with the proton exchange method exhibits strong linear dichroism and acts just like a polarizer.

In a fiber link which contains many optical components with different PDL values, the total PDL value depends on the SOPs of the light signal transmitted in the link and varies between a maximum and a minimum value. The maximum value is equal to the vector summation of the PDL values of all the components in the link. It is the difference between two insertion loss measurements of the link: the first one is when SOP of the light before each component is aligned with the minimum loss axis of the component and the second one is when SOP of the light before each component is aligned with the maximum loss axis. The minimum PDL value corresponds to the case that the SOPs of a light signal before all the PDL components are arranged such that the PDLs cancel one another out. The net residual PDL is the minimum PDL of the link.

The presence of PDL in a fiber link also complicates PMD compensation. When PDL is present in the fiber link, the link is no longer equivalent to a single retardation plate. Instead, it is retardation plates with a partial polarizer sandwiched in between. Any PMD compensation scheme therefore must take the effect of the partial polarizer into account, which may increase the complexity of the compensation arrangement significantly.

Like the PMD penalty, the PDL effect in a fiber link containing multiple PDL components separated by sections of single mode fiber is also time dependent. At any moment, the states of polarization in different sections of the link may be oriented favorably to allow a low link loss such that the detected optical power at the receiver is high enough to achieve acceptable bit-error rate (or signal-to-noise ratio). However at a different moment, the link loss may be too high to achieve quality transmission due to the external thermal or mechanical stress on the fiber that causes the states of polarization in different fiber sections to re-arrange. Note that unlike the pure PMD effect, the PDL cannot be compensated digitally even in the polarization multiplexed coherence detection systems, but can only be mitigated partially.

Polarization Dependent Gain (PDG)

The gain of an optical amplifier for the stronger polarization component is less than that for the weaker component (because the stronger component saturates the gain more) and the gain difference is called polarization dependent gain (PDG). One cause for the PDG is that the cross sections of the stimulated emission for different polarization states are different. This polarization hole burning always gives more gain to the weaker polarization component and thus tends to cause the polarization state to change with time. In addition, when the input SOP changes, the signal gain may increase temporarily, and then come down in a short period of time. Consequently, the polarization hole burning always encourages polarization fluctuations in a fiber laser system and thus causes mode-hopping and increases super-mode noise in a mode-locked laser.

An optical amplifier may, at the same time, exhibit PDL effect. For example, couplers and isolators are generally contained in an Er+ doped fiber amplifier (EDFA) and the presence of PDL in these components gives rise to the apparent PDL of the amplifier. Even in semiconductor optical amplifiers (SOA), the facets of the semiconductor chip are generally angle cleaved to prevent optical feedback into the amplifier. As discussed previously, these angled interfaces exhibit large PDL, which directly contributes to PDL of the SOA.

In a fiber optic link with many optical amplifiers and many components with PDL, the effect of PDG can be significant at some moments and negligible at other times. When a large number of optical amplifiers are cascaded in a long haul fiber link, the performance degradation caused by PDG is significant even though each amplifier may have a very small PDG ($\sim$0.1 dB). The performance degradation becomes even worse when PDG is combined with the PMD and PDL of the fiber and other components in the link.

Polarization Dependent Modulation (PDM)

External modulators, such as LiNbO3 based electro-optical modulators and semiconductor electro-absorption modulators, also exhibit polarization dependent modulation in that the modulation depth of signals with different polarization states are different. As a result, the amplitude of the received data bits varies when the state of polarization of light before entering the modulator varies due to the fluctuation of temperature or other external constraints on the fiber, resulting in bit-error-rate fluctuation.

To assist polarization alignment, most Ti-indiffused LiNbO3 modulators embed a polarizer at input or out of the waveguide and thus convert the PDM problem into a more easily identified PDL problem. The LiNbO3 modulators made with proton exchange process act like a polarizer themselves without the embedded polarizer. Thus one method of eliminating PDM effects uses a fast response dynamic polarization controller placed in front of the modulator to assure that light passing through the modulator has minimal loss.

Polarization Dependent Response (PDR)

Photonic integrated circuit (PIC) receivers generally consist of an input fiber pigtail and a semiconductor chip with waveguides, gratings, splitters, isolators, filters and photodetectors (PD) to convert the input light into photocurrents. The total conversion loss (CL) of these opto-electrical integrated devices is the loss associated with each optical path on the chip (including the coupling loss from the fiber pigtail to the chip, the insertion loss of the components in each optical path, the coupling loss to the PD, and the PD's conversion efficiency). Such loss is polarization sensitive and the difference between the maximum and minimum losses corresponding to all possible polarization states is defined as polarization dependent response.

12.2.1 Polarization Mode Dispersion (PMD) Measurement

Binary Magneto-optic Polarization Rotators Method

Traditionally, polarization related measurements have been relied on analog technologies [7], including rotating retarders [8, 9], rotating polarizers [10, 11], liquid crystal cells [12], phase modulators [13], and four-detector methods [14]. Such analog technologies suffer from inherently low repeatability and require complicated compensation techniques in order to achieve high accuracy measurements.

Here we describe a binary polarization measurement system comprised of a pair of polarization stage generator (PSG) and polarization state analyzer (PSA) made with binary magneto-optic polarization rotators. By taking the binary advantages of the rotators, all polarization related parameters, including the PMD values vs. wavelength, can be accurately measured. Unprecedented DGD and SOPMD measurement accuracies of 2.6 fs and 1.39 ps^2, respectively; repeatabilities of 0.022 fs, 0.28 ps^2, respectively; and resolutions of 1 fs, 0.005 ps^2, respectively, were achieved from 1480 to 1620 nm [14].

As shown in Figure 12.5, the binary polarization measurement system consists of a tunable laser (TLS), a binary MO PSG, a binary MO PSA, and a control personal computer (PC). The device under test (DUT) is placed between the PSG and PSA. The PSG and PSA each contain 6 binary MO polarization rotators, a polarizer, and a quarter wave plate. The PSA also contains a photodetector (PD) and a signal amplification circuit.

As described in [15], the MO rotators have the following attractive binary properties: when applying a positive magnetic field above a saturation field, the rotator rotates SOP by a precise angle around 22.5°. When applying a negative magnetic field beyond saturation, the rotator rotates SOP by a precise angle around −22.5°. Therefore, when two rotators rotate in the same direction, the net rotation is +45° or −45°. On the other hand, if the two rotators rotate in the opposite direction, the net SOP rotation is zero.

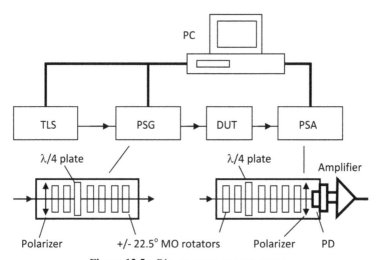

Figure 12.5   Binary measurement system.

By controlling the rotation directions of each rotator with the PC, the PSG can generate 6 distinctive polarization states and the PSA can accurately measure the state of polarization (SOP) and the degree of polarization (DOP) of light entering it by analyzing voltages generated in the photodetector using Muller matrix analysis. The binary PSA can also be self-calibrated to remove inaccuracies caused by the imperfections in components and workmanship, and therefore give superb measurement accuracies.

The PMD parameters of a fiber or optical components can be measured using Jones Matrix Eigen analysis or Muller Matrix method. With the frequency change of the optical signal which propagates through an optical component, the output state of polarization (SOP) rotates on the Poincare sphere around the principle state of polarization (PSP) vector Ω. The first-order PMD (DGD) of the DUT can be found as

$$\tau = |\Delta\theta(w)/\Delta\omega| \tag{12.4}$$

where $\Delta\theta(\omega)$ is the rotation angle of SOP along PSP and Δw is the angular frequency difference. The second-order can be calculated as

$$SOPMD = \frac{d\Omega(\omega)}{d\omega} = \frac{d\Delta\tau(\omega)}{d\omega}q(\omega) + \Delta\tau(\omega)\frac{dq(\omega)}{d\omega} \tag{12.5}$$

where $\mathbf{q}(\Omega)$ is a unit vector that points in the direction of the fast principle axis [14].

Method 1: Jones Matrix Eigenvalue Analysis

Generally, the polarization transfer matrix of optical components or fiber can be described by a 2x2 complex Jones transfer matrix Γ. The input and output polarization states can be related by

$$\boldsymbol{J}^{PSA} = c * \boldsymbol{\Gamma} \cdot \boldsymbol{J}^{PSG} \tag{12.6}$$

$$\begin{bmatrix} J_x^{PSA} \\ J_y^{PSA} \end{bmatrix} = c * \begin{bmatrix} \Gamma_{00} & \Gamma_{01} \\ \Gamma_{11} & 1 \end{bmatrix} \begin{bmatrix} J_x^{PSG} \\ J_y^{PSG} \end{bmatrix} \tag{12.7}$$

where J^{PSG} is the normalized Jones vector generated by PSG, J^{PSA} is the normalized Jones vector measured by PSA and the c* is a complex constant related to the absolute amplitude and absolute phase of light wave. It should be noted that the ratio of $\frac{J_x^{PSA}}{J_y^{PSA}}$ is

$$\frac{J_x^{PSA}}{J_y^{PSA}} = \frac{\Gamma_{00}^J J_x^{PSG} + \Gamma_{01}^J J_y^{PSG}}{\Gamma_{10}^J J_x^{PSG} + J_y^{PSG}} \tag{12.8}$$

and is independent of the constant c*. If three sets of vectors $\mathbf{J}^{PSG}$ and $\mathbf{J}^{PSA}$ are generated and measured, we can obtain the following equations

$$J_{0,x}^{PSG} J_{0,y}^{PSA} \Gamma_{00}^{J} + J_{0,y}^{PSG} J_{0,y}^{PSA} \Gamma_{01}^{J} - J_{0,x}^{PSG} J_{0,x}^{PSA} \Gamma_{10}^{J} = J_{0,y}^{PSG} J_{0,x}^{PSA}$$

$$J_{1,x}^{PSG} J_{2,y}^{PSA} \Gamma_{00}^{J} + J_{1,y}^{PSG} J_{1,y}^{PSA} \Gamma_{01}^{J} - J_{1,x}^{PSG} J_{1,x}^{PSA} \Gamma_{10}^{J} = J_{1,y}^{PSG} J_{1,x}^{PSA}$$

$$J_{2,x}^{PSG} J_{2,y}^{PSA} \Gamma_{00}^{J} + J_{2,y}^{PSG} J_{2,y}^{PSA} \Gamma_{01}^{J} - J_{2,x}^{PSG} J_{2,x}^{PSA} \Gamma_{10}^{J} = J_{2,y}^{PSG} J_{2,x}^{PSA}$$

$$(12.9)$$

let $k_{i,0} = J_{i,x}^{PSG} J_{i,y}^{PSA}$, $k_{i,1} = J_{i,y}^{PSG} J_{i,y}^{PSA}$ $k_{i,2} = -J_{i,x}^{PSG} J_{i,x}^{PSA}$, $k_{i,3} = J_{i,y}^{PSG} J_{i,x}^{PSA} (i = 0, 1, 2)$, then the Equation (12.9) can be simplified to be

$$k_{00} \Gamma_{00}^{J} + k_{01} \Gamma_{01}^{J} + k_{02} \Gamma_{10}^{J} = k_{03}$$

$$k_{10} \Gamma_{00}^{J} + k_{11} \Gamma_{01}^{J} + k_{12} \Gamma_{10}^{J} = k_{13} \qquad (12.10)$$

$$k_{20} \Gamma_{00}^{J} + k_{21} \Gamma_{01}^{J} + k_{22} \Gamma_{10}^{J} = k_{23}$$

So the Jones transfer matrix Γ of DUT can be easily calculated using

$$\Gamma_{00} = \frac{\begin{vmatrix} k_{03} & k_{01} & k_{02} \\ k_{13} & k_{11} & k_{12} \\ k_{23} & k_{21} & k_{22} \end{vmatrix}}{\begin{vmatrix} k_{00} & k_{01} & k_{02} \\ k_{10} & k_{11} & k_{12} \\ k_{20} & k_{21} & k_{22} \end{vmatrix}}, \Gamma_{01} = \frac{\begin{vmatrix} k_{00} & k_{03} & k_{02} \\ k_{10} & k_{13} & k_{12} \\ k_{20} & k_{23} & k_{22} \end{vmatrix}}{\begin{vmatrix} k_{00} & k_{01} & k_{02} \\ k_{10} & k_{11} & k_{12} \\ k_{20} & k_{21} & k_{22} \end{vmatrix}}, \Gamma{21} = \frac{\begin{vmatrix} k_{00} & k_{01} & k_{03} \\ k_{10} & k_{11} & k_{13} \\ k_{20} & k_{21} & k_{23} \end{vmatrix}}{\begin{vmatrix} k_{00} & k_{01} & k_{02} \\ k_{10} & k_{11} & k_{12} \\ k_{20} & k_{21} & k_{22} \end{vmatrix}}$$

$$(12.11)$$

For two adjacent two input optical frequencies, define the matrix:

$$T(\Delta\omega) = \Gamma(\omega_2)\Gamma(\omega_1)^{-1} \qquad (12.12)$$

where

$$\overline{\omega} = \frac{\omega_1 + \omega_2}{2} \qquad (12.13)$$

then the eigenvectors ρ_f and ρ_s are the polarization state principle (PSP) (they are related to the fast and slow axis), and the DGD can be calculated from the corresponding eigenvalues $\rho_s = a_s * \exp(i\tau_s)$ and $\rho_f = a_f * \exp(i\tau_f)$:

$$DGD(\overline{\omega}) = \tau(\omega) = |\tau_s - \tau_f| = \left| \frac{Arg(\rho_s/\rho_f)}{\omega_1 - \omega_2} \right| \qquad (12.14)$$

and arg(ρ_s/ρ_f) stands for the phase angle of (ρ_s/ρ_f). The wavelength dependent PMD vector $\overrightarrow{W}$ can be defined as

$$\overrightarrow{W}(\omega) = \tau(\omega)\overrightarrow{q}(\omega) \qquad (12.15)$$

where $\vec{q}(\omega)$ is the unit vector of the fast principal state of polarization. The second-order

PMD, defined as the frequency derivative of the PMD vector $\overrightarrow{W}$, can be calculated as

$$SOPMD = \frac{d\overrightarrow{W}(\omega)}{d\omega} = \frac{d\tau(\omega)}{d\omega}\vec{q}(\omega) + \tau(\omega)\frac{d\vec{q}(\omega)}{d\omega} \tag{12.16}$$

Method 2: Muller Matrix Measurement

Using the same Binary measurement system shown in Figure 12.5, let the Stokes vector of the ith output of PSG is

$$S_i^{PSG} = \begin{pmatrix} S_{0i}^{PSG} \\ S_{0i}^{PSG} \\ S_{0i}^{PSG} \\ S_{0i}^{PSG} \end{pmatrix} \tag{12.17}$$

the corresponding Stokes vector after passing through DUT measured by PSA is

$$S_i^{PSA} = \begin{pmatrix} S_{0i}^{PSA} \\ S_{1i}^{PSA} \\ S_{2i}^{PSA} \\ S_{3i}^{PSA} \end{pmatrix} = \begin{pmatrix} m_{00} & m_{01} & m_{02} & m_{03} \\ m_{10} & m_{11} & m_{12} & m_{13} \\ m_{20} & m_{21} & m_{22} & m_{23} \\ m_{30} & m_{31} & m_{32} & m_{33} \end{pmatrix} \begin{pmatrix} S_{0i}^{PSG} \\ S_{1i}^{PSG} \\ S_{2i}^{PSG} \\ S_{3i}^{PSG} \end{pmatrix} \tag{12.18}$$

then,

$$\mathbf{S}^{PSA} = \begin{pmatrix} S_{00}^{PSA} & S_{01}^{PSA} & S_{02}^{PSA} & S_{03}^{PSA} & S_{04}^{PSA} & S_{05}^{PSA} \\ S_{10}^{PSA} & S_{11}^{PSA} & S_{12}^{PSA} & S_{13}^{PSA} & S_{14}^{PSA} & S_{15}^{PSA} \\ S_{20}^{PSA} & S_{21}^{PSA} & S_{22}^{PSA} & S_{23}^{PSA} & S_{24}^{PSA} & S_{25}^{PSA} \\ S_{30}^{PSA} & S_{31}^{PSA} & S_{32}^{PSA} & S_{33}^{PSA} & S_{34}^{PSA} & S_{35}^{PSA} \end{pmatrix}$$

$$= \begin{pmatrix} m_{00} & m_{01} & m_{02} & m_{03} \\ m_{10} & m_{11} & m_{12} & m_{13} \\ m_{20} & m_{21} & m_{22} & m_{23} \\ m_{30} & m_{31} & m_{32} & m_{33} \end{pmatrix}$$

$$\begin{pmatrix} S_{00}^{PSA} & S_{01}^{PSA} & S_{02}^{PSA} & S_{03}^{PSA} & S_{04}^{PSA} & S_{05}^{PSA} \\ S_{10}^{PSA} & S_{11}^{PSA} & S_{12}^{PSA} & S_{13}^{PSA} & S_{14}^{PSA} & S_{15}^{PSA} \\ S_{20}^{PSA} & S_{21}^{PSA} & S_{22}^{PSA} & S_{23}^{PSA} & S_{24}^{PSA} & S_{25}^{PSA} \\ S_{30}^{PSA} & S_{31}^{PSA} & S_{32}^{PSA} & S_{33}^{PSA} & S_{34}^{PSA} & S_{35}^{PSA} \end{pmatrix} \tag{12.19}$$

$$= \mathrm{M} \cdot \mathrm{S}^{\mathrm{PSG}}$$

Consequently, the Mueller matrix of the DUT can be obtained from

$$\mathrm{M} = \mathrm{S}^{\mathrm{PSA}} \cdot (\mathrm{S}^{\mathrm{PSG}})^{\mathrm{T}} \cdot [\mathrm{S}^{\mathrm{PSG}} \cdot (\mathrm{S}^{\mathrm{PSG}})^{\mathrm{T}}]^{-1} \tag{12.20}$$

where $(\mathrm{S}^{\mathrm{PSG}})^{\mathrm{T}}$ is the transpose of matrix $\mathrm{S}^{\mathrm{PSG}}$ and $(\mathrm{S}^{\mathrm{PSG}} \cdot (\mathrm{S}^{\mathrm{PSG}})^{\mathrm{T}})^{-1}$ is the inverse of matrix $\mathrm{S}^{\mathrm{PSG}} \cdot (\mathrm{S}^{\mathrm{PSG}})^{\mathrm{T}}$. In formula (12.19), four pairs of $\mathrm{S}^{\mathrm{PSG}}$ and $\mathrm{S}^{\mathrm{PSA}}$ are enough to calculate the Muller matrix of DUT. In real system, six pairs of $\mathrm{S}^{\mathrm{PSG}}$ and $\mathrm{S}^{\mathrm{PSA}}$ can give more accurate results because more polarization states are used.

For two adjacent optical frequencies, define the matrix:

$$M_{\Delta}(\overline{\omega}) = M(\omega_2)M^{-1}(\omega_1) \tag{12.21}$$

then the complex PMD vector $\overrightarrow{W} = \overrightarrow{\Omega} + i\overrightarrow{\Lambda}$ can be found from the matrix of $\mathbf{M}_{\Delta}$,], where $\overrightarrow{\Omega}$ and $\overrightarrow{\Lambda}$ are the real and complex components of $\overrightarrow{W}$, respectively. DGD and PSP can be calculated from

$$DGD = Re(\sqrt{\overrightarrow{W} \cdot \overrightarrow{W}}) = \sqrt{\Omega^2 - \Lambda^2} \tag{12.22}$$

$$q(\omega)_{\pm} = PSP_{\pm} = \frac{\pm\overrightarrow{\Omega} + \overrightarrow{\Omega} \otimes \overrightarrow{\Lambda}}{\overrightarrow{\Omega} \cdot \overrightarrow{\Lambda}} \tag{12.23}$$

where q_+ and q_- are the unit vectors of the slow and fast PSP, respectively. The symbol "$\otimes$" in Equation (12.23) stands for cross product, and "$\cdot$" stands for inner product. As with the case using the JME method, the second order PMD can be calculated numerically using Equation (12.16), but with q replaced by q^- [14].

12.2.2 Polarization Dependent Loss (PDL) Measurement

Binary Magneto-optic Polarization Rotators Method
Using the same binary polarization measurement system depicted in Figure 12.5, high accuracy PDL measurement can be conducted with PDL

accuracies of 0.06 dB, repeatability of 0.034 dB, and resolutions of 0.01 dB, respectively [14].

Method 1: Jones Matrix Measurement Method (JMM)

For a given optical frequency, the PDL of a DUT can be calculated as

$$PDL = 10 * \log \left| \frac{r1}{r2} \right| \qquad (12.24)$$

Where, $r_{1,2} = \frac{m_{11}+m_{22}}{2} \pm \sqrt{(\frac{m_{11}+m_{22}}{2})^2 - m_{11}m_{22} + m_{12}m_{21}}$, are the eigenvalues of the matrix $M = (\Gamma^{-1})^* \cdot \Gamma = \begin{bmatrix} m_{11} & m_{12} \\ m_{21} & m_{22} \end{bmatrix}$, Γ^{-1} is the inverse of Jones transfer matrix Γ, and the star $*$ indicates the complex conjugate.

Method 2: Mueller Matrix Measurement Method (MMM)

For a given optical frequency, the PDL of a DUT can be obtained from the following formula [14],

$$PDL = -10 \times \log \left(\frac{P_{Min}}{P_{Max}} \right) = -10 \times \log \frac{m_{00} - \sqrt{m_{01}^2 + m_{02}^2 + m_{03}^2}}{m_{00} + \sqrt{m_{01}^2 + m_{02}^2 + m_{03}^2}}$$
$$(12.25)$$

Note that only the knowledge of the first row of the Muller matrix is sufficient to determine the PDL of an optical component, and therefore the setup of the PDL measurement can be simplified to have a photodetector to replace PSA in Figure 12.5. A multi-channel optical component analyzer based on Muller Matrix can be constructed using a tunable laser and a PSG, with multiple photodetectors, as shown in Figure 12.6, which is capable of performing simultaneous insertion loss (IL), polarization dependent loss (PDL), and optical power (P) measurements on multiple optical paths. It offers fast characterization of wavelength dependent optical parameters, which is an ideal solution for easy, accurate characterization of components and modules with multiple outputs, including DWDMs, ROADMs, AWGs and PLCs. Figure 12.6 shows the test setup to perform such measurement on multiport fiber optic compoenents:

Figure 12.7 shows the IL/PDL display taken with a commercial optical component analyzer while testing an 8 ports AWG.

For DWDM components, more information can be extracted which are crucial to characterize before using in the system. Figure 12.8 show the pass-band information such as ITU central, central wavelength, 3 dB BW, 10 dB BW, 20 dB BW, IL, flatness, PDL, cross-talk, adjacent cross-talk, and Non-adjacent cross-talk.

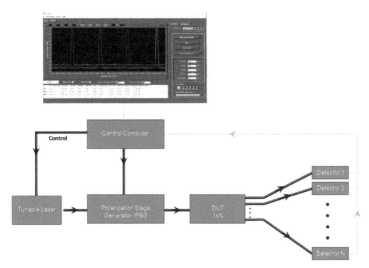

Figure 12.6   Setup diagram for PDL/IL measurement.

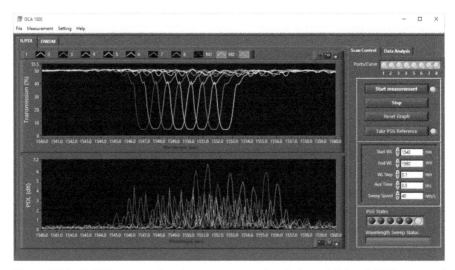

Figure 12.7   IL/PDL vs. wavelength measurement of an 8-channel C-band AWG measured with a 1540–1560nm wavelength sweep.

Polarization Scrambling Method

The PDL of an optical component is the maximum optical power transmittance change over all polarization states. By using a polarization generator or scrambler to generate either a deterministic or a pseudorandom subset of all

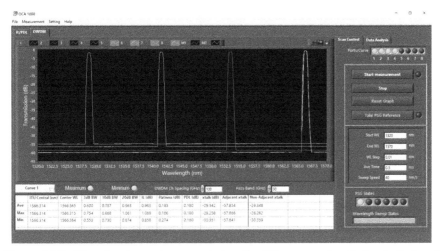

Figure 12.8 The plot shows the transmission (in dB) vs. wavelength for 4 displayed channels.

Figure 12.9 Polarization scrambling method.

possible states of polarization (SOPs) at the input light source, PDL can be calculated from the maximum and minimum power values obtained from the subset.

$$PDL = -10 \times \log(\frac{P_{Min}}{P_{Max}}) \qquad (12.26)$$

Max/min Search Method

The results obtained with polarization scrambling method may not be the exact maximum and minimum transmission of the component. It is highly dependent on the Poincaré sphere coverage, PDL range, scrambling frequency, scrambling pattern and the measurement time. Moreover, the Jones-matrix and Mueller-matrix methods measure the optical transmission at a set of fixed SOPs that do not generally coincide with the maximum and minimum

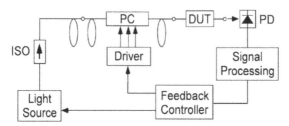

Figure 12.10 Block diagram of the automatic max–min transmittance search implementation (PC, fiber-squeezer-based polarization controller; DUT, device under test; PD, photode-tector; ISO, optical isolator).

transmissions. The PDL is calculated from the matrix elements that were obtained from the optical intensity measurements. Small measurement errors or circuit noise can affect the results significantly. An additional drawback is that most of such deterministic fixed SOP measurements require system or wavelength calibrations.

The Max/Min searching method is also based on changing the polarization states before the DUT to get the P_{max} and P_{min}. This method, however, has the advantage of offering an automatic and unambiguous determination of PDL by ensuring reaching the best and worst case transmission scenarios quickly. The max/min searching method employs an active feedback polarization control algorithm using a fiber squeezer polarization controller (PC) to systematically and rapidly search for the maximum and minimum transmittances without the need to measure all possible SOPs as shown in Figure 12.10. The direct measurement of maximum and minimum power transmittances assures calibration-free, highly accurate, high-speed measurement over a larger PDL range [16].

Figure 12.11 shows results taken with a commercial instrument that uses the min/max method. It has a wide PDL dynamic up to 45 dB range, and yet, still maintains high accuracy for DUT's with both high and low PDL values.

12.2.3 PDR Measurement of Receivers

The responsivity of photodetectors, defined as the ratio of the output photocurrent and the input optical power, also has slight dependency on the state of polarization, probably due to the stress built inside the semiconductor chip from dicing or other operations, although the chip itself is isotropic. We call such polarization dependence as polarization dependent response, or PDR. With the advances in Photonics Integrated Circuits (PIC), phtotodetectors

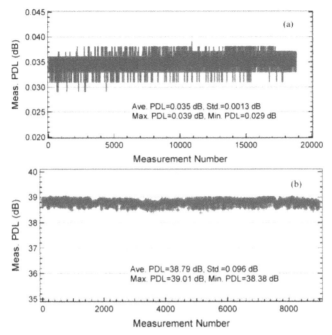

Figure 12.11 PDL measurement data and statistics for (a) an FC/APC–air interface and (b) an in-line fiber polarizer at 1550 nm.

can be integrated with others optical components, such as waveguides, filters, arrayed waveguide gratings (AWG), optical amplifiers, on a single chip to form a receiver, and hence the PDL of the other optical components, together with the PDR of the photodetector, will contribute to the final PDR of the receiver. It is therefore necessary to characterize the total PDR of the receiver. Figure 12.12 shows an example of such a 4-channel integrated receiver.

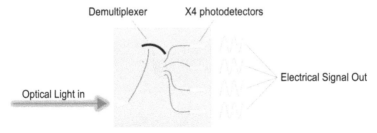

Figure 12.12 Simple representation of a 4-channel receiver PIC.

A Polarization Dependent Response meter (PDR meter) with an optical output port and electrical input port is used to measure and characterize, not only the total conversion loss (CL) of each optical path on the chip (including the coupling loss from the fiber pigtail to the chip, the insertion loss of the components in each optical path, the coupling loss to the PD, and the PD's conversion efficiency), but also the polarization dependence of the conversion loss.

The max and min search method (which is compliant with TIA/EIA-455-198), can simultaneously measure the PDR and CL of a PIC receivers the number can be different for different instruments and therefore is an ideal tool for characterizing the total responsivity and polarization dependence of opto-electrical integrated circuits with photodetector outputs.

The PDR meter can be combined with a tunable laser, as shown in Figure 12.13, to measure the polarization dependent response vs. wavelength which is needed in WDM communication systems.

12.2.4 PDL Measurement of Fiber Optic Link

In a multi-channel, WDM link which contains many optical components with different PDL values, the total PDL value depends on the SOPs of the light signal transmitted in the link and varies between a maximum and a minimum value. The maximum value is equal to the vector summation of the PDL values of all the components in the link. High PDL causes degradation in

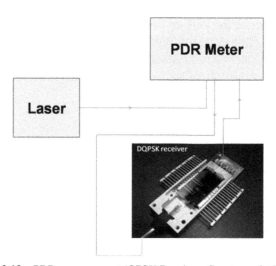

Figure 12.13 PDR measurement (QPSK Receiver: Courtesy of APACHE).

the system performance and may lead to service outage if it is not fully characterized. PDL can cause polarization cross-talk in polarization division multiplex signals, in addition to PDL-induced variation of optical signal to noise ratio (OSNR). PDL of fiber itself is time dependent, especially for old dark fibers. Its characteristics, such as PMD, IL, and PDL, may change and get worse over time. Moreover, the fact that several optical components are integrated on a single chip, it is becoming challenging to measure their PDL independently. There is a need to fully characterize the polarization dependent polarization dependent loss of the overall link of the overall link.

PDL Meter with special algorithm can be used to measure the overall polarization dependent loss through one specific channel of a multi-span, WDM link with different lengths, as shown in Figure 12.14. OSNR values and the expected propagation delay shall be taking into consideration when performing the measurement.

The presence of system noise affects the measured PDL. For a system with a finite SNR, the PDL measured by the instrument is:

$$PDL = PLD_0 + 10\log\left(\frac{SNR + (\frac{1+r^2}{2})}{SNR + (\frac{1+r^{-2}}{2})}\right) \qquad (12.27)$$

where PDL_0 is the PDL without system noise (in dB), r is a parameter defined by $PDL_0 = 10\log(1/r^2)$, $SNR = P_s/P_n$ (linear ratio, unitless), P_s is the total signal

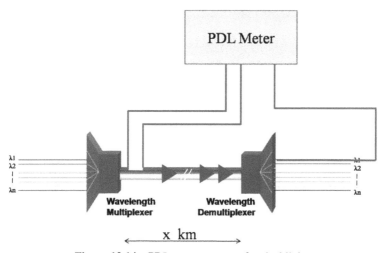

Figure 12.14   PDL measurement of optical links.

power (at the PDL Meter's detector), P_n is the total noise power (at the PDL Meter's detector). For large SNR and small PDL_0, $PDL \sim PDL_0$.

12.2.5 Measuring In-band OSNR by DOP Measurement

Polarization state analyzers made with binary magneto-optic (MO) polarization switches and one quarter-wave plate have fast response, superior repeatability, stability and very high accuracy and, thus, are used for optical performance monitoring [17]. The SOP/DOP of the signal can be obtained by measuring only four optical power levels at four polarization states, with a total response time of less than 1 ms. By measuring the instantaneous DOP value for each channel in a division-multiplexing (WDM) system, the in-situ optical OSNR can be measured.

For the PSA structure shown in Figure 12.15, a photodiode is required to detect the output optical power following the MO switch assembly. Assuming lossless optical transmission, the Mueller matrix of the PSA can be derived as shown below

$$
M(T) = \frac{1}{2}
\begin{pmatrix}
1 & \cos 2(\alpha+\beta)\cos 2(\gamma+\delta) & \sin 2(\alpha+\beta)\cos 2(\gamma+\delta) & \sin 2(\gamma+\delta) \\
1 & \cos 2(\alpha+\beta)\cos 2(\gamma+\delta) & \sin 2(\alpha+\beta)\cos 2(\gamma+\delta) & \sin 2(\gamma+\delta) \\
0 & 0 & 0 & 0 \\
0 & 0 & 0 & 0
\end{pmatrix}
$$

(12.28)

where α, β, γ, and δ are the relative rotation angles of each switch. If the Stokes vector $S = (S_0, S_1, S_2, S_3)$ represents the input polarization state, then the output optical power (S_0^{OUT}) is

$$
S_0^{OUT} = \frac{1}{2}[S_0 + \cos 2(\alpha+\beta)\cos 2(\gamma+\delta)S_1 + \sin 2(\alpha+\beta)\cos 2(\gamma+\delta) \\
S_2 + \sin 2(\gamma+\delta)S_3]
$$

(12.29)

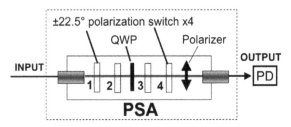

Figure 12.15   Polarization State Analyzer based on MO switches.

Define $\theta = \alpha + \beta$ and $\varphi = \gamma + \delta$, where $\alpha = \beta = \gamma = \delta = \pm 22.5°$ Equation x can be rewritten as

$$S_0^{OUT}(\theta, \varphi) = \frac{1}{2}[S_0 + S_1 \cos 2\theta \cos 2\varphi + S_2 \sin 2\theta \cos 2\varphi + S_3 \sin 2\varphi]$$

(12.30)

Since the possible combinations of double-stage rotation angles (θ, φ) can only be selected from combinations of 45°, 0° and −45°, by measuring the optical power at the four resultant polarization states, the SOP of the input signal can be obtained using the following equations [17]:

$$S_0 = S_0^{OUT}(45^0, 0^0) + S_0^{OUT}(-45^0, 0^0) \tag{12.31}$$

$$S_1 = 2S_0^{OUT}(0^0, 0^0) - S_0^{OUT}(45^0, 0^0) - S_0^{OUT}(-45^0, 0^0) \tag{12.32}$$

$$S_2 = S_0^{OUT}(45^0, 0^0) - S_0^{OUT}(-45^0, 0^0) \tag{12.33}$$

$$S_3 = 2S_0^{OUT}(0^0, 45^0) - S_0^{OUT}(45^0, 0^0) - S_0^{OUT}(-45^0, 0^0) \tag{12.34}$$

where $S_0^{OUT}(\theta, \varphi)$ is the detected optical power at different polarization switch conditions. The standard deviations SOP and DOP using this method are less than 0.1% and 0.07%, respectively, while the maximum deviations for both are 0.3%, with a resolution of 0.001; such high accuracy and repeatability are highly desirable features in both measurement and monitoring, and much higher than regular polarimeters (typically 2% DOP accuracy for low-cost polarimeters, while 1% for high accuracy analyzers).

OSNR monitoring using the DOP of the signal is a well-known technique based on the fact that the optical signal is polarized while the optical noise is not. The relationship between DOP and OSNR is as simple as

$$OSNR = \frac{DOP}{DOP - 1} \tag{12.35}$$

An optical filter (OF) is needed between the PSA and the photodiode to filter out the signal for WDM system. The OSNR monitoring result using the PSA is close to the measurements using the OSA. The DOP-based OSNR monitoring will be affected by the PMD of the link. To improve the accuracy, a demultiplexer (DEMUX) can be placed after the PSA instead of optical filter. Each WDM channel will then have a separate photodetector, instead of using only one photodetector inside the PSA; thus, all of the channels can share one PSA module and perform the monitoring function simultaneously. Integrated DEMUX and photodetector assemblies are commercially available and can make the configuration compact and cost-effective.

12.2.6 Polarization Emulation for Non-coherent and Coherent Systems

Coherent detection and polarization division multiplexing (PDM) has emerged as the key technology enabler for 40-Gbps, 100-Gbps, 200-Gbps and 400-Gbps networks in recent years because they can significantly increase the spectral efficiency of each channel and allow each channel to transmit high bit rate with a relatively small optical bandwidth. PDM doubles the spectral efficiency by combining two polarization channels of same bit rate and same wavelength [18]. On the other hand, coherent detection allows multiple levels of phase and amplitude modulations of lower rates to be multiplexed on the same wavelength channel. For example, PDM, together with quadrature-phase-shifted-keying (QPSK) modulation, enables 40 Gbps transmission with a bandwidth of a 10 Gbps direct channel. At such a small bandwidth, the impairments due to polarization mode dispersion (PMD) and chromatic dispersion (CD) are greatly reduced and will have less impact on the 40-Gbps transmission. In addition, because both the amplitude and phase information of the two orthogonal polarizations of a signal are preserved in a coherent detection system, PMD and CD compensations can be performed by digital signal processing (DSP) in the electrical domain to further extend system's PMD and CD tolerances. Note that PDL cannot be compensated digitally, however, can be mitigated partially with DSP.

An important task for a system deploying polarization multiplexing is to track the changes in polarization and separate the two polarization channels at the receiving end. Such a task can be accomplished optically by controlling polarization with a feedback signal. On the other hand, because of the preservation of both the phase and amplitude information of the two orthogonal polarizations of a signal when detected coherently, information in both polarization channels can be separated and obtained by digital signal processing (DSP). Such a digital polarization demultiplexing potentially has significantly reduced cost and size, compared with optical demultiplexing.

Coherent detection systems generally perform the following polarization control functions:

1. Polarization tracking and demultiplexing
2. PMD Compensation (PMDC)
3. PDL mitigation (PDLM)

These functions can be accomplished by different high speed DSP circuitries and algorithms in the transceivers as shown in Figure 12.16. For transceiver

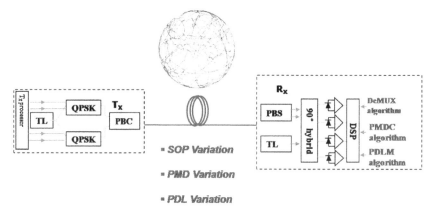

Figure 12.16 Pol-muxed fiber optic system that shows three main polarization related function of coherent receivers.

developers, how to evaluate the performances of different DSP circuitries and algorithms for these polarization control functions is critical to the success of their transceiver developments. For the system integrators using the transceivers, it is important to compare the performances of the transceivers from different vendors, including the three key polarization control functions. Finally, for network operators using the systems developed by different system vendors, evaluating and comparing the polarization performances from different vendors can help an operator making correct purchasing decisions.

Polarization Demultiplexing Related Tests
For polarization demultiplexing, the important testing parameters include SOP tracking speed, SOP recovery time, and SOP orthogonality between the two polarization channels. SOP tracking speed is defined as the highest speed of SOP (state of polarization) variations at which a system can still work properly without losing track. It has a unit of radians per second and quantitatively measures how well the demultiplexing circuitry and algorithm can track polarization variations of different patterns at different speeds. SOP recovery time is defined as the time required for a system to recover a loss of polarization track caused by an abrupt polarization change. It indicates how well the demultiplexing circuitry and algorithm work in response to sudden SOP jumps as shown in Figure 12.17.

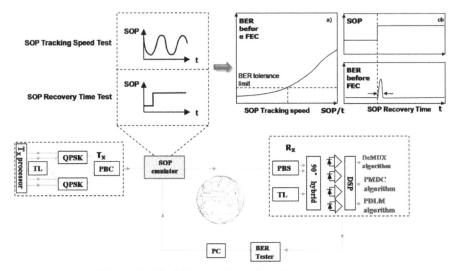

Figure 12.17　SOP related tests for coherent receivers.

PMD Related Tests

For PMD compensation, the important testing parameters are SOP tracking speed, SOP recovery time, PMD tracking speed, PMD recovery time, and PMD tolerance range. Similar to the definitions in polarization demultiplexing, SOP tracking speed is defined as the highest polarization variation at which the PMD compensation hardware and algorithm (to be referred as PMDC hereafter) can still function effectively in reducing PMD related signal distortions. SOP recovery time is defined as the time required for a PMDC to recover a loss of track caused by an abrupt polarization jump.

On the other hand, PMD tracking speed is defined as the highest PMD changing speed at which the PMD compensator is still effective in reducing PMD induced signal distortion. It is in unit of ps/s and a measure of how fast a PMDC can respond to PMD value changes. PMD recovery time is defined as the time required for a PMDC to recover a loss of track caused by an abrupt PMD value jump. It measures how well a PMDC work in response to sudden PMD value jumps. A fast PMD emulator is needed to perform these two tests.

PMD tolerance range is defined as the maximum PMD value in a transmission system with which data can be transmitted with bit error rate (BER) smaller than that required by system design. It is a measure how well the PMDC works in compensating PMD effects and extending system's tolerance

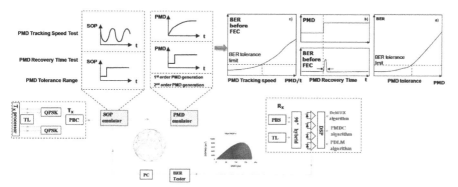

Figure 12.18 Test Setup for PMD's related tests of coherent receivers.

range on PMD. Figure 12.18 shows the setup used to characterize the PMDC function of coherent receivers. A fast polarization scrambler is needed before the PMD emulator if the signal is polarization multiplexed to average out the PMD effect among the two orthogonal channels. For non coherent systems, when polarization multiplex technique is not use, the effect of PMD on a system is highly dependent on the input SOP. To eliminate test uncertainties and increase test speed, the input SOP must be optimized and maintained against polarization fluctuations caused by external disturbances to ensure the worst effect PMD on the signal, by actively aligning the SOP to 45° deg. from the DGD's PSPs.

Test results can be used by network operators to compare systems made by different vendors and verify PMD related specifications promised by the vendors. They can also be used by system vendors to determine suitability of PMD algorithms, to tune algorithms, and for quality control screening of the transceivers. The bit-error rate (BER) of the system, or another performance indicator parameter such as power penalty, is monitored as the 1st order PMD (DGD) values generated by the PMD source and increased until the BER reaches the limits of the system. The corresponding DGD is the 1st order PMD tolerance of the system. Both the 1st and 2nd order PMD (SOPMD) values can also be increased as the BER of the system is measured and plotted. The system outage probability can be calculated from the data obtained.

PDL Related Tests
For PDL performances, the critical test parameters are SOP tracking speed, SOP recovery time, PDL tracking speed, PDL recovery time, and the

PDL tolerance range. SOP tracking speed is defined as the highest SOP variation speed at which the PDL mitigation hardware and algorithm can still function effectively in reducing PDL related transmission errors. SOP recovery time is defined as the time required for PDLM to recover a loss of track caused by an abrupt polarization change. PDL tracking speed is defined as the highest PDL changing speed (in unit of dB/s) at which PDLM can still function effectively in reducing PDL related transmission errors. PDL recovery time is defined as the time required for a PDLM to recover a loss of track caused by an abrupt PDL change. Finally, PDL tolerance range is defined as the maximum PDL value in a transmission system with which data can be transmitted with bit error rate (BER) smaller than that required by system design. It is a measure how well the PDLM works in mitigating PDL effects and extending system's tolerance range on PDL.

Figure 12.19 shows the setup used to characterize the PDLM function of coherent receivers. A fast polarization scrambler is, again, needed before the PDL emulator if Pol-Muxed signals are under test to evenly distribute the PDL effect among the two orthogonal channels. System vendors and operators need to know how much PDL a transceiver can tolerate, and therefore must conduct tolerance range testing. A system vendor can use test results to optimize hardware and algorithms, and to perform quality checks. Network operators can use PDL tolerance range testing to qualify vendors and to perform quality inspection of incoming equipment to ensure the performance of their systems.

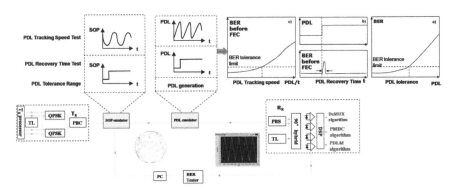

Figure 12.19   Test Setup for PDL's related tests of coherent receivers.

12.3 Optical Signal-to-Noise Ratio Measurement

Optical Signal-to-Noise Ratio (OSNR) is a key performance indicator of amplified optical networks to provide an assessment of the critical signal quality degradation caused by the amplified spontaneous emission (ASE).

Over the years, multi-wavelength amplified systems have evolved into dense wavelength division multiplexing (DWDM), incorporating add/drop filters or reconfigurable add/drop multiplexers (ROADMs) and more recently, into coherent-detection systems using combinations of polarization multiplexed (PM), multilevel-modulation formats, pulse shaping and channel grouping (e.g. super-channels), to maximize the throughput rate. The optical communication systems deployed to carry the massive amounts of data of data center interconnect (DCI) applications consequently rely on these high spectral density multi-level formats. More specifically, four-level amplitude modulation (PAM-4) direct detection is preferred for moderate link lengths while coherent-detection polarization multiplexed quadrature phase shift keying (PM-QPSK) and 16-ary quadrature amplitude modulation (PM-16QAM) are used for longer multiple spans transmissions with in-line amplification [19]. Measuring the OSNR to ensure that the system operates within the required parameters poses different but yet significant challenges in both of these use cases.

The following section briefly reviews the definition of OSNR, discusses available options to measure it and covers in more detail OSNR measurements performed with familiar and traditionally available optical spectrum analyzer (OSA). The specific challenges associated with the PAM-4 and DP-QPSK use cases more specifically pertinent do DCI applications are also presented.

For an optical transmission system, the OSNR of a given channel is defined as the ratio of total signal power to the ASE noise power spectral density within the optical spectrum of the signal [20]. For amplified optical networks, the impact of ASE noise measured by the OSNR is generally the dominant contributor to the quality of signal (QoS) and to the bit-error ratio (BER) performance of a channel. The configuration of early generations of amplified transmission links, with a limited number of channels and significant spacing between channels, allowed for a relatively simple spectral power interpolation measurement technique to determine the ASE level and thus measure OSNR [21]. Optical spectrum analyzers of various configurations[1]

---

[1]IEC 61280-2-9 describes grating-based, Michelson interferometer and Fabry-Perot as dominant technologies used for early commercially available OSAs.

have emerged as the preferred OSNR measurement tool among which the grating-based OSAs have become the test and measurement industry standard. Over the years, other measurement techniques[2] [22], mostly relying on delayed sampling [23] or high speed electrical conversion have been proposed but the OSA's ability to measure at the lower power levels available via taps of the network and to simultaneously measure multiple channels helped secure its position as the only International Electrotechnical Commission (IEC) standardized approach.

The continued increase in adoption of coherent optical transceivers has eased the measurement of several optical performance metrics such as chromatic dispersion (CD) and polarization mode dispersion (PMD) through the capabilities of the advanced digital signal processing (DSP) embedded in the coherent receiver. Coherent receivers also provide the electrical signal-to-noise ratio (SNR) where the electrical noise includes contributions from the ASE but also from other additive noise sources, such as distortions generated by fiber nonlinearity in long haul transmission systems which can be modeled as additive Gaussian noise [24, 25]. Since ASE and nonlinear noise can each degrade the received SNR in the same way, it is often useful to quantify these noise sources separately, especially when trying to identify the relative impact of the different impairments while troubleshooting. Thus, traditional OSNR where optical noise is defined as ASE remains a useful and relevant performance metric for coherent signals.

12.3.1 Measuring OSNR with an OSA

In many legacy systems (Figure 12.20), measuring OSNR was straight forward since WDM signals were not tightly spaced, allowing the measurement of optical ASE noise by interpolating its level from the spectral gaps between signals. The interpolation method simply consists in measuring the ASE level at the inter-channel position on either side of the channel and interpolating this value at the channel center. This can be done using the markers but most commercial OSAs have an integrated processing function to perform the measurement automatically in order to avoid the risk of common calculation errors resulting from differences between peak and integrated signal power and from OSA resolution bandwidth (RBW) considerations

---

[2]For a comprehensive review of direct and indirect OSNR measurement approaches, see [Zhenhua Dong, Faisal Nadeem Khan, Qi Sui, Kangping Zhong, Chao Lu, and Alan Pak Tao Lau, "Optical Performance Monitoring: A Review of Current and Future Technologies," *J. Lightwave Technol.* 34, 525–543 (2016)].

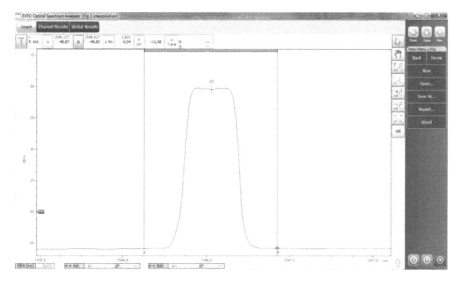

Figure 12.20 Interpolation OSNR measurement on wide channel spacing signal.

when subtracting noise from the signal and normalizing to the 0.1nm reference bandwidth. The calculation with markers for the trace of Figure 12.20 acquired with a 33pm RBW is done in Table 12.2 below using the traditional definition for OSNR as the ratio:

$$R = \frac{\int_{\lambda_1}^{\lambda_2} P_{sig}(\lambda)d\lambda}{B_r \cdot \rho} = \frac{P_{sig}}{B_r \cdot \rho} \qquad (12.36)$$

where:

$P_{\text{Sig}}(\lambda)$ is the time-averaged signal power spectral density, not including ASE, expressed in W/nm;

P_{Sig} is the total signal power, i.e. the integral of $P_{Sig}(\lambda)$ over wavelength, expressed in W;

ρ is the interpolated ASE power spectral density within the spectral range of the signal, independent of polarization, expressed in W/nm;

B_{r} is the reference bandwidth expressed in nm (usually 0.1 nm if not otherwise stated); and the integration range in nm from λ_1 to λ_2 is chosen to include the total signal spectrum.

and OSNR is usually expressed in dB as 10 log(R).

However, DWDM transmission systems and links with in-line filtering preclude the use of the interpolation method. As can be observed from Figure 12.21, such channels are relatively wide and densely packed so that the

Table 12.2   Interpolation OSNR calculation referring to the signal of Figure 12.20

	Wavelength [nm]	Power [dBm/dB]	Formula Used and Comments
Marker A	1546.127	−45.87	Marker refers to 0.033 RBW from OSA
Marker B	1546.927	−45.83	Marker refers to 0.033 RBW from OSA
Interpolated Noise at Channel Center	1546.527	−45.85	$10 \cdot \log_{10}\left[\dfrac{10^{\left(-\frac{45.87}{10}\right)} - 10^{\left(-\frac{45.83}{10}\right)}}{2}\right]$
Total Power in Channel		−11.08	$\int_{1546.127}^{1546.927} P_{Meas}(\lambda)$
Noise in Channel		−32.00	$10 \cdot \log_{10}\left[10^{\left(-\frac{45.85}{10}\right)} \cdot \frac{0.8}{0.033}\right]$ with 0.8nm channel width and 0.033nm OSA RBW
Signal Power (P_{Sig})		−11.12	$10 \cdot \log_{10}\left[10^{\left(-\frac{11.08}{10}\right)} - 10^{\left(-\frac{32}{10}\right)}\right]$
Noise in 0.1nm B_r (ρ*0.1)		−41.04	$10 \cdot \log_{10}\left[10^{\left(-\frac{45.85}{10}\right)} \cdot \frac{0.1}{0.033}\right]$ where B_r is 0.1nm and 0.033 the OSA RBW
OSNR		29.92	−11.12 − −41.04
OSA's IEC OSNR		29.92	OSA's interpolation built-in calculation

inter-channel measurement would be well above the actual ASE noise level and underestimate the OSNR level. The actual channel Off noise overlay trace also shows that the noise level outside the channel may not be representative of the in-channel noise and lead to overestimating the OSNR. Furthermore, filtered ASE imposed a refinement of the IEC OSNR measurement recommendation since the previous definition, which assumed spectrally uniform ASE beyond the channel limit, could lead to differences in OSNR penalty when comparing with BER transmission performance degradation with filtered signals [26]. The updated recommendation for a spectrally-integrated in-band OSNR, R_{int}, is calculated as follows:

$$R_{\text{int}} = \frac{1}{B_r} \int_{\lambda_1}^{\lambda_2} \frac{P_{sig}(\lambda)}{\rho(\lambda)} d\lambda \qquad (12.37)$$

where: $P_{Sig}(\lambda)$ is the time-averaged signal power spectral density, not including ASE, expressed in W/nm; $\rho(\lambda)$ is the ASE power spectral density, independent of polarization, expressed in W/nm; B_r is the reference bandwidth expressed in nm (usually 0.1 nm if not otherwise stated); and the integration range in nm from λ_1 to λ_2 is chosen to include the total signal

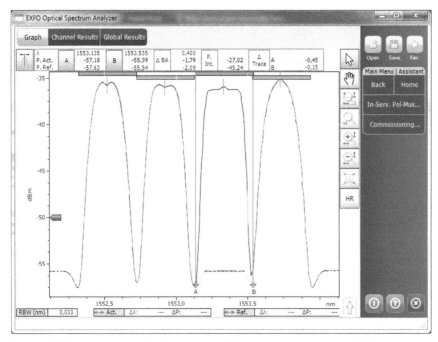

Figure 12.21 Systems precluding the use of interpolation: DWDM QPSK channels in 50GHz spacing with per-channel in-line filtering.

spectrum. $OSNR_{int}$ is usually expressed in dB as $10 \log(R_{int})$. Note that when the ASE is not carved by filtering, the spectrally-integrated OSNR yields the same result as the previous definition of Equation 12.36.

For these network configurations, an alternate method commonly referred to as the "Signal On/Off" method consists in temporarily turning off the channel of interest and measuring the in-band ASE noise at the channel center wavelength. This is generally done at commissioning before the system is used when turning off a channel does not cause actual traffic interruption and can be highly accurate provided that the elements in the signal path remain relatively stable during the process of turning the signal off and back on. Care must be taken to actually turn the signal Off at its source rather than blocking the path at a filtering node which would also remove the noise from the path. This may not be possible after commissioning when monitoring and management features are enabled as they may be programmed to detect a missing signal and block the corresponding channel. The On/Off method is generally provides a good measurement as long as a sufficient number WDM

Table 12.3   Signal On/Off method OSNR calculation referring to Figure 12.21

	Power [dBm/dB]	Formula Used and Comments
Total Power in Channel "On" Trace	−27.02	$\int_{1553.135}^{1553.535} P_{Meas}(\lambda)$
Total Power in Channel "Off" Trace	−45.24	$\int_{1553.135}^{1553.535} N_{ASE}(\lambda)$ "Off Trace"
Signal Power (P_{Sig})	−27.09	$10 \cdot \log_{10}\left[10^{\left(-\frac{27.02}{10}\right)} - 10^{\left(-\frac{45.24}{10}\right)}\right]$
Noise in 0.1nm B_r	−50.95	$\int_{1553.285}^{1553.385} N_{ASE}(\lambda)$ "Off Trace"
OSNR$_{\text{Max}}$	23.86	−27.09 − −50.95 i.e. from Equation 12.36
OSNR$_{\text{CCSA}}$	23.87	OSA built-in calculation (CCSA)
OSNR$_{\text{IECi}}$	23.93	OSA built-in calculation (IEC Equation 12.37)

channels are present such that the ASE power level does not significantly change when removing a channel[3]. Many commercial OSAs actually include tools to calculate OSNR$_{int}$ and to assist the user in making the multiple acquisitions measurements with typically N+1 traces, where N is the desired number of channels to measure. The On/Off calculation, done in accordance with the above-mentioned definitions, is detailed in Table 12.3 with reference to Figure 12.21.

Measuring the OSNR at commissioning allows the system operator to have the Start of Life (SOL) network conditions so they can ensure that the system is configured and operates within its design parameters. This information is required when QoS performance issues arise. At troubleshooting time, in order to determine if the OSNR measurement value obtained has degraded from its SOL condition, the troubleshooting decision will have to rely on a comparison to the "assumed design parameters" of the system if the actual SOL OSNR was not measured.

Once the system is in operation, it is generally no longer possible to use the Signal On/Off method as it would require interrupting or re-routing traffic. For conventional On/Off Keying (OOK) transmission systems, the ASE is generally depolarized and the signal strongly polarized, so that it is possible to discriminate the ASE from the signal without affecting service by exploiting these different polarization properties. A simple approach, known

---

[3]Chinese standard CCSA YD/T 2147-2010, Appendix A (non-normative) suggests a correction approach that takes into account local power changes around the channel under test for low channel counts (e.g. <10) or near the edges of the amplifier gain band.

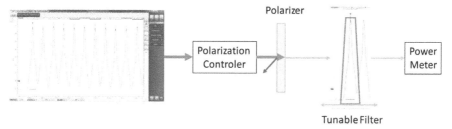

Figure 12.22 Polarization nulling measurement equipment.

as polarization nulling [27], consists in adjusting a polarizing filter "before" the OSA to substantially[4] block the polarized signal, until the signal power becomes negligible as compared to the ASE, and measure the remaining power which will represent $1/2$ of the total depolarized ASE level.

Many practical enhancements to the polarization nulling technique have been developed to overcome filter polarization extinction limitations and measurement errors arising from depolarization of the signal upon propagation due to the presence of polarization mode dispersion (PMD) for longer links or non-linear effects (NLE) in large channel-count amplified networks, especially when dispersion compensation elements are present [28, 29] Both PMD and NLE may induce partial depolarization of the signal which can no longer be completely blocked by the polarizing filter and, as a result, be wrongly measured as ASE contribution. For example, provided that the ASE spectral extent is wider than the signal's, the blocking filter's limited polarization extinction ratio and signal depolarization impacts can be mitigated by leveraging spectral discrimination to perform the measurement using at least two different RBW measurements. With reference to Figure 12.23, the calibrated (coupling loss factored in) power levels P_1, P_2 and P_3 are written as:

$$P_1 = (1 - \varepsilon)^* P_{\text{Sig}} + 0.5^* P_{\text{ASE}}{}^* B_1$$
$$P_2 = \varepsilon^* P_{\text{Sig}} + 0.5^* P_{\text{ASE}}{}^* B_1$$
$$P_3 = \varepsilon^* P_{\text{Sig}} + 0.5^* P_{\text{ASE}}{}^* B_2$$

where P_{Sig} is the signal power (mW), S is the power spectral density of ASE noise (mW/nm) and B_1 and B_2 (with $B_1 > B_2$) are the optical bandwidths

---

[4]The polarization extinction capability of the filter used (typically around 25dB) conditions the amount of polarized signal that passes through and could be wrongly measured as ASE contribution.

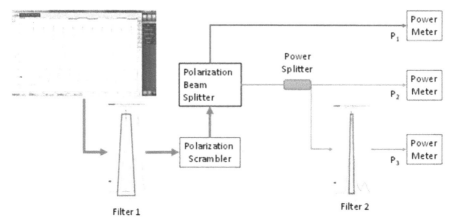

Figure 12.23   Improved polarization-nulling measurement setup.

in nm of the input filter (Filter 1) and bandpass filter (Filter 2) respectively. From P_1, P_2 and P_3, OSNR can be calculated as:

$$OSNR = \frac{P_{sig}}{(P_{ASE} \cdot B_r)} = \frac{P_1 + P_2}{P_2 - P_3} \cdot \frac{B_1 - B_2}{2 \cdot B_r} - \frac{B_1}{B_r} \qquad (12.38)$$

where B_r is the reference resolution bandwidth (typically 0.1nm) for the OSNR measurement. It is important to note here that the spectral extent of P_{Sig} is assumed to be entirely encompassed in B_2 of the narrower bandpass filter such that the same P_{Sig} power is assumed in P_2 and P_3 which would lead to measurement error for higher rate signals (see Table 12.4 below for an example).

Further improvements to the polarization-based techniques later made it possible to measure OSNR beyond the 30dB levels required for PAM-4 using OSAs equipped with a polarization scrambler and polarizing filter [30]. Repeated acquisitions while varying the input state of polarization (SOP) incident on the polarizing filter allows a polarization resolved spectral analysis. In this approach, a polarization-diversity OSA (shown in Figure 12.24) discriminates the unpolarized ASE noise from the polarized signal in order to obtain the signal's relative spectral power distribution. The calibrated sum and difference of the measured orthogonally-polarized spectra (// and ⊥) give:

$$P_{Meas}(\lambda) = [P_{Sig}(\lambda) + N_{ASE}(\lambda)]_{//} + [P_{Sig}(\lambda) + N_{ASE}(\lambda)]_{\perp}$$
$$= P_{Sig}(\lambda) + N_{ASE}(\lambda)$$

Table 12.4 Polarization-based compared OSNR calculations with Figure 12.25 signal

	Wavelength [nm]	Power [dBm/dB]	Formula Used and Comments
On/Off Noise at Channel Centerin 0.1nm B_r	1534.407	−52.06	$\int_{1534.357}^{1534.457} N_{ASE}(\lambda)$ on noise only "Off Trace" This is the actual noise level.
Total Power in Channel		−17.73	$\int_{1534.207}^{1534.607} P_{Meas}(\lambda)$
On/Off Total Noise in Channel		−46.04	$10 \cdot \log_{10}\left[10^{\left(-\frac{52.06}{10}\right)} \cdot \frac{0.4}{0.1}\right]$ where 0.4nm is the channel width and B_r is 0.1nm
Signal Power (P_{Sig})		−17.74	$10 \cdot \log_{10}\left[10^{\left(-\frac{17.73}{10}\right)} - 10^{\left(-\frac{46.04}{10}\right)}\right]$
$\text{OSNR}_{\text{On/Off}}$		34.32	$= -17.74 - -52.06$
OSA's IEC OSNR		32.44	OSA built-in interpolation calculation
Polarization Nulling Noise Cancellation		20.93	where ASE is $2 \cdot \int_{1534.207}^{1534.607} P_{Min}(\lambda)$ normalized to B_r 0.1nm
Improved Polarization Nulling Calculation		31.57	Applying Equation 12.38 with B_1 = 0.4nm and B_2 = 0.1nm
OSA's Polarized OSNR		34.42	OSA built-in calculation from Equation 12.39

which comprises the data-carrying signal and the ASE noise, and:

$$\Delta P_{\text{pol}}(\lambda) = |[P_{\text{Sig}}(\lambda) + N_{\text{ASE}}(\lambda)]_{//} - [P_{\text{Sig}}(\lambda) + N_{\text{ASE}}(\lambda)]_{\perp}|$$

respectively. With $|N_{\text{ASE}}(\lambda)_{//} - N_{\text{ASE}}(\lambda)_{\perp}| << |P_{\text{Sig}}(\lambda)_{//} - P_{\text{Sig}}(\lambda)_{\perp}|$, the difference spectrum provides a relative optical power distribution of the polarized signal. That is $\Delta P_{\text{pol}}(\lambda) \approx P_{\text{Sig}}(\lambda)/\kappa_0$ where κ_0 is a pro-portionality constant[5]. Provided that the spectral power distributions of the data-carrying signal and noise are different, which is generally the case in practice, a comparison of the difference $\Delta P_{\text{pol}}(\lambda)$ and sum $P_{\text{Meas}}(\lambda)$ spectra, done in a relative manner, allows the determination of the proportion of noise

---

[5]The number of independently and uniformly distributed SOP covered during the acquisition can also provide a statistical estimation of maximum and minimum values of transmitted signal through the polarizing filter and thus an estimation of κ_0. Adding an estimation of the effective differential group delay (DGD) obtained using scrambled state-of-polarization analysis (SSA) method can further reduce the PMD induced OSNR measurement errors.

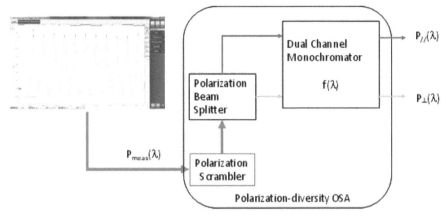

Figure 12.24   Schematic drawing of polarization-diversity OSA.

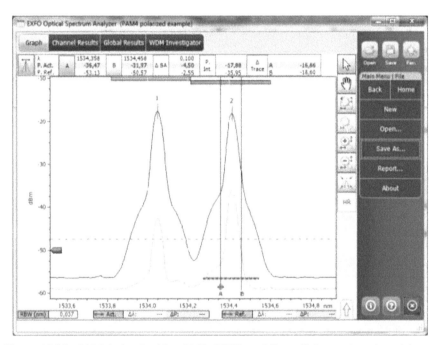

Figure 12.25   PAM-4 signal with >30dB OSNR and P_{Min} (light grey) as the minimum of 250 scrambled SOPs measurements through the polarization filter. The vertical markers indicate the approximate width of Filter 2 from Equation 12.38 and the red horizontal bar the OSA's built in calculation of noise based on Equation 12.39.

contributing to $P_{Meas}(\lambda)$. OSNR can thus to calculated by finding $N_{ASE}(\lambda)$ with:

$$\frac{P_{Meas}(\lambda) - N_{ASE}(\lambda)}{P_{Meas}(\lambda_p) - N_{ASE}(\lambda_p)} - \frac{\Delta P_{pol}(\lambda)}{\Delta P_{pol}(\lambda_p)} = 0 \qquad (12.39)$$

solving for portions of the spectrum (between λ_1 and λ_2) where the local variations in ASE noise power levels are small, i.e. where $|N_{ASE}(\lambda_1) - N_{ASE}(\lambda_2)| \ll |\Delta P_{pol}(\lambda_1) - \Delta P_{pol}(\lambda_2)|$ and where λ_p is arbitrarily selected as the signal's peak wavelength. Polarization-based OSNR measurements are compared with a PAM-4 signal example in Table 12.4 with reference to Figure 12.25. Note that for a meaningful comparison, the same measurement data was used for three compared techniques.

The basic polarization-nulling measurement for this high OSNR condition typical of single span PAM-4 transmission case produces a worse deviation from the true On/Off value than the interpolation method but improved polarization-based techniques (Equations 12.38 and 12.39) reduce or substantially eliminate the measurement error.

For coherent-detection systems, coherent optical transceivers have eased the measurement of several optical performance metrics through the capabilities of the advanced digital signal processing (DSP) embedded in the coherent receiver. Relying on the DSP, they can readily measure signal parameters such as chromatic dispersion (CD) and polarization mode dispersion (PMD). However, other impairments such as in-band crosstalk, passband narrowing due to optical filtering, and fiber nonlinear effects are more challenging to directly measure. Notwithstanding interoperability issues, the limited dynamic range available from monitoring taps may further preclude measurements with "golden receivers" along the link. The taps typically extract only a few percent of the power actually present on the link at that location and the use of high speed "golden receivers" would necessitate amplification and filtering[6]. The lower available powers at taps and the ability to measure multiple channels simultaneously have traditionally been the main advantages of OSA based techniques, which rely on low speed electronics in the kHz to MHz ranges and thus operate, even via taps, well above their sensitivity limits. The downside is that, at these low speeds, the equal power orthogonally-polarized PM signals appear depolarized and do not allow sufficient discrimination from the ASE to provide a reliable OSNR measurement based on the above polarization-based techniques.

---

[6]The external amplifier must be factored-in the measurement as it changes the OSNR which is no longer the actual value to be measured.

To date, no equivalently simple, cost-effective and accurate method has emerged for independently and non-intrusively measuring OSNR on PM coherent-detection systems. A number of non-service-affecting techniques have been proposed based on several different principles of measurement, which include, interferometric devices [31], characterization of beat noise [32], Stokes-space based polarization measurements ("polarization disc") [33], DSP-based algorithms [34], and more recently, a correlation-based technique using coherent-receivers with MHz speed electronics [35]. Each of these methods has its merits and drawbacks, including the limited dynamic range for measurement via taps, the need for additional specialized hardware or lack of applicability under certain system conditions, such as chromatic dispersion or fiber nonlinearity.

PM signals measured with a diversity OSA yield intrinsically insufficient polarization discrimination such that the level of $\Delta P_{pol}(\lambda)$ is small and there is no longer a spectral region where $|N_{ASE}(\lambda_1) - N_{ASE}(\lambda_2)| << |\Delta P_{pol}(\lambda_1) - \Delta P_{pol}(\lambda_2)|$. In parallel with the measurement where OSNR is required $P_{Meas}(\lambda)$, another technique called reference-Based [36] proposes to use a reference spectrum $P_{Ref}(\lambda)$ of the same signal at another location where the ASE noise level is known (or negligible) instead of using the unsuitably small $DP_{pol}(\lambda)$. This approach allows measurement of the OSNR non-intrusively via taps and on multiple channels simultaneously. In an ideal case (e.g. back-to-back measurement condition), $P_{Ref}(\lambda)$ can be compared with $P_{Meas}(\lambda)$ to determine the ASE level in much the same way as was demonstrated for polarization-based signals:

$$\frac{P_{Meas}(\lambda) - N_{ASE}(\lambda)}{P_{Meas}(\lambda_p) - N_{ASE}(\lambda_p)} - \frac{\Delta P_{ref}(\lambda)}{\Delta P_{ref}(\lambda_p)} = 0 \qquad (12.40)$$

solving for portions of the spectrum (between λ_1 and λ_2) where the local variations in ASE noise power levels are small, i.e. where $|N_{ASE}(\lambda_1) - N_{ASE}(\lambda_2)| << |P_{Ref}(\lambda_1) - P_{Ref}(\lambda_2)|$ and where λ_p is arbitrarily selected as the signal's peak wavelength.

In practice however, the received signal $P_{Sig}(\lambda)$, or $P_{Meas}(\lambda) - N_{ASE}(\lambda)$, differs from $P_{Ref}(\lambda)$. The two are nonetheless related since $P_{Sig}(\lambda) = \kappa(\lambda) * P_{Ref}(\lambda)$ where $\kappa(\lambda)$ is the link's spectral transfer function between the two locations, which comprises a linear net gain/loss profile $\kappa_L(\lambda)$ and, in a NL transmission regime, $\kappa_{NL}(\lambda)$ which represents the spectral deformation incurred by the signal. This spectral transfer function can be written as:

$$\kappa(\lambda) = \kappa_L(\lambda) \cdot \kappa_{NL}(\lambda) = \kappa_0 \cdot [1 + \Delta\kappa_L(\lambda)] \cdot [1 + \Delta\kappa_{NL}(\lambda)] \quad (12.41)$$

where κ_0 is now the constant transmission ratio between $P_{Sig}(\lambda)$ and $P_{Ref}(\lambda)$ at the signal's peak wavelength λ_p, $\Delta\kappa_L(\lambda)$ is the relative wavelength dependent linear transmission resulting from the gain/loss profile of the fiber, amplifiers and other network elements such as ROADMs, while $\Delta\kappa_{NL}(\lambda)$ is the relative wavelength dependent spectral deformation resulting from NL effects on the signal. From a detailed analysis of the spectra OSNR can be determined by using the optical spectrum difference (OSD) between $P_{Meas}(\lambda)$ and $P_{Ref}(\lambda)$ expressed as:

$$OSD(\lambda) = \frac{P_{Meas}(\lambda) - \kappa_0 \cdot P_{Ref}(\lambda)}{\kappa_0 \cdot P_{Ref}(\lambda_p)} = OSD_S(\lambda) + OSD_{ASE}(\lambda)$$

(12.42)

where

$$\begin{aligned}
OSD_S(\lambda) &= [P_{Sig}(\lambda)/P_{Sig}(\lambda_p)] - [P_{Ref}(\lambda)/P_{Ref}(\lambda_p)] \\
&= \Delta\kappa(\lambda) * [P_{Ref}(\lambda)/P_{Ref}(\lambda_p)] \\
OSD_{ASE}(\lambda) &= N_{ASE}(\lambda)/P_{Sig}(\lambda_p) \\
&= 1/OSNR_{ASE}(\lambda) \\
\Delta\kappa(\lambda) &= [\kappa(\lambda)/\kappa_0] - 1
\end{aligned}$$

"ASE only" OSNR can be calculated as:

$$OSNR_{ASE}(\lambda) = \frac{1}{OSD_S(\lambda)} = \frac{1}{OSD(\lambda) + OSD_S(\lambda)}$$

(12.43)

In the linear regime, $\Delta\kappa_{NL}(\lambda) = 0$ and the transfer function can be obtained by comparing the spectra at two spectral positions near the signal peak, e.g. within the central pass-band of the filters/ROADMs, where their wavelength dependent response is approximately constant, so that $\Delta\kappa_L(\lambda)$ could also be neglected. The resulting transfer function, practically independent of wavelength within the signal pass-band, can be written as:

$$\kappa_{0-Est} = \frac{P_{Meas}(\lambda_1) - P_{Meas}(\lambda_2)}{P_{Ref}(\lambda_1) - P_{Ref}(\lambda_2)} \approx \kappa_0$$

(12.44)

with λ_1, λ_2 chosen such that:

$$[N_{ASE}(\lambda_1) - N_{ASE}(\lambda_2)] << [P_{Meas}(\lambda_1) - P_{Meas}(\lambda_2)]$$

and the OSNR can be written as:

$$OSNR_{ASE}(\lambda) = \frac{\kappa_{0-Est} \cdot P_{Ref}(\lambda)}{P_{Meas}(\lambda) - \kappa_{0-Est} \cdot P_{Ref}(\lambda)}$$

(12.45)

yields the same as Equation 12.40 above. But in the nonlinear regime, NL spectral deformation of the signal is present, a significant $\Delta\kappa_{NL}(\lambda)$ contribution renders $\kappa_{0-Est} \neq \kappa_0$ and applying the Equations 12.44 and 12.45 leads to an estimated level:

$$OSNR_{Est}(\lambda) = OSNR_{ASE}(\lambda) + \Delta OSNR(\lambda) \qquad (12.46)$$

where $\Delta OSNR(\lambda)$ is a wavelength dependent difference determined by $\Delta\kappa_{NL}(\lambda)$. For weak NL effects, $\Delta\kappa_{NL}(\lambda) \approx 0$ and $\Delta OSNR(\lambda) \approx 0$, while for strong NL effects, a large $\Delta\kappa_{NL}(\lambda)$ leads to a larger $\Delta OSNR(\lambda)$. Without a priori knowledge of the ASE level, $\Delta\kappa_{NL}(\lambda)$ can be estimated indirectly from the ratio of the derivatives of $P_{Meas}(\lambda)$ and $P_{ref}(\lambda)$:

$$R(\lambda) = \frac{dP_{Meas}(\lambda)}{dP_{Ref}(\lambda)} = \kappa(\lambda) + P_{Ref}(\lambda) \cdot \frac{d\kappa(\lambda)}{dP_{Ref}(\lambda)} + \frac{dN_{ASE}(\lambda)}{dP_{Ref}(\lambda)} \quad (12.47)$$

$$\approx \kappa(\lambda) + P_{Ref}(\lambda) \cdot \frac{d\kappa(\lambda)}{dP_{Ref}(\lambda)}$$

where

$$dP_{Meas}(\lambda) = dP_{Meas}(\lambda)/d\lambda$$
$$d\kappa(\lambda) = d\kappa(\lambda)/d\lambda$$

and λ is chosen in a region from λ_1 to λ_2 such that $dN_{ASE}(\lambda) = dN_{ASE}(\lambda)/d\lambda << dP_{ref}(\lambda) = dP_{ref}(\lambda)/d\lambda$ and $dP_{ref}(\lambda) \neq 0$. From the properties (baud rate, spectral shaping) of the signal $P_{ref}(\lambda)$ an empirical relation applying to a wide range of signal types and system operating conditions can be determined that allows to infer $\Delta\kappa_{NL}(\lambda) = \kappa(\lambda)/\kappa_{0-1}$ from $\Delta R(\lambda) = R(\lambda)/R(\lambda_1) - 1$ and thus quantify $\Delta OSNR(\lambda)$. Figure 12.26 illustrates how the spectral signature of $\Delta R(\lambda)$ correlates with the spectral signature of $1 + \Delta\kappa_{NL}(\lambda)$ and provides an indication of the NLE conditions.

The application of this technique is shown for the setup of Figure 12.27 below and the corresponding spectrum shown in Figure 12.28.

The system power had been adjusted to optimize the BER conditions which entailed increasing the power level into the nonlinear regime. Upon transmission, the signal has thus suffered NL induced spectral deformation. In this example, an ASE OSNR level in excess of 23dB is measured to within 0.5dB of the On/Off measurement target value (ASE level also visible from the extinguished neighbor channels) despite significant NLE. A key advantage of this technique is thus that it allows the measurement of the ASE

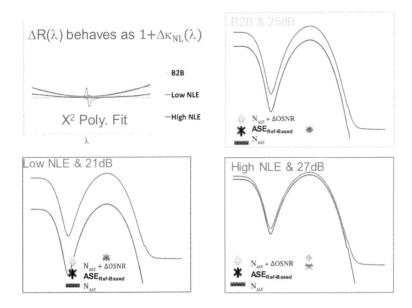

Figure 12.26 Shows how $\Delta OSNR(\lambda)$ of Equation 12.46 can be determined using $\Delta R(\lambda)$ from Equation 12.47 and an example of application for different NLE conditions.

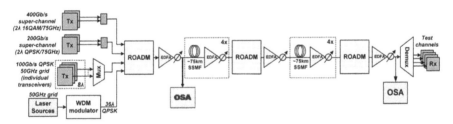

Figure 12.27 Description of setup used for the non-intrusive reference-based OSNR test.

noise contribution even in the presence of NLE induced noise when coherent receivers are generally not capable of distinguishing between such additive noise sources with approximately Gaussian statistical properties. This provides supplementary troubleshooting information when a degradation of the (electrical) SNR at the coherent receiver reveals performance issue.

The two highlighted channels of Figure 12.28 are within the central passband of the ROADM which was set to pass the bank of four channels together since the technique does not perform well for large baud rate signals passing through narrow filters/ROADMs or wavelength selective switches (WSS). For such conditions, spectral deformations add a dominant linear

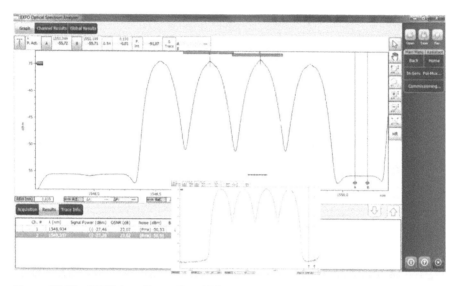

Figure 12.28 100Gb/s unfiltered PM-QPSK channels on the 50 GHz grid after transmission over eight spans of standard single-mode fiber with ~75 km per span and all-EDFA amplification (reference taken at beginning of link shown inset).

contribution $\kappa_L(\lambda)$ to $\kappa(\lambda)$ thus making it very difficult to infer $\Delta\kappa_{NL}(\lambda)$ since the reference signal $P_{Ref}(\lambda)$ is no longer representative of the received signal $P_{Sig}(\lambda)$.

This can be mitigated by using another reference for the received signal such as a measurement done at the receive location while doing commissioning of the system. At SOL, an initial $P_{SOL}(\lambda)$ measurement can be used to determine $P_{Sig}(\lambda)$ by leveraging the known initial OSNR value, for example obtained using the On/Off method. When at a later time a $P_{Meas}(\lambda)$ spectrum is acquired, while doing preventive maintenance or troubleshooting, it can be compared to the recalled $P_{Sig}(\lambda)$ spectrum so that their power spectral distribution could be analyzed. The processing is very similar but the approach is different in that the original $P_{SOL}(\lambda)$ measurement produces a reference $P_{Sig-SOL}(\lambda)$ which includes the link's initial $\kappa(\lambda)$ deformation. The measurement performance will thus be mainly limited by changes occurring on $P_{Sig}(\lambda)$ or $\kappa(\lambda)$ over time such as transmitted signal deformations (e.g. from DSP or modulator instabilities), relative drift between the signal and the filters' central wavelengths or filter transfer function variations (e.g. ROADM passband ripple changes). The technique has been demonstrated to be fairly robust for small to moderate changes, incurred on $P_{Sig}(\lambda)$, representative of a typical system operation [37]. A preventive maintenance strategy with repeated measurements, such as in a monitoring use case, would limit the changes that occur on $P_{Sig}(\lambda)$ or $\kappa(\lambda)$ between measurements (e.g. wavelength drifts of 2GHz do not usually happen abruptly). In that case, the reference can be updated regularly since the ASE level is correctly measured at time T and a new $P_{Sig-T}(\lambda)$ can be generated for further measurements.

12.4 Characterization of Optical Vector-Modulated Signals

Optical transmission systems for data center network have traditionally been based on binary modulation formats, most notably on-off-keying (OOK), either as non-return-to-zero (NRZ) or return-to-zero (RZ) format. These formats are straightforward to implement and provide excellent noise tolerance and low cost. However, with the continuous need to increase transmission capacity and distance, higher-order multilevel modulation formats and coherent detection have been increasingly used to achieve higher spectral efficiency. Coherent detection increases the receiver complexity, but it brings very important advantages. For example, digital coherent receivers can map the complete optical field in real time, which enables compensation of impairments such as chromatic dispersion (CD) and polarization mode dispersion (PMD) by linear digital signal processing (DSP) [38]. Coherent systems also enable polarization division multiplexing (PDM), as demultiplexing is performed entirely in the processing stage. This multiplies the capacity (i.e. spectral efficiency) by a factor of two without needing additional optical polarization control hardware in front of the receiver. Hence, 28 GBd PDM-QPSK carries 100 Gb/s with 12% forward error correction (FEC) overhead, and has become the industry consensus QPSK format [39] for 100 Gb/s single-carrier transmission. Furthermore, commercial systems having even greater spectral efficiency are gaining interest, which leads to the use of more complex modulation formats, such as M-ary quadrature amplitude modulation (e.g. 16-QAM [40] or 64-QAM [41]).

Well established techniques and standards exist to quantitatively test and characterize OOK hardware and systems. Examples include measurement of eye diagrams with an oscilloscope having predetermined filter response characteristics to quantify extinction ratio (ER), temporal jitter, and overshoot for example and standardized eye diagram masks to quantify margin before certain devices failure [42, 43]. With non-binary modulation formats, this becomes much more complex [44] and to date, no comprehensive standards exist that specify the parameters to characterize and how such quantification should be undertaken. This is further complicated when considering coherent transmission systems since there exists an intricate interplay between the hardware constrained performance and the DSP's impairment mitigation capability.

The error vector magnitude (EVM) has long been a common metric in wireless *electronic* digital communication systems [45], such as cellular telephony and satellite communications. With the advent of coherent optical systems, EVM is also been applied in fiber-optic communications [46], and, more recently, it was defined as a metric for quantifying the quality of an optical vector-modulated signal, e.g. a signal that is modulation in both phase and magnitude [47].

Section 12.4.1, introduces constellation and IQ diagrams for vector-modulated signals. The definition of EVM and metrics of root-mean-squared EVM (RMS-EVM) and time resolved EVM (TR-EVM) are presented in section 12.4.2. Analogous to SNR for OOK signal, relation between RMS-EVM, Q-Factor, OSNR and BER are covered in section 12.4.3. Finally, section 12.4.4 discusses characterization of signal quality and impairments for transmitters using complex optical modulation.

12.4.1 Constellation and IQ Diagrams for Vector-Modulated Signal

Constellation and IQ Diagrams:
A constellation diagram indicates the amplitude and phase of the signal at the decision point. This is the point in time when the signal must have the correct phase and amplitude value for error-free transmission. In OOK modulation, this corresponds to the decision threshold above which the receiver decodes the signal as a "1" instead of a "0". A cluster of points is displayed at each coding location, with a point for each detected symbol in a data pattern.

The IQ diagram displays the complete phase and amplitude transitions between transmitted vectors. It reflects directly the combined I and Q components of the signal at any sample time of the data acquisition. The traces on the diagram show the path of the signal vector over the data pattern where the transition traces depend on the modulation format (e.g. NRZ or RZ), as well as modulator configurations.

Figures 12.30, 12.31 and 12.32 show IQ diagrams for QPSK, 8-PSK and 16-QAM modulation formats [44].

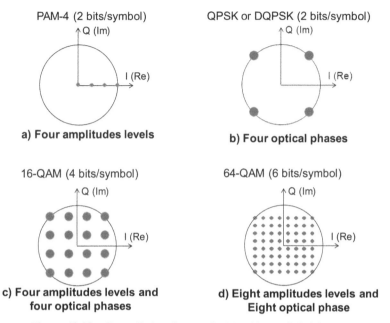

Figure 12.29 Constellation diagram for Non-binary digital format.

QPSK carries 2 bits/symbol

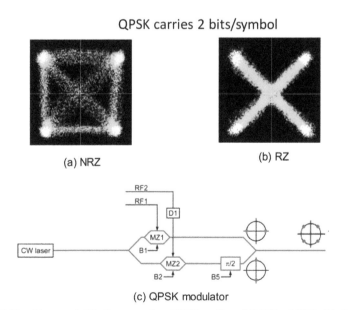

Figure 12.30 Measured IQ diagrams for QPSK coding (a)NRZ; (b)RZ; (c) Schematic drawing of I-Q modulator.

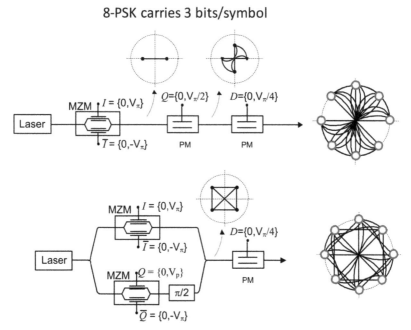

Figure 12.31   IQ diagrams for 8-phase coding (two configurations).

Polarization Multiplexing:

The phase modulation of a signal is demodulated by optical mixing and depends on the relative polarization of the two optical carriers. Since the incoming signal generally has an unknown and non-constant polarization, signals from two orthogonal polarization axes must be demodulated. Leveraging this doubling of the demodulation information, it is possible to detect signals from two carriers with orthogonal polarization, each carrying independent bit streams, and thus to double the transmission rate for a given wavelength channel. For such polarization multiplexed signals, two independent pairs of I and Q traces are present and two separate constellation or IQ diagrams are used.

12.4.2 Definitions of EVM, RMS-EVM and TR-EVM

Each transmitted symbol can be described by a vector with amplitude and phase, which codes a number of bits. Deviations from ideal modulation due to inherent impairments of transmitter and receiver, as well as impairments during transmission, impair the received vector with distortions and noise

16-QAM carries 4 bits/symbol

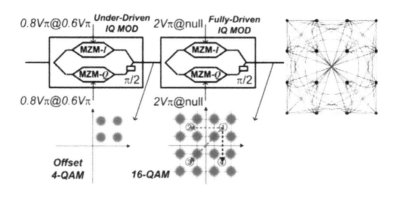

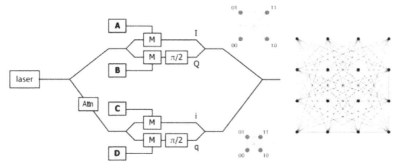

Figure 12.32 IQ diagrams for 16 QAM coding (two configurations).

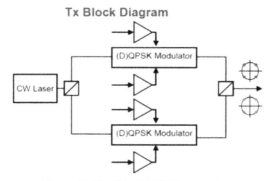

Figure 12.33 DP-(D)QPSK transmitter.

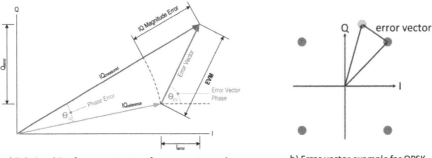

a) Relationship of error vector to reference vector and
measured signal vector in the constellation diagram

b) Error vector example for QPSK

Figure 12.34   Error Vector and Error Vector Magnitude (EVM).

resulting in a different vector location in the IQ diagram. As shown in Figure 12.34, the error vector is defined as the difference between an ideal constellation point and an actual received symbol in the I-Q plane, whose magnitude is the EVM.

In the mathematical terms the EVM is defined as

$$EVM(s) = |S_{meas}(s) - S_{ref}(s)|/|S_{refmax}| \qquad (12.48)$$

$$= \sqrt{Q_{error}{}^2(s) - I_{error}{}^2(s)} \qquad (12.49)$$

where S_{meas} is the measured vector for symbol number s, and S_{ref} is the reference vector given by the known symbol pattern. If the reference symbol pattern is unknown, the EVM can be computed in relation to the closest symbol, i.e. a decision directed approach. The EVM is normalized to the reference vector, S_{ref}, with largest magnitude.

For a vector-modulated signal that has been measured to give the time-dependent I and Q traces with sampling interval, T_s, the EVM of a particular measurement sample with index k is given by time-dependent I and Q traces with sampling interval, T_s, the EVM of a particular measurement sample with index k is given by

$$EVM(kT_s) = \sqrt{Q_{error}(kT_s)^2 + I_{error}(kT_s)^2} \qquad (12.50)$$

where

$$I_{error}(kT_s) = \alpha I_{meas}(kT_s) - I_{ref}^{r(k)}$$

$$Q_{error}(kT_s) = \alpha Q_{meas}(kT_s) - Q_{ref}^{r(k)}$$

and I_{ref} and Q_{ref} correspond to the reference symbol r(k) for the sample k, that is r(k) is an index pointing to the symbol with the reference vector for sample k.

In practical measurements the measured vector has arbitrary magnitude scaling with the receiver sensitivity, so normalization by a factor α of the measured vectors is required to make the error vector independent of the scaling. Detailed description and derivation of α can be found in [47], where it is defined as

$$\alpha = \frac{\sum_{s=1}^{n}(I^r_{ref}(s)^2 + Q^r_{ref}(s)^2)}{\sum_{s=1}^{n}(I^r_{ref}(s) \cdot I_{meas}(s) + Q^r_{ref}(s) \cdot Q_{meas}(s))} \qquad (12.51)$$

The EVM can be averaged over all symbols to obtain a single RMS value according to

$$EVM_{RMS} = \sqrt{\frac{1}{n}\sum_{s=1}^{n}EVM^2(s)} \qquad (12.52)$$

The EVM_{RMS} is commonly expressed in percentage of the magnitude of the longest reference vector. Its value is normally computed at the center of the symbol slot, or decision point, for example, with samples that lie within 10% of Ts. Figure 12.35 shows a measurement example of a NRZ 28GBd DP-QPSK. The white highlighting in IQ diagrams and in the eye diagrams of I-Q components indicate where the rms EVM measurement (average EVM_{RMS} of 7.455%) is taken. The rms error vector magnitude of a burst of n measured symbols can be used as a figure of merit for complex signals, summarizing the quality of the signal in one number [48].

The EVM concept can be strengthened by extending its application in the time domain and to the whole IQ diagram to take inter-symbol transitions into account. Using the general definition of the EVM each measured sample of the signal under test can be associated with an EVM value. Furthermore, each measured sample can be time stamped and given an accurate measurement time relative to the symbol slot. By combining the time information with the EVM information about each sample, we can define the time-resolved EVM as

$$EVM_{TR}(k,t) = |S_{meas}(k,t) - S_{ref}(s)|/|S_{ref}|_{max} \qquad (12.53)$$

with k as the index of each measured and analyzed sample from the burst of n samples and t as the acquisition time relative to the symbol slot [50]. The time-resolved EVM (TR-EVM) is defined exactly as the EVM, but

28GBd NRZ-DP-QPSK

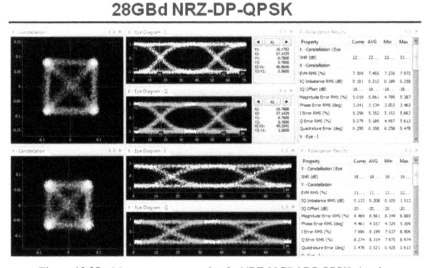

Figure 12.35   Measurement example of a NRZ 28GBd DP QPSK signal.

for all samples, not only the constellation points. Hence, also samples that appear in the transition area are compared with the ideal constellation point. Figure 12.36 (a) shows a typical NRZ-QPSK constellation diagram with 10kS, where two samples and their error vectors are highlighted. Figure 12.36 (b) shows the corresponding time-resolved EVM plot. The two highlighted samples are also shown; one in the symbol center and the other in the transition area. When the samples correspond to transitions passing

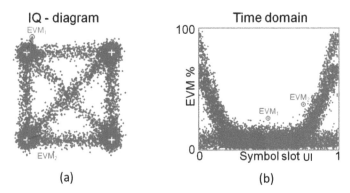

Figure 12.36   (a) NRZ-QPSK constellation diagram; (b) corresponding time-resolved EVM graph.

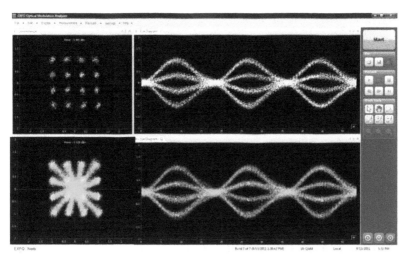

Figure 12.37   A measured 56GBd RZ-16-QAM signal.

through the origin of the constellation, the EVM reaches its maximum. Near the decision points, in the middle of the symbol slot, the EVM should be small, and the samples within [0.45, 0.55] UI are the samples that provide the EVM$_{RMS}$ metric. The advantage of TR-EVM is that all information is collected in one plot rather than from a complicated constellation diagram and numerous eye diagrams, which can become unwieldy for higher-order modulation formats having many bits per symbol, such as the 16-QAM format. Figure 12.37 shows the constellation diagram, IQ diagram and eye diagrams of I-Q components of a measured 56GBd RZ-16-QAM signal. The time-resolved EVM plot also indicates the margins that may be available against additional distortions accumulated in the transmission.

12.4.3 Relationships between EVM$_{RMS}$, Q-Factor, OSNR and BER

The Q-Factor metric is well established for on-off keying (OOK) optical systems. For such two-level signals, the Q factor is defined in the following way in [43]:

$$Q = \frac{|\mu_1 - \mu_2|}{\sigma_1 + \sigma_2} \tag{12.54}$$

Where μ_i are the mean values of the two levels, and σ_i their standard deviations, within a predefined time window around the sampling decision

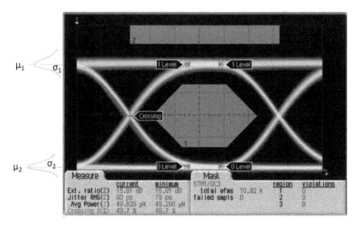

Figure 12.38   Example of eye diagram with mask.

point, as shown in Figure 12.38. To estimate BER from Q, marks and spaces in the detected photocurrent are assumed to be superimposed with additive white Gaussian noise (AWGN), the probability density of which is fully described by its mean and variance. A large Q leads to a small BER.

Using the above definition of Q factor and the definition of the EVM and assuming isotropic noise in the IQ plane the following relationship holds true for QPSK modulation, based on the normalized two-level I or Q-components with $|\mu1 - \mu2| = \sqrt{2}$ and $\sigma1 = \sigma2 = \text{EVM}_{\text{RMS}}$.

$$Q_{QPSK} = \frac{1}{\sqrt{2}EVM_{RMS}} \tag{12.55}$$

Extending this concept to higher multilevel modulation formats such as M-QAM leads to the following definition of Q-Factor for N-level signals

$$Q_{QPSK} = \frac{1}{N-1} \sum_{i=1....k-1} \frac{|\mu_i - \mu_{i+1}|}{\sigma_i + \sigma_{i+1}} \tag{12.56}$$

Again, assuming isotropic noise in the IQ plane the following relationship between EVM and Q-Factor holds true and for M-QAM, the relation of Q-Factor to EVM_{RMS} is

$$Q_{M-QAM} = \frac{1}{\sqrt{2} \cdot (\sqrt{M} - 1) \cdot EVM_{RMS}} \tag{12.57}$$

If the received optical field is perturbed by AWGN only, the EVM_{RMS} can be related to BER and to the optical signal-to-noise ratio (OSNR) [50, 51]. A small EVM_{RMS} corresponds to a small BER.

Table 12.5 Modulation Format-Dependent Factor p

Formats	B/Q/8 PSK	16-QAM	32-QAM	64-QAM
p	1	9/5	17/10	7/3

EVM_{RMS} can be estimated from the measured OSNR (e.g. measured with an optical spectrum analyzer) [50] and can be found as [51]:

$$EVM_{RMS} \approx \frac{1}{\sqrt{p}} \left[\frac{1}{OSNR} - \sqrt{\frac{96}{\pi(M-1)OSNR}} \sum_{i=1}^{\sqrt{M}-1} \gamma_i e^{-\alpha_i} \right.$$

$$\left. + \sum_{i=1}^{\sqrt{M}-1} \gamma_i \beta_i erfc(\sqrt{\alpha_i}) \right]^{1/2} \tag{12.58}$$

where $\alpha_i = 3\beta_i^2 OSNR/2(M-1)$, $\beta_i = 2i-1$, $\gamma_i = 1-i/M^{1/2}$ and erfc() is the complementary error function. Table 12.5 [48] shows the value of p, a modulation format dependent factor. OSNR in Equation 12.58 refers the signal power to the polarized ASE power in the actual optical receiver bandwidth B_O. Commonly the normalized or reference $OSNR_{ref}$ is defined as the total signal power to the unpolarized ASE power in a given noise equivalent bandwidth B_{ref}, where B_{ref} is typically 0.1nm or 12.5GHz at 1550nm [52, 53]. As a result, OSNR relates to $OSNR_{ref}$ as OSNR = $(B_{ref}/B_o)OSNR_{ref}$ for DP signal.

Here, basic assumptions are that system errors are mainly due to optical AWGN (neglecting electronic noise), that quadratic M-QAM signal constellations are investigated and that reception is non data-aided, i.e. the received data is not a priori known. For very noisy signals, non data-aided reception leads to an underestimated EVM because a received symbol could be closer to a neighbor constellation position than to its own targeted constellation point. In contrast, for data-aided reception where a predetermined sequence known to the receiver can be transmitted for measurement purposes, or larger OSNR conditions of non data-aided reception, the second and third square-rooted terms go to zero, and equation 12.58 simplifies to

$$EVM_{RMS} \approx \frac{1}{\sqrt{M} \cdot OSNR} \tag{12.59}$$

Representing ideal rms EVM value.

For data-aided reception, knowing EVM the BER can be approximated by [48, 50]

$$BER \approx \frac{2(1 - \frac{1}{\sqrt{M}})}{\log_2 M} erfc \left[\sqrt{\frac{3}{\sqrt{2}(M-1)pEVM^2}} \right] \qquad (12.60)$$

For (12.60), the same limitations as with (12.58) apply. If the EVM is not derived by evaluating (12.58) but measured directly, the influence of electronic noise is also included.

It is demonstrated experimentally and by simulations that the BER can be estimated from EVM data by this analytic relation [50, 51]. Strictly speaking, equation 12.60 is valid only for data-aided reception, but it was shown to be applicable for non data-aided reception when BER $< 10^{-2}$ [48].

From Equations 12.58, 12.59 and 12.60, both EVM and BER can be estimated from an OSNR measurement and both estimates are valid for systems limited by optical AWGN. Experiments with binary phase shift keying (BPSK), quadrature PSK (QPSK), 8PSK, 16QAM, 32QAM, and 64QAM modulation formats at symbol rates of 20 GBd and 25 GBd [48] showed good agreement when comparing measured OSNR, EVM and BER with calculated BER and EVM estimations.

12.4.4 Characterization of Transmitter Impairments

Figure 12.39 shows a typical optical DP-QPSK transmitter. A narrow linewidth continuous wave (CW) laser is used to limit the impacts of phase noise in the signal recovery process at the coherent receiver which eventually leads to bit errors. Each polarization is modulated in I/Q modulators with incoming QPSK data, consisting of four parallel binary streams (i.e. 2 bits/symbol/polarization). In order to optimize the transmitter and minimize penalties associated with any offset from the optimum setting, several delays and biases must be well designed or adaptively controlled. Some signal disturbances can be equalized to some extent in the receiver, whereas others will inevitably degrade the signal. The most common disturbances are directly drawn from the physical modulator architecture and may include for example I/Q data skew (D1 or D2), quadrature error (B5 or B6), and bias errors (B1 to B4).

Note that the transmitter scheme with I/Q modulators is general in the sense that it can handle almost any modulation format. For example 16-QAM can be realized by applying 4-level electrical drive signals, but other more complex transmitter architectures for 16-QAM can be devised to avoid

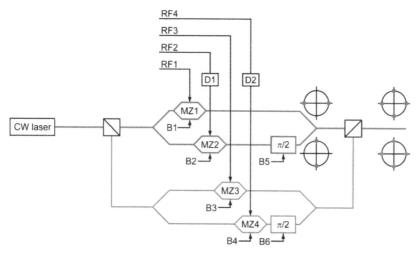

Figure 12.39 Dual-Polarization QPSK Transmitter.

multilevel electrical signals [54, 55]. The following section will discuss examples using QPSK transmitter impairments and their characterizations for simplicity but without loss of generality.

For characterization of such a transmitter, the most powerful tool is a coherent receiver that gives the complete optical field information. The sampling can be either real-time electrical [56], or coherent optical sampling [57]. Real-time sampling has the advantage of enabling standard receivers DSP, while optical sampling offers high bandwidth measurements of the true waveform under test with lower cost electronics. Figure 12.40 illustrates examples of different distortions in the IQ diagram, eye diagrams of I and Q components, and the corresponding shapes of the TR-EVM for NRZ-QPSK. A complete table is useful for reference, and not all effects are readily discernable. Figure 12.41 (a) shows an undistorted constellation. The table is then divided into different types of imperfections; modulator amplitude related imperfections such as bias-errors and RF signal level distortions (b)-(e), and timing-related distortions, such as jitter and skew (f)-(k). The table also includes signal chirp and signal SNR distortions (l)-(m). Amplitude distortions generally lead to shifts in the constellation points from the ideal, thereby increasing the EVM from the bottom in the middle of the UI and having significant impact on EVM_{RMS}. Timing distortions generally close the TR-EVM from the sides and reduce the margins towards mask violations, providing information which is not taken into account by the

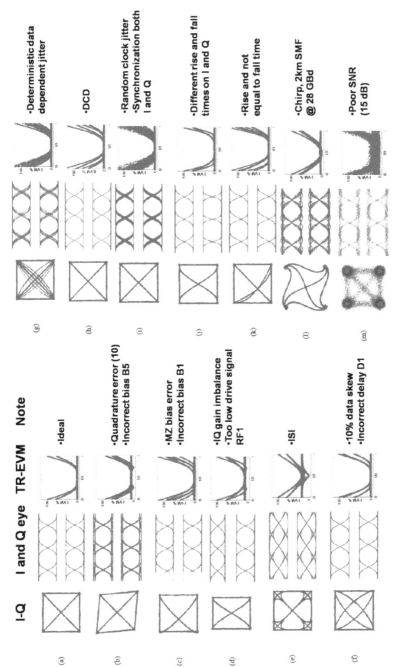

Figure 12.40　IQ diagrams, eye diagrams of IQ, and TR-EVM plots for different distortions.

Skew = 0 ps   Skew = 6 ps

Skew = 4 ps   Skew = 8 ps

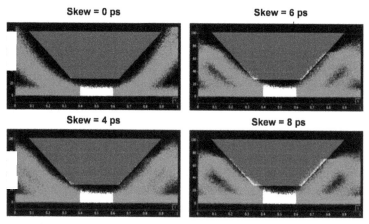

Figure 12.41 Exemplary mask with different I-Q timing skews.

EVM_{RMS} measurement. The quadrature error is minimized by optimizing bias B5 in Figure 12.39, which generally is more sensitive than the MZ biases. Note that ISI in Figure 12.40 (e) may be caused by the transmitter drive signals or transmitter bandwidth limitations, but it may also occur from receiver bandwidth limitations. High bandwidth detection is therefore primordial in order to obtain a true picture of the transmitted signal. Data skew in Figure 12.40 (f), which is one of the most important parameters to measure, leads to the characteristic opening in the phase diagram since the transitions do not occur at the same time for I and Q, and closes the TR-EVM from the sides. Deterministic data jitter (Figure 12.40 (g)) results in various transition routes in the phase diagram as well as jitter in the eye diagrams, and closure of the TR-EVM. Note that random symbol clock jitter (Figure 12.40 (i)) is not visible in the phase diagram since both I and Q transitions are shifted in time exactly the same way. They are however clearly visible in the eye diagrams and the EVM plot, particularly if the frequency of the jitter is too high to be tracked by the clock recovery algorithm. Variations in rise and fall times are common and lead to the typical distortions shown in Figure 12.40 (j) and Figure 12.40 (k). For example, dispersion-induced chirp gives rise to the well-known spiral transitions in Figure 12.40 (l). Poor SNR (Figure 12.40 (m)) can be caused by thermal noise due to low input power to the receiver, or optical ASE noise added to the signal, but it can also be optical or electrical noise coming directly from the transmitter. In general, a real signal suffers impairments resulting from combinations of all these distortions.

In summary, a majority of the signal distortions that may influence the quality of vector-modulated optical signals will strongly affect the transition region of the signals and hence change the appearance of the time-resolved EVM plot. Modulation distortions will show in the time-resolved EVM plot by both increasing the EVM in the symbol centers (higher EVM_{RMS}) and reducing the opening in the time domain. The time-resolved EVM plot will give additional information compared to EVM_{RMS} by better indicating margins to test failure or revealing reasons for test failures. It is thus natural to introduce a mask in analogy with traditional eye-diagram mask testing, which has been applied to transmitter production testing for many years in as discussed in section 12.4. A mask violation probability exceeding a certain level would lead to a failed test and the component under test would not pass. Figure 12.41 illustrates an exemplary TR-EVM mask for I-Q timing skews of 0ps, 4ps, 6ps and 8ps.

Figure 12.42 presents the time-resolved EVM for different modulation formats including: NRZ-BPSK, NRZ-QPSK, NRZ-8-APSK, and NRZ-16-QAM, and shows that a TR-EVM mask test is compatible with all vector-modulated signal formats. It should noted here that the ideal constellation points are given in Figure 12.42 by the known symbol pattern but a decision directed approach can be applied if the symbol pattern is unknown. Figure 12.43 illustrates TR-EVM for the same set of modulation formats as

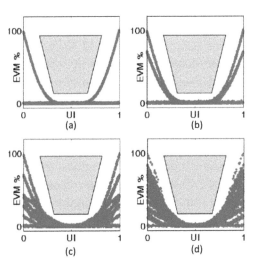

Figure 12.42   Typical TR-EVM plots for (a) NRZ-BPSK, (b) NRZ-QPSK, (c) NRZ-8APSK, and (d) NRZ-16QAM with a known reference symbol pattern.

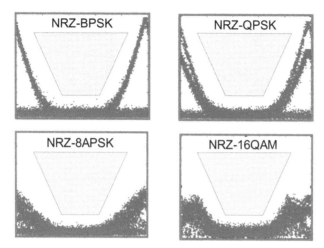

Figure 12.43 Typical TR-EVM plots for NRZ-BPSK, NRZ-QPSK, NRZ-8APSK and NRZ-16QAM with the decision directed reference approach.

for Figure 12.42 with the decision directed approach. The difference is small for QPSK but decreases the contrast (EVM_{TR} does not reach 100%) in the transition area for higher order modulation formats, such as 8-APSK and 16-QAM. Contrary to rms EVM, Figure 12.42 and Figure 12.43 show that mask testing using TR-EVM works as well for both the reference approaches.

Figure 12.44 shows a mask hits test example on a measured 28 Gbd QPSK signal with 8 ps skew, using (a) known symbol pattern reference approach. and (b) the decision directed reference approach.

Figure 12.45 and Figure 12.46 illustrate simulation results of a case study [58] for multi-impairment pass/fail testing of a 28 GBd DP-QPSK transmitter using TR-EVM plot and mask. Figure 12.45 the investigation results on

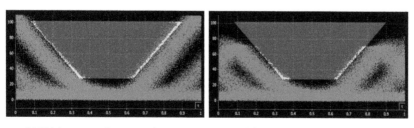

(a) With known sample pattern reference   (b) With direct decision reference

Figure 12.44 A mask hits test examples on a measured 28 Gbd QPSK signal with 8 ps I-Q timing skew.

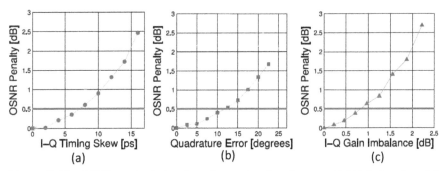

Figure 12.45   OSNR penalty in function of transmitter impairments: (a) I-Q timing skew, (b) quadrature error, (c) I-Q gain imbalance.

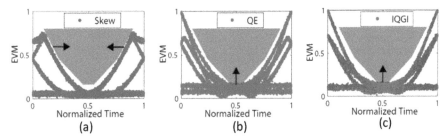

Figure 12.46   TR-EVM plots with (a) I-Q time skew (b) quadrature error (c) I-Q gain imbalance causing 0.5dB OSNR penalty.

OSNR penalties @BER of 10^{-3} with (different) transmitter impairments of I-Q timing skew (skew), quadrature error (QE) and I-Q gain imbalance (IQGI), by making Monte Carlo simulations finding the BER as a function of the OSNR for different amounts of the impairments. From these curves, it can obtain impairment level causing a given OSNR penalty, such as the horizontal red lines marking the limit of 0.5 dB. Table 12.6 shows corresponding impairment values to 0.5 dB OSNR at BER = 10^{-3}, obtained by interpolation in the Figure 12.45. It is desirable and useful to find a mask catching all impairments at approximately the same OSNR penalty (e.g. 0.5 dB). Figure 12.46 shows simulated TR-EVM plots and the common example mask with 0.5 dB OSNR penalty (or impairment values listed in Table 12.6).

Table 12.6   Allowed impairments for 0.5 dB OSNR penalty @BER 10^{-3}

skew	QE	IQGI
7.2 ps	11.9 deg.	0.83 (0.8dB)

The arrows in the figure indicate how the samples protrude into the mask for a further increased impairment level.

The design of proper compliance test masks for different vector-modulated optical signals is beyond the scope of this section. Mask definition (e.g. shape) would ideally be a topic of discussion for a specialized standards body, with strong input from both of network equipment and transmitter manufacturers.

12.5 Conclusion

A wide variety of data transmission technologies have been developed with optimized speed to cost ratios for different types of data centers in the past decade. To ensure the performance of the data transmission systems in operation, advanced tests and measurements are essential. In this chapter, we describe different advanced optical measurement technologies required for the successful deployment of optical fibers in data centers, including the measurements of polarization related parameters; the measurements of WDM channel bandwidth, crosstalk and ripple of the components and fiber links; the determination of in-band OSNR of the optical signals propagating in the systems, and the measurements of parameters relating to the vector modulated signals, such as error vector magnitude (EVM), bit-error-rate (BER), and OSNR. Finally, the emulation systems for determining the tolerances to different polarization related impairments in the systems, such as PDL, PMD, SOP variations, are also discussed.

Acknowledgements

We thank Dr. James Chen and Susan Wey form General Photonics for for their suggestions, and proof-reading of this chapter.

References

[1] Yao, X. S., "Controlling polarization related impairments." Lightwave Magazine, 2000.
[2] H. Kogelnik, R. Jopson, and L. Nelson, "Polarization-Mode Dispersion," in Optical Fiber telecommunications, Vol. IV-B, Systems and Impairments, I. P. Kaminow and L. Tingye ed., (Elsevier Science (USA).

[3] E. Lichtman, "Limitations imposed by polarization-dependent gain and loss on all-optical ultralong communication systems," J. Lightwave Technol. 13(5), 906–913 (1995).

[4] M. Chbat, "Managing polarization mode dispersion," Photonics Spectra, June 2000, pp. 100–104.

[5] S. Yao, "Combat Polarization Impairments with Dynamic Polarization Controllers", Lightwave Online articl., 2000.

[6] J. Tassé, W. Daab, "Why Coherent Detection Systems May Fail at Compensating for Polarization Mode Dispersion", white paper 063, 2015.

[7] R. A. Chipman, Polarimetry, Ch. 22 in Handbook of Optics, vol. II, 2nd Ed. M. Bass ed., McGraw-Hill, New York, 1995.

[8] D. H. Goldstein, "Mueller matrix dual-rotating retarder polarimeter," Appl. Opt. 31, 6676–6683 (1992).

[9] P. A. Williams, "Rotating-wave-plate stokes polarimeter for differential group delay measurements of polarization-mode dispersion," Appl. Opt. 38, 6508–6515 (1999).

[10] E. Dijkstra, H. Meekes, and M. Kremers, "The high-accuracy universal polarimeter," J. Phys. D 24, 1861–1868 (1991).

[11] P. A. Williams, A. H. Rose, and C. M. Wang, "Rotating-polarizer polarimeter for accurate retardance measurement," Appl. Opt. 36, 6466–6472 (1997).

[12] A. D. Martino, Y. Kim, E. Garcia-Caurel, B. Laude, and B. Drévillon, "Optimized Mueller polarimeter with liquid crystals," Opt. Lett. 28, 616–618 (2003).

[13] B. Wang, J. List, and R. Rockwell, "Stokes polarimeter using two photoelastic modulators," in Polarization Measurement, Analysis, and Applications V, D. H. Goldstein and D. B. Chenault, eds., Proc. SPIE 4819, 1–8 (2002).

[14] X. S. Yao, X. Chen, and T. Liu, "High accuracy polarization measurements using binary polarization rotators," Optics Express, Vol. 18, No. 7, pp. 6667–6685, 2010.

[15] X. S. Yao, L. Yan, and Y. Shi, "Highly repeatable all-solid-state polarization-state generator," Opt. Lett. 30(11), 1324–1326 (2005).

[16] Y. Shi, L. Yan, and X. S. Yao, "Automatic maximum-minimum search method for accurate PDL and DOP characterization," J. Lightwave Technol. 24(11), 4006–4012 (2006).

[17] L.-S. Yan, X. Yao, C. Yu, Y. Wang, L. Lin, Z. Chen, and A. E. Willner, "High-speed and highly repeatable polarization-state analyzer

for 40-Gb/s system performance monitoring," *IEEE Photon. Technol. Lett.*, vol. 18, no. 4, pp. 643–645, Feb. 2006.

[18] Bhandare, Sweety et al. "5.94-Tb/s 1.49-b/s/Hz (40/spl times/2/spl times/2/spl times/40 Gb/s) RZ-DQPSK polarization-division multiplex C-band transmission over 324 km." IEEE Photonics Technology Letters 17 (2005): 914–916.

[19] M. H. Eiselt, N. Eiselt, and A. Dochhan, "Direct-Detection Solutions for 100G and Beyond," in *Optical Fiber Communication Conference*, OSA Technical Digest (online) (Optical Society of America, 2017), paper Tu3I.3.

[20] Optical signal-to-noise ratio measurement for dense wavelength-division multiplexed systems: IEC 61280-2-9 (Edition 2.0 2009–02).

[21] Optical monitoring for dense wavelength division multiplexing systems: *ITU-T Recommendation G.697* (02/12).

[22] Zhenhua Dong, Faisal Nadeem Khan, Qi Sui, Kangping Zhong, Chao Lu, and Alan Pak Tao Lau, "Optical Performance Monitoring: A Review of Current and Future Technologies," J. Lightwave Technol. 34, 525–543 (2016).

[23] S. D. Dods, T. B. Anderson, K. Clarke, M. Bakaul, and A. Kowalczyk, "Asynchronous Sampling for Optical Performance Monitoring," in *Optical Fiber Communication Conference and Exposition and The National Fiber Optic Engineers Conference*, OSA Technical Digest Series (CD) (Optical Society of America, 2007), paper OMM5.

[24] F. Vacondio, O. Rival, C. Simonneau, E. Grellier, A. Bononi, L. Lorcy, J.-C. Antona, and S. Bigo, "On nonlinear distortions of highly dispersive optical coherent systems," *Optics Express* 20(2), pp. 1022–32 (2011).

[25] A. Carena, V. Curri, G. Bosco, P. Poggiolini, and F. Forghieri, "Modeling of the Impact of Nonlinear Propagation Effects in Uncompensated Optical Coherent Transmission Links," *J. Lightw. Technol.* 30(10), pp. 1524–1539 (2012).

[26] Fibre optic communication system design guides – Part 12: In-band optical signal-to-noise ratio (OSNR): *IEC TR 61282-12* (Edition 1.0 2016–02).

[27] M. Rasztovist-Wiech, M. Danner, and W.R. Leeb, "Optical signal-to-noise ratio measurement in WDM networks using polarization extinction," ECOC1998, pp. 549–550, (1998).

[28] J. H. Lee and Y.C. Chung, "Improved OSNR monitoring technique based on polarization-nulling method," Electron. Lett. 37(15) pp. 792–793 (2001).

[29] D. Gariépy, G. He, Y. Breton, B. Dery, and G. W. Schinn, "Novel OSA-based method for in-band OSNR measurement," *OFC*, paper JThA15, (2010).

[30] G. He, D. Gariépy, H. Chen, N. Cyr and G. W. Schinn, "In-band impairment measurement via Varied-SOP polarization-resolved Optical Spectrum Analysis," *International Conference on Advanced Infocom Technology 2011 (ICAIT 2011)*, Wuhan, China, 2011, pp. 1–8.

[31] M. R. Chitgarha, S. Khaleghi, W. Daab, M. Ziyadi, A. Mohajerin-Ariaei, D. Rogawski, M. Tur, J. D. Touch, V. Vusirikala, W. Zhao, A. E. Willner, "Demonstration of WDM OSNR Performance Monitoring and Operating Guidelines for Pol-Muxed 200-Gbit/s 16-QAM and 100-Gbit/s QPSK Data Channels" *OFC/NFOEC*, paper OTh3B.6, (2013).

[32] S. K. Shin et al., "A Novel Optical Signal-to-Noise Ratio Monitoring Technique for WDM Networks," *OFC*, paper WK6, (2000).

[33] Takashi Saida, Ikuo Ogawa, Takayuki Mizuno, Kimikazu Sano, Hiroyuki Fukuyama, Yoshifumi Muramoto, Yasuaki Hashizume, Hideyuki Nosaka, Shuto Yamamoto, and Koichi Murata, "In-band OSNR monitor with high-speed integrated Stokes polarimeter for polarization division multiplexed signal", *Optics Express* **20**(26), 177311 (2011).

[34] D. J. Ives, B.C. Thomsen, R. Maher, and S. Savory, "Estimating OSNR of equalized QPSK signal", *ECOC*, paper Tu.6.A.6, (2011).

[35] W. Moench and E. Loecklin, "Measurement of Optical Signal-to-Noise-Ratio in Coherent Systems using Polarization Multiplexed Transmission," *OFC*, paper Th2A.42, (2017).

[36] D. Gariépy, S. Searcy, G. He, and S. Tibuleac, "Non-intrusive measurement of polarization-multiplexed signals with spectral shaping and subject to fiber non-linearity with minimum channel spacing of 37.5GHz," Optics Express 24, 20156–20166 (2016).

[37] D. Gariépy, S. Searcy, M. Leclerc, P. Gosselin-Badaroudine, G. He, and S. Tibuleac, "Novel In-Service OSNR Monitoring Method for Reconfigurable Coherent Networks", *OFC*, paper W2A4, (2018).

[38] S. J. Savory, "Digital filters for coherent optical receivers," *Opt. Express*, vol. 17, no. 2, Jan. 2008, pp. 804–817.

[39] Optical Internetworking Forum, "100G Ultra Long Haul DWDM Framework Document".

[40] P. J. Winzer, A. H. Gnauck, S. Chandrasekhar, S. Draving, J. Evangelista, and B. Zhu, "Generation and 1,200-km transmission

of 448-Gb/s ETDM 56-Gbaud PDM 16-QAM using a single I/Q modulator," *Proc. ECOC 2010*, PD2.2.

[41] A. H. Gnauck, P. J. Winzer, A. Konczykowska, F. Jorge, J. Dupuy, M. Riet, G. Charlet, B. Zhu, and D. W. Peckham, "Generation and transmission of 21.4-Gbaud PDM 64-QAM using a high-power DAC driving a single I/Q modulator," *OFC 2011*, PDPB2.

[42] IEC-61282-2

[43] IEC-61282-8

[44] P. A. Andrekson, "Metrology of complex modulation formats," *OFC2011*, Tutorial, Paper OWN1.

[45] IEEE Standard for Wireless LAN Medium Access Control (MAC) and Physical Layer (PHY) Specifications: High-Speed Physical Layer in the 5 GHz Band, IEEE Standard 802.11a-1999.

[46] R. Schmogrow, D. Hillerkuss, M. Dreschmann, M. Huebner, M. Winter, J. Meyer, B. Nebendahl, C. Koos, J. Becker, W. Freude, and J. Leuthold, "Real-Time Software-Defined Multi Format Transmitter Generating 64QAM at 28 GBd.," *IEEE Photon. Technol. Lett.*, Vol. 22, 2010.

[47] IEC TR 61282-10

[48] R. Schmogrow, B. Nebendahl, M. Winter, A. Josten, D. Hillerkuss, S. Koenig, J. Meyer, M.Dreschmann, M. Huebner, C. Koos, J. Becker, W. Freude, and J. Leuthold, "Error Vector Magnitude as a Performance Measure for Advanced Modulation Formats", *IEEE Photon.Technol. Lett.* 24, 61–63, 2012.

[49] H. Sunnerud, M. Westlund, M. Sköld, and P. Andrekson, "Time-Resolved Error Vector Magnitude for Transmitter Mask Testing in Coherent Optical Transmission Systems," *OFC/NFOEC 2011* (Los Angeles), Paper JWA031.

[50] R. A. Shafik, M. S. Rahman, and A. H. M. R. Islam, "On the extended relationships among EVM, BER and SNR as performance metrics*," in Proc. 4th ICECE, 2006*, pp. 408–411.

[51] H. Arslan and H. A. Mahmoud, "Error vector magnitude to SNR conversion for nondata-aided receivers," *IEEE Trans. Wireless Commun.*, vol. 8, no. 5, pp. 2694–2704, May 2009.

[52] IEC 61280-2-9

[53] IEC TR 61282-12

[54] T. Sakamoto, A. Chiba, and T. Kawanishi, "50-Gb/s 16 QAM by a quad-parallel Mach-Zehnder modulator," *ECOC 2007*, paper PD2.8, 2007.

[55] G.-W. Lu, M. Sköld, P. Johannisson, J. Zhao, M. Sjödin, H. Sunnerud, M. Westlund, A. Ellis, and P. A. Andrekson, "40-Gbaud 16-QAM

transmitter using tandem IQ modulators with binary driving electronic signals," *Opt. Express*, vol. 18, issue 22, pp. 23062–23069, 2010.

[56] M. G. Taylor, "Measurement of phase diagrams of optical communication signals using sampled coherent detection," *Symp. Optical Fiber Measurements*, pp. 163–166, 2004.

[57] Henrik Sunnerud, Mats Sköld, Mathias Westlund, and Peter A. Andrekson "Characterization of Complex Optical Modulation Formats at 100 Gb/s and Beyond by Coherent Optical Sampling", *J. Lightwave Technol.*, vol. 30, pp. 3747–3759, 2012.

[58] Henrik Eliasson, Pontus Johannisson, Henrik Sunnerud, Mathias Westlund, Magnus Karlsson and Peter A. Andrekson "Transmitter mask testing for 28GBaud PM-QPSK", *ECOC 2013*, paper Tu.3.C.2, 2013.

13

Digital Signal Processing for Short-reach Optical Communications

Kangping Zhong[1], Xian Zhou[2], Jiahao Huo[2], Alan Pak Tao Lau[3], Chao Lu[4] and Li Zeng[5]

[1]MACOM Technology Solutions Inc, Shenzhen, China
[2]University of Science and Technology Beijing, Beijing, China
[3]Department of Electrical Engineering, The Hong Kong Polytechnic University, Hung Hom, Kowloon, Hong Kong SAR, China
[4]Department of Electronic and Information Engineering, The Hong Kong Polytechnic University, Hung Hom, Kowloon, Hong Kong SAR, China
[5]Huawei Technology Ltd, Shenzhen, China

13.1 Introduction

Fiber-optic communications traditionally focused on high-speed connections over vast distances across the globe. In recent years, demands for short-reach optical communications in data centers (DCs) have been driven by new applications such as Internet of Things (IoTs), Augmented/Virtual Reality, Cloud Computing, storage and other software-as-a-service (SaaS), and infrastructure-as-a-service (IaaS). Short-reach optical communications collectively refer to optical communication links in the hundreds of meters connecting one server to another within a DC to inter-DC links in the tens of kilometers as shown in Figure 13.1.

A recent Cisco report predicts that 86% of the global Internet traffic will be DC related in which 77% will be within a DC by 2020. Such an explosive growth for DC traffic is driven by content providers such as Facebook, Google, Amazon, and the like. Over the past few years, they have heavily invested in their own infra-structure and overtaken traditional telecommunication carriers as market drivers for optical transceivers, especially for short-reach links.

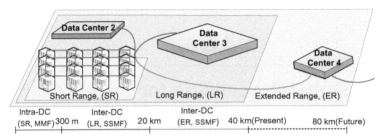

Figure 13.1   The length scale of short-reach optical communications for DCs can range from <300 m using a multi-mode fiber (MMF) to over 20 km using single-mode fibers with advanced modulation formats.

In terms of developments in optical transceiver technologies, perhaps the most significant advancement in the past decade is digital coherent systems in which all degrees of freedom of the optical carrier (amplitude, phase, and polarization) are exploited to encode information. With the powerful digital signal processing (DSP) at both the transmitter and the receiver sides, most of the transmission impairments like chromatic dispersion (CD) and polarization-mode dispersion (PMD) become linear and can be compensated easily and adaptively.

In order to mirror the achievements in long-haul coherent transceivers, the research community has put tremendous efforts to develop DSP strategies for short-reach systems in recent years. However, the sheer market size for short-reach transceivers renders cost the prime concern and prohibits direct application of the digital coherent system philosophy into short-reach links. Also, transceiver form factors are particularly important as port density, or the total number of physical connections and bit rate per area, is a key parameter determining the total size and energy usage of DC. These constraints impose unprecedented challenges in physical layer devices and necessitate innovative signaling, detection, and DSP strategies to realize next-generation DC for various applications aforementioned.

13.1.1 Challenges for Short-reach Optical Systems

13.1.1.1 Cost

For the low-cost short-reach transceivers' market, high-performance devices such as low-linewidth external cavity lasers (ECLs), lithium niobate (LiNbO$_3$) external modulators, thermoelectric cooling (TEC), LOs, and full coherent receivers are prohibited. Instead, components such as vertical-cavity surface-emitting lasers (VCSELs), directly modulated lasers (DMLs), electro-absorption modulators (EAMs), and low-bandwidth PIN make up the building blocks of most short-reach transceivers. Also, direct detection is much preferred over coherent detection. Simplified DSP is much desired and channel spacing (for WDM implementation) is much larger than standard 50-GHz international telecommunication union (ITU) grids because of the lack of TEC and wavelength locking mechanisms.

13.1.1.2 Form factor

While small form factors are desired for all types of optical transceivers in general, the form factor is imperative for DC applications as it significantly affects the port density, rack size which directly translates into the overall size of DCs, the amount of cooling/air conditioning required, and hence ultimately total energy usage.

13.1.1.3 Latency

For DC systems, latency tolerance is more stringent than long-haul systems because of super-computing and various other delay-sensitive Cloud applications. Also, propagation delay is so much smaller than long-haul links and there are fewer link components in short-reach systems to induce further signal delays. In this case, a typical FEC latency of 15–150 μs [1] becomes a major contributor of the overall system latency, and therefore complicated soft-decision FEC with iterative decoding techniques is out of consideration. Standard hard decision-forward error correction (HD-FEC) or innovative low-latency codes are more relevant for short-reach systems. This in turn imposes more stringent pre-FEC bit error rate (BER) and receiver sensitivity (RS) requirements.

13.1.2 Different Types of Short-reach Systems

13.1.2.1 Server-to-server or Intra-data-center links

Server-to-server or intra-DC interconnections are links below 300 m connecting one server to another within a DC. They constitute the biggest volume among all market segments in DC interconnects. They are currently dominated by VCSEL transmitters and MMF.

13.1.2.2 Inter data-center links

These are links below 20 km connecting one DC to another. The length scale is too long for MMF because of inter-modal dispersion and hence the standard single-mode fiber (SSMF) is the default choice. On the other hand, this length scale is still short enough that optical amplifiers are not preferred. Consequently, RS is an important parameter for system optimization. Advanced photo-diodes such as high-bandwidth avalanche photo-diodes (APDs) enable much lower RS than traditional PIN and are important research achievements in this area. In addition, as CD effects grow with the baud rate and distance and direct detection eliminates signal phase information and makes the system nonlinear, CD effects will need to be addressed by innovative signaling and DSP techniques. Alternatively, O-band transmissions are proposed to address such problems as CD effects are minimal at that wavelength range for SSMF.

13.1.2.3 Extended Reach Inter-data-center, Access, and Metro Links

These are links between 20 km and 80 km long and optical amplification is acceptable for some scenarios. However, direct detection receivers are still preferred over fully coherent receivers. To further increase the bit rate, polarization-multiplexed direct detection configurations such as Stokes-vector direct detection (SV-DD) receivers are being actively pursued by various groups worldwide. On the other hand, there are ongoing efforts to push the bit rate and transmission distance of O-band systems without optical amplification into this application domain. Furthermore, as full coherent transceivers are becoming cheaper, smaller, and more power efficient, extended range (ER) inter-data-center links may adopt fully coherent transceivers in the foreseeable future.

Table 13.1 summarizes the key features and limiting factors of the few types of short-reach systems aforementioned.

13.2 Modulation Formats for Short-reach Systems

Intensity modulation combined with direct detection (IM/DD) systems, which are technologically simple, cost effective, and small form-factor, becomes the most practical solution for short-reach systems [2].

Table 13.1   Different types of short-reach systems and their respective challenges

Type	Length Scale	Features and Limiting Factors	Remarks
Server-to-server, intra-data center, and intra-/inter-rack	<300 m	Inter-modal dispersion for MMF	
Inter-data center	<20 km	RS, CD	O-band transmissions as alternative
ER inter-data center, metro, and access	<80 km	CD	Fully coherent transceivers eventually?

Traditional optical interconnects implemented with simple non-return to zero – on–off-keying (NRZ-OOK) format and direct detection receivers without any special DSP (normally up to 25 Gbit/s) are no longer sufficient to support the increasing interface speed requirements. In order to scale the transmission speed of short-reach systems, high-order modulation formats including pulse amplitude modulation (PAM), carrier-less amplitude and phase modulation (CAP), and discrete multi-tone (DMT) modulation are adapted in 100-Gbit/s/lambda short-reach transmission systems.

13.2.1 Pulse Amplitude Modulation (PAM)

PAM is the simplest high-order modulation formats. Unlike the NRZ-OOK format, information is encoded on multiple amplitude levels such as PAM-4, PAM-8, or PAM-16. As the number of levels doubles, the RS will be reduced significantly. Figure 13.2 shows the BER vs. OSNR for 112-Gbit/s PAM-4, PAM-8, and PAM-16. It can be seen that a higher OSNR is required for higher order PAM signals that are more sensitive to channel impairments such as CD and require a more complex DSP. Therefore, PAM-4 is a more practical option considering the tradeoff between the system performance and the achievable bit rate. Recently, the IEEE 400GbE P802.3bs task force has adopted PAM-4 for DC interconnects [3].

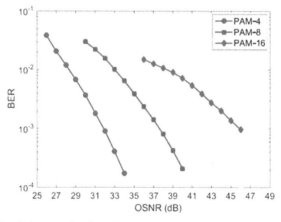

Figure 13.2 Simulation results for BER vs. optical signal-to-noise ratio (OSNR) for 112-Gbit/s PAM-4, PAM-8, and PAM-16.

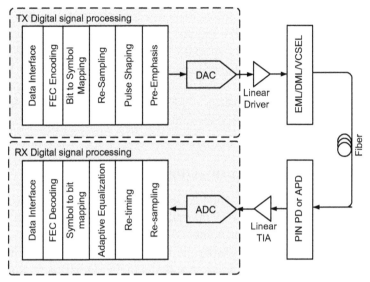

Figure 13.3   DSP configuration for PAM transmission systems.

Figure 13.3 shows the DSP configuration for PAM transmission systems. Pulse shaping and pre-emphasis through a time-domain finite impulse response (FIR) filter are employed to compensate for the DAC bandwidth limitations and transmitter nonlinearity such as the nonlinear characteristics of the modulator [4]. The signal is then loaded into the DAC to generate the electrical driving signal, boosted by a linear driver amplifier, and modulate the laser. The laser can be a DML, EAMs integrated with distributed feedback lasers (EMLs), or VCSELs. The received optical signal will be directly detected by PIN PDs or APDs. Compared to the receiver DSP of coherent systems, the difference of PAM-4 direct detection systems is the adaptive equalization using a feed-forward equalizer (FFE) and a decision feedback equalizer (DFE) to compensate for various transceiver and channel impairments [5, 6]. However, when the bandwidth of the concatenated components in the link is considerably lower than the signal bandwidth, significant ISI will be induced. Compensation of ISI by linear equalizers will greatly enhance the noise only in the high-frequency region which is not desirable. Instead, techniques such as MLSE can be employed. In this connection, Zhong et al. [7] proposed direct detection faster than Nyquist (DD-FTN) algorithm to overcome the drawbacks of conventional equalizers. More details about these algorithms will be discussed in the next section.

13.2.2 Carry-less Amplitude and Phase (CAP) Modulation

An alternative intensity modulation format that can provide high-order modulation is CAP modulation that enables the generation of QAM-type signals but with simpler implementation. In this case, baseband QAM signal with a frequency shift to a subcarrier is required. CAP uses an FIR filter to achieve this up-conversion and modulates the QAM signal to a real signal or demodulates the QAM signal from a received real signal. Although the shaping and matched filters in CAP system can be implemented using analog techniques [8], digital implementation is the preferred choice given today's DAC and ADC technologies [9].

The DSP structure of CAP is also shown in Figure 13.4. At the transmitter side, the in-phase (I) and quadrature (Q) components of the QAM signal are sent into two shaping filters with impulse responses $h_I(t) = g(t) \cdot \cos(2\pi f_c t)$ and $h_Q(t) = g(t) \cdot \sin(2\pi f_c t)$ that form a Hilbert transform pair. A square-root-raised-cosine shaping filter $g(t)$ with a roll-off coefficient of α is used as the baseband impulse response. The center frequency f_c is given by $(1 + \alpha) \cdot B/2 + \Delta f$, where Δf is the frequency offset in order to avoid spectrum aliasing. Figure 13.5(a) shows the time-domain response of the shaping filter. Figure 13.5(b) depicts the baseband QAM signal. After the shaping filter, the electrical spectrum of the generated CAP signal is shown in Figure 13.5(c). The signal is shaped to be a Nyquist-like signal and up-converted to an intermediate frequency (IF) of f_c. After the shaping filter, pre-emphasis is employed to compensate for the bandwidth limitation of

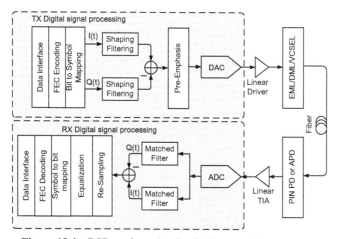

Figure 13.4 DSP configuration for CAP transmission systems.

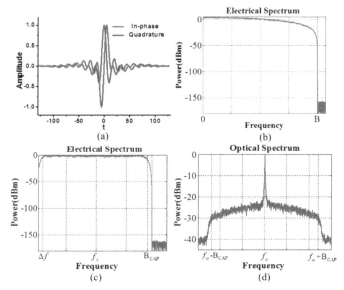

Figure 13.5   (a) Response of the shaping filter. (b) Electrical spectrum of the baseband signal. (c) Electrical spectrum of the electrical CAP signal. (d) Optical spectrum of the CAP signal.

DAC and transmitter laser nonlinearity with a time-domain FIR filter. The processed signal is then fed into a DAC to generate the electrical drive signal. Figure 13.5(d) shows the optical spectrum of the optical CAP signal. The optical signal after transmission would be directly detected by PIN PDs or APDs, sampled, and sent into two matched filters. The two matched filters are the time-reversed version of the shaping filters at the transmitter, i.e., $m_I(t) = g(-t) \cdot \cos(2\pi f_c t)$ and $m_Q(t) = g(-t) \cdot \sin(2\pi f_c t)$ in order to separate I and Q components. The separated I and Q signals are then combined to reconstruct the complex QAM signal in digital domain which is followed by a resampling function. As for complex high-order QAM signals, common adaptive equalization algorithms such as constant multi-modulus algorithm (CMMA) [10], multi-constant multi-modulus algorithm (MCMMA) [11], or decision directed least-mean-square (DD-LMS) can be used to compensate for various channel impairments.

13.2.3 Discrete Multi-tone (DMT) Modulation

Also known as direct detected orthogonal frequency division multiplexing (OFDM), DMT is another attractive modulation scheme for low-cost short-reach systems [12–14]. In contrast with OFDM, only half of the sub-carriers in DMT are coded with information, while the other half are the complex conjugate of the first half to ensure that the DMT signal is real for intensity modulation in IM/DD systems. Like all multi-carrier systems, DMT enables bit loading or power loading [15] that allows a flexible set of modulation formats for each subcarrier optimized with respect to the channel transfer function to maximize the overall bit rate or power margin. DMT has advantages such as high spectral efficiency, high tolerance to impairments, and flexible coding. The most popular bit loading method is Chow's algorithm [16], and Figure 13.6 shows the principle of bit loading. Before signal transmission, a probe signal would be transmitted and received to estimate the end-to-end SNR at each frequency subcarrier as shown by the blue curve. In this case, high-order modulation formats such as 128-QAM and 64-QAM can be coded in the subcarriers at low frequency which have high SNRs. 32-QAM, 16-QAM, and 8-QAM can be coded in the subcarriers at the mid-frequencies and QPSK or BPSK is coded in the subcarriers at the high frequency part to achieve a similar bit error ratio performance across all subcarriers. In this case, the bandwidth usage is maximized. The DSP configuration for DMT systems is also shown in Figure 13.7. After FEC encoding, the S/P function transfers the serial information bits to parallel blocks. Information bits are then mapped to QAM symbols with different modulation orders for each sub-carrier, followed by inverse fast Fourier transform (IFFT) to transform the signal to time domain. Cyclic prefix (CP) is

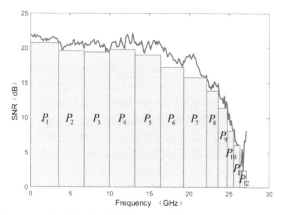

Figure 13.6 Bit or power loading for DMT transmission systems. The blue curve is the estimated SNR as a function of frequency and different amounts of signal powers (represented by the height of the green rectangles) and modulation format can be encoded in each subcarrier.

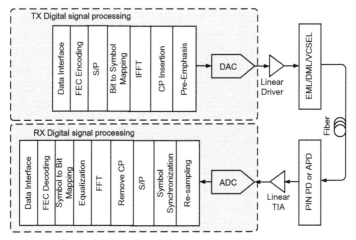

Figure 13.7   DSP configuration for DMT systems.

added to avoid inter-block interference and allow simple one-tap frequency-domain equalization. Pre-emphasis is added to compensate for the bandwidth limitation of DAC while transmitter laser nonlinearity effects are mitigated with a time-domain FIR filter. At the receiver side, the photo-detected signal is resampled and undergoes symbol frame synchronization, serial-to-parallel conversion, CP removal, and fast Fourier transform (FFT) to obtain the data symbols on each subcarrier. A frequency-domain equalizer is used to compensate for channel distortions followed by FEC decoding and symbol decisions.

13.2.4 Performance Comparison of Modulation Formats

Each modulation format has its own advantages and drawbacks; 100-Gb/s/lambda short-reach links are focused in this chapter. In [17], Zhong et al. compared the performance of these three modulation formats with typical commercial devices. The key parameters for different modulation formats are shown in Table 13.2. For PAM-4 signals, the baud rate is 56 Gbaud which gives a bit rate of 112 Gbit/s. The bandwidth is 31.36 GHz using a raised-cosine pulse shaping with a roll-off factor of 0.12. For CAP-16, the baud rate is 28 Gbaud and the root-raised-cosine pulse for both in-phase and quadrature components has a roll-off factor of 0.06. It should be noted that the roll-off factor for pulse shaping was obtained by a parameter sweeping in order to achieve the best performance for each modulation. In addition,

Table 13.2 Key parameters for 100-Gb/s PAM-4, CAP, and DMT short-reach systems

Modulation Format	Baud Rate	Roll-off Factor	Signal Bandwidth	No. of Subcarriers
PAM-4	56G	0.12	31.36 GHz	1
CAP (16 QAM)	28G	0.06	29.78 GHz	1
DMT (QPSK to 64 QAM)	NA	NA	31.5 GHz	256

CAP-16 has a slightly smaller roll-off factor than PAM-4 since a frequency offset of 0.1 GHz is used for CAP-16 here which would require the spectrum of CAP-16 to be slightly tighter for the same bandwidth. For the DMT signal, FFT size is 512 with Hermitian symmetry, CP is 8, and 217 subcarriers are coded with signals from QPSK to 64 QAM.

Figure 13.8 shows the simulation results for single-wavelength 112-Gbit/s IM/DD systems. Detailed information of the simulation setup can be found in [17]. Figure 13.8(a) shows the BER as a function of received optical power for different formats. FFEs adapted by DD-LMS are used for PAM-4 signals at the receiver. The performance of DMT is the best among the three formats. PAM-4 is slightly better than CAP-16. Receiver sensitivities of −6.95 dBm, −6.4 dBm, and −6.15 dBm at a 7% FEC overhead limit of 3.8e-3 are demonstrated for DMT, PAM-4, and CAP-16, respectively. Figure 13.8(b) shows the RS penalty as a function of transmitter bandwidth. The received optical power at BER = 3.8e-3 for a transmitter bandwidth of 35 GHz was used as the reference. It can be seen that PAM-4 is most sensitive to transmitter bandwidth. On the other hand, CAP-16 occupies less bandwidth with respect to PAM-4 for the same data rate due to high spectral efficiency and sharp pulse shaping, and hence it is more tolerant to transmitter bandwidth. DMT provides the best transmission performance because of the availability of bit loading. Figure 13.8(c) shows the RS penalty as a function of RIN for different formats. PAM-4 and CAP-16 have a similar tolerance to RIN. Moreover, as DMT waveforms resemble an analog signal with a higher peak-to-average power ratio (PAPR), they are more sensitive to RIN, ADC/DAC resolution among others. At 1 dB penalty, the tolerances to RIN are −147 dB/Hz, −143 dB/Hz, and −143 dB/Hz for DMT, PAM-4, and CAP-16, respectively.

Figure 13.9 shows the experimentally measured BER as a function of received optical power for 112-Gbit/s IM/DD transmission over 10 km with different modulation formats using a 25-Gbps device. It can be seen that PAM-4 slightly outperforms CAP-16. At received optical power regime above −5 dBm in which the BER is below the HD-FEC threshold,

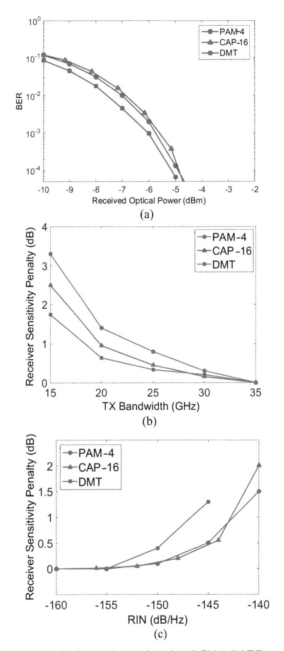

Figure 13.8   Simulation results for single wavelength 112 Gbit/s IM/DD system with different modulation formats: (a) BER vs. received optical power, (b) RS penalty vs. TX bandwidth, (c) RS penalty vs. RIN [17].

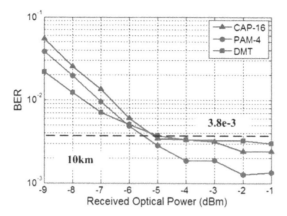

Figure 13.9 Experimental results for single-wavelength 112-Gbit/s IM/DD system with different modulation formats for 10-km transmission [17].

the performance of DMT is the worst among the three modulation formats and a BER floor of 3e-3 was observed for the DMT signal. This is due to the high PAPR of the DMT signal, limited 8-bit resolution of the DAC/ADC, and the nonlinearity of the driver amplifier for EML. Receiver sensitivities of −5.2 dBm, −5.1 dBm, and −5.6 dBm were obtained for CAP-16, DMT, and PAM-4, respectively.

13.2.5 Complexity Comparison of Modulation Formats

The required hardware (driver, transmitter, and receiver, etc.) for PAM-4, CAP-16, and DMT is similar. Instead, the biggest difference between the modulation formats lies in the transmitter and receiver DSP. The DSP complexity (including DACs and ADCs) is a critical factor that affects cost, power consumption, and hence practical implementation. While all three formats share a few common DSP blocks, major differences are highlighted.

First, PAM-4 is the simplest among the three formats in terms of signal generation. Pulse shaping is an option to reduce the signal bandwidth but not necessary. For demodulation, only threshold decision is required for the PAM-4 signal. However, for CAP-16, a pair of shaping filters is required to combine the I and Q components of the complex QAM signal to generate a real CAP-16 signal. At the receiver side, a pair of matched filters is required to separate the I and Q components from the received CAP-16 signal. The shaping and match filters can be realized by a digital FIR. But to achieve more spectrally efficient rectangle spectral shaping, a long FIR with a large number of taps is needed. DMT, which encodes information on frequency subcarriers and requires the IFFT function for DMT signal generation at the transmitter and the FFT function for demodulation at the receiver. In addition, bit loading and power loading are required for system optimization, which also increase the DSP complexity. For signal generation and demodulation, PAM-4 is the simplest, DMT requires most DSP resource, and CAP-16 lies in between. Moreover, the equalization complexities of the three different modulation formats are compared in terms of real multipliers per bit; 301 and 43 real multipliers per bit are required for PAM-4 with time-domain equalization (TDE) and frequency-domain equalization (FDE), respectively; 198 (TDE) and 40 (FDE) real multipliers per bit are required for CAP-16. For DMT, only 21 real multipliers per bit are required. Therefore, the computational complexities of PAM-4 and CAP-16 are similar while DMT is simpler as it needs only one tap equalizer. More in-depth discussions about the computational complexity can be found in [17]. Last but not least, the ADC/DAC resolution requirements are also compared. In theory, a resolution of 2 bits is sufficient for PAM-4 while DMT needs a much higher resolution due to its analog-like nature and the difficulty to fully suppress PAPR effects [18–20]. DMT also requires good linear electrical amplifiers. In this regard, CAP-16 is in between PAM-4 and DMT and is less demanding on ADC/DAC than DMT but more than PAM-4.

Table 13.3 shows the comparison of different modulation formats in terms of RS, RIN, TN, requirement on DAC/ADC, complexity of equalization, and complexity of signal generation and demodulation. Taken into account various performance attributes, we believe that PAM-4 is the most desirable choice for high-speed short-reach transmission systems as it is relatively simple to implement with good performance. The industry is also focusing on PAM-4 as new optical modules and chipsets based on PAM-4 are being produced in recent years as PAM-4 has been chosen as a standard for 400G Ethernet [3]. Meanwhile, the performance of CAP is good but it does not demonstrate unique advantages that outperform other formats in practical

Table 13.3 Comparison of different modulation formats. BW: bandwidth and EQ: equalization

Modulation Format	RS	RIN	BW	ADC/DAC	EQ	DSP Complexity
PAM-4	Fair	Good	Bad	Good	Fair	Fair
CAP-16	Fair	Good	Fair	Fair	Fair	Fair
DMT	Good	Bad	Good	Bad	Good	Good

application scenarios. DMT performs best in terms of RS, bandwidth tolerance, thermal noise, and complexity of equalization. However, complexity of generation and demodulation, bit loading, high requirements on ADC/DAC and linearity of the transmitter and the receiver, and poor tolerance on RIN standardization difficulties are issues to be solved for practical implementation. Finally, comparisons of higher order PAM and CAP modulation formats can be found in [21].

Table 13.4   Recent experimental demonstrations of short-reach systems using PAM-4, CAP-16, and DMT formats

Modulation Format	Bit Rate (Gb/s)	Number of Wavelengths	Distance (km)	References
PAM4	100	1	1	[22]
PAM4	140	1	20	[7]
PAM4	500	4	2	[23]
PAM4	214	1	2	[24]
CAP16	102	1	10	[27]
CAP16	400	4	20	[28]
Half-cycle SCM-16QAM	608	4	10	[32]
DMT	100	1	10	[33]
DMT	468	4	40	[34]
DMT	560	4	2	[35]

13.2.6 Recent Experiment on High-Speed Short-reach Transmission Systems

Based on those high-order modulation formats, researchers have already demonstrated high-speed short-reach optical transmission systems. Some recent high-speed short-reach transmission experiments are summarized here. Rodes et al. [22] demonstrated the first single-wavelength 100-Gbit/s transmission using PDM-PAM4 in 2012. Zhong et al. [7, 23] demonstrated single-channel PAM-4 signal transmission with a bit rate up to 140 Gbit/s and a four-lane 500-Gbit/s PAM-4 transmission system using commercially available 25-GHz EML-TOSA and direct detection. Most recently, Kanazawa et al. [24] demonstrated a record-high 107-Gbaud PAM-4 IM/DD system using an ultra-broadband EML. CAP-16 and other high-order CAP signals were first demonstrated for short-reach applications by Tao et al. [25, 26]. Employing multi-band techniques, single-lane 102-Gbit/s and four-lane 400-Gbit/s multi-band CAP transmissions were also demonstrated [27, 28].

Another related format called half-cycle M-QAM Nyquist subcarrier modulation (SCM) also attracts much interest [29–31] and four-lane 608 Gbit/s using a half-cycle 16-QAM Nyquist-SCM signal using 25-GHz EML and direct detection [32] were demonstrated. For DMT signals, a single-channel 101-Gb/s DMT signal transmission over 10 km was achieved with a 64-GSa/s arbitrary waveform generator (AWG) and a DML [33]. More recently, 4×117 Gb/s DMT signals have been successfully transmitted over 40 km of SSMF [34]. A four-channel 560-Gbit/s 128 QAM-DD-OFDM transmission over 2 km of SSMF was also demonstrated [35]. Table 13.4 summarizes recent demonstrations of PAM-4, CAP-16, and DMT short-reach systems.

13.3 Digital Signal Processing for Short-reach Systems

In short-reach links, the most important transmission impairments are bandwidth limitations of the transmitter and the receiver, nonlinearity of EML/DML and distortions due to interaction with CD, real signaling, and direct detection as the transmission distance increases. Few key DSP blocks such as FFE as equalizer have been briefly mentioned in the last section. In this section, the DSP algorithms to compensate for these impairments are introduced.

13.3.1 Feed-forward Equalizer (FFE)

A feed-forward equalizer is a widely used equalizer for linear impairments' compensation. Figure 13.10 shows the block diagram of FFE. The FFE output is expressed as

$$z[k] = \sum_{i=0}^{N-1} w_i E_r((k-i)T) = \mathbf{w} \cdot \mathbf{E_r}^T[k] \qquad (13.1)$$

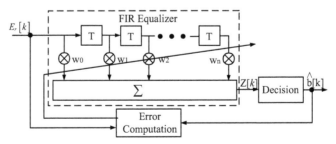

Figure 13.10   Block diagram of FFE.

where $z[k]$ is the equalizer output, $\mathbf{w} = [w_0 \ w_1 \ w_2 \ \cdots \ w_{N-1}]$ is the vector of tap weights, N is the number of taps, and $\mathbf{E_r}[k] = [E_r(kT) \ E_r((k-1)T) \ \cdots \ E_r((k-N+1)T)]$. The tap weights can be adaptively updated by many different strategies and interested readers are referred to [36] and [37] for a more in-depth treatment of adaptive DSP for optical communications. Here, we will focus on one of the most common adaptive signal processing techniques – the DD-LMS algorithm. In this case, we define the error $\varepsilon[k] = \hat{b}[k] - z[k]$ between $z[k]$ and the decided symbol $\hat{b}[k]$, the cost function $J(\mathbf{w}) = (\varepsilon[k])^2 = (\hat{b}[k] - \mathbf{w} \cdot \mathbf{E}_\mathbf{r}^T[k])^2$ and use the iterative method of stochastic gradient descent [37] to obtain the best filter configuration that minimizes $J(\mathbf{w})$. If we start our iteration with the signals at time k_0T, the $n + 1^{\text{th}}$ update of the filter tap weights is given by

$$\mathbf{w}^{(n+1)} = \mathbf{w}^{(n)} + \mu \left. \frac{\partial J}{\partial \mathbf{w}} \right|_{\mathbf{w}^{(n)}} = \mathbf{w}^{(n)} + \mu \varepsilon[k_0 + n]\mathbf{E_r}[k_0 + n] \quad (13.2)$$

where μ is the step size. It should be noted that FFE can operate at symbol rate sampling or higher. For symbol rate FFE, a filter matched to the channel is required prior to sampling and filtering. If FFE is operated at 2 samples per symbol, the FFE can be the matched filter itself. In this case, T equals $T_S/2$ while the tap weights are updated only at time slots $2T$, $4T$, $6T$, etc. We simulate a 56-Gbaud PAM-4 system with various amounts of low-pass filtering using fourth-order Bessel filters and study the corresponding FFE characteristics. Figure 13.11 shows the frequency response of FFE for the 56-Gbaud PAM4 system with different bandwidths. The number of taps

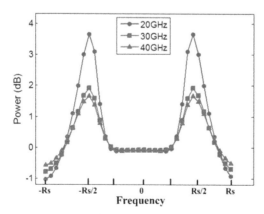

Figure 13.11 Frequency response of FFE for a 56-Gbaud PAM-4 system with different transceiver component bandwidths.

is 51 for FFE. It can be seen that FFE boosts the power of the high-frequency components that undergo large losses due to the system bandwidth limitations.

13.3.2 Decision Feedback Equalizer (DFE)

A decision feedback equalizer (DFE) is an alternative equalizer for channel impairments. Figure 13.12 shows the block diagram of DFE. Different from FFE, the input of DFE is the symbols after decision $\hat{b}[k]$ as shown in Figure 13.12. DFE is usually operated at 1 sample per symbol. In this case, the signal compensated by the DFE output before decision is given by

$$z[k] = E_r(kT) - \sum_{i=0}^{N-1} w_i \hat{b}[k-i] \qquad (13.3)$$

If we start our DD-LMS iteration with the signals at time $k_0 T$, the $n+1^{\text{th}}$ updates for the tap weights are given by

$$\mathbf{w}^{(n+1)} = \mathbf{w}^{(n)} + \mu \varepsilon[k_0 + n]\mathbf{Z}[k_0 + n] \qquad (13.4)$$

where $\mathbf{Z}[n] = [z[n]z[n-1]\cdots z[n-(N-1)]]$ and $\varepsilon[n] = \hat{b}[n] - z[n]$. It should be noted that FFE is simple and does not suffer from feedback delays but is ineffective in equalizing spectral nulls. DFE on the other hand can successfully equalize spectral nulls by pole insertion but it may become unstable and suffer from decision error propagation. Nonetheless, since the CD-induced channel impulse response contains both pre-cursor (future symbols) and post-cursor (past symbols) ISI, the best choice for equalization is a combination

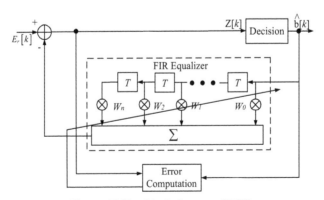

Figure 13.12   Block diagram of DFE.

of a feed-forward and a feedback structure equalizer. Recently, Tomlinson–Harashima pre-coding, a transmitter-side DFE that does not suffer from error propagation, is shown to outperform DFE in various PAM-4 short-reach transmissions scenarios [38].

13.3.3 Direct Detection Faster-than Nyquist (DD-FTN)

While FFE and DFE are widely used, it is well known that they severely enhance the in-band noise in bandwidth-limited optical transmission systems. This can be easily illustrated by the frequency response of FFE shown in Figure 13.10. As the equalizer tries to boost the power of the high-frequency components to compensate for the low-pass effects, it also amplifies the noise power at high-frequency components. Noise enhancement results in an SNR degradation of the equalized signal. To address such problems, we proposed a DD-FTN algorithm [7] and the schematic diagram is shown in Figure 13.13.

The DD-FTN consists of an FFE adapted by LMS, a digital post filter, and a maximum likelihood sequence estimation (MLSE) function. The post filter is placed after the equalizer for in-band noise suppression. The response of the post filter has a transfer function $H(z) = 1 + \alpha z^{-1}$. The tap coefficient α is introduced to facilitate the optimization of frequency response of the post filter. Figure 13.14(a) shows the frequency response of post filter far different tap coefficients. It can be seen that larger tap coefficients result in lower bandwidth, sharper roll-off, and larger attenuation at high frequencies. Normally, α increases with signal baud rate/bandwidth. The post filter suppresses the noise but induces ISI, which is further addressed by MLSE. Although DD-FTN includes two more DSP blocks compared to conventional

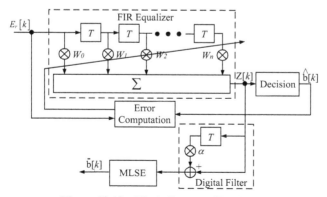

Figure 13.13 Block diagram of DD-FTN.

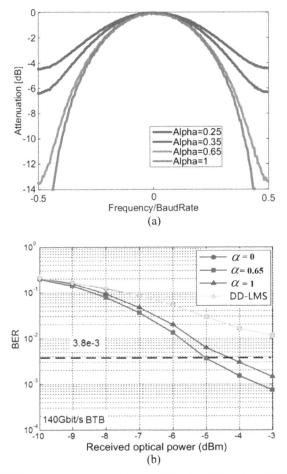

Figure 13.14   (a) Frequency response of the post filter for different α. (b) BER vs. received optical power (ROP) for a 70 GBaud PAM-4 IM/DD system with DD-FTN [7].

equalizers, the complexity increments are quite small. The post filter is only a delay and addition. As studied in part E of [39], MLSE requires $3M^2$ additions, M^2 multiplications, and $M(M-1)$ comparisons for M-ary PAM signals. However, approaches such as state grouping are proposed to merge several states for trellises originally with a large number of states [40, 41]. In this case, the state count can be halved to merely two for the PAM-4 signal, which significantly reduces the complexity. In the meantime, we have found that a memory length of 20 symbols is sufficient to ensure good detection performance. Therefore, the overall complexity of DD-FTN is not high compared to other DSP approaches.

Figure 13.14(b) shows the BER vs. received optical power for a 140-Gbit/s PAM-4 IM/DD transmission experiment with and without DD-FTN [7]. The BER with DD-LMS only is also shown (green curve). With a tap coefficient of $\alpha = 0$ (blue curve), the performance is the same with that of DD-LMS only (hence the blue and green curves overlap with each other), which exhibits the worst performance. For a received optical power of -4 dBm, a BER of 1.7E-2 is obtained. BER decreases significantly when tap coefficients $\alpha = 1$ and $\alpha = 0.65$ are used. For a received optical power of -4 dBm, BERs of 3E-3 and 1.6E-3 are obtained for $\alpha = 1$ and $\alpha = 0.65$, respectively. The best performance is obtained with an optimal tap coefficient $\alpha = 0.65$. Receiver sensitivities at the 7% FEC limit of 3.8E-3 are -5 dBm and -4.3 dBm for the cases of $\alpha = 0.65$ and $\alpha = 1$, respectively.

13.3.4 Volterra-series Based Nonlinear Equalizer (VNLE)

While linear distortions can be efficiently compensated using FFE, DFE, or DD-FTN, nonlinear distortions such as modulation chirp, saturated power amplification, nonlinear modulation characteristics of lasers or modulators, and square law detection for the signal with CD [42–44] are difficult to compensate effectively. One of the most common algorithms to compensate for nonlinear effects is VNLE. Since most of the nonlinear impairments in IM/DD transmission systems such as saturated power amplification, non-linearity from modulators, and CD induced ISI are primarily second-order nonlinearities, second-order VNLE is introduced in this paper. It should be noted that VNLE can be implemented based on FFE or DFE. Here, we focus on the VNLE based on FFE. Figure 13.15 shows the schematic diagram of VNLE with input $E_r[k]$ and output

$$
\begin{aligned}
Z[k] &= Z_L[k] + Z_{NL}[k] \\
&= \sum_{i=0}^{N_L-1} w_{L,i} E_r((k-i)T) \\
&\quad + \sum_{i=0}^{N_{NL}-1} \sum_{j=0}^{N_{NL}-1} w_{NL,ij} E_r((k-i)T) E_r((k-j)T) \quad (13.5)
\end{aligned}
$$

where $z_L[k]$ and $z_{NL}[k]$ are the linear and nonlinear equalized outputs, respectively, $w_{L,i}$ is the tap weight for the linear part of VNLE, $w_{NL,ij}$ is

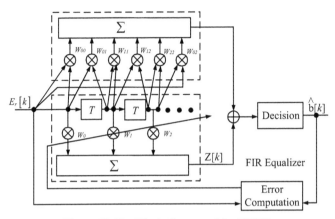

Figure 13.15 Block diagram of the VNLE.

the tap weight for the nonlinear part of VNLE, N_L is the number of taps for the linear part of VNLE, and N_{NL} is the number of taps of the nonlinear part of VNLE. Representing the filter taps as vectors $\mathbf{w}_L$ and matrix $\mathbf{w}_{NL}$ and if we start our DD-LMS iteration with the signals at time $k_0 T$, the $n + 1^{th}$ tap weights' update will be given by

$$\mathbf{w}_L^{(n+1)} = \mathbf{w}_L^{(n)} + \mu\varepsilon[k_0 + n]\mathbf{E}_r[k_0 + n] \tag{13.6}$$

$$\mathbf{w}_{NL}^{(n+1)} = \mathbf{w}_{NL}^{(n)} + \mu\varepsilon[k_0 + n]\mathbf{E}_r[k_0 + n](\mathbf{E}_r[k_0 + n])^T \tag{13.7}$$

where $\varepsilon[n] = \hat{b}[n] - z[n]$. Zhang et al. [42] employed VNLE in an SSB-DMT system and an average SNR improvement of 2 dB across subcarriers for a 100G SSB-DMT signal is obtained by using VNLE. It should be noted that for systems with high carrier to signal power ratio (CSPR), the signal component is smaller, and hence the overall system response will be more linear, rendering VNLE less effective.

13.4 Polarization Division Multiplexed Transmission for Short-reach Systems

Polarization division multiplexing with coherent optical detection (PDM-CO) systems provides three degrees of freedom, including intensity, phase, and polarization, which can encode four-dimensional (4D) information (namely, two complex signals) to improve the spectral efficiency of systems significantly. However, the previously discussed IM/DD systems can support only one degree of freedom, hence 1D modulation only. For short-reach systems limited by direct detection and other low-cost implementation constraints, innovative transceiver configurations and DSP are actively being pursued to exploit such additional degrees of freedom inherent in polarization multiplexing. One of the key research directions in this regard is SV-DD transceivers.

13.4.1 Stokes-vector Direct Detection (SV-DD) Receiver

The polarization of an optical signal can be described in a vector notation by separately describing the electric field vector of the x and y components as Jones vector, as

$$\mathbf{E} = \begin{bmatrix} E_x \\ E_y \end{bmatrix} \tag{13.8}$$

where E_x and E_y denote the electric fields of X and Y polarizations, respectively. The Jones space is a two-dimensional complex space where each dimension represents an orthogonal state of polarization (SOP) of light, providing total 4D field. However, to acquire full 4D information in Jones space, a dual-polarization coherent receiver is required, as discussed in Section 3.5. Besides Jones space, an alternative representation to describe the SOP of an optical signal is the Stokes space representation which expresses the SOP in optical powers instead of electric field. The transmitted Stokes vector consists of four Stokes parameters, $\mathbf{S} = [S_0, S_1, S_2, S_3]^T$. Here, the electric field vectors of the signal in Jones space can be converted into Stokes space,

$$S_0 = |E_x|^2 + |E_y|^2, \ S_2 = 2\text{Re}\{E_x \cdot E_y^*\}$$
$$S_1 = |E_x|^2 - |E_y|^2, \ S_3 = 2\text{Im}\{E_x \cdot E_y^*\} \tag{13.9}$$

where Re{} and Im{} denote the real part and the imaginary part of a complex number, respectively, and ()* denotes the complex conjugate. The total power S_0 can also be represented by: $S_0 = \sqrt{S_1^2 + S_2^2 + S_3^2}$. In this case, the same field can be equivalently represented using a 3D real-valued vector. But, 1D information loses in comparison of the Jones vector, because the relative phase between two polarizations substitutes for their absolute phases in Stokes space. Thus, the signal representation is in 3D Stokes space (S_1, S_2, S_3), which can be directly detected by SV-DD receiver (SV-R) shown in Figure 13.16 [45–47].

Figure 13.16 summarizes the four architectures of the SV-DD receiver. For all the architectures, the signal is first split with a polarization beam

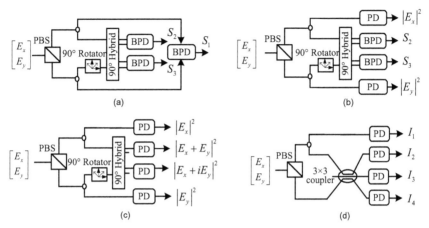

Figure 13.16   Options for the architecture of the SV-DD receiver.

splitter (PBS) into two outputs of X and Y, respectively. For the hybrid-based SV-DD receivers, both X and Y are split with 1×2 couplers, providing four output ports. In Figure 13.16(a), the outputs of branches 1 and 4 are fed into a balanced PD (BPD) directly, whose output is the first component of SV: $S_1 = |E_x|^2 - |E_y|^2$. Ports 2 and 3 are fed into a 90° hybrid with two BPD receivers, and their outputs are $\mathrm{Re}\{E_x \cdot E_y^*\}$ and $\mathrm{Im}\{E_x \cdot E_y^*\}$, which are exactly the S_2 and S_3. For simplicity, the scaling constants for couplers are not considered, and the responsivities of photodiodes are assumed to be 1. Thus, the linear transformation between the three outputs and Stokes parameters S_1, S_2, and S_3 is expressed as

$$
\begin{bmatrix} S_1 \\ S_2 \\ S_3 \end{bmatrix} = \begin{bmatrix} 1 & 0 & 0 \\ 0 & 1 & 0 \\ 0 & 0 & 1 \end{bmatrix} \cdot \begin{bmatrix} |E_x|^2 - |E_y|^2 \\ 2\mathrm{Re}\{E_x \cdot E_y^*\} \\ 2\mathrm{Im}\{E_x \cdot E_y^*\} \end{bmatrix} \tag{13.10}
$$

Here, the total optical power S_0 can be obtained by $S_0 = \sqrt{S_1^2 + S_2^2 + S_3^2}$. Instead of using a balanced PD to receive the S_1, the split orthogonal components can be detected by using two single-ended PDs, respectively. Then, S_0 and S_1 can be reconstructed in digital domain. The linear transformation between the four outputs and Stokes parameters can be addressed as

$$
\begin{bmatrix} S_0 \\ S_1 \\ S_2 \\ S_3 \end{bmatrix} = \begin{bmatrix} 1 & 0 & 0 & 1 \\ 1 & 0 & 0 & -1 \\ 0 & 1 & 0 & 0 \\ 0 & 0 & 1 & 0 \end{bmatrix} \cdot \begin{bmatrix} |E_x|^2 \\ 2\mathrm{Re}\{E_x \cdot E_y^*\} \\ 2\mathrm{Im}\{E_x \cdot E_y^*\} \\ |E_y|^2 \end{bmatrix} \tag{13.11}
$$

The above two receiver structures can obtain the SV components straight-forwardly. But in fact, the SV-DD receiver can be generalized by any three or four detections of polarization states as long as they are non-singular superposition of the SV components (S_0, S_1, S_2, S_3) [48]. Another SV-DD option is shown in Figure 13.16(b) [46]. After a PBS and couplers, ports 1 and 4 are launched into the PD directly, resulting in the outputs of $|E_x|^2$ and $|E_y|^2$. Ports 2 and 3 are fed into the 90° hybrid. But, two pairs of outputs of the hybrid, only two outputs are fed into the PDs to provide the outputs of $|E_x + E_y|^2$ and $|E_x + iE_y|^2$, respectively. In this case, the four outputs provide a linear transformation to the Stokes space expressed by:

$$
\begin{bmatrix} S_0 \\ S_1 \\ S_2 \\ S_3 \end{bmatrix} = \begin{bmatrix} 1 & 0 & 0 & 1 \\ 1 & 0 & 0 & -1 \\ -1 & 1 & 0 & -1 \\ -1 & 0 & 1 & -1 \end{bmatrix} \cdot \begin{bmatrix} |E_x|^2 \\ |E_x + E_y|^2 \\ |E_x + iE_y|^2 \\ |E_y|^2 \end{bmatrix}
\tag{13.12}
$$

Figure 13.16(c) reduces the PD amount to four, but the optical hybrid has still to use, which will induce optical power loss. To further reduce the structure and improve power efficient, a 3×3 coupler can be applied to the SV-DD receiver, as shown in Figure 13.16(d). In this case, the linear transformation can be computed as:

$$
\begin{bmatrix} S_0 \\ S_1 \\ S_2 \\ S_3 \end{bmatrix} = \begin{bmatrix} 1 & 1 & 1 & 1 \\ 3 & -1 & -1 & -1 \\ 0 & -\sqrt{2} & 2\sqrt{2} & -\sqrt{2} \\ 0 & -\sqrt{6} & 0 & \sqrt{6} \end{bmatrix} \cdot \begin{bmatrix} I_1 \\ I_2 \\ I_3 \\ I_4 \end{bmatrix}
\tag{13.13}
$$

The SV-DD receiver can recover the signal in the Stokes space from 1D up to 3D, regardless of any transmitter structure or modulation format. Based on S_0 and S_1, the intensity information of X and Y polarizations can be recovered; while based on S_2 and S_3, and inter-polarization phase information can be recovered.

Based on the SV-DD receiver, 3D information can be recovered from the directly detected signal. In the next two sections, different types of modulation schemes based on the PDM-DD system using the SV-DD receiver are introduced.

13.4.2 2-Dimensional (2D) PDM-DD System Based on SV-DD Receiver

Figure 13.17 illustrates two typical transmitter architectures of 2D modulation. In Figure 13.17(a), two IM signals are modulated on two orthogonal

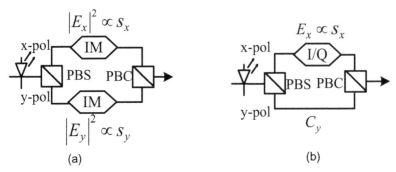

Figure 13.17   Transmitter architectures for 2D modulation. (a) Two 1D IM signals. (b) 2D complex signal + carrier.

polarizations [49, 50]. In this case, S_0 and S_1 are required to recover the intensity of the two polarizations as

$$
\begin{bmatrix} S_0 \\ S_1 \\ S_2 \\ S_3 \end{bmatrix} = \begin{bmatrix} |E_x|^2 + |E_y|^2 \\ |E_x|^2 - |E_y|^2 \\ 2\mathrm{Re}\{E_x \cdot E_y^*\} \\ 2\mathrm{Im}\{E_x \cdot E_y^*\} \end{bmatrix} \Rightarrow \begin{aligned} |E_x|^2 &= \frac{1}{2}(S_0 + S_1) \\[2mm] |E_y|^2 &= \frac{1}{2}(S_0 - S_1) \end{aligned}
\tag{13.14}
$$

Due to fiber polarization variation, the receiver SV components using the SV-DD receiver will not be aligned with the transmitted Stokes space. The SOP recovery is a crucial issue for the PDM-DD system. Since the SV-DD-based direct detection channel is a linear channel, the arbitrary modulation formats, which satisfy to modulate in 3D Stokes space, can be de-multiplexed based on the SV-DD receiver in theory. However, the detailed DSP operation will be related with the signal modulation format. For the two IM-based single-carrier PDM systems [51], a 1×1 single-input and single-output (SISO) FIR filter and a 3×1 multiple-input and single-output (MISO) bank of FIR filters are used to de-rotate polarization and mitigate the inter-symbol interference (ISI) simultaneously, as shown in Figure 13.18.

Since the zeroth SV component is insensitive to polarization rotation, the received S_0' can be filtered using a 1×1 SISO filter $\mathbf{h}_{00}$ with N real-valued taps to provide the estimate of the transmitted S_0, denoted by $\hat{S}_0 = \mathbf{h}_{00} \otimes S_0'$. Here, the superscript '$\wedge$' stands for the estimated value. Also, the estimated $\hat{S}_1$ can be obtained by employing a 3×1 MISO bank of FIR filters $\mathbf{h}_{11}, \mathbf{h}_{21}$, and $\mathbf{h}_{31}$ to remove the polarization rotation and mitigate ISI, namely,

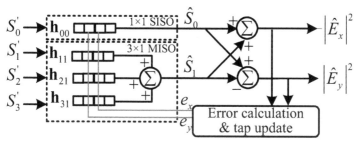

Figure 13.18 The de-multiplexing and equalization for 2D PDM-PAM4.

$\hat{S}_1 = \mathbf{h}_{11} \otimes S_1' + \mathbf{h}_{21} \otimes S_2' + \mathbf{h}_{31} \otimes S_3'$. The taps of these SISO and MISO filters can be updated using the sophisticated LMS algorithms as follows,

$$
\begin{aligned}
h_{00}^{i+1} &= h_{00}^i + \mu(e_x + e_y)S_0' \\
h_{11}^{i+1} &= h_{11}^i + \mu(e_x - e_y)S_1' \\
h_{21}^{i+1} &= h_{21}^i + \mu(e_x - e_y)S_2' \\
h_{31}^{i+1} &= h_{31}^i + \mu(e_x - e_y)S_3'
\end{aligned}
\qquad (13.15)
$$

where μ is the step size parameter and $e_{x(y)}$ are the error signals of the x- (y-) polarization, which can be calculated based on training sequence (TS) initially, and switching to decision-directed mode at steady state. Then, the intensity of the two polarizations can be recovered based on the outputs of the two filters.

The 2D IM-based transmitter can also load the real-valued OFDM, namely, DMT on two polarizations. In terms of OFDM, the de-multiplexing and equalization operation will be converted from time domain to frequency domain, performed after FFT. Since polarization rotation can be regarded as frequency independent, the 1×1 SISO and 3×1 MISO equalizers can be used to retrieve S_0 and S_1 for each subcarrier in frequency domain. Here, the coefficients of the frequency-domain equalizers can be estimated by using pilot subcarriers [52]. However, the 2D PDM-IM systems have the same property as conventional 1D IM-DD systems, which are susceptible to the CD-induced frequency-selective power fading impairment.

Another 2D modulation scheme in Stokes space [see Figure 13.17(b)] is to modulate a complex signal on only x-polarization using an IQ modulator, and a constant carrier is sent on y-polarization, denoted by C_y. Obviously, instead of S_0 and S_1, S_2 and S_3 are required to recover the complex signal, as follows

$$\begin{bmatrix} S_0 \\ S_1 \\ S_2 \\ S_3 \end{bmatrix} = \begin{bmatrix} |E_x|^2 + |E_y|^2 \\ |E_x|^2 - |E_y|^2 \\ 2\mathrm{Re}\{E_x \cdot E_y^*\} \\ 2\mathrm{Im}\{E_x \cdot E_y^*\} \end{bmatrix} \Rightarrow E_x C_y^* = \frac{1}{2}(S_2 + iS_3) \qquad (13.16)$$

This 2D PDM-DD system can be seen as a self-coherent detection, which realizes a linear mapping of channel from optical domain to electrical domain. Therefore, it effectively avoids the CD-induced power fading. However, the SOP recovery is also necessary. Owing to the use of IQ modulator, a straight-forward method can be used for SOP recovery. At the transmitter, three sets of training symbols with the Jones space representation of [0 1], [1 1], and [i 1] can be sent as shown in Figure 13.19, which correspond to three basis time-interleaved vectors, i.e., [−1 0 0], [0 1 0], and [0 0 1], in Stokes space, respectively.

Then, the estimation of the first column of polarization RM, denoted by **R**, is given by

$$\underbrace{\begin{bmatrix} S_1' \\ S_2' \\ S_3' \end{bmatrix}}_{Detected\ TS} = \underbrace{\begin{bmatrix} h_{11} & h_{12} & h_{13} \\ h_{21} & h_{22} & h_{23} \\ h_{31} & h_{32} & h_{33} \end{bmatrix}}_{\mathbf{R}:\ Channel\ RM} \cdot \underbrace{\begin{bmatrix} -1 \\ 0 \\ 0 \end{bmatrix}}_{TS} \Rightarrow \begin{bmatrix} \hat{h}_{11} = -S_1' \\ \hat{h}_{21} = -S_2' \\ \hat{h}_{31} = -S_3' \end{bmatrix} \qquad (13.17)$$

and the second and third columns of **R** are estimated by the second and third pilot symbols in a similar fashion. The output signal of the SOP recovery stage can be described as

$$\underbrace{\begin{bmatrix} \hat{S}_1 \\ \hat{S}_2 \\ \hat{S}_3 \end{bmatrix}}_{Aligned\ signal} = \underbrace{\begin{bmatrix} \hat{h}_{11} & \hat{h}_{12} & \hat{h}_{13} \\ \hat{h}_{21} & \hat{h}_{22} & \hat{h}_{23} \\ \hat{h}_{31} & \hat{h}_{32} & \hat{h}_{33} \end{bmatrix}^{-1}}_{\hat{\mathbf{R}}:\ Estimated\ RM} \cdot \underbrace{\begin{bmatrix} S_1' \\ S_2' \\ S_3' \end{bmatrix}}_{Detected\ signal} \qquad (13.18)$$

where in principle matrix $\hat{\mathbf{R}}^{-1}$ is an approximation of the inverse of the channel RM-**R**. In fact, the channel RM can also be determined by two angle parameters

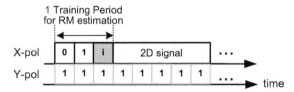

Figure 13.19　The signal structure for the SOP recovery in PDM-SC systems.

$$\mathbf{R} = \begin{bmatrix} h_{11} & h_{12} & h_{13} \\ h_{21} & h_{22} & h_{23} \\ h_{31} & h_{32} & h_{33} \end{bmatrix} = \underbrace{\begin{bmatrix} \cos\theta & -\sin\theta\cos\varepsilon & \sin\theta\sin\varepsilon \\ \sin\theta & \cos\theta\cos\varepsilon & -\cos\theta\sin\varepsilon \\ 0 & \sin\varepsilon & \cos\varepsilon \end{bmatrix}}_{(Mueller\ Matrix)}$$

(13.19)

whose corresponding equivalent rotation matrix in Jones space is

$$\begin{bmatrix} e^{j\varepsilon/2}\cos^\theta/_2 & -e^{-j\varepsilon/2}\sin^\theta/_2 \\ e^{j\varepsilon/2}\sin^\theta/_2 & e^{-j\varepsilon/2}\cos^\theta/_2 \end{bmatrix}$$

(13.20)

where θ and ε denote the random polar angle and azimuth angle, respectively, which can be blindly estimated according to the normal vector (v_1, v_2, v_3) of the least squares plane (LSP) [53]. Here, the LSP is the plane that minimizes the sum of the distances between the plane and the received Stokes vectors. Then, after determining the normal vector (v_1, v_2, v_3) of the LSP, the polar angle θ can be estimated by

$$\hat{\theta} = \tan^{-1}(\sqrt{v_2^2 + v_3^2}, v_1)$$

(13.21)

with the phase delay estimate

$$\hat{\varepsilon} = \tan^{-1}(v_3/v_1)$$

(13.22)

and the rotation matrix $\mathbf{R}$ can be obtained accordingly. This blind estimation algorithm can work for all modulation formats and offer performance close to the training symbol-based method. Then we multiply the inverse of $\mathbf{R}$ to estimate the transmitted SV, and the received complex signal can be obtained by

$$E_x' = \hat{S}_2 + i\hat{S}_3$$

(13.23)

Subsequently, appropriate equalizers for single-carrier or multi-carrier signals will be used to compensate both the additional ISI resulting from the low-pass filtering and other channel distortions.

13.4.3 3-Dimensional (3D) PDM-DD System Based on SV-DD Receiver

To maximize the spectral efficiency, 3D modulation can be supported by using the SV-DD receiver. Figure 13.20 illustrates a 3D modulation

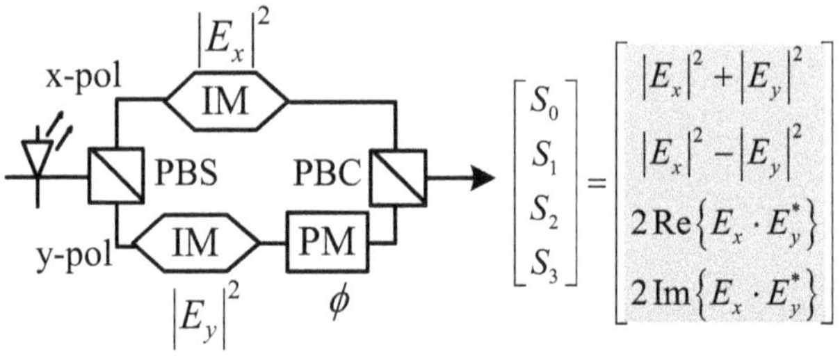

Figure 13.20   Transmitter structures for 3D modulation with two 1D IM signals and a 1D inter-polarization phase (IP).

transmitter, where IM is followed by a phase modulator (PM) in one of polarizations to further modulate the phase difference between two orthogonal polarizations, namely, an inter-polarization phase [54].

Since all the dimensions in Stokes space are utilized to load data information, the SV-DD receiver will use all the SV components to recover the 3D information. In this case, the SOP recovery and equalization become more complex, which require a 4 × 4 MIMO equalizer to realize. There is an exponential increase in computational complexity due to the use of the 4 × 4 MIMO equalizer. In order to reduce the tap number of the 4 × 4 MIMO equalizer, some multi-stage DSP implementations were proposed. In [54], the first stage is a 4 × 4 MIMO filter bank which is responsible for inverting the polarization rotation and roughly mitigating the ISI. The second stage comprises four SISO FIRs that are used to remove any residual ISI uncompensated by the first stage. Another implementation involves three-stage DSP procedures in [55]. The first stage of the DSP filters all four received waveforms independently and mitigates the ISI. The second stage comprises a 4 × 4 MIMO by using the four filtered outputs of the first stage for polarization demultiplexing. The third stage mitigates the residual transmitter side ISI by applying three independent SISO FIRs. Both DSP approaches perform equally but the three-stage DSP can track the SOP faster as the receiver-side ISI mitigation and polarization de-multiplexing are separated. This approach also requires a smaller computational complexity.

13.5 Conclusion

In this chapter, we have reviewed various DSP techniques that enable high capacity transmissions for short-reach systems. Advanced modulation formats and direct detection are typically used due to cost constraints.

DSP techniques are essential to overcome component bandwidth limitations, impairments introduced by nonlinear modulation characteristics and CD in the transmission channel. Various approaches for realizing direct detected polarization division multiplexed transmissions are promising techniques for further increase in system capacity. Future efforts on joint optimization of optical components and DSP techniques may become a key area of growth. This may also enable gradual introduction of many coherent detection functionalities into future short-reach optical communication systems.

Acknowledgments

The authors want to acknowledge research fund support from the National Natural Science Foundation of China NSFC, 61671053, 61435006, 61370191, the Hong Kong Government General Research Fund PolyU 152248/15E, the project 1-ZVGB of Huawei Technologies Co. Ltd and Fundamental Research Funds for the Central Universities FRF-BD-17-015a.

References

[1] V. Bobrovs, S. Spolitis, and G. Ivanovs, "Latency causes and reduction in optical metro networks." Proc. *SPIE* 9008, 90080C-1, 2013.

[2] J. L. Wei *et al.*, "400 Gigabit Ethernet using advanced modulation formats: Performance, complexity, and power dissipation." *Commun. Mag.,* vol. 53, no. 2, pp. 182–189, Feb. 2015.

[3] IEEE 802.3, [online]. Available: http://www.ieee802.org/3

[4] J. W. Zhang *et al.,* "Time-domain digital pre-equalization for band-limited signals based on receiver-side adaptive equalizers," *Opt. Exp.,* vol. 22, no. 17, pp. 20515–20529, Aug. 2014.

[5] J. W. Man *et al.*, "A low-cost 100 GE optical transceiver module for 2 km SMF interconnect with PAM4 modulation." in Proc. *OFC*, San Francisco, CA, USA, 2014, Paper M2E.7.

[6] F. Breyer *et al.*, "Comparison of OOK- and PAM-4 modulation for 10 Gbit/s transmission over up to 300 m polymer optical fiber." in Proc. *OFC*, San Diego, CA, USA, 2008, Paper OWB5.

[7] K. P. Zhong *et al.*, "140 Gb/s 20 km transmission of PAM-4 signal at 1.3um for short reach communications." *Photon. Technol. Lett.*, vol. 27, no. 16, pp. 1757–1761, Aug. 2015.

[8] J. L. Wei *et al.*, "Experimental demonstration of optical data links using a hybrid CAP/QAM modulation scheme," *Opt. Lett.*, vol. 39, no. 6, pp. 1402–1405, Mar. 2014.

[9] J. Wei *et al.*, "400 Gigabit Ethernet using advanced modulation formats: performance, complexity, and power dissipation." in *IEEE Commun. Mag.*, vol. 53, no. 2, pp. 182–189, Feb. 2015.

[10] X. Zhou and J. Yu, "Multi-level, multi-dimensional coding for high-speed and high-spectral-efficiency optical transmission," in *J. Lightw technol.*, vol. 27, no. 16, pp. 3641–3653, Aug. 2009.

[11] Y. Wang *et al.*, "High Speed WDM VLC system based on multi-band CAP64 with weighted pre-equalization and modified CMMA based post-equalization," in *IEEE Commun. Lett.*, vol. 18, no. 10, pp. 1719–1722, Oct. 2014.

[12] C. Xie *et al.*, "Single-VCSEL 100-Gb/s short-reach system using discrete multi-tone modulation and direct detection." in Proc. *OFC*, Los Angeles, CA, USA, 2015, Paper Tu2H.2.

[13] F. Li *et al.*, "Optimization of training sequence for DFT-spread DMT signal in optical access network with direct detection utilizing DML," *Opt. Exp.*, vol. 22, no. 19, pp. 22962–22967, Oct. 2014.

[14] W. Yan *et al.*, "100 Gb/s optical IM-DD transmission with 10G-class devices enabled by 65G Samples/s CMOS DAC Core," in Proc. *OFC*, Anaheim, CA, USA, 2013, Paper OM3H.1.

[15] S. Randel *et al.*, "Advanced modulation schemes for short-range optical communications." in *J. Select. Topics in Quantum Electronics*, vol. 16, no. 5, pp. 1280–1289, Oct. 2010.

[16] P. Chow *et al.*, "A practical discrete multitone transceiver loading algorithm for data transmission over spectrally shaped channels," *Trans. Commun.*, vol. 43, no. 11, pp. 773–775, June, 1995.

[17] K. P. Zhong *et al.*, "Experimental study of PAM-4, CAP-16, and DMT for 100 Gb/s short reach optical transmission systems." *Opt. Exp.*, vol. 23, no. 2, pp. 1176–1189, Feb. 2015.

[18] F. Li *et al.*, "Optimization of training sequence for DFT-spread DMT signal in optical access network with direct detection utilizing DML." *Opt. Exp.*, vol. 22, no. 19, pp. 22962–22967, Nov. 2014.

[19] W. A. Ling *et al*, "112 Gb/s transmission with a directly-modulated laser using FFT-based synthesis of orthogonal PAM and DMT signals." *Opt. Exp.*, vol. 23, pp. 19202–19212, Dec. 2015.

[20] L. Nadal *et al.*, "Low complexity PAPR reduction techniques for clipping and quantization noise mitigation in direct-detection O-OFDM systems." *Optical Fiber Technol.*, vol. 20, no. 3, pp. 208–216, Oct. 2014.

[21] J. Shi *et al,.* "Comparison of 100G PAM-8, CAP-64 and DFT-S OFDM with a bandwidth-limited direct-detection receiver," *Opt. Exp.*, vol. 25, no. 26 pp. 32254–32262. Dec. 2017.

[22] R. Rodes *et al.*, "100 Gb/s single VCSEL data transmission link." in Proc. *OFC*, Los Angeles, CA, USA, 2012, Paper PDP5D.10.

[23] K. P. Zhong *et al.*, "Experimental demonstration of 500Gbit/s short reach transmission employing PAM4 Signal and direct detection with 25 Gbps device." in Proc. *OFC*, Los Angeles, CA, USA, 2015, Paper Th3A.3.

[24] S. Kanazawa *et al.*, "Transmission of 214-Gbit/s 4-PAM signal using an ultra-broadband lumped-electrode EADFB laser module." in Proc. *OFC*, Anaheim, CA, USA, 2016, Paper Th5B.3.

[25] L. Tao *et al.*, "Experimental demonstration of 10 Gb/s multi-level carrier-less amplitude and phase modulation for short range optical communication systems." *Opt. Exp.*, vol. 21, no. 5, pp. 6459–6465, Mar. 2013.

[26] L. Tao *et al.*, "High order CAP system using DML for short reach optical communications." *Photon. Technol. Lett.*, vol. 26, no. 13, pp. 1348–1351, Jul. 2014.

[27] M. I. Olmedo *et al.*, "Multiband carrier-less amplitude phase modulation for high capacity optical data links" *J. Lightw. Technol.*, vol. 32, no. 4, pp. 798–804, Feb. 2014.

[28] T. Zuo *et al.*, "O-band 400 Gbit/s client side optical transmission link." in Proc. *OFC*, San Francisco, CA, USA, 2014, Paper M2E.4.

[29] J. C. Cartledge *et al.*, "100 Gb/s Intensity Modulation and Direct Detection." *J. Lightw. Technol.*, vol. 32, no. 16, pp. 2809–2814, Aug. 2014.

[30] M. S. Erkilinç *et al.*, "Nyquist-shaped dispersion-precompensated subcarrier modulation with direct detection for spectrally-efficient WDM transmission." *Opt. Exp.*, vol. 22, no. 8, pp. 9420–9431, Apr. 2014.

[31] M. S. Erkilinç *et al.*, "Spectrally efficient WDM Nyquist pulse-shaped 16-QAM subcarrier modulation transmission with direct detection." *J. Lightw. Technol.*, vol. 33, no. 15, pp. 3147–3155, Aug. 2015.

[32] K. P. Zhong *et al.*, "Experimental demonstration of 608 Gbit/s short reach transmission employing half-cycle 16QAM Nyquist-SCM signal and direct detection with 25 Gbps EML." *Opt. Exp.*, vol. 24, no. 22, pp. 25057–25067, Oct. 2016.

[33] W. Z. Yan *et al.*, "100 Gb/s optical IM-DD transmission with 10G-class devices enabled by 65 Gsamples/s CMOS DAC core." in Proc. *OFC*, Anaheim, CA, USA, 2013, Paper OM3H.1.

[34] T. Tanaka *et al.*, "Experimental demonstration of 448-Gbps+ DMT transmission over 30 km SMF." in Proc. *OFC*, San Francisco, CA, USA, 2014, Paper M2I. 5.

[35] F. Li *et al.*, "Demonstration of four channel CWDM 560 Gbit/s 128QAM-OFDM for optical interconnection." in Proc. *OFC*, Anaheim, CA, USA, 2016, Paper W4J.2.

[36] A. P. T. Lau *et al.*, "Advanced DSP Techniques Enabling High Spectral Efficiency and Flexible Transmissions: Toward Elastic Optical Networks," Special Issue on Advanced Digital Signal Processing and Coding for Multi-Tb/s Optical Transport, *IEEE Signal Processing Magazine*, vol. 31, no. 2, pp. 82–92, Mar. 2014.

[37] J. G. Proakis and D. G. Manolakis, *Digital Signal Processing: Principles, Algorithms and Applications, 4th edition,* Pearson Education, 2007.

[38] R. Rath, *et al.*, "Tomlinson–Harashima Precoding For Dispersion Uncompensated PAM-4 Transmission With Direct-Detection." *J. Lightw. Technol.*, vol. 35, no. 18, pp. 3909–3917, Sept. 2017.

[39] J. Li, E. *et al.*, "Approaching Nyquist Limit in WDM Systems by Low-Complexity Receiver-Side Duobinary Shaping." *J. Lightw. Technol.*, vol. 30, no. 11, pp. 1664–1676, June, 2012.

[40] M. V. Eyuboglu and S. U. H. Qureshi, "Reduced-state sequence estimation for coded modulation of intersymbol interference channels." *IEEE Journal on Selected Areas in Communications*, vol. 7, no. 6, pp. 989–995, Aug. 1989.

[41] S. Olcer, "Reduced-state sequence detection of multilevel partial-response signals." *Trans. Commun.*, vol. 40, no. 1, pp. 3–6, Jan. 1992.

[42] L. Zhang, *et al.*, "C-band Single Wavelength 100-Gb/s IM-DD Transmission over 80-km SMF without CD compensation using SSB-DMT." in Proc. *OFC*, Los Angeles, CA, USA, 2015, Paper Th4A.2.

[43] W. Yan, *et al.*, "80 km IM-DD Transmission for 100 Gb/s per Lane Enabled by DMT and Nonlinearity Management." in Proc. *OFC*, San Francisco, CA, USA, 2015, Paper M2I.4.

[44] J. Zhang, *et al.*, "An Efficient Hybrid Equalizer for 50 Gb/s PAM-4 Signal Transmission Over 50 km SSMF in a 10-GHz DML-Based IM/DD system." in Proc. *CLEO*, San Jose, CA, USA, 2017, Paper SF1L.1.

[45] D Che, *et al.*, "Stokes vector direct detection for linear complex optical channels". *J. Lightw. Technol.*, vol. 33, no. 3, pp. 678–684, Feb. 2015.

[46] D. Che *et al.,* "Stokes vector direct detection for short-reach optical communication", *Opt. Lett.,* vol. 39, no. 11, pp. 3110–3113, Jun. 2014.

[47] W. Shieh, H. Khodakarami, D Che, "Polarization diversity and modulation for high-speed optical communications: architectures and capacity". *APL Photonics*, vol. 1, no. 4, pp. 040801, Jul. 2016.

[48] D Che, "Coherent optical short-reach communications", https://minerva-access.unimelb.edu.au/handle/11343/129708

[49] M. Osman *et al.,* "1λ× 224 Gb/s 10 km transmission of polarization division multiplexed PAM-4 signals using 1.3μm SiP intensity modulator and a direct-detection MIMO-based receiver." presented at European Conference on Optical Communications, Cannes, France, 2014. PDP4.4.

[50] X. Zhou *et al.,* "Polarization-Multiplexed DMT with IM-DD using 2×2 MIMO processing based on SOP estimation and MPBI elimination." *Photon. J.*, vol. 7, no. 6, pp. 1–12, Mar. 2015.

[51] M. Morsy-Osman *et al.,* "224-Gb/s 10-km transmission of PDM PAM-4 at 1.3 μm using a single intensity-modulated laser and a direct-detection MIMO DSP-based receiver." *J. Lightw. Technol.*, vol. 33, no. 7, pp. 1417–1424, Apr. 2015.

[52] D. Che, Y. Feng, W. Shieh, "200-Gb/s polarization-multiplexed DMT using stokes vector receiver with frequency-domain MIMO." presented at the Optical Fiber Communication Conference, Los Angeles, CA, USA, 2017, Paper Tu3D.4.

[53] D. Che, S. William. "Polarization demultiplexing for Stokes vector direct detection." *J. Lightw. Technol.*, vol. 34, no. 2, pp. 754–760, Jun. 2016.

[54] C. Mathieu et al., "1 λ, 6 bits/symbol, 280 and 350 Gb/s direct detection transceiver using intensity modulation, polarization multiplexing, and inter-polarization phase modulation." presented at Optical Fiber Communications Conference, Los Angeles, California, USA, 2015, Paper Th5B.

[55] C. Mathieu *et al.,* "Digital signal processing for dual-polarization intensity and interpolarization phase modulation formats using Stokes detection." *J. Lightw. Technol*, vol. 34, no. 1, pp. 188–195, Feb. 2016.

14

Multi-dimensional Polarization Modulation

Di Che[1], An Li[2], Xi Chen[3], Qian Hu[4] and William Shieh[1]

[1]The University of Melbourne, Melbourne, Australia
[2]Futurewei Technologies, Santa Clara, CA, USA
[3]Nokia Bell Labs, Holmdel, New Jersey, USA
[4]Nokia Bell Labs, Stuttgart, Germany

Intensity modulation (IM) with direct detection (DD) dominates commercial short-reach optical interconnects, due to its simple architecture, low cost and power consumption. Parallel IM-DD optical links with coarse wavelength division multiplexing (CWDM) have become mature solutions for 100G short-reach transport. A promising approach of further upgrading the data rate per wavelength is the multi-dimensional modulation and detection exploiting polarization diversity. While polarization multiplexing (POL-MUX) of two complex-modulated signals has been widely deployed in modern coherent transceivers enabled by digital MIMO equalization [1, 2], it is not yet implemented in short-reach applications, impeded by the high transceiver expense. This chapter reviews the recent research efforts that aim at transforming POL-MUX technologies to be suitable for datacenter connectivity.

The polarization state changes randomly during fiber transmission due to the time-varying birefringence distributed along the fiber [3]. Without the active optical polarization control to align the polarization states between the transmitter and receiver, receivers should perform the detection with polarization diversity [1, 2] to avoid any information loss of the POL-MUX signal. Section 14.1 introduces the optical detection methods with polarization diversity, consisting of both the coherent detection in Jones space by the local oscillator, and the direct detection in Stokes space. The simplifications of POL-MUX can be performed through two perspectives: the transmitter and the receiver. Section 14.2 retains the coherent receiver, and focuses on the transmitter simplification from the sophisticated nested-external-modulators to the directly modulated lasers (DML), the most widely deployed optical component in short-reach interconnect. In particular, it reveals a remarkable phenomenon of direct modulation (DM), that the conventionally "detrimental" DM chirp [4] can be converted to a transmission benefit in coherent systems. Section 14.3, on the other hand, mainly focuses on the receiver simplification, namely, the DD Stokes vector receiver (SVR) that achieves the polarization diversity without a local laser. For the first time, it endows DD receivers with the capability of digital polarization recovery, a distinct characteristic in modern coherent receivers. By varying the modulation formats in Stokes space, the SVR can be applied either to medium-reach applications over several hundred kilometers, or optical interconnect below tens of kilometers. Furthermore, Section 14.4 addresses both perspectives, combining the POL-MUX DM with DD SVR that offers the simplest POL-MUX architecture ever with digital polarization recovery. The wavelength of the two DMLs can even be free-running, which enables an uncooled transceiver with much lower power consumption.

By direct intensity detection, multimode fiber can serve as the medium for very-short-reach interconnects, where the signal suffers from severe modal dispersion. However, when taking into account the polarization as one additional modulation dimension, the receiver must detect all the spatial and polarization modes and perform massive multi-input-multi-output (MIMO) for mode demultiplexing [5, 6]. Such scheme is currently too complicated for datacenter applications. As a result, we only focus on the POL-MUX inside the single-mode fiber.

14.1 Optical Detection with Polarization Diversity

14.1.1 The Need of Polarization-Diversity Detection

Although named as single-mode fiber (SMF), SMF supports two polarization modes during propagation. These two modes inevitably experience mode coupling that is usually random due to the uncontrollable birefringence distributed along the fiber. It is well-known, that the polarization state inside the SMF can be represented as a 2-D complex-valued (4-D real-valued) Jones vector, or a 3-D real-valued Stokes vector, as expressed by Equation (14.1) [7]:

$$|s\rangle = \begin{bmatrix} X \\ Y \end{bmatrix} \quad \hat{s} = \begin{bmatrix} S_0 \\ S_1 \\ S_2 \\ S_3 \end{bmatrix} = \begin{bmatrix} |X|^2 + |Y|^2 \\ |X|^2 - |Y|^2 \\ 2\mathrm{Re}(X \cdot Y^*) \\ 2\mathrm{Im}(X \cdot Y^*) \end{bmatrix} \tag{14.1}$$

where X/Y stand for the field of the 0/90-degree linear polarization, Re/Im stands for the real/imaginary part of the signal, and superscript "$*$" is the conjugate. It is noted that the total optical power S_0 is not an independent dimension in Stokes space because $S_0 = \sqrt{S_1^2 + S_2^2 + S_3^2}$, but it is convenient to include S_0 for the analysis of POL-MUX IM signal because $|X|^2 = (S_0 + S_1)/2$ and $|Y|^2 = (S_0 - S_1)/2$.

Corresponding to the Jones and Stokes vectors, the inter-polarization coupling can be characterized by the 2×2 channel matrix in Jones space, or the 3×3 rotation matrix in Stokes-space [7]. Although the short-reach fiber links normally have negligible polarization mode dispersion (PMD) that is frequency dependent, the frequency-independent polarization rotation brings extra trouble for polarization demultiplexing. To reveal this issue vividly, we pick up a special polarization modulation format (the system details can be referred to [8]): in Jones space, we send a 2-D complex modulated signal (S) at X-POL, and keep Y-POL as a constant carrier (C); converting this modulation to Stokes space via Equation (14.1), the modulated signal can be retrieved from $S_2 = 2Re(SC^*)$ and $S_3 = 2Im(SC^*)$. A notable feature of this modulation is that it does not fully utilize the degree of freedom in either 4-D Jones space or 3-D Stokes space. Because of the non-modulation constraint on Y-POL, the signal in Stokes space is restricted onto a surface instead of filling the entire 3-D space, as shown by Figure 14.1. The surface

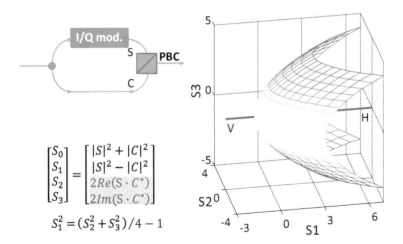

Figure 14.1 Self-coherent polarization modulation [8] and its signal distribution in Stokes space. PBC: polarization beam combiner; S/C: signal/carrier; H/V: horizontal/vertical polarization in Jones space; the H-V line aligns with the direction of S_1 axis.

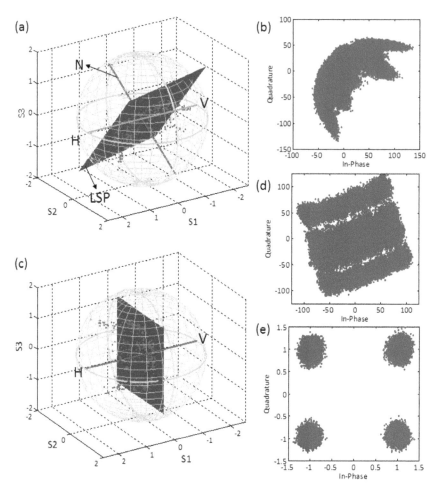

Figure 14.2   4-QAM signal distribution in Stokes space with an arbitrary polarization rotation: (a) raw received data; (c) data after polarization de-rotation. Projections of the 3-D Stokes-space distribution to the 2-D S_2-S_3 plane: (b) without polarization de-rotation; (d) with de-rotation; (e) with de-rotation and equalization. LSP: least-square plane of the 3-D distribution [8]; N: normal vector of LSP.

equation is $S_1^2 = (S_2^2 + S_3^2)/4 - 1$, assuming a normalized carrier (namely, $C = 1$) [8]. Using the 4-QAM at X-POL, we depict the signal distribution in Stokes space after an arbitrary polarization rotation in Figure 14.2(a).

Considering the modulation has 2-D degree of freedom, the question is whether the receiver can retrieve the signal properly by only 2-D detection, in other words, the projection of the 3-D distribution to the $S2$-$S3$ plane.

Figure 14.2(b) shows the projection without any polarization control. Obviously, the 4-QAM constellation has been completely distorted due to the discard of *S1* dimension after arbitrary rotation. To avoid such a problem, proper polarization de-rotation should be performed before the projection operation, as shown by Figure 14.2(c). The corresponding *S2-S3* projection is illustrated in Figure 14.2(d). Compared with Figure 14.2(b), this 2-D constellation becomes much regular, contaminated only by the sampling offset and constant phase offset. After equalization, the transmitted 4-QAM constellation at X-POL is shown in Figure 14.2(e).

The need of polarization de-rotation before the projection is critical for any polarization modulation system. The discussion above can also be extended to POL-MUX IM systems, where two intensity signals are modulated on $|X|^2$ and $|Y|^2$, equivalently, S_0 and S_1. The demodulation is equivalent to a pair of 1-D projections to $|X|^2$ and $|Y|^2$. Without polarization control, the direct square-law detection of the two polarizations along the receiver polarization axes would result in channel fading, or even channel singularity (the complete information loss of one polarization), depending on the unpredictable polarization rotation. This phenomenon leads to two types of receivers for polarization demultiplexing (POL-DEMUX), as shown by Figure 14.3. In terms of Figure 14.3(a), the polarization is actively controlled before optical detection, namely, the polarization is adaptively de-rotated in optical domain. As a result, the optical receiver only needs to recover the Jones or Stokes vector partially (e.g. S_2 and S_3 for the modulation in Figure 14.1). This partial detection is equivalent to the projection operation. Alternatively, without optical polarization control, the optical detection in Figure 14.3(b) must achieve polarization diversity, namely, retrieve the complete set of the 4-D Jones vector or the 3-D Stokes vector, to enable the subsequent MIMO equalization for digital polarization de-rotation. After MIMO, the projection can be performed in digital domain.

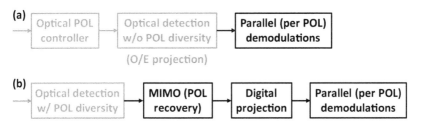

Figure 14.3 Two types of POL-DEMUX receiver: (a) polarization control in optical domain; (b) polarization recovery by digital MIMO equalization.

14.1.2 Automatic Polarization Control

The critical part of the POL-DEMUX in Figure 14.3(a) is the optical polarization control. The initial motivation of polarization control emerged in late 1980s for coherent detection, where the polarization should be aligned between the signal and local oscillator [9, 10]. However, the invention of erbium-doped fiber amplifier (EDFA) in 1990s solves the power budget issue of WDM transmissions, which overshadows the receiver sensitivity advantage of coherent detection. The polarization control technologies came soon after. With the popularization of optical chromatic dispersion compensation, polarization mode dispersion (PMD) became a nonnegligible transmission impairment, leading to a resumed interest on polarization control for PMD compensation [11, 12]. Finally, in early 2000s when WDM channel resources were exhausted, polarization became exploited as a further degree of freedom of the electromagnetic field inside the fiber to increase the spectral efficiency per wavelength, namely, the POL-MUX [13, 14].

In early POL-MUX systems, DD receivers lack the capability of polarization-diversity detection and DSP-enabled MIMO. Therefore, the control of polarization state should be performed in optical domain to avoid improper projection. Because the receiver has a fixed polarization axis, the polarization state before the receiver polarization splitting should maintain stable to well align with the local polarization axis. It is important that such axis matching is maintained continuously, irrespective of variations of the input state, referred as the so-called endless polarization control [10]. Besides realizations of endless polarization control, another key consideration for POL-DEMUX is the tracking speed. At incipient POL-MUX demonstrations, polarization controller is adjusted manually [15], which is not practical to be deployed to the field transmissions. The early 2000s witnessed an enthusiastic pursuit for POL-MUX systems, where the POL-DEMUX is realized by an automated approach [13, 14], with an example shown in Figure 14.4. Partial optical signals are tapped out as the input of a feedback circuit. The circuit determines the drive signal of the polarization controller, which adjusts polarization state based on the input. This circuit can be completely analog [16], or digital to achieve a faster processing speed [13, 14, 17, 18]. The optical POL-DEMUX no doubt simplifies the receiver, because it only needs to recover a partial set of Jones or Stokes vector (i.e. the O/E projection). However, in addition to its bulkiness, the low tracking speed prevents its applications to modern optical transmission systems.

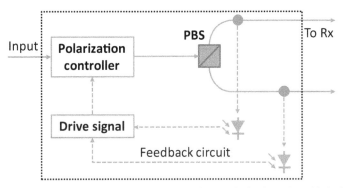

Figure 14.4 A typical setup for automated optical polarization demultiplexing. PBS: polarization beam splitter; Rx: receiver.

14.1.3 Polarization-Diversity Detection in Jones Space

With the rapid cost drop of receiver hardware and DSP chips during the past decade, digital POL-DEMUX as shown by Figure 14.3(b) becomes a preferable option, which is now a remarkable feature for modern optical transmission systems. The configuration in Figure 14.3(b) can be further divided into two types: the polarization-diversity detection in either Jones or Stokes space. Because Jones space contains the phase information of both polarizations, the Jones-space signal detection is generally the coherent detection which employs a local oscillator (LO) to provide the phase reference. A typical structure for the dual-polarization coherent receiver is shown by Figure 14.5 [19]. With the full optical field of both polarizations, the polarization can be recovered by the 2×2 complex-valued MIMO equalization in Jones space.

14.1.4 The Barrier of Self-Polarization Diversity

For short-reach applications like the datacenter connectivity, cost becomes one primary consideration for transceiver design. The dual-polarization coherent receiver is usually regarded to be too sophisticated for these cost-sensitive applications. As the counterpart, direct detection is preferred due to its natural advantage – the simplicity. In general, direct detection is referred to any detection method without the reference from an LO. This not only saves the expense of one laser, but more importantly, realizes a completely colorless receiver; namely, there is no need to match the wavelengths of the two lasers between transmitter and receiver. This avoids the sophisticated

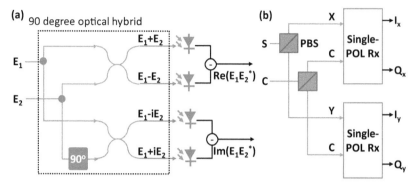

Figure 14.5 (a) Single-polarization coherent receiver; (b) dual-polarization coherent receiver. PBS: polarization beam splitter; S: dual-polarization signal; C: carrier from LO; Re/Im: real/imaginary part; X/Y: the pair of orthogonal polarizations in Jones space; I/Q: in-phase/quadrature part.

wavelength alignment as well as carrier phase recovery [20, 21] at receiver; and in consequence, relaxes the requirements of both laser linewidth and frequency stability at transmitter. Such characteristics are critical for short-reach networks with asymmetric nodes, where the remote unit is usually equipped with low-cost transceivers without cooling to save the power consumption.

To achieve the direct detection with polarization diversity, one attempt is to send a reference carrier from the transmitter to avoid the extra laser at receiver [22, 23]. For conventional coherent detection with local oscillator (LO), the LO polarization is fine controlled and then split to generate a pair of orthogonal carriers in Jones space with the same intensity [19]. Nevertheless, the polarization state of the carrier that travels with the signal is not known at receiver. As a consequence, the receiver can no longer generate the proper pair of carriers. At the local Jones space, the two carriers would have different intensity during most time, which breaks the unitary condition of the 2×2 channel matrix, equivalent to the effect of polarization dependent loss (PDL). One extreme case is the complete carrier fading of one local axis which results in channel singularity. This is well known as the carrier fading of POL-MUX direct detection. A modified (but still unworkable) attempt, instead of splitting the incoming carrier directly, is to generate its orthogonal counterpart through a polarization rotator (e.g. Faraday rotator mirror) to form a pair of orthogonal polarization bases at receiver, namely, the self-polarization diversity [24]. However, it is not generally valid to transform any arbitrary polarization into its orthogonal state by a static rotator [25, 26]. Below, we briefly prove this claim by an arbitrary polarization in Jones space E and a rotator with the fixed Jones matrix of F. The self-polarization-diversity should satisfy

$$\forall E : E^H F E = 0 \tag{14.2}$$

where "H" stands for Hermitian conjugate. As a unitary matrix, F can be can be diagonalized as $F = P^H D P$, where P is a diagonalization unitary matrix and D is a diagonal matrix with the diagonal elements η_1 and η_2 (i.e. $D = [\eta_1 \; 0; ; 0 \; \eta_2]$). Substituting F to Equation (14.2), we reach

$$E_1^H D E_1 = 0 \quad \rightarrow \quad \eta_1 |E_{1X}|^2 + \eta_2 |E_{1Y}|^2 = 0 \tag{14.3}$$

where $E_1 = P E = [E_{1X} \; E_{1Y}]$, an arbitrary polarization without loss of generality. For arbitrary E_1, the only solution of Equation (14.3) is $\eta_1 = \eta_2 = 0$, which is unattainable for any rotator.

14.1.5 Polarization-Diversity Detection in Stokes Space

In general, it is not possible to regenerate an orthogonal basis in Jones space by only the transmitted carrier without active optical polarization control. Therefore, a direct detection receiver should achieve the detection with polarization diversity in the isomorphic space of Jones space, namely, the Stokes space [7]. Stokes vector is born to be directly detected [27, 28], because it characterizes the polarization states by the intensity of three orthogonal polarizations in Stokes space, expressed as Equation (14.1). Figure 14.6(a) illustrates a receiver structure that can recover a Stokes vector defined by Equation (14.1) straightforward [29]. In fact, Stokes vector receiver (SVR) can be generalized by any 3 or 4 detections of polarization states as long as they are non-singular superposition of the Stokes components (S_0, S_1, S_2, S_3). Figure 14.6(b–d) provides a few more configurations [25, 30, 31].

SVR can be regarded as a hybrid receiver of coherent and non-coherent detection. From S_0 and S_1, the non-coherent information (intensity of X and Y) can be recovered; while from S_2 and S_3, the coherent information between X and Y can be recovered. In the following sections, we will introduce the applications of both the coherent receiver in Figure 14.5, and the Stokes vector receiver in Figure 14.6 for datacenter connectivity, following the conceptual signal processing flow in Figure 14.3(b), namely, the digital POL-DEMUX.

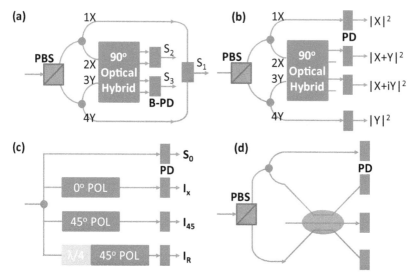

Figure 14.6 Various optical architectures for Stokes vector receiver. PBS: polarization beam splitter; PD: photodiode; B-PD: balanced photodiode; POL: polarizer; λ/4: quarter-wave plate; I: intensity; Figure (d) contains a 3x3 coupler to mix the two polarizations.

14.2 Direct Modulation with Coherent Receiver

In early 1990s, the commercialization of 2.5-Gb/s submarine optical fiber links using direct modulation (DM) and direct detection (DD) drove the development of Internet as a global phenomenon. However, the detrimental frequency chirp [4] of directly modulated lasers (DML) impeded its application to the subsequent 10-Gb/s evolution. Nowadays, industry offers the 100 Gb/s per channel long-haul solution by external modulation (EM) and coherent detection (COHD), to avoid the DM chirp effect. EM-COHD transceivers employing the dual-polarization QPSK modulations [32, 33] have already been commercialized. However, for medium-reach applications, there exists an incentive to find a compromise between the DM-DD and EM-COHD. Especially, EM faces the following shortcomings when applied to medium reach applications: (1) the price is 2 to 3 orders higher than DML, making it cost-inefficient; (2) the high insertion loss of EM reduces the power budget, which is not suitable for passive optical networks without extra amplification; (3) the bulky combination of laser and modulator prevents them from integrating to the compact transmitter optical sub-assembly (TOSA). As a result, there emerged the revival of interest in DML as a replacement of EM transmitter in coherent system [34–39].

14.2.1 The Intensity-only POL-MUX-DM Coherent System

In general, a POL-MUX DM transmitter consists of two DMLs for POL-MUX that exploits 2-D modulation degrees of freedom in Jones space, namely, the optical intensity of the two polarizations. The coherent receiver still needs to perform 4-D detection to guarantee a proper projection as we analyzed in Part 1. The 4-D information is processed for fiber dispersion compensation and POL-DEMUX. After then, the coherent receiver discards the phase information, and makes decision only by optical intensity. One unique feature for the POL-DEMUX of DM signal is the modulus-based equalization [40], because the baseband signal is modulated only on the optical intensity. Namely, for either polarization, the cost-function should be:

$$ J(\boldsymbol{w}) = E\left[\left(|z_k|^2 - m_k\right)^2\right] \quad z_k = \boldsymbol{w}^H \begin{bmatrix} x_k \\ y_k \end{bmatrix} \qquad (14.4) $$

where w is the equalizer weight, E is the expectation operation, z_k is the equalizer output, x_k and y_k are the equalizer inputs of the received X and Y polarizations, respectively, and the subscript k is the sampling sequence. m_k is the discrete modulus of the intensity signal that can be acquired from either the training sequence or the symbol decision (namely, the decision-directed mode). In contrast, the cost function for the constellation-based equalization, like the most popular least mean square (LMS) algorithm [40], is:

$$ J(\boldsymbol{w}) = E\left[|z_k - c_k|^2\right] \quad z_k = \boldsymbol{w}^H \begin{bmatrix} x_k \\ y_k \end{bmatrix} \qquad (14.5) $$

where c_k is the complex-valued constellation point. Compared with LMS algorithm, the update error factor of multi-modulus algorithm (MMA) changes from $(z_k - c_k)$ to $(|z_k|^2 - m_k)z_k$. The cost function of MMA can be upgraded to the least square criterion for faster convergence but with more complicated recursive algorithms [41].

By coherent detection, the achievable transmission distance of POL-MUX DM signal can be elongated significantly to multiple fiber spans, owing to the digital dispersion compensation. In this application regime, the optical amplified spontaneous emission (ASE) noise becomes the dominant noise source in the transmission system.

Conventionally in direct detection system, DM PAM signals are modulated with equal intensity-spacing. However, this gives rise to a unique issue for DM coherent system. Namely, higher intensity level after the square-law function suffers from larger noise variance, because the ASE noise is

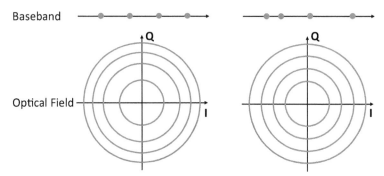

Figure 14.7 Constellation of POL-MUX DM PAM-4 signal. Left: equal-intensity-spacing; right: equal-amplitude-spacing.

added linearly to the optical field, as shown by Figure 14.7(left). To address such an issue, the transmitter can perform constellation shaping, like the geometric shaping in Figure 14.7(right) to stretch the effective Euler distance between higher intensity levels, or the probabilistic shaping to assign higher probability to lower power level [42]. Alternatively, an optical filter may be applied before the receiver to perform the frequency-modulation to amplitude-modulation (FM-AM) conversion that realizes a similar effect of geometric shaping, as investigated in [36].

Because the decision of POL-MUX DM signal does not rely on the phase, there is no need to perform carrier recovery at receiver. Consequently, the linewidth requirement of both lasers at transmitter and receiver can be relaxed. For example, [43] use the ultra-low-cost VCSELs for both the transmitter laser and the local oscillator. However, the intensity-only decision wastes the information of the phase dimension. Compared between two 4-level modulations, the IM PAM-4 has a large OSNR sensitivity gap with the complex modulation (CM) QAM-4 signal. This significantly limit the number of optical amplification spans. It would be desirable to take advantage of the phase domain to enhance the performance of DM coherent systems.

14.2.2 Complex DM Model

Since its birth, DMLs have been exploited as versatile optical transmitters. (i) The linear relationship between the injection current and the laser output power enables IM; (ii) the injection current perturbs the refractive index of the laser active region, resulting in a wavelength deviation simultaneously with the IM, namely, the frequency chirp [4]. Chirp sweeps the optical frequency, endowing DMLs the capability of angle modulation [44], such as frequency-shift-keying and differential phase-shift-keying, revealed more than 30 years ago. The intensity and frequency responses of DMLs are strongly coupled, as derived by the laser diode rate equation [4]. Limited by the DD receiver, DMLs used to be exploited only as 1-D (intensity or angle) transmitters by suppressing or simply ignoring the other modulation. For example, when performing DM angle modulation, the intensity variation is suppressed by various approaches such as the small-signal modulation with a bias much higher than the DML threshold or short-pulse modulation to change the instantaneous optical frequency [44].

Opposite to the traditional 1-D DM, a 2-D DM signal can be recovered by the advanced digital coherent detection. This distinguished DM concept is named as complex DM (CDM) [45, 46], because it contains 2-D information of both intensity and angle modulations. CDM uses the same DM transmitter in very-short-reach interconnects. In contrast, its achievable transmission distance without optical dispersion management is beyond 1000 km [45, 46], due to the superior OSNR sensitivity. The principle of CDM can be revealed by the following approximation of chirp model:

$$\Delta f = \frac{\alpha}{4\pi} \left(\frac{d}{dt} \ln I(t) + \kappa \cdot I(t) \right) \tag{14.6}$$

where Δf is the chirp induced frequency shift, α/κ is the laser transient/adiabatic chirp coefficient, and $I(t)$ is the laser output intensity. The equation indicates that the frequency shift of DML can be determined by intensity, in other words, $\Delta f(t)$ contains extra information of $I(t)$. Therefore, $\Delta f(t)$ can be combined with $I(t)$ to enhance the symbol decision of IM.

At digital domain, the differential phase between the adjacent sampling point is a good estimation of the instantaneous frequency. The crucial technique of CDM is the joint demodulation of intensity and differential phase with the maximum likelihood sequence estimation (MLSE) [46]. CDM decoder performs 2-tap MLSE taking the intensity I_t and differential phase $\Delta \varphi_t$ as inputs. We define two data sets: (i) the state $\{S_t\}$ (each state corresponds to an intensity level); (ii) the transition $\{\chi_t | \chi_t \triangleq (S_t, S_{t-1})\}$. The crucial task for MLSE is to calculate the transition probability $P(I_t, \Delta \varphi_t | S_t, S_{t-1})$ in order to determine the most reliable transitions. $P(I_t, \Delta \varphi_t | S_t, S_{t-1})$ is the conditional probability of observing $(I_t, \Delta \varphi_t)$ when the transition (S_t, S_{t-1}) is sent. Although I_t and $\Delta \varphi_t$ are correlated with each other, to simplify the calculation, we regard I_t and $\Delta \varphi_t$ as 2 independent random variables and decompose the 2-D joint distribution into 2 1-D distributions:

$$P(I_t, \Delta\varphi_t | S_t, S_{t-1}) = P(I_t | S_t) \cdot P(\Delta\varphi_t | S_t, S_{t-1}) \qquad (14.7)$$

It is assumed that the intensity channel has no ISI so that $P(I_t | S_t, S_{t-1}) = P(I_t | S_t)$, where $P(I_t | S_t)$ represents the IM model and $P(\Delta\varphi_t | S_t, S_{t-1})$ is the chirp modulation model. Both probabilities can be estimated either by the transition distance on the 2-D complex plane assuming a Gaussian-noise channel [45], or by the statistical probability estimation more precisely [46].

14.2.3 100-Gb/s CDM Transmission Over 1600-km SMF

To better explain the CDM principle and advantage, here we review a 100-Gb/s CDM transmission experiment [47], with the setup shown by Figure 14.8. The baseband 25-GBaud PAM-4 signal is Nyquist-pulse shaped by a raised cosine filter with roll-off of 0.01. Two independent 1549.44-nm distributed feedback (DFB) lasers serve as the light sources of dual polarizations. Their frequency offset is tuned to be less than 0.02 nm by adjusting the temperature. The two outputs are polarization combined and then launched into a recirculating loop consisting of 1 span of 80-km standard single mode fiber (SSMF) and an EDFA to compensate the loop loss of 17 dB. At receiver, an external cavity laser serves as the local oscillator (LO), which is fed into a dual-polarization coherent receiver together with the signal. The receiver offline DSP is shown by the inset (right) of Figure 14.8. The intensity-only decision is immediately made after the polarization demultiplexing for performance comparison.

We first observe the measured distributions. The 4 intensity distributions $(I_t | S_t)$ in Figure 14.9(a) present Gaussian shapes. Because the optical signal is modulated with equal-intensity spacing while the ASE noise is added to the signal field, PAM level with high intensity suffers from larger noise after the square-law detection. Figure 14.9(b) illustrates 16 differential phase distributions $(\Delta\varphi_t | S_t, S_{t-1})$. Each curve cluster represents the statistics of the current state S_t; within each cluster, the 4 curves stand for 4 previous states, respectively. Conventionally, DM PAM-4 constellations after coherent detection shows 4 rings with arbitrary absolute phase [36]. In contrast, the differential phase distributions in Figure 14.9(b) presents regular relations with the optical intensity, which coincide well with the explanation above.

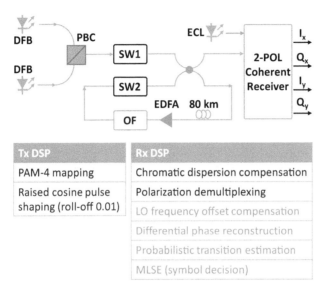

Tx DSP	Rx DSP
PAM-4 mapping	Chromatic dispersion compensation
Raised cosine pulse shaping (roll-off 0.01)	Polarization demultiplexing
	LO frequency offset compensation
	Differential phase reconstruction
	Probabilistic transition estimation
	MLSE (symbol decision)

Figure 14.8 Experiment setup of the 100-Gb/s POL-MUX DM coherent system. Inset (left) transmitter DSP; (right) receiver DSP. DFB: distributed feedback laser; PBC: polarization beam combiner; SW: (optical) switch; OF: optical filter; ECL: external cavity laser; I/Q: in-phase/quadrature; x/y: X/Y polarization.

We evaluate the system performance. The back-to-back OSNR sensitivity is illustrated in Figure 14.10(a). MLSE decoding achieves about 7 dB OSNR gain over the intensity-only hard decoding. This no doubt supports much more optical amplification spans, as shown by Figure 14.10(b), a transmission distance extension from 400 km to 1600 km. With the identical DM-coherent transceiver by virtue of only a little DSP increment (the 2-tap MLSE), CDM successfully elongates the DM transmission distance to the long-haul range beyond 1000 km. Compared with the external modulation based coherent systems, CDM brings a huge cost reduction from the combination of laser and nested external modulators to 2 DMLs (for POL-MUX); while at receiver, CDM relaxes the linewidth requirement of the local oscillator, considering its demodulation utilizes the differential phase which avoids the carrier phase recovery.

Because CDM is enabled by the frequency chirp, it gives rise to a contradictory yet innovative perspective of DML design and manufacture. Instead of minimizing the chirp impact in conventional DM-DD transceivers over the past 30 years, it may demand even higher chirp parameters, especially

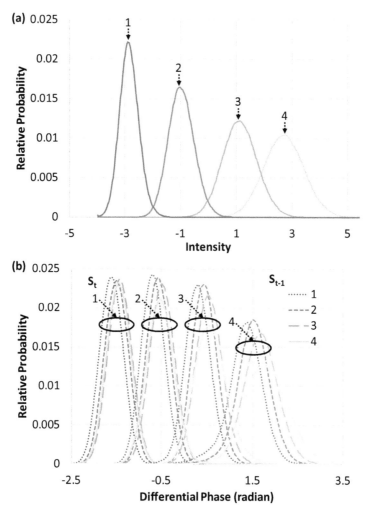

Figure 14.9 Probability density functions of the 25-Gbaud DM PAM-4 at 24-dB OSNR: (a) intensity; (b) differential phase. $S_{t/t-1}$: current/previous sampling point.

for higher order modulation [48] and larger baud-rate [49]. Investigations have indicated more than 5-dB OSNR penalty due to the lack of chirp in CDM systems. It is very promising to develop the CDM system close to the theoretical OSNR sensitivity by customized DML chirp in the near future.

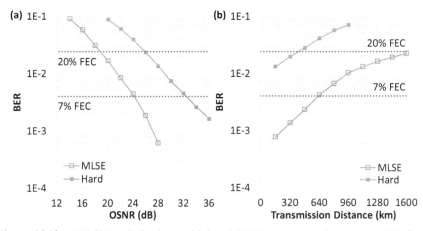

Figure 14.10 100 Gb/s polarization-multiplexed PAM-4 system performance. (a) Back-to-back OSNR sensitivity; (b) BER versus SSMF transmission distance. 7%-HD/20%-SD FEC threshold: 4e-3/2.4e-2; Hard: Intensity-only decision.

14.3 Polarization Modulation in Stokes Space

14.3.1 Stokes-space Modulation

Intensity modulation with direct detection (IM-DD) dominates the commercial optical short-reach interconnects, due to its simple implementation and low expense [50, 51]. Parallel IM-DD pipelines using coarse WDM (CWDM) has been mature solutions for 100G short-reach transport. A promising approach of further upgrading the data rate per channel is the multi-dimension modulation and detection by exploiting polarization diversity. POL-MUX has been widely adopted in modern coherent transceivers by signal representation in Jones space [1, 2]. However, Stokes vector [7] is better suitable for cost-sensitive short-reach applications, because it contains only optical intensity along orthogonal polarizations in Stokes space which naturally fits for DD. By signal modulation in Stokes space, the signal can be directly detected by various Stokes vector receivers [29–31] introduced in Section 14.1.5. To explain the necessity of polarization diversity DD, Figure 14.1 introduces a kind of Stokes-space modulation (SSM), namely, the self-coherent SSM which places the signal and the constant carrier onto two orthogonal polarizations in Jones space [8]. In fact, SSM is a much wider concept beyond the self-coherent concept. Because Stokes space has three degrees of freedom, SSM supports the modulation dimension from 1-D up to 3-D. Figure 14.11(a)

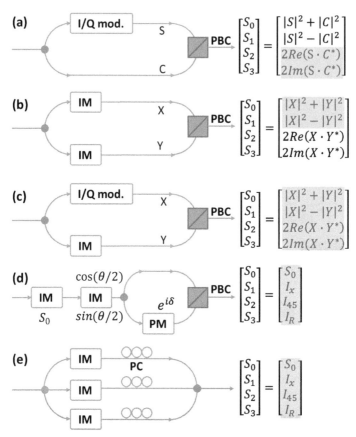

Figure 14.11 Transmitter structure for SSM using (a) self-coherent complex modulation; (b) POL-MUX intensity modulation (IM); (c) X-POL complex modulation and Y-POL intensity modulation; (d) 3-D SSM with polar coordinates (S_0, θ, δ); (e) 3-D SSM by polarization controller (PC). PM: phase modulator; PBC: polarization beam combiner. The shaded elements of the Stokes vector are modulated with signals.

depicts the self-coherent SSM [8, 29], which is a 2-D SSM format. Another 2-D SSM in Figure 14.11(b) sends the polarization-multiplexed (POL-MUX) IM signal [52], which is equivalent to the signal modulation on S_0 and S_1. By Stokes-space representation, the reason becomes clear for the channel fading issue of the 2-D receiver that recovers $|X|^2$ and $|Y|^2$ (equivalently, S_0 and S_1). When the polarization state changes, the modulated information inside $|X|^2$ and $|Y|^2$ would leak to S_2 and S_3. Without the complete detection of Stokes vector, the information loss is inevitable.

To maximize the spectral efficiency, SSM can at most perform 3-D modulation [25, 31, 53, 54]. As shown in Figure 14.11(c), the X polarization is complex modulated while Y is intensity modulated. SVR recovers the complex value X and intensity $|Y|$. Another 3-D modulation in Stokes space uses the polar coordinates (S_0, θ, δ) in Figure 14.11(d). The first intensity modulator (IM) modulates the signal intensity S_0. The second IM splits the signal into two output ports with an arbitrary splitting ratio of $tan(\theta\,/\,2)$. The phase modulator gives the phase difference δ between the split signals. Signals after the polarization beam combiner (PBC) forms the Stokes vector in Poincare sphere:

$$\begin{bmatrix} S_1 \\ S_2 \\ S_3 \end{bmatrix} = S_0 \cdot \begin{bmatrix} \cos\theta \\ \sin\theta\cos\delta \\ \sin\theta\sin\delta \end{bmatrix} \tag{14.8}$$

Alternatively, the 3-D SSM can be realized by three static polarization controllers as shown by Figure 14.11(e). It is noted that the 3-D SSM formats from Figure 14.11 (c) to (e) can be linearly transformed with each other, and is not limited to these three implementations.

14.3.2 Universal MIMO Equalization in Stokes Space

The detection of polarization-modulated signal using SVR contains 3 steps as illustrated in Figure 14.12: (i) the linear transformation of the received signal to the standard definition of Stokes vector; (ii) the 3×3 MIMO in Stokes space; (iii) the projection of the 3-D received signals to proper polarization axes. The first step is related to the selected structure of SVR, which normally conducts 3 or 4 intensity detections as illustrated in Figure 14.6. The transformation matrix is fixed regardless of the polarization variation. The third step is related to the SSM formats as presented in Figure 14.11. The crucial task is step 2. The fiber channel with random polarization variation can be characterized by a real-value 3×3 rotation matrix (RM). Namely, the polarization recovery in Stokes space is a completely a linear operation regardless of the modulation format. The universal MIMO concept for SVR can be interpreted from the following three aspects:

1. The MIMO can be applied to any 2-D or 3-D polarization modulation formats, as long as the signal can be converted to Stokes space uniquely. The MIMO equalizer always has 3 inputs corresponding to 3 received Stokes elements; and the outputs are determined by the modulation dimensions at transmitter, as illustrated at the right side of Figure 14.12.

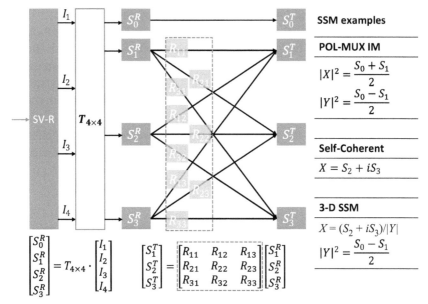

Figure 14.12   Digital polarization recovery for SVR. The Stokes-space modulation (SSM) at the right side are referred to the experiment demonstrations as: POL-MUX IM [5, 6]; Self-Coherent [24, 25]; 3-D SSM [26, 27]. The superscript R/T stands for received/transmitted Stokes vectors.

2. The MIMO can be performed either in time domain, or frequency domain due to the linearity of DFT operation and MIMO equalization, namely [55],

$$DFT\left(RM_{3\times3} \cdot \begin{bmatrix} S_1 \\ S_2 \\ S_3 \end{bmatrix}\right) = RM_{3\times3} \cdot \begin{bmatrix} DFT(S_1) \\ DFT(S_2) \\ DFT(S_3) \end{bmatrix} \quad (14.9)$$

It is noted that the above equation is accurate when the PMD is negligible. This is very practical for an SSM format that involves frequency domain modulations such as the OFDM.

3. The MIMO is a generalized concept which can be realized by various mature MIMO algorithms, such as least square method [56], steep-decent adaptive algorithm [40], or even blind search.

Categorized by the application distance, the various SSM formats can be divided into two types: the self-coherent CM [29] and the multi-dimension IM [52, 53]. The unique characteristic that distinguishes between them is whether it supports the digital dispersion compensation, in other words, the optical field recovery with phase diversity.

14.3.3 Self-coherent SSM

The self-coherent concept is normally applied to only single-polarization system, because traditional POL-MUX self-coherent systems suffer from the carrier fading issue explained in Section 14.1.4. In general, the single-polarization self-coherent system can be divided into two categories: the single-sideband (SSB) modulation and double-sideband (DSB) modulation. To recover the optical field from the intensity-only 1-D detection, the signal should be modulated with SSB, where the optical carrier is out of the signal spectrum. Assuming the baseband complex signal has the bandwidth of B, the corresponding minimum sampling rate at receiver should be at least $2B$ according to the Nyquist sampling theorem. The doubling bandwidth is consistent with the fact that the 2-D field signal at transmitter is transferred by square-law detection into a 1-D intensity signal without information loss. Given an SSB signal $S(t)$, there is a determinant relation between its in-phase (I) and quadrature (Q) parts:

$$S_Q(t) = HT[S_I(t)] \tag{14.10}$$

where HT stands for Hilbert transform. Under polar coordinates, Equation (14.10) can be converted to the relation between the intensity $I(t) = |S(t)|^2$ and the phase $\varphi(t)$:

$$\varphi(t) = HT[\log I(t)] \tag{14.11}$$

Thus, the SSB signal field recovery can be categorized into two types. (i) Using the I-Q relation in Equation (14.10), the major limitation is that DD cannot recover the pure in-phase part, contaminated by the signal-to-signal beat noise (SSBN). SSBN can be separated from the signal by a frequency gap between signal and carrier (i.e. offset-SSB [57]), which sacrifices half of the spectral efficiency. Alternatively, the gapless methods [58] suppress SSBN by high carrier-signal power ratio (CSPR) and digital SSBN cancellation algorithms. (ii) With the nonlinear logarithm in Equation (14.11), the intensity-phase relation can reconstruct the signal phase from the photo-current directly [59–61], but a large CSPR is still required to guarantee the positive intensity envelope, which is also interpreted as the minimum-phase condition [61].

The advantage of the self-coherent SSB is the 1-D detection that is the simplest receiver to recover an optical field. On the other hand, it brings obvious drawbacks:

1. It doubles the receiver bandwidth requirement, leading to an asymmetric requirement between transmitter and receiver. It is impractical for commercial products to develop transmitter and receiver with different upgrade speed. As a result, the receiver bandwidth would become the ceiling for the achievable capacity of SSB system.

2. Without frequency gap, SSB system requires high CSPR to minimize the 2nd-order noise impact. This no doubt reduces the system OSNR sensitivity, in other words, reduces the number of optical amplification spans.

3. The SSB modulation condition is strict for transmitter design. Because the optical carrier should be located out of the signal spectrum, the signal is normally generated with strong pulse shaping to form a rectangular-like spectrum, so that the receiver bandwidth could be minimized to the theoretical $2B$.

If the signal and carrier is properly separated before detection, they can be fed into a single-polarization coherent receiver in Figure 14.5(a) to recover the linear field directly [22, 62]. This enables the DSB modulation and detection that avoid the SSB drawbacks. However, it usually wastes the optical spectral efficiency to guarantee a strict signal-carrier separation. Taking one step further, as a DSB approach, the self-coherent CM in Stoke space successfully resolves signal-carrier separation problem by place the carrier on one polarization that is orthogonal to the signal. For the first time, it realizes 100% spectral efficiency for both the optical channel and the electronic receiver, with reference to the single-polarization coherent detection. Looking at the Stokes space signal: (i) the full optical field information can be retrieved from the 2nd and 3rd components of the Stokes vector, and (ii) the 2nd-order nonlinearity term is completely lumped into the 1st component. This orthogonal separation of the signal and its 2nd-order term in Stokes space significantly decreases the optimum system CSPR to 0 dB in theory (i.e. the same signal and carrier power), and consequently, enhance the system OSNR sensitivity. Compared with other self-coherent subsystems, the advantages of self-coherent CM in Stoke space are summarized below:

1. It realizes a linear complex optical channel similar to coherent detection. The field recovery is performed by pure linear MIMO without any

nonlinear operation like the iterative SSBN cancellation or logarithm in Equation (14.11).

2. The 0-dB optimal CSPR results in only 6-dB OSNR sensitivity gap between the Stokes vector direct detection and the POL-MUX-CM coherent system. This is the narrowest gap that can be achieved by direct detection ever.

3. The linear optical field recovery enables various DSP to compensate fiber transmission impairments, among which the dispersion compensation is the most critical one to elongate the achievable distance of direct detection system.

14.3.4 Multi-Dimensional IM in Stokes Space

Besides the self-coherent SSM that realizes a linear optical channel, SSM is also applied to the multi-dimensional IM to double or even triple the spectral efficiency of short-reach interconnect below tens of kilometers where the dispersion impairment can be negligible. For any 3-D SSM format, the Stokes-space MIMO can be realized either by the 3×3 adaptive equalization, or by special polarization training pattern like the 3 orthogonal basis vectors in Stokes space [8] that estimates the channel matrix considering the transmitter owns full capability to generate any desired 3-D pattern.

One notable SSM is the 2-D POL-MUX IM illustrated in Figure 14.11(b). Because the transmitter has no access to control the optical phase of both polarizations, it becomes unattainable to estimate the 3×3 channel matrix by 3-D training patterns directly. In this case, the single-carrier POL-MUX IM (e.g. PAM [52]) can perform 3×1 adaptive equalization to retrieve the transmitted S_1, and then recovers $|X|^2$ and $|Y|^2$ from S_0 ang S_1. However, for POL-MUX DMT signal, both polarization training and adaptive equalization at time domain becomes unattainable. It becomes imperative to move the Stokes-space MIMO to frequency domain [55]. The polarization rotation can be performed after DFT as expressed by Equation (14.9). Namely, after receiving the Stokes vectors, the digital receiver first performs DFT, and then estimate RM in frequency domain based on signal constellations. To simplify the frequency-domain POL-DEMUX, equalizer coefficients can be estimated by only a few pilot subcarriers, and other subcarriers directly use these coefficients to retrieve the transmitted S_0 and S_1. Similar MIMO could be generalized to other OFDM SSM systems, like the self-coherent SSM mentioned above.

14.4 Noncoherent Polarization Multiplexing

14.4.1 Degree of Coherence in POL-MUX Transmitter

We have analyzed various POL-MUX systems through the above few sections. Among them, the DML based POL-MUX transmitter is distinguished from the external modulator: it has much lower cost and power consumption that makes it attractive for short-reach applications, but it lacks the capability of controlling either wavelength or phase between the two polarizations. To better analyze the DML-based POL-MUX system, we categorize POL-MUX transmitters by the degree of coherence (between the two polarizations) into five levels in a descending order in the following Table 14.1.

Level 1 is currently the most widely utilized POL-MUX scheme because of its straightforward realization of coherence. In the state-of-the-art coherent transceivers, digital polarization recovery normally requires certain degree of coherence, because it performs MIMO equalization by the field signal in Jones space that utilizes the phase information of dual polarizations [40]. Level 1 is suitable for these coherent transceivers in nature. However, this blocks the pathway of DMLs becoming low-cost POL-MUX transmitters for short-reach applications. To realize a POL-MUX scheme compatible with the parallel DM-DD, the transmitter must select the degree of coherence from

Table 14.1   Degree of Coherence in Polarization-Multiplexed Optical Transmitters

1	One laser source [52, 55]	Light from the laser is split and modulated by external modulators.
2	Injection-locked lasers [34, 35]	DM lasers (DML) can maintain high phase coherence by injection-locking with an additional master laser. This offers high coherence not only suitable for POL-MUX (e.g. 3-D IM in the right inset), but even for the I/Q modulation.
3	Lasers with wavelength alignment [36, 47]	DML are combined without phase coherence, but their wavelength is well aligned (e.g. by temperature control). In DM coherent systems, DML wavelengths should be aligned to that of the local oscillator.
4	Lasers with fixed-grid wavelengths [63]	Each laser should stay within its grid (without spectra overlapping). Using POL (instead of wavelength) DEMUX, this can realize MUX of 4 DM signals at most.
5	Lasers with no wavelength control [64]	There is no coherence constraint on any POL-MUX laser (the spectra of the 2 polarizations can be completely or partially overlapped, or completely detached).

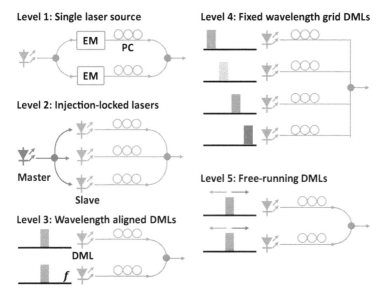

Figure 14.13 Degree of coherence in polarization-multiplexed optical transmitters.

level 2 to 5 that support DM. Level 2 can achieve high coherence among DMLs, but it requires an extra high-performance master laser to lock the slave DMLs. We have discussed the POL-MUX CDM system that belongs to Level 3 in Section 14.2. Although it relies on sophisticated coherent receiver, its application distance up to 1000 km well deserves such expense. In this section, we introduce the principle of level 4 and 5, where the DMLs inside the POL-MUX transmitter are completely noncoherent with each other, namely, achieving the noncoherent (NC) POL-MUX. In essence, level 4 is identical to WDM, but it offers an interesting perspective of using POL-DEMUX instead of wavelength DEMUX to separate parallel DM signals. The most practical and meaningful design for short-reach applications is the degree of coherence at level 5, because it uses free-running DMLs without temperature control, and thus offers a smooth upgrade from parallel DM-DD to their POL-MUX counterpart.

14.4.2 Noncoherent POL-MUX Schemes

The outputs of 2 free-running lasers can be modelled in Jones space by:

$$\begin{bmatrix} X \\ Y \end{bmatrix} = \begin{bmatrix} |X| \exp(i\varphi_x + i2\pi\Delta ft) \\ |Y| \exp(i\varphi_y) \end{bmatrix} \tag{14.12}$$

where the intensity $|X|$ and $|Y|$, the phase φ_x and φ_y are freely modulated, and $\triangle f$ stands for an arbitrary frequency offset. By Jones-Stokes conversion:

$$
\begin{bmatrix} S_0 \\ S_1 \\ S_2 \\ S_3 \end{bmatrix} = \begin{bmatrix} |X|^2 + |Y|^2 \\ |X|^2 - |Y|^2 \\ 2\,|X|\,|Y|\,\mathrm{Re}(\exp(i\varphi_x - i\varphi_y + i2\pi\triangle ft)) \\ 2\,|X|\,|Y|\,\mathrm{Im}(\exp(i\varphi_x - i\varphi_y + i2\pi\triangle ft)) \end{bmatrix}
\tag{14.13}
$$

We pick up two crucial points from Equation (14.13):

1. $[S_0, S_1]$ contains no angle components, in other words, it only contains baseband modulations on optical intensity;
2. When $\triangle f$ is larger than the baseband signal bandwidth, low-pass filters (LPF) can be cascaded after the SVR to completely filter out the cross-beating components ($[S_2, S_3]$) regardless of the DML phase behavior, without impairing the baseband intensity modulation on S_0 and S_1.

When the DML wavelengths are completely detached, even if the polarization variation randomly mixes $[S_0, S_1]$ and $[S_2, S_3]$ during fiber transmission, the low-pass filters can separate the useful baseband modulation on $[S_0, S_1]$ from the cross-beating noise at receiver. As a result, at transmitter, $[S_2, S_3]$ can be regarded as two nullified dimensions, which have potential to be exploited as two extra modulation dimensions. For monochromatic light, the Stokes space has at most 3 dimensions. This is the fundamental theory that supports single-wavelength 3-D SSM. However, taking into account the noncoherent feature indicated by Equation (14.13), the degree of freedom for polarization modulation is extend to four [63]. To explain this 4-D modulation, we derive from a 3-D basis polarization set for transmitter: 0-degree linear POL, 45-degree linear POL and right-circular POL, and add one more dimension as 90-degree POL, expressed via Jones vectors as

$$
J_T = A_1 e^{i\phi_1} \begin{bmatrix} 1 \\ 0 \end{bmatrix} + A_2 e^{i\phi_2} \begin{bmatrix} 0 \\ 1 \end{bmatrix} + \frac{A_3 e^{i\phi_3}}{\sqrt{2}} \begin{bmatrix} 1 \\ i \end{bmatrix} + \frac{A_4 e^{i\phi_4}}{\sqrt{2}} \begin{bmatrix} 1 \\ 1 \end{bmatrix}
\tag{14.14}
$$

where A stands for the amplitude of each polarization and ϕ is the angle component. Obviously, beyond 2-D NC-POL-MUX with wavelength offset, each polarization should convey more than one wavelength (i.e. X-POL: 1, 3 and 4; Y-POL: 2, 3 and 4). Using Jones-Stokes conversion and discarding all the cross-beating components through the low-pass filters, the corresponding

Stokes vector is

$$S_T = \begin{bmatrix} 1 & 1 & 1 & 1 \\ 1 & -1 & 0 & 0 \\ 0 & 0 & 0 & 1 \\ 0 & 0 & -1 & 0 \end{bmatrix} \begin{bmatrix} A_1^2 \\ A_2^2 \\ A_3^2 \\ A_4^2 \end{bmatrix} \qquad (14.15)$$

Compared with 2-D DM, while $[S_2, S_3]$ is non-used at transmitter in Equation (14.13), they are loaded with intensity modulations of A_3^2 and A_4^2 in Equation (14.15). The DML number with WDM cannot further increase beyond 4, because Equation (14.15) only provides 4 equations, and these 4 equations are independent with each other only if the 4 channels are completely offset in frequency domain.

We now draw our attention to the most practical NC-POL-MUX at level 5 by Figure 14.14. According to the analysis of Equation (14.13), we label the baseband components (BC, $[|X|^2, |Y|^2]$ or $[S_0, S_1]$) as the blue spectra, and the cross-beating components (CC, $[S_2, S_3]$) as green. Assuming the baseband bandwidth as B (namely, the optical bandwidth is $2B$), the wavelength offset is classified into 3 types:

1. $\triangle f = 0$ (Figure 14.14a). Both BC and CC stay within the baseband at transmitter (Tx). At receiver (Rx) after polarization rotation, each Stokes element is a linear superposition of BC and CC, with the spectra constrained within the baseband.
2. $\triangle f > 2B$ (Figure 14.14d). CC spectra are completely detached from the baseband as analyzed in Section IV.A. Each Stokes element at Rx would be a linear superposition of BC $[|X|^2, |Y|^2]$ only. LPFs can filter out CC, and Rx bandwidth requirement is B.
3. $0 < \triangle f \leq 2B$ (Figures 14.14b and 14.14c). At Tx, the CC spectra reside in both baseband and out-band. At Rx, each Stokes element has the spectrum exceeding the baseband. Seemingly, this requires at most $3B$ Rx bandwidth (corresponding to $\triangle f = 2B$) to recover the full set of Stokes vector. However, the Stokes space offers a 3×3 linear MIMO channel, which maintains BC's spectra within the baseband regardless of the polarization rotation. The LPFs with bandwidth of B only filter out the out-band CC, but never impair BC.

The principle of level-5 NC-POL-MUX is fully revealed:

1. There is no constraint on wavelength or phase between the two polarizations at transmitter.

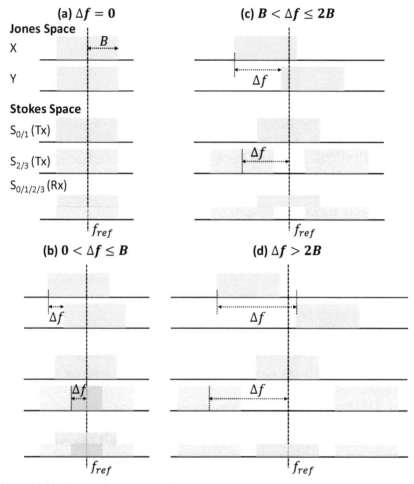

Figure 14.14 Spectra in Jones and Stokes space with varying laser frequency offset ($\triangle f$): (a) $\triangle f = 0$; (b) $0 < \triangle f \leq$ B; (c) $B < \triangle f \leq$ 2B; (d) $\triangle f >$ 2B.

2. A universal receiver for any wavelength offset is realized by a modified SVR required only the baseband bandwidth B, where the LPFs should be inserted after the SVR photodiodes while before the digital detection to avoid sampling aliasing.

By a pair of 16-GHz DFB lasers, a >100-Gb/s NC-POL-MUX system is demonstrated in Ref. [64]. For the first time, POL-MUX is realized by DM-DD transceivers like the transmitter/receiver optical assembly (TOSA/ROSA) instead of the sophisticated coherent transceivers. NC-POL-MUX transmitter can be uncooled which gets rid of the power-hungry temperature control; and its receiver requires only the baseband bandwidth which achieves high electrical spectral efficiency. NC-POL-MUX enables POL-MUX as a powerful solution to double the DM-DD bit-rate.

14.5 Summary

This chapter reviews a variety of simplified digital POL-MUX transceivers tailored for short-reach optical communications. For interconnects where fiber dispersion impact is negligible, direct-detection (DD) Stokes-vector receiver (SVR) is a powerful optical subsystem to support multi-dimensional intensity modulations (IM) through polarization diversity, that doubles or even triples the spectral efficiency per wavelength for parallel IM-DD pipelines. In particular, the noncoherent (NC) modulation in Stokes space enables a pair of uncooled directly modulated lasers (DML) as the POL-MUX transmitters, which offers a simple transmitter with low energy consumption. For single or a-few spans fiber transmission, the self-coherent modulation in Stokes space linearizes the DD optical channel, endowing DD-SVR the capability of digital dispersion compensation. The 6-dB OSNR sensitivity gap leads to the DD closest to coherent detection. For distance up to 1000-km level, coherent detection should be the ultimate choice. By coherent detection, complex DM (CDM) offers a completely new perspective of the traditional DM transmitter, and significantly enhances the OSNR sensitivity of DM coherent system. These technologies will no doubt accelerate the application of multi-dimensional polarization modulation to various layers of short-reach applications.

References

[1] Han, Y., and Li, G. (2005). Coherent optical communication using polarization multiple-input-multiple-output. *Opt. Exp.* 13, 7527–7534.
[2] Shieh, W., Yi, X., Ma, Y., and Tang, Y. (2007). Theoretical and experimental study on PMD-supported transmission using polarization diversity in coherent optical OFDM systems. *Opt. Exp.* 15, 9936–9947.
[3] Foschini, G. J., and Poole, C. D. (1991). Statistical theory of polarization dispersion in single mode fibers. *J. Lightwave Technol.* 9, 1439–1456.
[4] Tucker, R. (1985). High-speed modulation of semiconductor lasers. *J. Lightwave Technol.* 3, 1180–1192.

[5] Ryf, R., Randel, S., Gnauck, A. H., Bolle, C., Sierra, A., Mumtaz, S., et al. (2012). Mode-division multiplexing over 96 km of few-mode fiber using coherent 6 × 6 MIMO processing. *J. Lightwave Technol.* 30, 521–531.

[6] Li, A., Al Amin, A., Chen, X., and Shieh, W. (2011). "Reception of mode and polarization multiplexed 107-Gbps CO-OFDM signal over a two-mode fiber," in *Proceedings of the Optical Fiber Communication Conference*, Los Angeles, CA, Paper PDPB8.

[7] Gordon, J. P., and Kogelnik, H. (2000). "PMD fundamentals: Polarization mode dispersion in optical fibers," in *Proceedings of the National Academy of Sciences*, 97, 4541.

[8] Che, D., and Shieh, W. (2016). Polarization demultiplexing for Stokes vector direct detection. *J. Lightwave Technol.* 34, 754–760.

[9] Walker, N. G., and Walker, G. R. (1990). Polarization control for coherent communications. *J. Lightwave Technol.* 8, 438–458.

[10] Noé, R., Heidrich, H., and Hoffmann, D. (1988). Endless polarization control systems for coherent optics. *J. Lightwave Technol.* 6, 1199–1207.

[11] Heismann, F., Fishman, D. A., and Wilson, D. L. (1998). "Automatic compensation of first-order polarization mode dispersion in a 10 Gbps transmission system," in *Proceedings of the European Conference on Optical Communication*, Madrid, Spain, 529–530.

[12] Bülow, H. (2000). "PMD mitigation techniques and their effectiveness in installed fiber," in *Proceedings of the Optical Fiber Communication Conference*, Baltimore, MD, Paper ThH1.

[13] Bigo, S., Frignac, Y., Charlet, G., Idler, W., Borne, S., Gross, H., et al. (2001). "10.2Tbit/s (256x42.7Gbit/s PDM/WDM) transmission over 100km TeraLight^{TM} fiber with 1.28bit/s/Hz spectral efficiency," in *Proceedings of the Optical Fiber Communication Conference*, Anaheim, CA, paper PD25.

[14] Hecker, N. E., Gottwald, E., Kotten, K., Weiske, C. J., Schopflin, A., Krummrich, P. M., et al. (2001). "Automated polarization control demonstrated in a 1.28 Tbit/s (16x2˜40G bit/s) polarization multiplexed DWDM field trial," in *Proceedings of the European Conference on Optical Communication*, Amsterdam, paper Mo.L.3.1.

[15] Hill, P. M., Olshansky, R., and Burns, W. K. (1992). Optical polarization division multiplexing at 4 Gbps. *IEEE Photon. Technol. Lett.* 4, 500–502.

[16] Heismann, F., and Whalen, M. S. (1992). Fast automatic polarization control system. *IEEE Photon. Technol. Lett.* 4, 503–505.

[17] Koch, B., Hidayat, A., Zhang, H., Mirvoda, V., Lichtinger, M., Sandel, D., et al. (2008). Optical endless polarization stabilization at 9 krad/s with FPGA-based controller. *IEEE Photon. Technol. Lett.* 20, 961–963.

[18] Martinelli, M., Martelli, P., and Pietralunga, S. M. (2006). Polarization stabilization in optical communications systems. *J. Lightwave Technol.* 24, 4172–4183.

[19] Ip, E., Lau, A. P. T., Barros, D. J. F., and Kahn, J. M. (2008). Coherent detection in optical fiber systems. *Opt. Express* 16, 753–791.

[20] Pfau, T., Hoffmann, S., and Noé, R. (2009). Hardware-Efficient Coherent Digital Receiver Concept With Feedforward Carrier Recovery for M-QAM Constellations. *J. Lightwave Technol.* 27, 989–999.

[21] Ip, E., and Kahn, J. M. (2007). Feedforward carrier recovery for coherent optical communications. *J. Lightwave Technol.* 25, 2675–2692.

[22] Schmidt, B. J. C., Zan, Z., Du, L. B., and Lowery, A. J. (2010). 120 Gbit/s over 500-km using single-band polarization-multiplexed self-coherent optical OFDM. *J. Lightwave Technol.* 28, 328–335.

[23] Qian, D., Cvijetic, N., Hu, J., and Wang, T. (2010). 108 Gbps OFDMA-PON with polarization multiplexing and direct detection. *J. Lightwave Technol.* 28, 484–493.

[24] Xie, C, (2008). "PMD insensitive direct-detection optical OFDM systems using self-polarization diversity," in *Proceedings of the Optical Fiber Communication Conference*, San Diego, CA, Paper OMM2.

[25] Shieh, W., Khodakarami, H., and Che, D. (2016). Polarization diversity and modulation for high-speed optical communications: architectures and capacity. *APL Photonics* 1:040801.

[26] Peng, W.-R., Feng, K.-M., and Willner, A. E. (2008). "Direct-detected polarization division multiplexed OFDM systems with self-polarization diversity," in *Proceedings of the 21st Annual Meeting of the IEEE Lasers and Electro-Optics Society*, Acapulco, Paper MH3.

[27] Betti, S., De Marchis, G., and Iannone, E. (1992). Polarization modulated direct detection optical transmission systems. *J. Lightwave Technol.* 10, 1985–1997.

[28] Benedetto, S., Gaudino, R., and Poggiolini, P. (1995). Direct detection of optical digital transmission based on polarization shift keying modulation. *IEEE J. Select. Areas Commun.* 13, 531–542.

[29] Che, D., Li, A., Chen, X., Hu, Q., Wang, Y., and Shieh, W. (2015). Stokes Vector Direct Detection for Linear Complex Optical Channels. *J. Lightwave Technol.* 33, 678–684.

[30] Che, D., Li, A., Hu, Q., Chen, X., and Shieh, W. (2015). *"Implementing Simplified Stokes Vector Receiver for Phase Diverse Direct Detection,"* in *Proceedings of the Optical Fiber Communication Conference*, Los Angeles, CA, Paper Th1E.4.

[31] Kikuchi, K., and Kawakami, S. (2014). Multi-level signaling in the Stokes space and its application to large-capacity optical communications. *Opt. Express* 22, 7374–7387.

[32] Fludger, C. R. S., Duthel, T., van den Borne, D., Schulien, C., Schmidt, E.-D., Wuth, T., et al. (2008). Coherent equalization and POLMUX-RZ-DQPSK for robust 100-GE transmission. *J. Lightwave Technol.* 26, 64–72.

[33] Charlet, G., Renaudier, J., Mardoyan, H., Tran, P., Pardo, O. B., Verluise, F., et al. (2009). Transmission of 16.4-bit/s capacity over 2550 km using PDM QPSK modulation format and coherent receiver. *J. Lightwave Technol.* 27, 153–157.

[34] Liu, Z., Kakande, J., Kelly, B., O'Carroll, J., Phelan, R., Richardson, D. J., et al. (2014). Modulator-free quadrature amplitude modulation signal synthesis. *Nat. Commun.* 5:5911.

[35] Fontaine, N. K., Xiao, X., Chen, H., Huang, B., Neilson, D. T., Kim, K. W.et al. (2016). *"Chirp-Free Modulator using Injection Locked VCSEL Phase Array,"* In *42nd European Conference on Optical Communication*, Dusseldorf, Germany, paper Th3A.3.

[36] Xie, C., Spiga, S., Dong, P., Winzer, P. J., Bergmann, M., Kgel, B., et al. (2014). *"Generation and Transmission of a 400-Gbps PDM/WDM Signal Using a Monolithic 2x4 VCSEL Array and Coherent Detection,"* In *Optical Fiber Communication Conference*, San Francisco, CA, paper Th5C.9.

[37] Hu, Q., Che, D., Wang, Y., Li, A., Fang, J., and Shieh, W. (2015). Beyond amplitude-only detection for digital coherent system using directly modulated laser. *Opt. Lett.*, 40(12), 2762–2765.

[38] Che, D., Hu, Q., Yuan, F., and Shieh, W. (2015). *"Enabling Complex Modulation Using the Frequency Chirp of Directly Modulated Lasers,"* In *European Conference on Optical Communication*, Valencia, Spain, paper Mo.4.5.3.

[39] Cano, I. N., Lerín, A., and Prat, J. (2016). DQPSK directly phase modulated DFB for flexible coherent UDWDM-PONs. *IEEE Photon. Technol. Lett.* 28, 35–38.

[40] Savory, S. J. (2008). Digital filters for coherent optical receivers. *Opt. Express* 16, 804–817.

[41] Haykin, S. O. (2013). *Adaptive Filter Theory*, 5th edition. London: Pearson.

[42] Eriksson, T. A., Chagnon, M., Buchali, F., Schuh, K., Ten Brink, S., and Schmalen, L. (2017). 56 Gbaud probabilistically shaped PAM8 for data center interconnects," in *Proceedings of the 43rd European Conference on Optical Communication*, Gothenburg, Paper Tu2D.4.

[43] Xie, C., Spiga, S., Dong, P., Winzer, P., Andrejew, A., Kögel, B., et al. (2014). "All-VCSEL based 100-Gbps PDM-4PAM coherent system for applications in metro networks," in *Proceedings of the 40th European Conference on Optical Communication*, Cannes, Paper P4.3.

[44] Kobayashi, S., Yamamoto, Y., Ito, M. and Kimura, T. (1982). Direct frequency modulation in AlGaAs semiconductor lasers. *IEEE J. Quantum Elect.* 18, 582–595.

[45] Che, D., Yuan, F., Hu, Q., and Shieh, W. (2016). Frequency chirp supported complex modulation of directly modulated lasers. *J. Lightwave Technol.* 34, 1831–1836.

[46] Che, D., Yuan, F., and Shieh, W. (2017). Maximum likelihood sequence estimation for optical complex direct modulation. *Opt. Exp.* 25, 8730–8738.

[47] Che, D., Yuan, F., and Shieh, W. (2017). "100-Gbps complex direct modulation over 1600-km SSMF using probabilistic transition estimation," in *Proceedings of the Optical Fiber Communication Conference*, Los Angeles, CA, Paper M3C.5.

[48] Che, D., Yuan, F., and Shieh, W. (2016). Towards high-order modulation using complex modulation of semiconductor lasers. *Opt. Express* 24, 6644–6649.

[49] Che, D., Yuan, F., and Shieh, W. (2017). "Adiabatic chirp impact on the OSNR sensitivity of complex direct modulation: an experiment investigation," in *Proceedings of the Optical Fiber Communication Conference*, Los Angeles, CA, paper Th2A.47.

[50] Yan, W., Tanaka, T., Liu, B., Nishihara, M., Li, L., Takahara, T., et al. (2013). "100 Gbps optical IM-DD transmission with 10G-class devices enabled by 65 GSamples/s CMOS DAC Core," in *Proceedings of the Optical Fiber Communication Conference*, Anaheim, CA, paper OM3H.1.

[51] Zhang, H., Fu, S., Man, J., Chen, W., Song, X., and Zeng, L. (2014) "30km downstream transmission using 4×25Gbps 4-PAM modulation with commercial 10Gbps TOSA and ROSA for 100Gbps-PON,"

in *Proceedings of the Optical Fiber Communication Conference*, San Francisco, paper M2I.3.

[52] Morsy-Osman, M., Chagnon, M., Poulin, M., Lessard, S., and Plant, D. V. (2014). "1λ × 224 Gbps 10 km transmission of polarization division multiplexed PAM-4 signals using 1.3 μm SiP intensity modulator and a direct-detection MIMO-based receiver," in *Proceedings of the 40th European Conference on Optical Communication*, Cannes, Paper PD4.4.

[53] Chagnon, M., Osman, M., Patel, D., Veerasubramanian, V., Samani, A., and Plant, D. (2015). "1 λ, 6 bits/symbol, 280 and 350 Gbps direct detection transceiver using intensity modulation, polarization multiplexing, and inter-polarization phase modulation," in *Proceedings of the Optical Fiber Communication Conference*, Los Angeles, CA, Th5B.2.

[54] Hoang, T. M., Sowailem, M., Osman, M., Paquet, C., Paquet, S., Woods, I., et al. (2017). "280-Gbps 320-km transmission of polarization-division multiplexed QAM-PAM with Stokes vector receiver," in *Proceedings of the Optical Fiber Communication Conference*, Los Angeles, CA, paper W3B.4.

[55] Che, D., Yuan, F., and Shieh, W. (2017). "200-Gbps polarization-multiplexed DMT using Stokes vector receiver with frequency-domain MIMO," in *Proceedings of the Optical Fiber Communication Conference*, Los Angeles, CA, paper Tu3D.4.

[56] Golub, G. H., and Reinsch, C. (1970). Singular value decomposition and least squares solution. *Numer. Math.* 14, 403–420.

[57] Lowery, A. J., and Armstrong, J. (2006). Orthogonal-frequency-division multiplexing for dispersion compensation of long-haul optical systems. *Opt. Express* 14, 2079–2084.

[58] Peng, W.-R., Wu, X., Feng, K.-M., Arbab, V. R., Shamee, B., Yang, J.-Y., et al. (2009). Spectrally efficient direct-detected OFDM transmission employing an iterative estimation and cancellation technique. *Opt. Express* 17, 9099–9111.

[59] Voelcker, H. (1966). Demodulation of single-sideband signals via envelope detection. *IEEE Trans. Comm. Tech.* 14, 22–30.

[60] Schuster, M., Randel, S., Bunge, C. A., Lee, S. C. J., Breyer, F., Spinnler, B., et al. (2008). "Spectrally efficient compatible single-sideband modulation for OFDM transmission with direct detection," *IEEE Photon. Technol. Lett.*, 20(9), 670-672.

[61] Mecozzi, A., Antonelli, C., and Shtaif, M. (2016). Kramers–Kronig coherent receiver. *Optica* 3, 1220–1227.

[62] Chen, X., Che, D., Li, A., He, J., and Shieh, W. (2013). Signal-carrier interleaved optical OFDM for direct detection optical communication. *Opt. Express* 21, 32501–32507.

[63] Estarán, J., Usuga, M. A., Porto, E., Piels, M., Olmedo, M. I., and Tafur-Monroy, I. (2014). "Quad-polarization transmission for high-capacity IM/DD links," in *Proceedings of the European Conference on Optical Communication*, Cannes, Paper PD4.3.

[64] Che, D., Fang, J., Khodakarami, H., and Shieh, W. (2017) "Polarization multiplexing without wavelength control," in *Proceedings of the European Conference on Optical Communication*, Gothenburg, paper Th1F.5.

15

High-speed Flexible Coherent Optical Transport Network

Tiejun J. Xia and Glenn A. Wellbrock

Verizon, Richardson, TX, USA

15.1 Introduction

In over 30 years, the optical transport network (OTN), the only physical network which actually carries the Internet data from the senders to the receivers, has experienced significant growth in both channel speed and link capacity. During the time period, the network has gone through several major upgrades as technology of optical transmission and networking continuously keeps advancing. In the middle of 1980s, early-stage fiber transmission system emerged as the first equipment for OTN ever introduced to network service providers, i.e., network carriers. The data rate on a pair of strands of fiber at that time was a few hundred Mbps for the system typically [1]. In the later 1980s, synchronous optical network (SONET as an example) began to play its role in carriers' transport networks [2]. SONET systems provided several hundred Mbps capacity per channel at the beginning and quickly moved to supply 10 Gbps per channel (OC-192) in the middle of the 1990s. Then, 40-Gbps channel (OC-768) was introduced to further increase channel capacity. As the required optical channel capacity became even larger when entering the new century, a new OTN standard, OTN, was introduced in the early 2000s. The new standard includes channels with capacities of 2.5 Gbps (OTU1), 10 Gbps (OTU2), 40 Gbps (OTU3), 100 Gbps (OTU4), and beyond [3, 4].

A pair of fibers in OTN usually carried only one optical channel in early days. Silica-based telecom fiber typically has a huge optical bandwidth, 10–50 THz, in the near infrared spectrum, depending on what tolerance of transmission attenuation is counted. An optical channel normally occupies only a tiny optical bandwidth compared with the total useable bandwidth of the fiber. For example, a 10-Gbps OC-192 channel has a only 7-GHz optical bandwidth. Therefore, using a strand of fiber to support just one channel is not economical, particularly in situation where an optical link requires transport capacity much higher than what a channel can provide. Then, the technology of loading multiple optical channels on to a single strand of fiber, wavelength division multiplexing (WDM), was introduced in the middle of 1990s, aiming to increase optical bandwidth utilization [5]. The number of channels, which use different wavelengths, on a strand of fiber has grown from a few at the beginning to about a hundred at the end of the first decade of the new century, when dense WDM (DWDM) became the standard in OTNs. In DWDM systems, channels are spaced at 50 or 100 GHz uniformly, except systems capable of flexible channel [6]. Using DWDM technology dramatically increased fiber capacity by allowing multiple channels traveling on the same fiber simultaneously. As the channel speed increases and with help of the DWDM technology, optical fiber is able to provide an unprecedented data bandwidth which surpasses bandwidths any other transmission media are ever able to provide. For example, a 100-channel DWDM transport system with 100 Gbps per channel can easily support a 10-Tbps total capacity.

Optical channels are added (at its transmitter end) and dropped (at its receiver end) at optical network nodes for loading and unloading data traffic from and to data equipment, such as routers or digital switches. Optical nodes also switch travel directions of channels moving toward their destinations. Introduction of DWDM technology provides opportunity for adding, dropping, or switching channels at optical nodes flexibly. Reconfigurable optical add/drop multiplexer (ROADM) is an optical switch device which controls the travel path of a channel in a transport network. When a channel needs to bypass a node, ROADM directly switches the travel direction of the channel in the optical domain and there is no need to go through a digital switch without involvement of a digital switch, which was a common device to switch channel directions before DWDM time. That means optical channels can skip optical–electrical–optical (oeo) conversion if the contents of the channels do not need to be touched at the node [7]. In recent years, more advanced DWDM technologies have been developed to enrich the features of ROADMs. The transport network becomes even more flexible with the enhanced ROADMs, which can support channels with any wavelength (colorless), switch a channel to any direction of a node (directionless), allow neighboring channels at a add/drop module to have even same wavelengths (contentionless), and support flexible spectral grid (CDC-F).

The CDC-F ROADM will improve the performance of the transport network significantly [8, 9].

The latest major transport network upgrade is the introduction of coherent optical transmission and detection that happened around 2010. Coherent optical technology significantly boosts the capability of the OTN once again. With coherent detection, full signal information of an optical channel, including its amplitude, phase, and polarization, is received and processed. That is a huge contrast with the situation in traditional optical transmission with direction detection, in which only the intensity of the optical signal is received and processed in most cases. With full information of the received signal, a coherent receiver's ability to recover the integrity of original data sent by the transmitter is much more powerful than that of a non-coherent receiver. As a rule of thumb, the performance of a coherent system, considering its data bandwidth capacity and reach distance, is about one order of magnitude better than that of a non-coherent system. Typical commercially available coherent transport systems in the current generation use polarization-multiplexed quadrature phase-shifted keying modulation format (PM-QPSK) for 100-Gbps channels and polarization multiplexed 16-ary quadrature amplitude modulation (PM-16QAM) format for 200-Gbps channels with an approximately 32-GBd symbol rate [10, 11].

In the next stage, OTN will become more powerful with the assistance of several new technologies. Higher orders of modulation and higher symbol rates are examples of the new technologies. Channels with higher capacity, such as 400 Gbps or more, will be available to accommodate data ports with a port speed of 400 gigabit Ethernet (GbE) or higher. Photonic integration technology continues to cut power consumption, reduce footprint, and increase front panel port density. Using higher symbol rates will significantly cut the number of components needed in a system so as its cost. Agnostic digital switches embedded in transport equipment will make the original transport layer (Layer 0 or 1) to be able to perform upper layer's functions, such as switching data packets. The unified multi-layer control plane will optimize resource utilization of the network by orchestrating network assets in various layers and make the network more reliable. By integrating all the new technologies together, a unified transport network will serve the Internet data demand with a high growth rate well in the near future. Figure 15.1 shows the evolution of service providers' OTN.

The very purpose to continuously improve OTN is to adequately serve ever growing Internet traffic demands. Every year new Internet-related applications drive users to consume more data bandwidths. One indicator

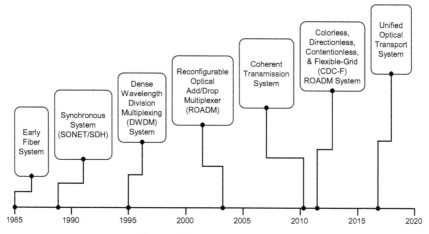

Figure 15.1   Evolution of OTN.

of required transport network capacity increase is the growth of end-user demand for data services. Figure 15.2a shows the end-user demand of data traffic in North America since the very beginning of the Internet [12–23]. The demand has increased for almost 10 orders of magnitude. Before 1995, the Internet was mainly used by researchers in academic institutes, so the total traffic amount was very tiny. After the Internet became a publicly accessible resource, it has undergone a tremendous expansion. For example, the annual traffic growth rate for both 1995 and 1996 was more than 800%! (see Figure 15.2b). Emails, website surfing, document sharing, e-commerce, enterprise data sharing, video uploading and downloading, and wireless data applications have been endlessly pushing transport networks to provide more capacity. Looking forward, new applications such as Internet of Things (IoT), virtual reality, smart cities, datacenter (DC) connections, and 5G wireless services [24–26] will no doubt require even higher channel speeds and fiber capacities in the near future.

Interconnection traffic index [27] is another indicator of Internet traffic growth. As the end-user demands grow, traffic passing through network interconnection increases accordingly. Figure 15.3 shows the latest result of an analysis of the growth of North America installed bandwidth capacity for interconnection from 2016 to 2020. Its annual growth rate is about 40%.

The transport network infrastructure, through which almost all the Internet data are delivered, is undergoing a deep makeover driven by various new applications, advanced with new technologies, and pressured by cost

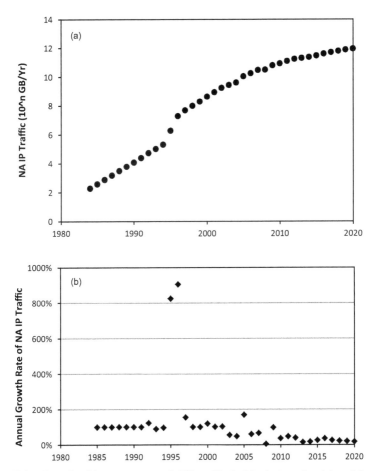

Figure 15.2 Growth of internet protocol (IP) traffic in North America (a) and its annual growth rates (b) [12–23].

reduction requirements. No matter how the infrastructure changes, however, the OTN must continue to provide (1) high enough link capacity to support overall data traffic between network nodes and (2) large enough transport channels to accept/deliver data streams from/to interfaces of data equipment (routers, switches, etc.). Therefore, maintaining the transport network to be able to always meet the demands of Internet's ever growing traffic is crucial. This chapter focuses on discussion about how the coherent OTN will meet or surpass this requirement.

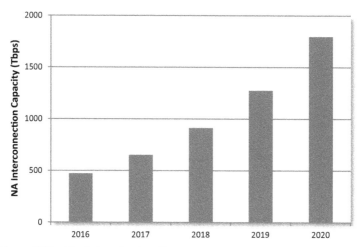

Figure 15.3   North America installed interconnection bandwidth capacity [27].

15.2 Why Optical Coherent Transmission?

Up to later 2000s, most optical channels in transport network were modulated with the simplest format, on–off keying (OOK), and most channels were running at a 10-Gbps channel speed or lower. As the available optical bandwidth on fiber seemed unlimited, people had the luxury to just implement the simple scheme yet were able to meet data traffic demands then. As the channel speed was increasing to cope with larger data flow required to support data traffic growth, however, the capacity supported by the 10-Gbps OOK channels was no longer enough. One approach to increase the channel speed is to increase the signal symbol rate of a channel. 40-Gbps channels based on 40-GBd OOK format [28] and 100-Gbps channels based on 100-GBd OOK format [29] were tested to increase the channel speed for four folds and ten folds than that of a 10-Gbps channel. The 40-Gbps OOK approach resulted in a few commercial products but the 100-Gbps OOK approach has never been commercialized. The reason why increasing channel speed by directly increasing the signal symbol rates is not an easy way to increase transport capacity is that the working bandwidth of commercially available semiconductor-based optical components, as the key building blocks of optical transport equipment, is approaching its limit. Therefore, when designing optical channels with a data rate higher than 10 Gbps, people began to consider more sophisticated modulation formats, rather than sticking to the simplest OOK format. Using various multi-level modulation formats is the main direction that was considered to increase channel capacity without increasing symbol rates. High-order modulation formats can provide high data rates with limited component bandwidths, as each symbol is able to carry multiple data bits. For example, a host of DWDM differential QPSK 40-Gbps channels successfully traveled over 4,500-km standard single mode fiber (SSMF) [30]. In this experiment, each QPSK symbol represents two data bits, and therefore, the symbol rate of each channel is 20 GBd, which is half of the data rate.

Two approaches are commonly used to increase the number of bits of a modulated symbol: pulse-amplitude modulation (PAM) and quadrature amplitude modulation (QAM). PAM signals are modulated only with their amplitudes while QAM signals are modulated with both their amplitudes and phases. An OOK signal is the simplest PAM signal, which only has two amplitude levels: zero and one. A PMA-4 signal has four equally spaced amplitude levels [31, 32]. For PAM signals, only the signal intensity needs to be detected at the receiver. The DQPSK signal mentioned above is an example of QAM signals. The modulation of a QAM signal is represented by a constellation diagram, dots in which represent the amplitude and phase of the signal. A PAM signal can be detected directly for its intensity without concerning of its phase. Direct detection is used for differential QAM signals, for example, the DQPSK signal, as well. Since only the phase differences between adjacent symbols need to be measured in differential QAM signals, usually a passive delay-and-interference device installed at the receiver is used to convert the phase difference into intensity variations, which are measured by photodetectors [33, 34]. Modulation with orders higher than PAM-4 or QPSK can potentially further increase the number of data bits a symbol presents. However, direct detection is not the best way to detect phase-modulated signals, especially for high-order and sophisticated phase-modulated signals. Hence, better detection technology needs to be developed to improve performance of optical fiber transmission.

It is certainly helpful to obtain information of a received optical signal as complete as possible since more information helps to better recover the original signal sent by the transmitter. The phase information of an optical signal gets lost when it is received with direct detection. To reserve both amplitude and phase information of an optical signal, coherent detection is needed [35]. In coherent detection, an optical continuous-wave signal from a local laser, working as a local optical oscillator, is mixed with the incoming data signal in an interferometer. Optical intensities at the output ports of the interferometer reveal the value of the amplitude of the incoming signal as well as its phase relative to the phase of the local laser. Since photodetectors can measure only the intensity of an optical wave, not its phase, the phase information of the incoming signal must be measured indirectly with the help of the interferometer.

The optical phase detection theoretically is similar to that in radio frequency (RF) communications. To retrieve the phase information of an incoming RF signal a phase-locked loop (PLL) must be established between a local RF oscillator and the incoming signal. This scheme, however, is not practical for optical coherent detection. The optical wave used in optical communications has a frequency about 200 THz. There is no commercial electronic PLL that is fast enough to lock the phase of a local laser to that incoming optical data signal. It is possible to lock the phase of a local laser to that of the laser in a remote transmitter optically but the method is not practical to commercial products [36]. To deal with the difficulty of the optical PLL, instead of attempting to lock the phase of a local laser to that of a transmitted signal, the local laser in a coherent receiver runs freely, as long as its optical frequency is very close to that of the transmitted signal. Since the two frequencies are very close, beating between the two signals in the interferometer varies slowly and can be detected. The slow beating can be processed and phase relationship between the local laser and the incoming signal can be determined for any particular moment. In this way, even though the local laser is not locked to the incoming signal, the detected phase relationship variation plays a similar role to an optical PLL. By including the slow variation in signal processing in a receiver, the phase information of the incoming signal can then be determined. If the frequency of the local laser is significantly different from that of the incoming signal, the beating will change too fast to be detected. This characteristic of coherent detection has an important merit. When multiple DWDM optical waves arrive at a coherent receiver, its detector picks up only the channel with a frequency very close to its local laser and ignores all other channels as noises. That means for a DWDM transport link, a coherent receiver does not need an optical filter in front of the detectors to receive a particular channel, since all non-relevant channels are ignored automatically. With coherent detection, higher order modulation formats, which pack more data bits into a signal symbol, can have similar performance to that with lower orders of modulation formats by non-coherent detection. That can support high data rates without increasing symbol rates when limited component working bandwidth is a constraint. For example, an 8QAM symbol can represent three data bits while a 32QAM symbol five data bits.

Another dimension of optical signal propagation in fiber, polarization, can be considered to further increase the channel capacity. Two optical signal streams traveling in fiber, with the same optical frequency, may ride on the other independent polarization state without problem. Traditionally, optical channels use only one polarization. For example, OC-192 channels usually use only one polarization mode of fiber. Coherent transmission makes it feasible to use both polarization modes to carry different data in the same channel in a standard, non-polarized fiber. Two polarization multiplexed (PM) data streams are detected at two orthogonal polarization states. Even though the two polarization states at the receiver normally do not align with the two original polarization states at the transmitter, DSP in the receiver is able to rotate polarization states in the digital domain and separate the two data streams [37].

Using both polarization modes immediately doubles the capacity of a channel, regardless what order of its modulation format is. For instance, a PM-QPSK 100-Gbps channel needs only a 25-GBd symbol rate (without considering overhead), as a QPSK symbol represents two data bits and using both polarization modes leads to four data bits represented by the same symbol. Figure 15.4 shows a comparison of direct detection and coherent detection for an exemplary 100-Gbps channel with certain overhead, where the 100-Gbps OOK non-coherent channel has a 107-GBd symbol rate since each symbol represents only one data bit, while the 100-Gbps PM-QPSK coherent channel has a 32-GBd symbol rate as each symbol carries four data bits.

With coherent detection, the full information of an optical signal, including the amplitude, phase, and polarization of its electromagnetic (EM) field, is detected and received. During its propagation from the transmitter to the receiver, an optical signal is inevitably impacted with linear and non-linear influences imposed by the fiber, the elements in the transmission line, and the interaction between the signal and the fiber. The level of the impact normally depends on how far an optical travels from the transmitter to the receiver in the fiber, which ranges from a few feet to thousands of miles. The impact causes signal impairments and reduces the recoverability of the signal at the receiver. Therefore, the influence of transmission line accumulated during signal propagation must be removed or mitigated so the data carried by the signal can be recovered. The linear influences, such as chromatic dispersion (CD) or polarization mode dispersion (PMD), are only related to fiber characteristics; they can be treated as a lump sum effect at the end of the propagation. Since the full EM field of the received signal is known, the lump

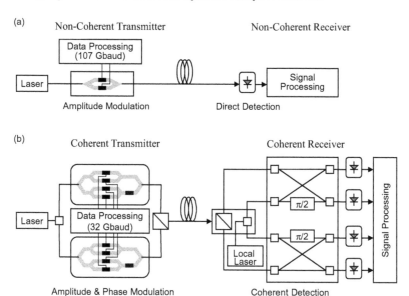

Figure 15.4 Comparison of direct detection and coherent detection for exemplary a 100-Gbps channel.

summed linear effects can be removed with powerful digital signal processing (DSP), without concern about the detail how the linear effect is generated at each individual fiber section along the transmission line. Therefore, if linear influences are the only adverse effects, then with coherent detection, the reach of an optical channel would purely depend on the noise level at the receiver that is characterized by the optical signal-to-noise ratio (OSNR). The random noise produced by optical power amplifiers along the transmission line is hard to compensate and mitigate. The more the amplifiers in the transmission line, the more the accumulated random noise. The reach of a signal would be determined by a point where the signal could no longer be recovered from the noise. In conventional non-coherent transport systems, because only the signal intensity, not its full EM field, is received, linear impact cannot be cleared with DSP. Therefore, the linear effects are usually compensated with additional optical devices, such as CD compensators or PMD compensators. One benefit of coherent detection is that these linear compensators are no longer needed. By removing CD compensators, the signal travel latency caused by the compensators is eliminated and signal delay decreases [38, 39].

Besides linear effects, non-linear effects of a transmission line must be mitigated as well. Removing non-linear impact on optical signal is more complicated than removing linear impact. The strength of non-linear effects on an optical channel at a particular section of the transmission link depends not only on fiber characteristics, such as CD, PMD, and other parameters, but also the amplitude, phase, and polarization of the channel and its neighboring channels traveling in the same fiber. The total non-linear effect to an optical signal at the receiver is an accumulated result of the non-linear effect happening at each transmission section. If all characteristics of the fiber link are known, the properties of a received channel can be predicted with a numerical calculation with a non-linear optical fiber propagation equation also involving all other channels in the fiber, theoretically. The problem is that the full characteristics of a fiber link are usually not stable quantities and that of the neighboring channels are usually unknown. That makes accurate calculation nearly impossible. Therefore, non-linear effects are very difficult to completely compensate. Various methods have been proposed and tested to compensate for non-linear effects as much as possible [40]. Back propagation is a widely used technique to remove intra-channel non-linear effect. A received optical signal is used as the initial condition for a numerical calculation, in which the received signal propagates in the same fiber link backward to the transmitter to discover the original data pattern [41]. Other methods, such as using a pilot tone, are used to estimate and remove non-linear effects generated by neighboring channels [42]. After the introduction of coherent detection, noise generated by optical amplifiers and fiber non-linear effects are the two major hurdles left for further improving optical transmission performance.

It must be noted that when loaded onto two orthogonal polarization modes, two individual data streams with the same optical frequency actually form only one optical wave in the fiber. The resulted polarization state of each symbol of the wave is determined by the phase relationship between two corresponding symbols originally coming from the two data streams. One can imagine that when checking the polarization state of the combined wave, the polarization state jumps from symbol to symbol. That leads to an average degree of polarization (DOP) of a channel to be zero. Since the strength of non-linear effects from neighboring channels is polarization dependent, the zero-averaged DOP helps to reduce random impact from neighboring channels. In short period of time, however, the DOP of a neighboring channel can still have a non-zero value, depending on data patterns. To reduce the impact of the short period non-zero DOP effect, a multi-dimensional data recoding has been proposed to minimize the polarization dependence of other channels in a DWDM coherent system [43].

Compared with direct detection, coherent detection makes optical channels much more resilient to fiber impairments, caused by both linear and non-linear effects of the fiber. That leads to much higher fiber capacity and much better reach distance compared with previous generation transport equipment. It is expected that most optical channels for transport networks will use coherent detection soon, except channels serving very short distances. Various orders of modulation formats have been considered in channel designs. Right now, many deployed transport networks use PM-QPSK for 100-Gbps channels and PM-16QAM for 200-Gbps channels [44, 45].

Just like the traditional transport system, the coherent transport system supports DWDM channel arrangements as well. Thanks to high-order modulation formats and polarization multiplexing, spectral efficiency, which measures how much data are packed into a given optical bandwidth, is much higher in coherent systems than that in traditional direct detection systems. As a comparison, an OC-192 channel (10 Gbps) normally occupies a 50-GHz optical bandwidth in commercial systems. The same optical bandwidth is used to house a 100-Gbps PM-QPSK channel in coherent commercial systems, leading to a 10x spectral efficiency enhancement. High spectral efficiency contributes to high fiber capacity directly. A DWDM coherent link readily supports 10 or 20 Tbps of fiber capacity when 100 channels of PM-QPSK or PM-16QAM are installed for long haul distances or regional/metro distances [46, 47].

15.3 Optical Transport Network with Coherent Transmission

Using coherent detection is not the only significant improvement of OTN in recent years. The industry has also introduced another important technology to advance the transport network: fully flexible reconfiguration of traveling paths of channels. The conventional ROADM has been used to provide some flexibility in the optical layer, such as switching the traveling direction of channels at nodes to avoid unnecessary oeo conversion or to control a channel whether it needs to be terminated at a node or pass through it. The functions of the conventional ROADM, however, are limited. It may only be able to add a channel from a particular assigned port in its add/drop module based on the wavelength of the channel. A channel from an add port may only be able to exit the node at a particular direction. From a perspective of transport network management, ideally, an optical channel should be able to be assigned with any color (wavelength) and any traveling path to reach its destination if necessary. Also, the channel management should be automatically controlled by software without human intervening. Therefore, the limitation of conventional ROADMs should be removed to reach the goal of fully automatic transport network management [8].

To reach the goal of full flexibility for OTN control, the concept of "colorless, directionless, and contentionless (CDC)" has been introduced to improve the performance of ROADMs [9]. Wavelength-tunable transmitters are available, so the color of a channel can be adjusted to meet the requirement of a particular networking circumstance. As the wavelength of a channel is no longer a fixed value, an add/drop port of a ROADM must be able to accommodate a channel with any possible wavelength, i.e., the port must be "colorless." Also, since reconfiguration of channels in a transport network is inevitable due to traffic pattern changes, an optical node should be able to switch a channel initiated from one of its add/drop modules to any direction facing the network. In other words, a ROADM should provide a "directionless" service to channels. In addition, another limitation must be removed. Channels connected to an add/drop module of a conventional ROADM are forbidden to have the same color since wavelength contention happens in the module when two or more channels have the same color. This constraint is overcome in a "contentionless" ROADM. The CDC ROADM is the basic enabler of full flexibility at optical layer for transport network management.

Several technologies have helped to make the CDC ROADM a reality. The wavelength selective switch (WSS), based on liquid crystal (LC) or LC on silicon technology, is a device which is able to select channels according to wavelength flexibly. The multi-cast switch (MCS), mainly based on planar lightwave circuit technology, is a device which is able to directionlessly and contentionlessly connect add/drop ports and WSSs in the CDC ROADM. Normally add/drop units of a CDC ROADM are modularized and each ROADM has a host of add/drop modules. Inside an MCS, all channels dropped to a particular add/drop module are broadcasted to all drop ports of the module. The beauty of coherent detection is that the local laser in the receiver of a transponder, which is attached to a particular add/drop port, only picks up the channel which wavelength is very close to that of the local laser and ignores all other channels in the broadcasted signals and a tunable filter in front of the detector for selecting a channel is not needed, as mentioned

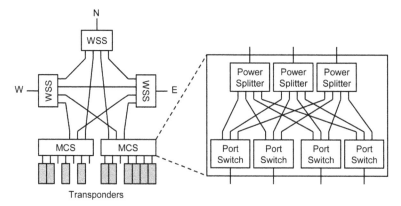

Figure 15.5   A typical design of a CDC ROADM.

in the previous section. Figure 15.5 shows a schematic diagram of a typical CDC ROADM design.

While the current coherent transport network serves data ports on routers or switches with port speeds up to 100 GE, the transport network is preparing to support a new data port speed standard of 400 GbE and beyond. Even though a single-carrier PM-QPSK 100-Gbps coherent channel is able to reach almost any distance, a channel with a higher data rate but still using a single optical carrier is expected to have limitation in reach distance. To increase the data rate of an optical channel with a single carrier, one must raise its modulation level if its baud rate is kept the same. The higher a channel's modulation level, the higher its OSNR is required at the receiver, leading to a shorter traveling distance. One approach to increase the data rate of a channel without scarifying its reach distance is to use multiple optical carriers for the channel. Since optical channels in conventional transport networks almost exclusively use only one optical carrier, the multi-carrier optical channel is called "super-channels" [48, 49]. For example, a 400-Gbps super-channel, which is able to carry 400GbE data as its payload, can be constructed with four optical carriers. Each optical carrier is modulated with PM-QPSK and provides 100-Gbps capacity. The 400-Gbps super-channel should be able to reach the same distance as a 100-Gbps PM-QPSK channel does.

It must be noted that introducing super-channels is a turning point in transport network history in terms of improvement of spectral efficiency when a higher data rate is introduced. Spectral efficiency (SE) is a parameter representing how much capacity a given optical spectrum can provide, which is directly related to how much capacity a pair of strands of fiber can support.

Let's use a standard 50-GHz channel spacing as an example. If the channel occupying the space has a data rate of 10 Gbps, for example, an OC-192 channel, then the SE is 0.2 b/s/Hz. If the channel data rate is enhanced to 40 Gbps, the SE is also enhanced for four times. If the channel data rate becomes 100 Gbps, the SE gets another 2.5x gain. When the channel data rate is further increased to 400 Gbps with a super-channel configuration as described above, however, the SE does not improve at all if all four optical carriers keep 50 GHz as carrier spacing. This fact reflects that the capacity of the SSMF is approaching its limit, if no feasible new technologies to be developed.

Introduction of super-channel, though, does provide a new opportunity for further improving spectral efficiency slightly. Since all optical carriers belong to a super-channel have the same traveling path, they will be switched altogether as a group at network nodes. Therefore, no guard band, normally reserved for switching individual channels, needs to be reserved between the carriers. The saved guard bands offer extra spectral bandwidth for transporting more data. It is estimated that 20–30% of SE improvement is realistic when super-channels are deployed. Figure 15.6 shows an exemplary comparison of four standard 100-Gbps channels and a four-carrier 400-Gbps super-channel. Each of the 100-Gbps channels occupies a 50-GHz bandwidth and the central frequency of each channel is assigned to a 50-GHz grid. The 400-Gbps channel occupies a 150-GHz bandwidth, no guard band is reserved between the carriers, and the central frequency of the super-channel is assigned to 193.3625 THz, which is not on a commonly seen DWDM grid. The 400-Gbps super-channel has an SE of 2.7 b/s/Hz while the 100-Gbps channel has an SE of 2 b/s/Hz.

Removing guard bands between optical carriers in super-channels to further increase the SE brings in new challenges for the transport network. As shown in Figure 15.6, since the central frequency may not sit on a conventional spectral grid, ROADMs must have a flexible grid to support super-channels. In conventional transport networks, the practical channel spacing is 50 GHz or its multiple times (except in submarine systems) and the central wavelength of each channel is locked on a 50-GHz grid. Introduction of super-channels means that the channel spacing can no longer be a fixed value and channels can no longer be assigned to a conventional grid. The spectral bandwidth of a super-channel will depend on the number of optical carriers it contains and the modulated bandwidth of the carriers. In short, ROADMs must support flexible spectral arrangement to accommodate super-channels. That imposes another requirement to ROADMs: flexible grid, in

(a) Four 100-Gbps channels   Guard Band

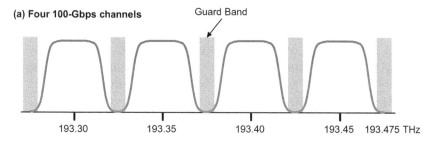

(b) A 400-Gbps super-channel

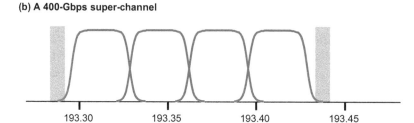

Figure 15.6   Exemplary spectra of four standard 100-Gbps channels and a 400-Gbps super-channel.

addition to the CDC requirements. With the capability of flexible grid-added CDC ROADMs, they are then called CDC-F ROADM [8].

15.4 What's Next for Optical Transport Network?

Until a few years ago, the development of OTN technologies had been mainly focused on enhancing channel speeds and fiber capacities to meet the fast growing Internet data traffic demands. Since C-band and L-band have the lowest optical attenuation in fiber and mature optical amplifiers, most transport equipment has been designed to use the two bands. As the industry anticipated that the practical data capacity of single-core/single-mode fiber, which had been widely used in transport networks, was approaching its limit, the approach of space-division multiplexing (SDM) was introduced and has been extensively studied since. Multi-core fiber [50] and multi-mode fiber [51] developed for transport network have shown their potential to further increase the data capacity of a single strand of fiber by 10–100 times. Figure 15.7 shows the record capacity thru the years in transport technology development [52–75]. The highest capacity reported so far for

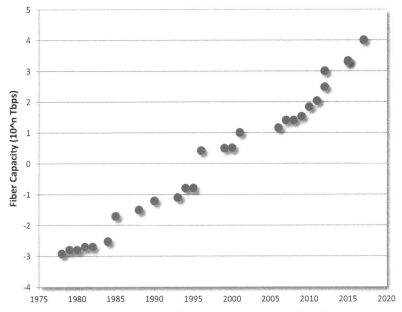

Figure 15.7 *Record fiber capacity along years [52–75].*

single-core/single-mode fiber is about 70 Tbps [69] and that for SDM enabled fiber is more than 10 Pbps [75].

The next stage of transport network development, however, is higher capacity; smaller footprint, easier control, and lower cost are high priorities. To increase the capacity, using higher orders of modulation format is a natural choice. Modulators used in coherent transport equipment are able to generate various orders of modulation formats with desired amplitudes and phases for optical signals. For example, 538-Gbps PM-64QAM single-carrier channels are tested to obtain high SE [76]. Of course, the higher a modulation order, the higher the required OSNR at a receiver. Unless new amplifiers with lower noise, such as phase sensitive amplifier [77], become available in the near future, higher order modulation schemes most likely will serve shorter distance transmission or help lower order modulation setups to optimize their transmission performance, such as in the case of constellation shaping [78].

Even though 32 GBd is viewed as a norm symbol rate in currently installed coherent transport networks, higher symbol rates are coming. A doubled symbol rate, 64 GBd, and other symbol rate settings between the two have been considered for the next-generation transport equipment development. The main benefit by raising symbol rates is to reduce the number of components required in the system. That will result in reduced equipment size, power consumption, and cost. A transport platform equipped with multiple symbol rates also provides flexibility in channel setup and for network management, as the symbol rate is able to be adjusted to best fit a particular networking circumstance. For example, a 400-Gbps coherent channel can be a super-channel with four carriers, each of which is modulated at 32 GBd with PM-QPSK. It can also be a super-channel with two carriers, each of which is modulated at 32GBd with PM-16QAM. It yet can be an ordinary channel with only one carrier, which is modulated at 64 GBd with PM-16QAM [79]. The option of multiple baud rates provides flexibility for path adaptive transmission as well, in which a channel's data rate reflects the condition of its fiber path [80–83]. It must be pointed out, though, increasing symbol rate usually does not help SE or fiber capacity enhancement. Even though the number of optical carriers needed for a super-channel decreases as the symbol rate of each carrier increases, the modulated bandwidth of each carrier is broadened as well, and therefore, the resulted spectral efficiency has almost no change.

The concept of super-channel has been used for channels with data rates higher than 100 Gbps, although the channels may be able to use just a single optical carrier in certain circumstances. As super-channels contain multiple optical carriers, when the data from a client port are loaded into a transport channel, they must be redistributed onto different carriers anyway, so that it seems reasonable that the capacity of a transport channel no longer needs to be strictly linked to that of the client port, contrast to a situation that the capacity of a transport channel is the same as that of a client port in a conventional transponder. Therefore, the capacity of transport channel does not need to be locked with client port speed settings. For example, when the client port data rate is 400 GbE, the line side may use a 400-Gbps channel or a 600-Gbps super-channel to carry the client data. The digital switch silicon in transport node allows the data to be redistributed.

Traditional digital switching fabrics in optical nodes are mainly for mapping and aggregating data streams from client ports into transport channels. TDM data switching has been performed by separate digital cross-connect equipment. Packet data switching is performed by separate Ethernet or MPLS switch equipment. In the next-generation transport system, however, optical nodes will have much stronger digital switching capability. Multi-purpose switch fabrics housed in transport gear will perform both TDM and packet data switching, in addition to mapping and aggregation. Figure 15.8 shows a

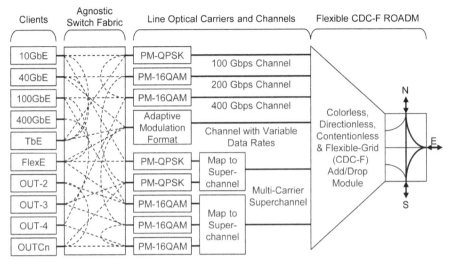

Figure 15.8 Example of the next-generation optical transport node design [84].

schematic diagram of the next-generation optical node design, where CDC-F ROADM and agnostic switch silicon are the two key building blocks [84].

A telecom network is a two-layer structure in the simplest view: the data layer and the transport layer. Traditionally, the two layers are independent and each has its own management system. One trend of transport network development is to merge the control of the two layers and strengthen communications between the two layers, so they will work together more closely. A unified control platform communicates with both layers and allocates resources in both layers more efficiently. That makes networks more easily controlled by software and is the essence of the software defined network (SDN) concept [85]. Software automatically changes network settings based on clients' demands and network conditions. For example, the data rate, modulation order, and symbol rate of an optical channel will be "elastic" to meet data transport demand timely. The concept of elastic optical network (EON) is under heavy investigation now [86].

15.5 Coherent Transport Technology Development by Network Operators

Network operators or network service providers are customers and users of network equipment. A lot of time they also actively participate in new technology development by working closely with their equipment and system vendors. Before introducing new technology into networks, service providers or operators often conduct field trials to verify the performance of prototypes, which are built on the new technology. Field trials are carried out in real network environment, which may be different from that in the R&D labs of equipment developers. Field trials of transport equipment are particularly valuable since fiber infrastructure usually presents more challenges than other network infrastructure does. Field trials give service providers realistic expectation for the performance of new systems, which will be deployed in the network soon. Issues found in trials are sent back to equipment vendors for product improvement [87]. The following are a few examples of field trials for coherent transport networks in past several years.

One of the benefits of coherent transport systems is that they are highly resilient to PMD. In an early coherent transmission field trial, a high data rate channel survives fiber routes with very high PMD values. It is demonstrated that the transmission remains error free (with FEC) for 107-ps instantaneous DGD in a 73-km field fiber route located in the east Texas region. The result shows that the dual subcarrier PM-QPSK channel exhibits much higher PMD tolerance than 10G OOK channels. The experiment also verified the ability of coherent channels to characterize fiber DGD values accurately to a first order. The combination of PM-QPSK, coherent receiver, and DSP shows the potential to mitigate PMD and CD effectively for networks that need to support high data rates [88]. A similar technology trial with a dual carrier 100G coherent channel over field fiber link of 769 km in a UK network shows that the coherent 100G is attractive for dispersion compensation module-free WDM systems. Its high PMD tolerance makes older fiber in outside plant useable again [89].

The first field trial of a single-carrier 100G coherent channel was reported in 2008 [90, 91]. In the trial transmission of a 111-Gbps channel combined with two 43-Gbps channels and eight 10.7-Gbps channels on a 50-GHz grid, over 1,040-km field fiber and two ROADMs was demonstrated. No Raman amplification was utilized. The interplay between the channels with different data rates has been analyzed. The performance of a 111-Gbps channel is optimized by carefully choosing the power levels of its neighboring channels. PMD tolerance tests on the 111-Gbps channel are performed by adding lumped multi-order PMD emulators in the beginning and the end of the transmission line. The results show that the 111-Gbps channel is able to sustain at least 23-ps mean DGD for error-free operation, owing to strong digital processing in the coherent receiver. The trial results confirmed the suitability of 111-Gbps PM-QPSK for multi-rate operation in existing systems on presently deployed fiber infrastructures.

Since the purpose of the transport network is to serve data traffic, it is important to demonstrate that high-speed coherent channels can support data traffic end to end. A field experiment just for this purpose was conducted in 2010. The trial demonstrates that an end-to-end 100G transport of native IP packet traffic over 1520-km field deployed fiber is realized with multi-suppliers' 112-Gbps single-carrier real-time coherent PM-QPSK DWDM transponder, 100GE router cards, and 100G CFP interfaces for the first time. The trial shows the feasibility of interoperability between multi-suppliers' equipment for 100G transport. This field trial, which fully emulated a practical near-term deployment scenario, confirmed that all key components needed for deployment of 100GbE technology were maturing [92]. Another field trial with a similar purpose was reported in the same year that a real-time single-wavelength coherent 100G PM-QPSK channel upgrade of a field system was demonstrated. In the trial, an error-free operation with FEC was achieved over installed 900-km and 1,800-km fiber links [93].

400G and 1T are two important channel speeds which match data port rates of 400GbE and potential TbE in the near future. To prepare supporting 400GbE and TbE, a field trial was conducted with 400G and 1T coherent channels with 100G neighboring channels. It is the first field experiment with mixed line-rate transmission of 100 Gbps, 450 Gbps, and 1.15 Tbps performed over a record distance of 3,560-km field fiber with EDFAs only. Both the 1.15-Tbps channel and the 450-Gbps channel are multi-subcarrier super-channels generated using all-optical OFDM technique with PM-QPSK modulation of each subcarrier. A filter-free coherent receiver was used in the trial to select the channel of interest by tuning the local oscillator frequency. This trial proves that super-channels are able to provide higher channel capacity and higher spectral efficiency than the existing 100-Gbps system without sacrificing the required long haul distance. A 1.1-Tbps super-channel with DP-16QAM modulation format is also tested in the trial as well to prove that higher spectral efficiency is possible for shorter distance requirements using the same line system [94, 95].

Demonstrating feasible high fiber capacity is one of main goals of field trials. One of such tests has set a record fiber capacity by using only C-band for terrestrial backbone networks. The team successfully performed a 21.7-Tbps field fiber transmission experiment. A total of 22 optical super-channels with a flexible band WDM and 1,503 km of field installed fiber with EDFAs and Raman amplifiers were used. All channels are multi-subcarrier super-channels generated using the all-optical OFDM technique with PM-8QAM or PM-QPSK modulation at each subcarrier. A novel dynamic modulation format selector is also used to combat uneven OSNR distribution within the transmitted spectrum. A coherent receiver was used to select the subcarriers of interest by tuning the local oscillator frequency. This trial proves that it is feasible to transport more than 20 Tbps on an installed standard fiber infrastructure at long haul distances [96, 97].

Using both C- and L-band, fiber capacity can be doubled compared with the situation where only C-band is used. Another field experiment just shows that the team successfully performed the high capacity field transmission trials with both C- and L-band: 40.5 Tbps for a long haul distance of 1,822 km and 54.2 Tbps for a regional distance of 634 km. Different modulation formats are used based on transmission conditions. PM-8QAM Nyquist carrier modulation is used for long haul distance and PM-16QAM is employed for regional distance. Adaptive FEC overheads are applied to guarantee every super-channel has achieved the BER values bellow HD-FEC threshold while the maximum channel capacity is obtained. The results are the highest field capacities for long-haul and regional distances in terrestrial networks achieved so far. These trials prove that it is feasible to have the installed fiber infrastructures support over 40 Tbps for national backbone networks and near 60 Tbps for regional/metro networks [98, 99].

It is worthwhile to investigate if PM-16QAM is able to support long-haul distance in the deployed fiber network as PM-QPSK is able to. A field trial for this purpose was reported several years ago. In the trial, a transmission experiment of eight dual-carrier 400G PM-16QAM channels with 50 GHz over a long haul distance of 1,500 km in field with aged fiber was successfully conducted. All-distributed Raman amplification and high-gain FEC codes play the key roles in this test. The result shows that PM-16QAM signals can be supported by existing long haul fiber networks by reducing OSNR degradation during signal propagation and by increasing FEC coding gain. This experiment demonstrates that to double the fiber capacity and spectral efficiency of the current industry norm of carriers' long haul network is feasible by using PM-16QAM signals [100].

Besides field trials for capacity enhancement mainly, network service providers and network operators also conduct other field trials to verify and demonstrate new technologies, which are continuously reshaping transport networks.

In a high symbol rate field trial, a super-channel with 28 optical carriers is constructed. Each carrier is modulated at 64 GBd with a PM-QPSK format. The super-channel traveled for nearly 400 km in field [101]. The experiment proves that 64-GBd transmission is mature enough to be used in the transport network. Modulation formats higher than 16QAM have been tested in real networks as well. In one field trial, 64QAM and 36QAM, in addition to 16QAM, have been considered. A super-channel with four optical carriers, each of which is modulated with the formats, is studied. Various symbol rates are used to search for optimal conditions in terms of reach distance and spectral efficiency. Different constellation shaping is applied to adjust bit distributions on the constellation diagram. The study results show that shaped constellation distributions give better spectral efficiencies than that obtained with uniform distributions for given reach distances [102]. A submarine field trial with constellation shaped 64QAM to increase spectral efficiency is also reported [103].

Unified control of multiple layers including transport network is another meaningful topic for field trials. In one experiment, DC operation is able to request inter-DC light paths dynamically from transport network operators via an orchestration platform to meet its traffic demands timely. The dynamic nature of the operation provides a lot of benefits to network service providers and their clients, such as to release and reassign unused bandwidth in time, flatten network load peaks, and improve the overall network resource utilization. The trial also proves that the SDN-based network control reduces lead-time to provide services to customers [104]. In another reported field trial, management of 400GbE via long-distance 400-Gbps transport lines is tested. The data service can be created, terminated, or rerouted with an SDN-based control plane [105]. The concept of EON has been experimented in field as well. The trial shows optical path with a required bandwidth can be set up in a multi-vendor, multi-domain network with bandwidth variable transponders. Interoperability between different vendors' transport equipment is successfully demonstrated [106].

The field trials discussed above reflect network service providers and network operators' continuous efforts to improve transport networks to meet the ever growing demands of data traffics.

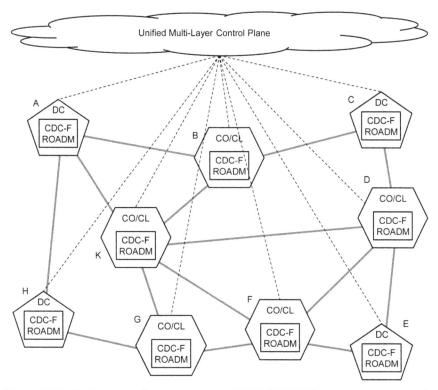

Figure 15.9   Architecture of next-generation CDC-F ROADM-based transport networks. DC – data center, CO – central office, and CL – co-location.

15.6 Datacenter Connections and Coherent Transport Networks

As data traffic continues growing with high growth rates, DCs are being built almost everywhere. For DC connections, the coherent transport network mainly serves inert-DC connections. The inter-DC connection has become one of the most important data networking services that telecom carriers provide, while large DC operators have also built transport links to connect their DCs directly. Figure 15.9 shows a typical service provider's transport network which connects DCs and other telecom facilities, such as central offices (COs) and colocations (CLs). CDC-F ROADM-based mesh architecture has become an established foundation of the next-generation transport network [8].

Mega Internet service companies require very high-capacity but dedicated simple point-to-point optical links between their DCs in metropolitan areas, bypassing multi-purposed transport network to save costs. Based on the typical link distance in metro areas and the predicted dominant router port capacity facing externally in the near future, a new interface standard for the very need, 400ZR, has been proposed in 2016 [107, 108]. It will be a single-carrier coherent 400-Gbps DWDM channel with about 60-GBd symbol rate and PM-16QAM modulation format. The reach distance is up to 120 km. Routers in different locations can be directly connected with the device. If a 100-GHz channel spacing is considered, a pair of fibers is able to support up to 20-Tbps capacity in the C-band for the direct DC–DC connection. For network service providers, the 400ZR interface can also serve high-capacity connections from CO to customer premises or small hub locations without the need of installing ROADM equipment at the remote location.

Unified network control plane, as shown in Figure 15.9, collects service request of applications and critical information of both data networking layer and transport networking layer. It orchestrates resources in the layers based on service demands and optimizes network settings for end-to-end data transportation, including that between DCs. One of the main advantages of implementing a unified management platform is that network service providers are able to respond to data traffic pattern changes quickly. The control software can quickly set up new links, turn off links no longer in use, and reassign idle capacity to parties who request more capacity.

The idea to use a unified control plane is not new, though, since its benefits are obvious. The reason that the control of transport network has not been quickly integrated into a software-based grand control plane is due to the special characteristics of transport networks. Unlike the data networking layer, where connections between network nodes are just virtual links, connections between transport nodes are tangible fiber links. Optical signals literally travel from one node to another on fiber routes. Hence, the performance of the connection highly depends on the physical condition of the route. Because the physical condition of fiber routes may vary due to changes in the environmental, weather, or fiber cable conditions, parameters governing optical signal transmission in fiber may change as well. That can directly impact the performance of deployed transport systems. Fortunately, equipped with flexible channel-setting capability, the next-generation transport equipment is not only more resilience to adverse physical link condition changes but also able to run at a condition code to its optimal state. Application of the concept of SDN to transport network is getting mature.

To maximize the fiber capacity of transport networks, many next-generation transponders are built with variable symbol rates and adjustable modulation orders. This transport equipment provides opportunities to network operators that every optical link maintains allowable maximum capacity at all times to reach the goal of having highest utilization of the fiber infrastructure. This capacity-maximizing process will include two steps: initial channel set-up optimization and continuous channel performance optimization. The following is an example.

Assume a 400-Gbps is requested between DC A and DC H referred to Figure 15.9. A constellation shaped PM-32QAM-based 100-Gbps channel with a 64-GBd symbol rate is chosen during the initial channel set-up. After the channel is in operation, the transponder continuously monitors transmission performance of the channel, such as the pre-FEC error rate, the current link margin, etc. When the unified control plane detects that the channel's performance is "too good," that means the channel has a link margin higher than a preset required value, the control plane adjusts the baud rate, the modulation order, and/or the constellation shaping to increase the SE of the channel that leads to a higher channel capacity, 500 Gbps for instance. On the other hand, if the control plane detects a margin drop below the preset value, it adjusts the settings of the channel again to lower its SE and capacity, for instance, to 350 Gbps. In this way, the channel always runs at the highest possible capacity it is able to offer at any time.

Flexible transport network equipment can guarantee not only maximized capacity for channels but also traffic pattern optimization at the optical layer. For example, there is an established transport link connecting DC A and DC C via optical node B in Figure 15.9. If the data capacity requested by the DCs is a small fraction of that of an OTN channel, the data are carried with other data in the channel and experience payload switching with the switch fabric in node B. When the data demand between the DCs grows to a point, at which a dedicated new OTN channel makes sense, then the new channel can be set up between the DCs. The new channel will pass through node B via its CDC-F ROADM and bypass the digital switch fabric. The released capacity in the original OTN channel can serve other applications and the digital switch's bandwidth for the data can be reused.

The next-generation transport network has a meshed architecture at the optical layer. An optical mesh network provides new opportunities for data traffic protection. Let's use the diagram of Figure 15.9 again for example. Assume a channel travels from node A to node G via node K. If the fiber cable between node K and node G is cut, the unified control plane can quickly adjust traveling directions of the channel in ROADM A and ROADM G and build a new optical path A-H-G to resume the data flow between the two end nodes. In this data transport recovery process, all actions are taken in the optical layer and client data port connections are intact.

The few examples presented above show some benefits of the next-generation coherent transport networks for network service providers and operators. More advantages will be surely explored in the near future.

15.7 Conclusions

As the Internet traffic continues to grow, OTNs have to pick up the pace to serve more capacity for various applications, such as cloud computing, DC connections, and 5G wireless connections. Flexible coherent transport equipment for the next-generation transport networks supports multiple baud rates, modulation orders, and data rates to meet various data transport demands. This new technology guarantees transport links to always be running at the optimal condition, in terms of offering highest possible capacity. CDC-F ROADM-based transport network provides new ways to automate the network and protect data traffic. Unified multi-layer network management platforms not only orchestrate resources to be used in an optimized way but also give network operators the ability to adjust bandwidth flexibly and automatically to fit dynamic traffic pattern changes.

References

[1] Kaminow, I. P., and Li, T. (eds). (2002). *Optical Fiber Telecommunications IV B System and Impairments*. New York: Academic Press, 1–16.

[2] Freeman, R. L. (1999). *Fundamentals of Telecommunications*. New York: John Wiley & Sons, 511–532.

[3] International Telecommunication Union (2016). "Interfaces for the optical transport network," ITU-T G.709/Y.1331.

[4] Gorshe, S. (2017). "Beyond 100G OTN Interface Standardization," *in Proceedings of the Optical Fiber Communications Conference and Exhibition (OFC)*, 1–62.

[5] Stern, T. E., and Bala, K. (2000). *Multiwavelength Optical Networks*. Upper Saddle River, NJ: Addison-Wesley, 661–678.

[6] Wright, P., Lord, A., and Nicholas, S. (2012). "Comparison of optical spectrum utilization between flexgrid and fixed grid on a real network topology," in *Proceedings of the Optical Fiber Communication Conference (OFC/NFOEC)*, OTh3B.5.

[7] Soole, J. B. D., Chand, N., Earnshaw, M. P., Kojima, K., Pafchek, R., Ling, M., et al. (2002). "Multipurpose reconfigurable optical add-drop multiplexer (ROADM)," *in Proceedings of the 28th European Conference on Optical Communication (ECOC)*, PD3.3.

[8] Wellbrock, G. A., and Xia, T. J. (2015). "True value of flexible networks," *in Proceedings of the Optical Fiber Communications Conference and Exhibition (OFC)* (Los Angeles, CA:IEEE), M3A1.

[9] Gringeri, S., Basch, B., Shukla, V., Egorov, R., and Xia, T. J. (2010). "Flexible architectures for optical transport nodes and networks," *in Proceedings of the IEEE Communications Magazine, 48, 40.*

[10] Salsi, M., Mardoyan, H., Tran, P., Koebele, C., Dutisseuil, E., Charlet, G., et al., (2009). "155x100Gbit/s coherent PDM-QPSK transmission over 7,200km," *in Proceedings of the 35th European Conference on Optical Communication (ECOC)*, PD2.5.

[11] Gnauck, A. H., Winzer, P. J., Chandrasekhar, S., Liu, X., Zhu, B., and Peckham, D. W. (2010). "10× 224-Gbps WDM transmission of 28-Gbaud PDM 16-QAM on a 50-GHz grid over 1,200 km of fiber," *in Proceedings of the National Fiber Optic Engineers Conference (OFC/NFOEC)*, PDPB8.

[12] Cisco (2012). "Cisco Visual Networking Index," *Private Communication, Cisco White Paper.*

[13] Cisco (2008). "Global IP Traffic Forecast and Methodology, 2006–2011," *Cisco White Paper.*

[14] Cisco (2008). "Cisco Visual Networking Index: Forecast and Methodology, 2007–2012," *Cisco White Paper.*

[15] Cisco (2009). "Cisco Visual Networking Index: Forecast and Methodology, 2008–2013," *Cisco White Paper.*

[16] Cisco (2010). "Cisco Visual Networking Index: Forecast and Methodology, 2009–2014," *Cisco White Paper.*

[17] Cisco (2011). "Cisco Visual Networking Index: Forecast and Methodology, 2010–2015," *Cisco White Paper.*

[18] Cisco (2012). "Cisco Visual Networking Index: Forecast and Methodology, 2011–2016," *Cisco White Paper.*

[19] Cisco (2013). "Cisco Visual Networking Index: Forecast and Methodology, 2012–2017," *Cisco White Paper.*

[20] Cisco (2014). "Cisco Visual Networking Index: Forecast and Methodology, 2013–2018," *Cisco White Paper*.

[21] Cisco (2015). "Cisco Visual Networking Index: Forecast and Methodology, 2014–2019," *Cisco White Paper*.

[22] Cisco (2016). "Cisco Visual Networking Index: Forecast and Methodology, 2015–2020," *Cisco White Paper*.

[23] Cisco (2017). "Cisco Visual Networking Index: Forecast and Methodology, 2016–2021," *Cisco White Paper*.

[24] OND (2017). "Verizon to offer 5G pre-commercial services to select customers in 11 U.S. markets by mid-'17," OND [Accessed: February 23, 2017].

[25] OND (2017). "AT&T's fixed 5G trial expands to more cities," OND [Accessed: August 31, 2017].

[26] OND (2017). "DT activates its first pre-standard 5G connection in Berlin," OND, [Accessed: September 5, 2017].

[27] Equinix (2017). "Global Interconnection Index," Equinix [Accessed: August 14, 2017].

[28] Bogner, W., Gottwald, E., Schopflin, A., and Weiske, C. J. 40 Gbit/s unrepeatered optical transmission over 148 km by electrical time division multiplexing and demultiplexing. *Electron. Lett.* 33, 2136–2137.

[29] Schuh, K., Junginger, B., Lach, E., and Veith, G. "1 Tbit/s (10x107 Gbit/s ETDM) NRZ transmission over 480 km SSMF," *in Proceedings of the Optical Fiber Communication Conference (OFC/NFOEC)*, PDP23.

[30] Jansen, S. L., van der Borne, D., Schopflin, A., de Waardt, H. (2005). "26 x 42.8-Gbit/s DQPSK transmission with 0.8-bit/s/Hz spectral efficiency over 4,500-km SSMF using optical phase conjugation," *in Proceedings of the 31st European Conference on Optical Communication, 2005 (ECOC)*, Th4.1.5.

[31] Chang, Y. F. (2016). "Link performance investigation of industry first 100G PAM4 IC chipset with real-time DSP for data center connectivity," *in Proceedings of the Optical Fiber Communication Conference (OFC)*, Th1G.2.

[32] OND (2017). "MACOM debuts 100G single lambda module scalable to 400G," OND [Accessed: September 7, 2017].

[33] Winzer, P. J., Raybon, G., Song, H., Adamiecki, A., Corteselli, S., Gnauck, A. H., et al. (2008). 100-Gb/s DQPSK transmission: From laboratory experiments to field trials. *J. Lightwave Technol.* 26, 3388–3402.

[34] Xia, T. J., Wellbrock, G., Lee, W., Lyons, G., Hofmann, P., Fisk, T., et al. (2008) "Transmission of 107-Gb/s DQPSK over Verizon 504-km commercial lambdaXtreme transport system," *in Proceedings of the Optical Fiber Communication/National Fiber Optic Engineers Conference, 2008 (OFC/NFOEC)*, NMC2.

[35] Ly-Gagnon, D. S., Tsukamoto, S., Katoh, K., and Kikuchi, K. (2006). "Coherent detection of optical quadrature phase-shift keying signals with carrier phase estimation. *J. Lightwave Technol.* 24, 12.

[36] Cao, X. D., Jiang, M., Liang, Y., Ahn, K. H., Xia, T. J., Lou, J. W., et al. (1996). "Using two synchronized fiber lasers to drive an all-optic logic gate for 100 Gbps network applications," *in Proceedings of the Conference on Lasers and Electro-Optics (CLEO)*, CPD13.

[37] Zhang, C., Mori, Y., Igarashi, K., Katoh, K., and Kikuchi, K. (2009). "Demodulation of 1.28-Tbit/s Polarization-multiplexed 16-QAM Signals on a Single Carrier with Digital Coherent Receiver," *in Proceedings of the Optical Fiber Communication Conference (OFC/NFOEC)*, OTuG3.

[38] Kahn, J. M., and Ip, E. (2009). "Principle of digital coherent receivers for optical communications," *in Proceedings of the Optical Fiber Communication-incudes post deadline papers (OFC/NFOEC)*, OTuG5.

[39] Taylor, M. G. (2010). "Algorithms for Coherent Detection," *in Proceedings of the Optical Fiber Communication (OFC), collocated National Fiber Optic Engineers Conference, 2010 Conference on (OFC/NFOEC)*, OThL4.

[40] Oliveira, J. B., Pessoa, L. M., Salgado, H., and Darwazeh, I. (2010). "Signal processing techniques for transmission impairments compensation in optical systems," *in Proceedings of the 12th International Conference on Transparent Optical Networks (ICTON)*, We.A3.2.

[41] Behrens, C., Killey, R. I., Savory, S. J., Chen, M., and Bayvel, P. C. (2010). "Benefits of digital backpropagation in coherent QPSK and 16QAM fibre links," *in Proceedings of the Asia Communications and Photonics Conference and Exhibition (ACPCE)*, 359.

[42] Zhao, Y., Tao, Z., Oda, S., Aoki, Y., and Hoshida, T. (2017). "Pilot based cross phase modulation power estimation," *in Proceedings of the Optical Fiber Communications Conference and Exhibition (OFC)*, W1G.2.

[43] Shiner, A. D., Reimer, M., Borowiec, A., Gharan, S. O., Gaudette, J., Mehta, P., et al. (2014). Demonstration of an 8-dimensional modulation format with reduced inter-channel nonlinearities in a polarization multiplexed coherent system. *Optics Express* 22, 20366–20374.

[44] Charlet, G., Renaudier, J., Mardoyan, H., Tran, P., Pardo, O. B., Verluise, F., et al. (2009). Transmission of 16.4-bit/s capacity over 2550 km using PDM QPSK modulation format and coherent receiver. *J. Lightwave Technol.* 27, 2009, 153–157.

[45] Alfiad, M. S., Kuschnerov, M., Jansen, S. L., Wuth, T., van den Borne, D., and de Waardt, H. (2010). "11 x 224-Gbps POLMUX-RZ-16QAM transmission over 670 km of SSMF With 50-GHz channel spacing," *in Proceedings of the IEEE Photonics Technology Letters*, 1150.

[46] Salsi, M., Koebele, C., Tran, P., Mardoyan, H., Bigo, S., and Charlet, G. (2010). "80×100-Gbit/s transmission over 9,000 km using erbium-doped fibre repeaters only," *in Proceedings of the 36th European Conference and Exhibition on Optical Communication (ECOC)*, We.7.C.3.

[47] Renaudier, J., Bertran-Pardo, O., Mardoyan, H., Tran, P., Charlet, G., Bigo, S., et al. (2012). "Spectrally efficient long-haul transmission of 22-Tbps using 40-Gbaud PDM-16QAM with coherent detection," *in Proceedings of the Optical Fiber Communication Conference (OFC/NFOEC)*, OW4C.2.

[48] Rahn, J. T., Kumar, S., Mitchell, M., Malendevich, R., Sun, H., Wu, K. T., et al. (2012). "250Gbps real-time PIC-based super-channel transmission over gridless 6000km terrestrial link," *in Proceedings of the Optical Fiber Communication Conference (OFC/NFOEC)*, PDP5C.5.

[49] Gavioli, G., Torrengo, E., Bosco, G., Carena, A., Curri, V., Miot, V., et al. (2010). "Investigation of the impact of ultra-narrow carrier spacing on the transmission of a 10-Carrier 1Tbps superchannel," *in Proceedings of the Optical Fiber Communication (OFC/NFOEC)*, OThD3.

[50] Sakaguchi, J., Awaji, Y., Wada, N., Hayashi, T., Nagashima, T., Kobayashi, T., et al., (2011). "Propagation characteristics of seven-core fiber for spatial and wavelength division multiplexed 10-Gbit/s channels," *in Proceedings of the Optical Fiber Communication Conference and Exposition (OFC/NFOEC), 2011 and the National Fiber Optic Engineers Conference (OFC/NFOEC)*, OWJ2.

[51] Zhu, B., Taunay, T. F., Yan, M. F., Fishteyn, M., Oulundsen, G., and Vaidya, D. (2010). "70-Gbps multicore multimode fiber transmissions for Optical data links," *in Proceedings of the IEEE Photonics Technology Letters*, 22, 1647–1649.

[52] Yamada, J., Saruwatari, M., Asatani, K., Tsuchiya, H., Kawana, A., Sugiyama, K., et al., (1978). High-speed optical pulse transmission at 1.29-μm wavelength using low-loss single-mode fibers. *J. Quantum Electron.* 14, 791–800.

[53] Yamada, J. I., Machida, S., Kimura, T., and Takata, H. (1979). Dispersion-free single-mode fibre transmission experiments up to 1.6 Gbit/s. *Electron. Lett.* 15, 278–279.

[54] Yamada, J. I., Machida, S., and Kimura, T. (1981). 2 Gbit/s optical transmission experiments at 1.3 μm with 44 km single-mode fibre. *Electron. Lett.* 17, 479–480.

[55] Olsson, N. A., Logan, R. A., and Johnson, L. F. (1984). Transmission experiment at 3 Gbit/s with close-spaced wavelength-division-multiplexed single-frequency lasers at 1.5 μm. *Electron. Lett.* 20, 673–674.

[56] Olsson, N. A., Hegarty, J., Logan, R. A., Johnson, L. F., Walker, K. L., Cohen, L. G., et al. (1985). 68.3 km transmission with 1.37 Tbit km/s capacity using wavelength division multiplexing of ten single-frequency lasers at 1.5 mm. *Electron. Lett.* 21, 105–106.

[57] Lin, C., Kobrinski, H., Frenkel, A., and Brackett, C. A. (1988). Wavelength-tunable 16 optical channel transmission experiment at 2 Gbit/s and 600 Mbit/s for broadband subscriber distribution. *Electron. Lett.* 24, 1215–1217.

[58] Toba, H., Oda, K., Nakanishi, K. E. N. J. I., Shibata, N., Nosu, K., Takato, N., et al. (1990). A 100-channel optical FDM transmission/distribution at 622 Mbps over 50 km. *J. Lightwave Technol.* 8, 1396–1401.

[59] Chraplyvy, A. R., Gnauck, A. H., Tkach, R. W., and Derosier, R. M. (1993). "8×10 Gbps transmission through 280 km of dispersion-managed fiber," *in Proceedings of the IEEE Photonics Technology Letters*, 5, 1233–1235.

[60] Chraplyvy, A. R., Gnauck, A. H., Tkach, R. W., and Derosier, R. M. (1994). "160-Gbps (8 x 20 Gbps WDM) 300-km transmission with 50-km amplifier spacing and span-by-span dispersion reversal," *in Proceedings of the Optical Fiber Communication Conference (OFC)*, PD19.

[61] Tkach, R. W., Derosier, R. M., Gnauck, A. H., Vengsarkar, A. M., Peckham, D. W., Zyskind, J. L., et al. (1995). "Transmission of eight 20-Gbps channels over 232 km of conventional single-mode fiber," *in Proceedings of the EEE Photonics Technology Letters*, 7, 1369–1371.

[62] Yano, Y., Ono, T., Fukuchi, K., Ito, T., Yamazaki, H., Yamaguchi, M., et al. (1996). "2.6 terabit/s WDM transmission experiment using optical duobinary coding," *in Proceedings of the 22nd European Conference on Optical Communication (ECOC)*, ThB.3.1.

[63] Scheerer, C., Glingener, C., Farbert, A., Elbers, J. P., Schopflin, A., Gottwald, E., et al. (1999). 3.2 Tbit/s (80×40 Gbit/s) bidirectional WDM/ETDM transmission over 40 km standard single mode fibre. *Electron. Lett.* 35, 1752–1753.

[64] Nielsen, T. N., Stentz, A. J., Rottwitt, K., Vengsarkar, D. S., Chen, Z. J., Hansen, P. B., et al. (2000). "3.28 Tbps (82×40 Gbps) transmission over 3×100 km nonzero-dispersion fiber using dual C- and L-band hybrid Raman/erbium doped inline amplifiers," *in Proceedings of the Optical Fiber Communication Conference*, PD23.

[65] Bigo, S., Frignac, Y., Charlet, G., Idler, W., Borne, S., Gross, H., et al. (2001). "10.2 Tbit/s (256x42.7 Gbit/s PDM/WDM) transmission over 100 km TeraLight^{TM} fiber with 1.28 bit/s/Hz spectral efficiency," *in Proceedings of the Optical Fiber Communication Conference (OFC)*, PD25.

[66] Sano, A. (2006). "14-Tbps (140 x 111-Gbps PDM/WDM) CSRZ-DQPSK transmission over 160 km using 7-THz bandwidth extended L-band EDFAs," *in Proceedings of the ECOC*, Th4.1.1.

[67] Gnauck, A. H., Charlet, G., Tran, P., Winzer, P., Doerr, C., Centanni, J., et al. (2007). "25.6-Tbps C+L-band transmission of polarization-multiplexed RZ-DQPSK signals," *in Proceedings of the Optical Fiber Communication conference (OFC)*, PDP19.

[68] Zhou, X., Yu, J., Huang, M. F., Shao, Y., Wang, T., Magill, P., et al. (2009). "32 Tbps (320 × 114 Gbps) PDM-RZ-8QAM transmission over 580km of SMF-28 ultra-low-loss fiber," *in Proceedings of the National Fiber Optic Engineers Conference (OFC/NFOEC)*, PDPB4.

[69] Sano, A., et al. (2010). "69.1-Tbps (432 × 171-Gbps) C- and extended L-band transmission over 240 km Using PDM-16-QAM modulation and digital coherent detection," *in Proceedings of the Optical Fiber Communication Conference (OFC/NFOEC)*, PDPB7.

[70] Sakaguchi, J., Awaji, Y., Wada, N., Kanno, A., Kawanishi, T., Hayashi, T., et al. (2011). "109-Tbps (7x97x172-Gbps SDM/WDM/PDM) QPSK transmission through 16.8-km homogeneous multi-core fiber," *in Proceedings of the Optical Fiber Communication Conference (OFC/NFOEC)*, PDPB6.

[71] Sakaguchi, J., Puttnam, B. J., Klaus, W., Awaji, Y., Wada, N., Kanno, A., et al. (2012). "19-core fiber transmission of 19x100x172-Gbps SDM-WDM-PDM-QPSK signals at 305Tbps," *in Proceedings of the Optical Fiber Communication Conference (OFC/NFOEC)*, PDP5C.1.

[72] Takara, H., Sano, A., Kobayashi, T., Kubota, H., Kawakami, H., Matsuura, A., et al. (2012). "1.01-Pb/s (12 SDM/222 WDM/456 Gbps) crosstalk-managed transmission with 91.4-b/s/Hz aggregate spectral efficiency," *in Proceedings of the European Conference and Exhibition on Optical Communication (ECOC)*, Th.3.C.1.

[73] Puttnam, B. J., Luís, R. S., Klaus, W., Sakaguchi, J., Mendinueta, J. M. D., Awaji, Y., et al. (2015). "2.15 Pb/s transmission using a 22 core homogeneous single-mode multi-core fiber and wideband optical comb," *in Proceedings of the European Conference on Optical Communication (ECOC)*, PDP 3.1.

[74] Soma, D., Igarashi, K., Wakayama, Y., Takeshima, K., Kawaguchi, Y., Yoshikane, N., et al. (2015). "2.05 Peta-bit/s Super-Nyquist-WDM SDM transmission using 9.8-km 6-mode 19-core fiber in full C band," *in Proceedings of the European Conference on Optical Communication (ECOC)*, PDP 3.2.

[75] Soma, D., Wakayama, Y., Beppu, S., Sumita, S., Tsuritani, T., Hayashi, T., et al. (2017). "10.16 Peta-bit/s Dense SDM/WDM transmission over low-DMD 6-mode 19-core fibre across C+L band," *in Proceedings of the European Conference on Optical Communication (ECOC)*, Th.PDP.A.1.

[76] Kobayashi, T., Sano, A., Matsuura, A., Yoshida, M., Sakano, T., Kubota, H., et al. (2011). "45.2Tbps C-band WDM transmission over 240km using 538Gbps PDM-64QAM single carrier FDM signal with digital pilot tone," *in Proceedings of the European Conference and Exposition on Optical Communications (ECOC)*, Th.13.C.6.

[77] Corcoran, B., Olsson, S. L., Lundström, C., Karlsson, M., and Andrekson, P. (2012). "Phase-sensitive optical pre-amplifier implemented in an 80km DQPSK link," *in Proceedings of the National Fiber Optic Engineers Conference (OFC/NFOEC)*, PDP5A.4.

[78] Bertignono, L., Pilori, D., Nespola, A., Forghieri, F., and Bosco, G. (2017). "Experimental comparison of PM-16QAM and PM-32QAM with probabilistically shaped PM-64QAM," *in Proceedings of the Optical Fiber Communication Conference (OFC)*, M3C.2.

[79] Zhu, Y., Li, A., Peng, W. R., Kan, C., Li, Z., Chowdhury, S., et al. (2017). "Spectrally-efficient single-carrier 400G transmission enabled by probabilistic shaping," *in Proceedings of the Optical Fiber Communications Conference and Exhibition (OFC)*, M3C.1.

[80] Kobayashi, T., Nakamura, M., Hamaoka, F., Shibahara, K., Mizuno, T., Sano, A., et al. (2017). "1-Pb/s (32 SDM/46 WDM/768 Gbps) C-band dense SDM transmission over 205.6-km of single-mode heterogeneous

multi-core fiber using 96-Gbaud PDM-16QAM channels," ," *in Proceedings of the Optical Fiber Communications Conference and Exhibition (OFC)*, Th5B.1.

[81] Schuh, K., Buchali, F., Idler, W., Eriksson, T. A., Schmalen, L., Templ, W., et al. (2017). "Single carrier 1.2 Tbit/s transmission over 300 km with PM-64 QAM at 100 Gbaud," *in Proceedings of the Optical Fiber Communications Conference and Exhibition (OFC)*, Th5B.5.

[82] Wolf, S., Zwickel, H., Kieninger, C., Kutuvantavida, Y., Lauermann, M., Lutz, J., et al. (2017). "Silicon-organic hybrid (SOH) IQ modulator for 100 GBd 16QAM operation," *in Proceedings of the Optical Fiber Communication Conference (OFC)*, Th5C.1.

[83] Lange, S., Wolf, S., Lutz, J., Altenhain, L., Schmid, R., Kaiser, R., et al. (2017). "100 GBd intensity modulation and direct detection with an InP-based monolithic DFB laser Mach-Zehnder modulator," *in Proceedings of the Optical Fiber Communication Conference (OFC)*, Th5C.5.

[84] Wellbrock, G. (2017). "White Box Optics?" *in Proceedings of the Optical Fiber Communication Conference (OFC)*, M1E, White Box Optics: Will It Kill or Encourage Innovation?

[85] Wellbrock, G., and Xia, T. J., "SDN enabled transport," *in Proceedings of the FOE 2013*, 10–2.

[86] Tang, F., Li, L., Chen, B., Bose, S. K., and Shen, G. (2017). "Mixed channel traffic grooming in shared backup path protected IP over elastic optical network," *in Proceedings of the Optical Fiber Communications Conference and Exhibition (OFC)*, W1I.1.

[87] Xia, T. J. (2013). "Near term terabit transmission field trial opportunities," *in Proceedings of the Optical Fiber Communication Conference and Exposition and the National Fiber Optic Engineers Conference (OFC/NFOEC)*, NW4E.6.

[88] Xia, T. J., Wellbrock, G., Pollock, M., Lee, W., Peterson, D., Doucet, D., et al. (2009). "92-Gbps field trial with ultra-high PMD tolerance of 107-ps DGD," *in Proceedings of the National Fiber Optic Engineers Conference* (OFC/NFOEC), NThB3.

[89] Zhou, Y. R., Smith, K., Wilkinson, M., Payne, R., Lord, A., Bennett, T., et al. (2012). "Coherent 100G field trial over installed fiber links: Investigating key network scenarios and applications," *in Proceedings of the National Fiber Optic Engineers Conference (OFC/NFOEC)*, NTh1I.2.

[90] Xia, T. J., Wellbrock, G., Peterson, D., Lee, W., Pollock, M., Basch, B., et al. (2008). "Multi-rate (111-Gbps, 2x43-Gbps, and 8x10.7-Gbps) transmission at 50-GHz channel spacing over 1040-km field-deployed

fiber," *in Proceedings of the 34th European Conference on Optical Communication (ECOC)*, Th.2.E.2.

[91] Alfiad, M. S., Kuschnerov, M., Wuth, T., Xia, T. J., Wellbrock, G., Schmidt, E. D., et al. (2009). "111-Gbps transmission over 1040-km field-deployed fiber with 10G/40G neighbors," *in Proceedings of the IEEE Photonics Technology Letters*, 21, 615–617.

[92] Xia, T. J., Wellbrock, G., Basch, B., Kotrla, S., Lee, W., Tajima, T., et al. (2010). "End-to-end native IP data 100G single carrier real time DSP coherent detection transport over 1520-km field deployed fiber," *in Proceedings of the National Fiber Optic Engineers Conference (OFC/NFOEC)*, PDPD4.

[93] Birk, M., Gerard, P., Curto, R., Nelson, L., Zhou, X., Magill, P., et al. (2010). "Field trial of a real-time, single wavelength, coherent 100 Gbit/s PM-QPSK channel upgrade of an installed 1800km link," *in Proceedings of the Conference on Optical Fiber Communication (OFC), collocated National Fiber Optic Engineers Conference (OFC/NFOEC)*, PDPD1.

[94] Xia, T. J., Wellbrock, G. A., Huang, Y. K., Ip, E., Huang, M. F., Shao, Y., et al. (2011). "Field experiment with mixed line-rate transmission (112-Gbps, 450-Gbps, and 1.15-Tbps) over 3,560 km of installed fiber using filterless coherent receiver and EDFAs only," *in Proceedings of the Optical Fiber Communication Conference and Exposition (OFC/NFOEC) and the National Fiber Optic Engineers Conference*, PDPA3.

[95] Huang, Y. K., Ip, E., Xia, T. J., Wellbrock, G. A., Huang, M. F., Aono, Y., et al. (2012). Mixed line-rate transmission (112-Gbps, 450-Gbps, and 1.15-Tbps) over 3,560 km of installed fiber with filterless coherent receiver. *J. Lightwave Technol.* 30, 609–617.

[96] Xia, T., Wellbrock, G. A., Huang, Y. K., Huang, M. F., Ip, E., Ji, P. N., et al. (2012). "21.7 Tbps field trial with 22 DP-8QAM/QPSK optical superchannels over 1,503-km of installed SSMF," *in Proceedings of the National Fiber Optic Engineers Conference (OFC/NFOEC)*, PDP5C.6.

[97] Huang, Y. K., Huang, M. F., Ip, E., Mateo, E., Ji, P. N., Qian, D., et al. (2013). "High-capacity fiber field trial using Terabit/s all-optical OFDM superchannels with DP-QPSK and DP-8QAM/DP-QPSK modulation. *J. Lightwave Technol.* 31, 546–553.

[98] Xia, T. J., Wellbrock, G., Tanaka, A., Huang, M. F., Ip, E., Qian, D., et al. (2013). "High capacity field trials of 40.5 Tbps for LH distance of 1,822 km and 54.2 Tbps for regional distance of 634 km," *in Proceedings of the National Fiber Optic Engineers Conference (OFC)*, PDP5A.4.

[99] Huang, M. F., Tanaka, A., Ip, E., Huang, Y. K., Qian, D., Zhang, Y., et al. (2014). Terabit/s Nyquist superchannels in high capacity fiber field trials using DP-16QAM and DP-8QAM modulation formats. *J. Lightwave Technol.* 32, 776–782.

[100] Xia, T. J., Wellbrock, G. A., Huang, M. F., Zhang, S., Huang, Y. K., Chang, D. I., et al. (2014). "Transmission of 400G PM-16QAM channels over long-haul distance with commercial all-distributed Raman amplification system and aged standard SMF in field," *in Proceedings of the Optical Fiber Communications Conference and Exhibition (OFC)*, Tu2B.1.

[101] Zhou, Y. R., Smith, K., West, S., Johnston, M., Weatherhead, J., Weir, P., et al. (2016). "Field trial demonstration of real-time optical superchannel transport up to 5.6 Tbps over 359 km and 2 Tbps over a live 727 km flexible grid link using 64GBaud software configurable transponders," *in Proceedings of the OFC*, Th5C.1.

[102] Idler, W., Buchali, F., Schmalen, L., Lach, E., Braun, R. P., Böcherer, G., et al. (2017). Field trial of a 1 Tbps super-channel network using probabilistically shaped constellations. *J. Lightwave Technol.* 35, 1399–1406.

[103] Cho, J., Chen, X., Chandrasekhar, S., Raybon, G., Dar, R., Schmalen, L., et al. (2017). "Trans-atlantic field trial using probabilistically shaped 64-QAM at high spectral efficiencies and single-carrier real-time 250-Gbps 16-QAM," *in Proceedings of the Optical Fiber Communication Conference (OFC)*, Th5B.3.

[104] Szyrkowiec, T., Autenrieth, A., Gunning, P., Wright, P., Lord, A., Elbers, J. P., et al. (2013). "First field demonstration of cloud datacenter workflow automation employing dynamic optical transport network resources under OpenStack and OpenFlow Orchestration," *in Proceedings of the European Conference on Optical Communication (ECOC)*, PD4.F.1.

[105] Nelson, L. E., Zhang, G., Padi, N., Skolnick, C., Benson, K., Kaylor, T., et al. (2017). "SDN-controlled 400 GbE and-to-end service using a CFP8 client over a deployed, commercial flexible ROADM system," *in Proceedings of the Optical Fiber Communication Conference (OFC)*, Th5A.1.

[106] De Dios, O. G., Casellas, R., Paolucci, F., Napoli, A., Gifre, L., Annoni, S., et al. (2015). "First demonstration of multi-vendor and multi-domain EON with S-BVT and control interoperability over Pan-European testbed," *in Proceedings of the European Conference on Optical Communication (ECOC)*, PDP.4.1.

[107] OIF (2016). OIF launches coherent transmission projects. *Lightwave* [Accessed: December 1, 2016].

[108] OIF (2017). The OIF's 400ZR coherent interface starts to take shape. *OIF* [Accessed: June 23, 2017].

16

Ultra-low-power SiGe Driver-IC for High-speed Electro-absorption Modulated DFB Lasers

Jung Han Choi

Fraunhofer Heinrich-Hertz-Institute, Berlin, Germany

16.1 Introduction

Explosion of data and IP traffic in wired and wireless services trigger data centers to increase their capacity to handle huge amount of digital contents. According to the survey by Cisco [1], in 2016, global IP traffic was 1.2 ZB per year or 96 EB per month (EB = one billion gigabytes [GB]). By 2021, global IP traffic will reach 3.3 ZB per year, or 278 EB per month. The global IP traffic will increase by five times from 2016 to 2021 showing 24% Compound Annual Growth Rate (CAGR). PC-originated traffic will grow at a CAGR of 10%, while TVs, tablets, smartphones, and machine-to-machine modules will have traffic growth rates of 21%, 29%, 49%, and 49%, respectively [1]. To cope with tremendous data traffic, the Ethernet standards have been increasingly challenged and updated. Various future standards have been actively discussed by several organizations, like IEEE, OIF, ITU-T, MSA, and COBO; 100-gigabit Ethernet (GbE) was standardized in 2010 featuring with a data rate of 4×25.8 Gbps using a wavelength of 1.3 μm. At the end of 2017, it is expected that 200 GbE and 400 GbE in IEEE are supposed to be ratified as a standard. Right now, 100 Gbit/s/λ PAM-4 using 1.3 μm seems the most attractive candidate for four-lane 400-GbE systems.

Electro-absorption modulated DFB laser (EML) is an attractive optical component for low-cost and small footprint 400 G transmitter optical subassembly (TOSA) in the IEEE GbE Standard 802.3bs [2]. There has been reports about four-array EMLs supporting flip-chip bonding [3] and showing high-power optical output [4]. Also, there has been a report [5, 6]

659

that an InGaAlAs-MQW 1.3 μm EML integrates a matching component at the electro-absorption modulator (EAM). A 3-dB bandwidth of 40 GHz, a dynamic extinction ratio (ER) of 9.5 dB at 56 GBd for back-to-back (B2B), and large-signal optical eye openings up to 56 GBd were reported exploiting the EML with an integrated matching in [5].

At a data rate of 56 GBd, less parasitic between the EAM and the driver integrated circuit (IC) is very significant for a good signal integrity. Without the matching element at the EAM, the termination resistance of 50 Ω shall be packaged at the RF-interposer, which in general generates more parasitics than the 50 Ω integrated EML. It moreover leads to additional dc-power consumption due to the EAM reverse biasing. The latter may be avoided at the expense of enhanced RF packaging efforts, by further adding a series blocking capacitor at the cold side of the 50 Ω termination on the RF-interposer.

In this chapter, we present a 56 GBd SiGe driver-IC for the Fraunhofer Heinrich-Hertz Institute (*hereafter* HHI) EML with integrated matching network, its chip-on-carrier (CoC), and transmission experiments [7]. Details about the driver-IC designs and its RF package for the lowest power TOSA are described. The driver-IC is co-designed with respect to the EML. The operation principle of the matching-integrated EML is exploited to reduce the driver-IC power and to achieve the matching. It is noted that the driver-IC as a standalone component could not produce the low-power TOSA. The lowest power and the best signal integrity TOSA could be achieved when it is co-designed with the matching-integrated EML, resulting in the good electro-optical (EO) eye.

16.2 IC Design for Low Power Consumption

16.2.1 Design Requirements

Electrical specifications are determined according to the operating speed, the required voltage for the EML, and the packaging. For example, for the 56 GBd PAM-4, the electrical bandwidth of 45 GHz of the driver-IC can be considered including packaging parasitics due to wire bonding between the EML chip and the driver-IC. It is noted that the bandwidth degradation due to package parasitics was implicated defining the required driver-IC bandwidth. Differential electrical input and single-ended output configurations were assumed in the driver-IC architecture since the preceding IC, digital to analog converter (DAC), may have differential output signals. The differential signal

Table 16.1 Example of driver-IC requirements for 56 GBd TOSA

Parameter	Symbol	Min	Typ	Max	Unit
Bandwidth	BW		45		GHz
Power	P			250	mW
Data Rate	DR		56		GBd
Rise/fall time (10%–90%)	tr/tf		8		ps
Single-ended Output Signal	VOUT		1600		mVpp
Input interface	VIN, P, VIN, N	Differential, DC-coupled			—
Output interface	VOUT	Single-ended, DC-coupled			—

scheme is attractive, in that it shows better signal integrity, less crosstalk interference, and less EMI, especially at high baud-rates > 25 GBd. The output voltage of the driver-IC about 1.6 V_{pp} was considered. One significant aspect with respect to the TOSA is to isolate the thermal heat of the driver-IC from the EML device, since the output wavelength of the EML device is very sensitive to temperature variations. Therefore, it is mostly required to design a good thermoelectric cooling (TEC) below the EML to provide thermal decoupling capability. For that purpose, the HHI develops the TOSA module, which isolates the thermal energy of the driver-IC from the EML by separating the motherboard. For example, the EML sits on-TEC and the driver-IC is off-TEC. Table 16.1 presents one typical example of driver-IC requirements for 56 GBd TOSA.

16.2.2 IC Architectures for Low Power Consumption

The patented single-ended HHI EML with integrated matching network is presented in Figure 16.1 [8]. The EML is composed of a DFB laser and the EAM. The matching component is integrated at the EAM junction to provide 50 Ω impedance matching to the driver-IC. The matching component is realized with the meander line, and thus can be represented with a resistance and an inductance in series. For example, the resistance and the inductance are extracted 78 Ω and 22.4 pH, respectively. The equivalent circuit is shown in Figure 16.2. It includes lossy 50 Ω coplanar waveguide, which can be expressed using R, L, G, and C distributed components and their values are simulated and given in [8].

The low-power driver-IC in Figure 16.3 is designed to drive the matching-integrated EML. The basic idea to achieve the lowest power consumption in TOSA will be the following. First, the broadband RF-coil is considered to

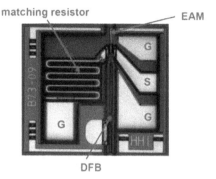

Figure 16.1   Electro-absorption modulated laser with integrated matching network [8].

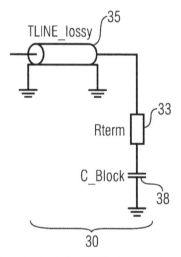

Figure 16.2   Equivalent circuit of the EAM's integrated matching termination [8].

decrease the supply voltage to the driver-IC. When the broadband RF-coil is located at the RF-hot side of the output matching resistance at the TWA, one can decrease the supply voltage, resulting in the lower power consumption. This idea as shown in Figure 16.3 was submitted for patent. Second, the generated photocurrent (I_{photo}) at the EAM junction can be exploited for the TWA power reduction. It flows into the TWA as shown in Figure 16.3, and the current consumption at the TWA's supply terminal can be decreased. It is noted that no photocurrent flows into the matching network at the EAM when the RF-coil is connected to the output resistance at the TWA. Third, the reverse bias at the EAM's cathode can also contribute the DC current

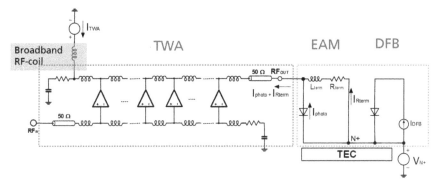

Figure 16.3 The concept of co-designed EML driver-IC for the lowest power consumption in the TOSA module.

(I_{Rterm}) through the matching resistance. This current also can flow into the TWA leading to the reduction of the current consumption at the TWA's supply terminal.

In fact, the power saving can only be achieved when the driver-IC architecture is simultaneously considered with the optical device. The objective of this approach is to obtain the lowest possible power consumption maintaining the best signal integrity. We call this strategic approach the co-design technique.

The TOSA power consumption can be considerably decreased when the driver-IC is located off-TEC while the EML still stays onto the TEC. The mechanical separation of these two is challenging, since a short distance between the IC and the EML is required for high-speed operation. We developed a novel EO package, which enables a thermal separation between EML and IC, resulting in a significant reduction in the TOSA power consumption without compromising high-frequency electrical performance. The impedance-matched flexible RF interconnection line can be exploited to enable the distance of the driver-IC and the matching-integrated EML increase. Figure 16.4 illustrated this TOSA package concept.

Since the RF impedance of the flexible transmission line is matched to the driver-IC and the EML, the signal integrity is not deteriorated. If, optionally, the matching network is integrated at the flexible transmission line, then the EML will not include the matching network. This idea is submitted for patent by the HHI, as well.

From the circuit diagram in Figure 16.5, one can see that EML driver-IC has two functionalities: receiving differential PAM-4 signals from DAC

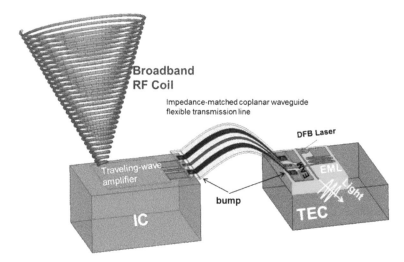

Figure 16.4   The RF flexible transmission line between the driver-IC and the impedance-matched EML enables to increase the distance between the two, maintaining the signal integrity.

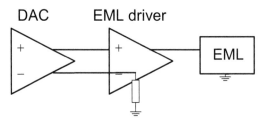

Figure 16.5   Circuit chain diagram: high-speed DAC (differential to differential), EML driver (differential in single-ended out), and EML devices (single-ended).

and terminating one input channel, and amplifying one RF channel signal to drive the EML. In general, differential driving signal for the EML is not appropriate, since the doping concentration of the EML InP substrate is high and single-layer capacitors (SLCs) to offer RF ground to the substrate are connected at the CoC. When the output differential signals are optionally combined to generate the single-ended output signal employing a broadband balun, then energy efficiency can be enhanced by two. However, to author's date, such broadband balun using active device and having low insertion loss is not yet developed. Therefore, one of signal outputs from the DAC is simply terminated at the TWA using the 50 Ω matching resistance, leading to the deterioration of the power efficiency.

The challenges of EML driver-IC design are high-speed operation and good linearity to support PAM-4 application, as well as low power consumption. To achieve those purposes, the traveling-wave amplifier (TWA) topology was chosen. The limiting amplifier is not appropriate since the multilevel amplification, e.g., PAM-4, is required for the EML. It is needed for the driver-IC to have a flat frequency response and a linear phase characteristic. The TWA is a simple circuit in its topology and has singled-ended input/output interface, which implicates simpler packaging design together with the single-ended EML.

One needs to consider the followings designing an EML driver-IC using the TWA configuration. For example, the broadband stability of the designed amplifier is required. The output voltage from the TWA is large-signal having ultra broadband spectral components from DC. The stability factor shall be above 1 throughout whole frequency bands, unless unwanted resonance may happen. It shall be noted that the packaging parasitic components for high-speed operations also need to be considered together with the driver-IC, especially bearing in mind of fabrication tolerances of each package component. If the stability factor has an influence due to packaging parasitic, then the driver-IC itself must be robust enough considering the package tolerances. In addition, very good input and output reflections are of significance to assure a good signal integrity. Any reflected signal from the load colliding with the outgoing signals from the driver-IC would degrade the resultant output signal. In practice, the reflection shall be reduced < -20 dB at the low-frequency band and about -15 dB at higher frequency. For example, for 56 GBd operation, $S_{11}(S_{22})$ is < -20 dB up to 10–15 GHz, and shows about -10 to -15 dB up to 25–30 GHz.

For those purposes, the followings are considered in the design phase. The broadband stability is obtained by using the cascode transistor configuration. Good input and output reflections are achieved by adding an input emitter follower stage before the cascode stage. It further improves the overall stability of the amplifier. It should be noted that the large-signal bandwidth of the TWA is reduced compared to the small-signal one. When amplifiers operate in large-signal domain, the bandwidth reduces due to the non-optimum power matching and the device model, which in fact is not the large- but the small-signal transistor model. In general, the transistor model used in the design is obtained under the small-signal condition, but is frequently used for the large-signal IC design. To overcome and avoid the device modeling issues, one can include power- and frequency-dependent model parameters in the device model. Including those nonlinearity behaviors, nowadays, one can use X-parameter for the transistor. However, this model is complicated and large, since it includes model parameters only at specific frequency ranges of interest. This topic is believed beyond the scope of this book chapter, and hence further discussions will not be made.

16.2.3 Driver-IC Design

16.2.3.1 Unit-cell design

In the TWA design, the number of fingers of the transistor is varied to accomplish the design specification. Figure 16.6 presents a proposed cascode unit cell. It improves the stability, isolates the input and output circuits, and has the capability to handle high output voltage swings. Furthermore, the optimization of the input and output matching networks can be done independently due to the reduced Miller capacitance. An emitter follower stage at the input of the cascode cell is added to reduce the input capacitance, and thus enhances the stability, as well as S_{11}. A 25 Ω resistor is added at the base of common base stage to protect the circuit against any parasitic inductance due to base interconnections. A capacitor of 80 fF and a resistor of 50 Ω are added at the common base stage to increase the input impedance

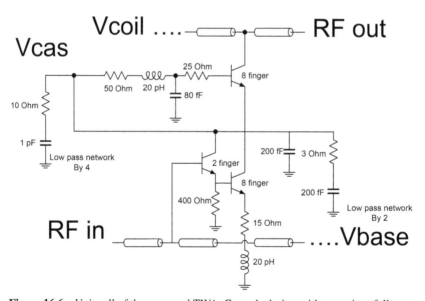

Figure 16.6 Unit cell of the proposed TWA. Cascode design with an emitter follower.

leading to high-voltage swing division between common emitter and common base stages.

Considering the output voltage swing of 1.6 V_{pp} and 50 Ω matching at the EML, the total DC loading to the TWA will be 25 Ω (due to the back terminated 50 Ω from the TWA and 50 Ω from the EML device). Then, the minimum current handling capability of TWA should be at least larger than 32 mA DC. Five amplification stages of the TWA are designed considering the gain flatness, transmission line losses, and chip area. Due to the linearity consideration and the transistor speed degradation at high collector current, the transistor size is increased to 8 fingers to handle 45 mA DC current with 1.125 mA/finger capability. The loading line of the eight-finger transistor is plotted together with DC transfer curves in Figure 16.7.

The TWA circuit diagram is shown in Figure 16.8, and the corresponding voltage supplies are also indicated. V_{coil} should be connected to the broadband RF-coil. It is located at the front of the back termination output resistor to avoid extra DC power dissipation. V_{base} should be bonded after input termination resistor, which is due to that the EML driver-IC has to be DC coupled with the preceding DAC chip for a proper voltage gain. V_{ctext} is used to extend the low cutoff frequency of the back termination network. V_{cas} is the bias voltage for the common base transistor at the cascode stage. All relevant bias voltages are documented in Table 16.2.

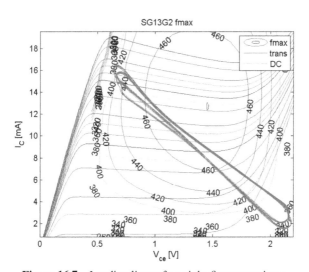

Figure 16.7　Loading lines of an eight-finger transistor.

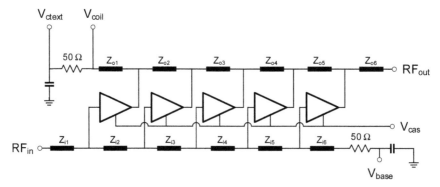

Figure 16.8   TWA circuit diagram with bias voltages.

Table 16.2   Descriptions of voltage supplies in the TWA circuit diagram

Supply	Descriptions
V_{base}	Base bias for emitter follower (input buffer stage)
V_{cas}	Base bias for common base (Cascode stage)
V_{coil}	Collector bias for common base (Cascode stage)
V_{ctext}	Bonded to single-layer capacitance to decrease the low cutoff frequency.

16.2.3.2 Circuit Simulations

Transmission lines between each amplification stage have different characteristic impedances and electrical lengths. They are optimized using the Keysight ADS software with respect to the lowest group delay ripple and flat overall frequency responses. Initial performance at the schematic level considering the small-signal response and time-domain eye diagram is estimated using input and output ideal bias-tees.

Based on the optimization values of ideal transmission lines shown in Table 16.3, small-signal S-parameters are simulated to examine the frequency characteristics. The simulation results are given in Figure 16.9. A 3 dB cutoff frequency is more than 60 GHz and reflections are below -15 dB up to 60 GHz. Group delay distortion is <1 ps up to 100 GHz. A stability factor larger than one is obtained for the whole frequency range of simulation, 160 GHz. Time-domain simulation results at 56 GBd are presented at

Table 16.3   Transmission line electrical length and the characteristic impedances

Z_{i1}	Z_{i2}	Z_{i3}	Z_{i4}	Z_{i5}	Z_{i6}	Input Line Length
$61.5\ \Omega$	$63.0\ \Omega$	$56.25\ \Omega$	$57.75\ \Omega$	$60.75\ \Omega$	$50.25\ \Omega$	$28.65°$
Z_{o1}	Z_{o2}	Z_{o3}	Z_{o4}	Z_{o5}	Z_{o6}	Output Line length
$54.0\ \Omega$	$65.25\ \Omega$	$50.25\ \Omega$	$50.25\ \Omega$	$63.0\ \Omega$	$53.25\ \Omega$	$35.47°$

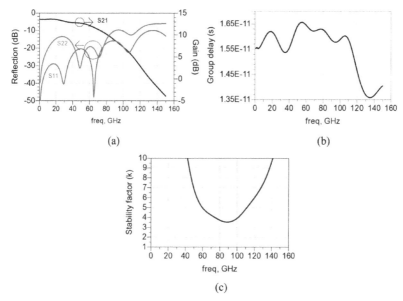

Figure 16.9 Frequency responses for the ideal schematic level of TWA: (a) S_{21}, S_{11}, S_{22} (b) group delay, and (c) stability factor.

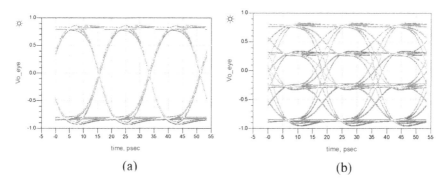

Figure 16.10 Simulated (a) OOK and (b) PAM-4 eye diagrams at the schematic level for the TWA, with an input swing of 320 mV$_{pp}$ of the Pseudorandom binary signal (PRBS) signal.

Figure 16.10. For both the NRZ and PAM-4 signals, clear eye-opening results are obtained.

The single-ended output driver-IC is fabricated using 0.13 µm SiGe BiCMOS technology with $f_T = 300$ GHz and $f_{max} = 500$ GHz. The driver is a linear amplifier with an integrated internal back-termination output resistance, 50 Ω. The driver is composed of five cascode amplification stages, which are electrically distributed to sum up output signals at each amplification stage constructively. The power supply is provided using a broadband RF-coil at the front of the internal output matching resistance, and therefore, the voltage drop at the matching resistance can be avoided.

16.3 Co-design and Electro-optical Simulation

16.3.1 Low-power CoC Design

Figure 16.11 illustrates the electrical diagram of the CoC, illustrating the DC bias schemes for the DFB, EAM, and the driver-IC. It also presents external SLCs, which were exploited to provide AC ground. The RF input pad-configuration at the EML is ground–signal–ground. In fact, the ground pads are connected to the N+ layer at the EAM, which is biased with V_{N+}. It means that ground pads of the EAM (N+) have a DC potential and functions as an AC ground. They should be connected to the output pads of the driver-IC. Therefore, the output ground pads of the driver-IC must be isolated from the DC ground of the driver-IC. The output ground pads at the driver-IC shall include shunt capacitors (C_{int}) in order to provide AC ground to the N+ contact. In fact, it is of significance to provide shunt capacitors to satisfy the low-cutoff frequency specification. Therefore, an extra capacitor ($C_{ext,IC}$) is connected using the SLC. Also, to provide N+ contact with AC ground, it is useful to include an extra shunt SLC capacitor ($C_{ext,EAM}$) for the millimeter-wave applications.

The supply voltage to the TWA is provided by the V_{TWA} through the RF-broadband coil. It is noted that the small end of conical coil directly should be down on the pad, and keeps the lead as short as possible. Also, it is preferred

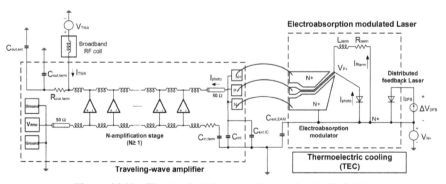

Figure 16.11   The detailed electrical diagram for the CoC [7].

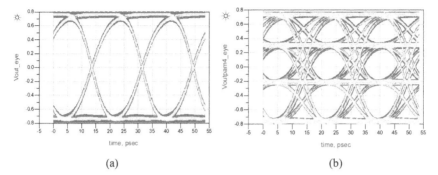

Figure 16.12 Electrical time-domain simulations after PEX including EM simulation results for (a) NRZ and (b) PAM-4 at 56 GBd.

to solder the SLC ($C_{out,ext}$) nearest to the $C_{out,term}$ due to the same reason above. It is important to keep the distance of the SLC and the corresponding pad as closer as possible, since it can generate unwanted LC resonance.

The power-on process of the CoC is the following. In general, the optical device shall be turned on first, and then the remaining electronics are switched on. In the proposed CoC, since the p+ contact of the EAM is electrically connected through the V_{TWA} (Figure 16.11) and also the EAM shall be reverse biased, it is always necessary to supply the voltage at the N+ contact first and then provide the corresponding DC voltage to the V_{TWA}. After that, the DFB current (I_{DFB}) can be given with respect to V_{N+}. V_{cas} and V_{base} in Figures 16.6 and 16.8 at the TWA can be turned-on, sequentially. To save the power consumption, it is useful to supply the laser DC current (I_{DFB}) with regard to V_{N+} reducing the voltage magnitude. The power-off process can be reversed to the power-on process.

The driver-IC power was 84.4 mW and the EML consumed 146.2 mW. The overall power consumption of the fabricated CoC was 230.6 mW. By co-designing the driver-IC combined with the EML, the power consumption of the driver-IC can be drastically reduced below 100 mW. By achieving such low-power consumption IC, one can minimize the thermal influence from the IC to the EML. This is one of advantages by doing the co-design of the driver-IC for the optical device. It should be noted that the signal integrity should be maintained even though the co-design is carried out to save the power consumption. In case the longer distance between the driver-IC and the EML is required due to the mechanical fabrication reasons, then the impedance-matched flexible transmission line can be employed, which has been submitted for patent by the Fraunhofer HHI.

16.3.2 Co-simulation of Driver-IC with EML

16.3.2.1 Electrical Time-domain Simulations

The driver-IC is simulated in the time domain at a line rate of 56 GBd both for NRZ and PAM-4. The load to the driver-IC is assumed to be ideal 50 Ω. After doing the full layout, the input and the output transmission lines are detached from the layout and are exported to the EM simulation tool, Keysight ADS Momentum, for the S-parameter simulation. The back-end-of-line information is given from the foundry service provider and is imported into the same simulation tool for the EM simulation. The simulation results, e.g., broadband S-parameters (or alternatively broadband Spice-model), are again plugged into the Cadence simulation testbench. The layout excluding transmission lines above will be parasitic extracted, called parasitic extraction (PEX) process, using the Assura software in Cadence. In the testbench, the transistor model including layout parasitics after the PEX and the S-parameters are combined and properly connected for the complete simulation. The results are presented in Figure 16.12 for both OOK and PAM-4 at 56 GBd. We observe very clear eye opening. In the PAM-4 eye waveform, the eye-height at the upper level is slightly decreased because the high output voltage level starts to be saturated. This phenomenon can be eliminated by increasing the 1 dB compression point (P_{1dB}).

16.3.2.2 Electro-optical Time-domain Simulations

The EO simulations have been carried out using the developed large-signal nonlinear EML model. The EML model was imported and combined with the parasitic extracted driver-IC block in the testbench. EO results are displayed in Figure 16.13. In the OOK eye, the EO shows higher jitter characteristic, which is attributed to the reduced overall 3 dB frequency response. In the testbench, the wire-bonding of 100–150 pH for the interconnection between the driver-IC and the EML was assumed. In the simulation setup, random noise generation is not activated, which means that the increased jitter is deterministic, and mostly stems from the frequency-dependent inter-symbol interference (ISI). For the PAM-4 EO eye, the eye-height at the upper level is more compressed than the electrical PAM-4 eye. This is due to the nonlinear frequency response of the EML. In fact, the proposed driver-IC does not include the eye-height compensation method to obtain equal eye-heights.

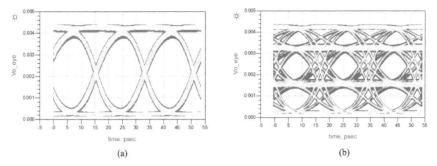

Figure 16.13 EO time-domain simulations combining the driver-IC testbench with the large-signal EML model for (a) NRZ and (b) PAM-4 at 56 GBd.

16.4 Measurements

16.4.1 EO Measurements

The fabricated CoC is presented in Figure 16.14. As discussed in earlier sections, the driver-IC and the EML are thermally separated, e.g. on- and off-TEC, and details are illustrated in the right figure of Figure 16.14. The EML stays onto diamond, which is supported by AlN board. A temperature sensor using negative temperature coefficient material (NTC) is placed near the EML device and is connected to the TEC controller. The driver-IC and the EML is connected using 20 μm diameter Au wire-bond. SLC arrays are placed nearest to the driver-IC to bypass low-frequency noises of DC-biases and provide AC-ground to the driver-IC, especially at the DC supply and ground.

The OOK EO results are presented in Figure 16.15 using the PRBS sequence length of $2^{31} - 1$. The driver-IC output was measured about 1.4 V_{pp}. The optical extinction ratios of 6.6 and 5.1 were obtained for 40 and 56 GBd, respectively. Very clear OOK eye-openings were obtained using a

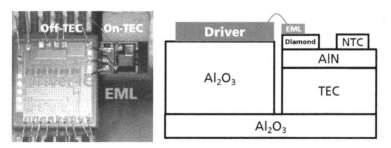

Figure 16.14 The fabricated CoC. Only EML stays on-TEC and driver-IC is off-TEC [7].

40 GBd   56 GBd

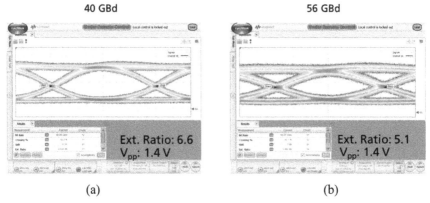

(a)   (b)

Figure 16.15   Measured EO eye at (a) 40 GBd and (b) 56 GBd. EML includes an integrated matching (optical wavelength: 1300 nm) [7].

28 GBd   32 GBd

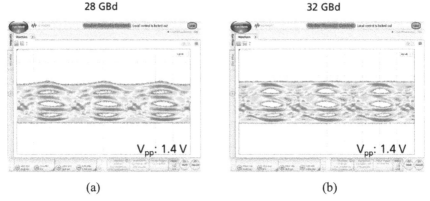

(a)   (b)

Figure 16.16   Measured EO PAM-4 eye at (a) 28 and (b) 32 GBd. EML includes an integrated matching (optical wavelength: 1300 nm) [7].

digital communication analyzer (DCA). The measurement temperature was 45°C and the DFB DC current was 70 mA.

PAM-4 signals at line rates of 28 and 32 GBd are generated using a 2-bit DAC module (SHF88120A) and are fed into the driver-IC. The measured PAM-4 optical eyes are given in Figure 16.16. The output voltage and the bandwidth shall be increased and the linearity is also to be enhanced in the revision tape-out to get the 56 GBd PAM-4 eye. It should also be noted that the nonlinearity of the EML affects the PAM-4 eye, which can be observed at the lower-eye of the PAM-4 eye in Figure 16.16(b), thus equalized electrical signal either using DSP or equalizer-integrated driver-IC to compensate for the nonlinearity will be considered.

The power consumption for the driver-IC and the EML remains the same compared with the OOK measurements. The TOSA power efficiency of 3.59 pJ/bit was obtained from these experiments excluding the additional TEC power consumption. The driver-IC efficiency corresponds to 1.3 pJ/bit. When only the EML is mounted on the TEC and the required TEC power ranges from 150 mW to 200 mW, we estimate about 6–7 pJ/bit energy efficiency of the developed TOSA.

16.4.2 Transmission Experiments [7]

The experimental setup for the transmission experiments about the NRZ signal is depicted in Figure 16.17. PRBS is generated using the bit pattern generator (SHF12103A) and fed into the driver-IC using the on-wafer RF-probe adjusting the input amplitude. The input signal to the driver-IC was configured to have 350 mV$_{pp}$. The light output from the CoC was coupled to a standard single-mode fiber (SSMF) span of 0, 20, and 40 km. The fiber is tilted with respect to the angle of the output facet of the EML. The x-, y-, and z-positions of the fiber are found out measuring the optical output from the EML. The optical output after span of fiber length is then divided into two. One goes to the optical spectrum analyzer and the other is amplified using the erbium-doped-fiber amplifier to compensate for the insertion loss of the fiber. An optical variable attenuator follows for the output power adjustment. The amplified optical signal is again divided into two paths. One is direct captured by a digital communication analyzer (Keysight DCA-X 86100D) with a 65 GHz analog bandwidth module (86116C). The other was converted into electrical signal using the high-speed photodiode module (u2t, XPDV 2320R). This electrical signal was amplified (SHF 804 EA) to fit the signal to the input range of the error analyzer (SHF 11100B).

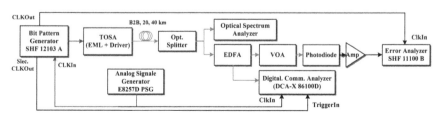

Figure 16.17 Experimental setup for the transmission. Blue lines indicate optical signal paths [7].

The photodiode and the following electrical amplifier have bandwidths of 50 and 45 GHz, respectively. It should be noted that the cascading of the bandwidth-limited components has an influence on the bit error rate (BER) measurement results. Despite the noise filtering effect thanks to the bandwidth reduction, the ISI inevitably increases. The de-embedding of these accumulated effects by off-the-shelf components is not considered in this book chapter.

The bit error analyzer is synchronized with the bit pattern generator. Analog signal generator (E8257D PSG) supplies the clock signal and distributes it to the digital communication analyzer (DCA-X 86100D) and to the bit pattern generator. The digital communication analyzer is also synchronized with the bit pattern generator using the trigger signals. The bit error analyzer (SHF 11100 B) receives the clock signal from the bit pattern generator (SHF 12103A) to count bit errors [7].

The EO OOK eyes for B2B were measured at 45°C for 28, 32, and 56 GBd, respectively, using PRBS sequence length $2^{31} - 1$ at 45°C. Results are displayed in Figure 16.18. Very good OOK eye could be obtained using the developed TOSA.

The power consumption for the driver-IC and the EML was 84 and 146 mW, respectively. The DC current for the DFB laser was 70 mA. The ER and the crossing rate of the OOK optical eye at 56 GBd were measured about 5.4 dB and 50%, respectively [7].

The reverse bias voltage of the EAM can be adjusted to vary the crossing of the optical OOK eye. It is noted that the reduced reverse bias at the EAM can also decrease the photocurrent.

The optical output power was about −1 dBm (B2B). The generated photocurrent at the EAM was about 12 mA, flowing back into the driver

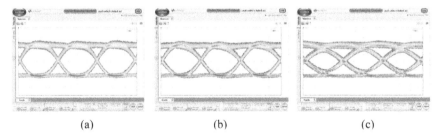

(a)   (b)   (c)

Figure 16.18   Electro-optical OOK eye (B2B) for (a) 28 GBd, (b) 32 GBd, and (c) 56 GBd [7].

and thus reducing the driver power consumption by about 27 mW. This can happen, since the RF broadband coil provides almost 0 Ω in DC. This DC photo-current is actually evenly distributed to each amplification stage at the TWA. It is noted that the output impedance of each amplification stage shall be equal to each other. There can also be a DC current through the internal matching resistance at the EAM. It adds to the photocurrent and contributes to the reduction in the current consumption at the driver-IC as well, reducing the external driver-IC power consumption. If the broadband RF-coil is not used for the biasing of the EML driver-IC, then DC supply voltage at the TWA will be supplied through the output impedance 50 Ω of the TWA at the expense of higher supply voltage. It is also mentioned that if the matching network at the EAM includes series capacitance to the resistance, then all photocurrent also flows down to the TWA, showing the same effect when the broadband RF-coil is used for DC biasing to the TWA.

OOK transmission experiments at 28, 32, and 56 GBd for B2B, 20, and 40 km were carried out by measuring the BER. In Figure 16.19, the error-free operation at BER of 10^{-9} was achieved for B2B for the EML with the matching by extrapolating the measured curves. When the received optical power is > -10 dBm, then a hard decision forward error correction (HD-FEC) threshold of 3.8×10^{-3} at line rates of 28 GBd and 32 GBd was achievable. At 56 GBd, the received optical power shall be larger than -4.5 dBm for below HD-FEC performance [7].

For the EML without matching, the BER was significantly higher and results were reported in [7]. The results show that below the soft decision forward error correction (SD-FEC) threshold of 1.9×10^{-2} was possible even though the EML does not include the matching inside.

From this experiment, it is clear that the EML with the matching is a good choice to achieve the lower BER, simultaneously saving the power consumption. Alternatively, the RF-interposer can be considered, sacrificing signal integrity and module cost. The matching component can be hybrid integrated onto the RF-interposer between the driver-IC and the EML (without the matching). However, RF parasitics from the off-the-shelf matching component, e.g., broadband 50 Ω, onto the RF-interposer cannot be avoided.

Figure 16.20 displays the measured OOK BER varying transmission distances, B2B, 20, and 40 km (EML with the matching element) at 45°C [7]. Transmission experiments at 32 GBd over fiber lengths up to 40 km of SSMF show BERs below the HD-FEC threshold. For 56 GBd, we achieved the HD-FEC threshold up to 20 km. Over 40 km at 56 GBd we could not achieve reasonable BER results and also the received power at the detector

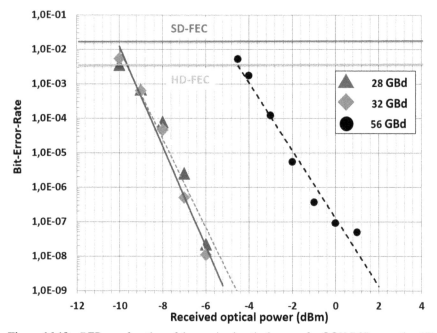

Figure 16.19   BER as a function of the received optical power for OOK B2B operation [7].

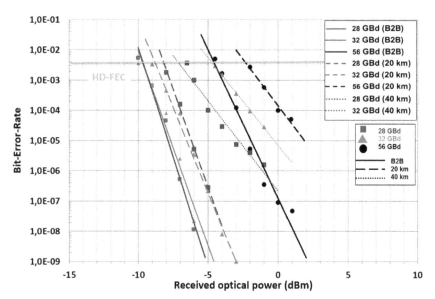

Figure 16.20   BER measurement (B2B, 20 km, and 40 km, OOK) for line rates of 28, 32, and 56 GBd [7].

shall be above 2 dBm to get the BER curves. Using the current measurement setup, it was difficult to increase the received power above 2 dBm.

16.5 Conclusion and Perspective

In this chapter, we present the low-power and small-footprint EML TOSA with a SiGe driver-IC combined with the matching-integrated EML. Using the proposed TOSA module, we could demonstrate a power efficiency of 3.59 pJ/bit up to 64 Gbps. The ideas to reduce the power were discussed in this book chapter in detail, as well as in [7]. They are repeated here to emphasize how the co-design of the driver-IC with the EML could enhance the TOSA power efficiency:

1. The DFB laser light that is emitted into the EAM generates a photocurrent. This EAM photocurrent is fed back to the driver-IC and thus reduces the current consumption at the power for the IC.
2. The total power consumption of the TOSA can be halved avoiding putting the driver-IC on the TEC. We developed a novel EO package, which enables a thermal separation between EML and IC, resulting in a significant reduction of the TOSA module power consumption without compromising high-frequency electrical performance.
3. The broadband RF coil can reduce the supply voltage to the driver-IC and leads to lower power consumption.

Details about the module in terms of DC and RF operations are presented in this chapter, focusing on how the generated photo-current can save the power consumption at the driver-IC and how matching element at the EML plays a role. An RF-interposer technology between the driver-IC and the EML, if the EML does not include matching components, can not be avoided for a TOSA package. The RF-interposer in that case will integrate matching elements and microwave transmission lines, presumably suffering from the deteriorated signal integrity due to RF parasitics from off-the-shelf components soldering and multiple wire-bonding.

Using the developed TOSA we achieved an error free operation up to a BER of 10^{-9}. OOK transmission experiments up to 40 km over SSMF show a BER below the HD-FEC threshold for 32 Gb/s over 40 km and 56 Gb/s over a 20 km SSMF fiber length. It is also measured that the BER for the EML without the integrated matching was worse than its counterpart but was still below SD-FEC threshold [7].

Acknowledgments

The author would like to thank Dr. Lei Yan at Sicoya GmbH, Berlin, Germany, for his design work. He is grateful to Prof. Dr. Martin Schell for his support of this work and also to Dr. Martin Möhrle and Dr. Ute Troppenz in the Laser group at the Fraunhofer Heinrich-Hertz Institute for their fruitful discussions and providing the EML laser. Dr. Heinz-Gunter Bach's proof-reading and patent work are highly appreciated.

References

[1] Cisco (2017). *Cisco Visual Networking Index: Forecast and Methodology, 2016–2021*, White Paper.

[2] [Online]. Available at: http://www.ieee802.org/3/bs/

[3] Kanazawa, S., Fujisawa, T., Ishii, H., Takahata, K., Ueda, Y., Iga, R., et al. (2015). High-speed (400 Gbps) eight-channel EADFB laser array module using flip-chip interconnection technique. *IEEE J. Sel. Top. Quant. Electron*, 21, 439–447.

[4] Kanazawa, S., Yamazaki, H., Nakanishi, Y., Fujisawa, T., Takahata, K., Ueda, Y., et al. (2016). "Transmission of 214 Gbit/s 4-PAM signal using an ultra-broadband lumped-electrode EADFB laser module," *in Proceedings of the OFC2016*, Th5B.3.

[5] Troppenz, U., Narodovitch, M., Kottke, C., Przyrembel, G., Molzow, W. D., Sigmund, A., et al. (2014). "1.3 μm electroabsorption modulated lasers for PAM4/PAM8 single channel 100 Gbps," *in Proceedings of the IPRM*, Th-B2-5.

[6] Klein, H., Bornholdt, C., Przyrembel, G, Sigmund, A., Molzow, W.-D., Bach, H.-G., et al. (2013). "56 Gbit/s InGaAlAs-MQW 1300 nm electroabsorption-modulated DFB-lasers with impedance matching circuit," *in Proceedings of the ECOC 2013*, TH.1.B.5.

[7] Choi, J.-H., Gruner, M., Bach, H.-G., Theurer, M., Troppenz, U., Möhrle, M., et al. (2017). "Ultra-low power SiGe driver-IC for high-speed electroabsorption modulated DFB lasers," *in Proceedings of the OFC 2017,* Th3G.3.

[8] US Patent (2017). Terminating impedance circuit for an electroabsorption modulator, *US 9678369 B2*.

Index

About the Editor

Dr. Frank Chang is an expert of photonic IC technologies and optical networks in research/development, and technology innovation. He currently is Principal Engineer – Optics at Inphi's CTO Optics Office for Optics Interconnect since April 2013 after over 11 years' service at Vitesse Semiconductors. He leads the optical system engineering efforts for physical layer IC platform and chipset solutions involving high speed drivers, TIAs and PAM4 PHYs, FECs for various 100/400G optical applications. He has about 20 years of working experience in the optical networking and communication IC industry. Prior to Vitesse, he held various senior project, architectural and management positions at Cisco/Pirelli, Mahi Networks, and JDS Uniphase. He has authored or co-authored over 100 peer-reviewed journal and conference articles, 4 book chapters and given numerous invited talks in the field. In addition, he is currently the OFC'18 N5 Chair for Market watch, network operators and data center summit to be held in San Diego and Industry Forum & Exhibit (IF&E) Co-chair for Globecom 2019 to be held in Hawaii. He was the IF&E chair to run the full industry program for Globecom 2015 in San Diego.

Dr. Chang obtained his Ph.D. in Optoelectronics from the Ecole Polytechnique, University of Montreal, Canada for his research thesis on ultrashort optical pulse generation of 1550nm tunable solid-state lasers. He is Sr. Member of IEEE/LEOS and OSA Fellow.